Classroom Companion: Business

The Classroom Companion series in Business features foundational and introductory books aimed at students to learn the core concepts, fundamental methods, theories and tools of the subject. The books offer a firm foundation for students preparing to move towards advanced learning. Each book follows a clear didactic structure and presents easy adoption opportunities for lecturers.

S. M. Riad Shams • David M Brown • Kimberley Hardcastle

Sustainable Marketing

Strategic Marketing for People, Planet and Profit

Springer

S. M. Riad Shams
Newcastle Business School
Northumbria University
Newcastle upon Tyne, UK

David M Brown
Edinburgh Business School
Heriot-Watt University
Edinburgh, UK

Kimberley Hardcastle
Newcastle Business School
Northumbria University
Newcastle upon Tyne, UK

ISSN 2662-2866 ISSN 2662-2874 (electronic)
Classroom Companion: Business
ISBN 978-3-031-79129-1 ISBN 978-3-031-79130-7 (eBook)
https://doi.org/10.1007/978-3-031-79130-7

This Springer imprint is published by the registered company Springer Nature Switzerland AG
The registered company address is: Gewerbestrasse 11, 6330 Cham, Switzerland

To my niece and daughter, Sharika and Riayana—RS

Love and thanks to my wife, Tannya, and our children, Lee and Dylan. As with so much academic writing, I produced my contributions during stolen family time—which is neither sustainable nor people-oriented. Even published advocates of sustainable business have much room for improvement!—DB

To my family and friends,
Your love and encouragement have been my guiding light. This work is a testament to your belief in me.—KH

Preface

Thank you for taking the time to read this preface and explore our views about sustainable marketing. You are our target audience. As marketers, we aim to share our experience about sustainable marketing with you, to empower and inspire you to achieve your career goals as a sustainable marketer, so that together we can contribute to making our planet Earth a better place to live for us and future generations.

Today's consumers are more attuned to social and environmental issues than ever before. An increasing number of consumers are eager to show loyalty to brands that enhance their societal and environmental impact. In response, sustainable marketing strategies have emerged as powerful tools for brands seeking to align with the values of socially and environmentally conscious consumers.

Sustainable marketing aims to give back to society and the environment, so that the societal and environmental goals of a brand can underpin its commercial competitive advantage. In this context, a sustainable marketer's mission is to apply marketing principles to create value for the triple bottom line, namely People, Planet, and Profit, while inspiring consumers and other stakeholders, eying on reinforcing diversity, equity, and inclusion of our global community for shared sustainable development.

Adopting a strategic view, each chapter of this core sustainable marketing textbook clarifies the application of different cross-disciplinary and cross-functional business theories to demonstrate how marketers can create value for People, Planet, and Profit. The traditional, contemporary, and embryonic marketing principles and their applications are discussed in a simplified manner; these are supported by practical case studies to exhibit how sustainable marketers can:

- Contribute to addressing the grand challenges associated with climate change, economic instability, geo-political uncertainty, and inequity, while underpinning their profit-making ventures responsibly.
- Contribute to the global pursuit of the 17 United Nations' Sustainable Development Goals (SDGs).

To nurture your learning experience, each chapter presents a Preliminary Vignette, Manager's Corners, (Sustainable) Marketing in Action, Web Exercises, Review Questions, Discussion Questions, and Multiple-Choice Questions. This core textbook serves as an essential resource, empowering and inspiring the next generation of sustainable marketers. It prepares you to pursue your career goals with a commitment to fostering a positive and sustainable impact on society, the environment, the economy, organizations, and yourselves.

We believe that our marketing dynamics are an ever-evolving phenomenon, where exciting things occur every day! In comparison to other established areas in marketing (e.g., principles of marketing, marketing management, strategic marketing, brand management, supply chain management, or sales management), sustainable marketing is a relatively new area in marketing thought and practice, and is yet at its embryonic stage. Considering the number of texts available in other areas of marketing, it is, therefore, one of the very few initial core textbooks on sustainable marketing for undergraduate and postgraduate teaching. Our approach to sustainable marketing helps us set this text apart from the competition. Here are the features that make this text special:

Our Applied Approach

Marketing is an applied field in social science. Thus, marketing as an academic discipline and as a subject of practice has been progressing based on the contribution of reflective marketing academics and practitioners. In this context, we adopt an experiential learning and teaching (L&T) approach. The constructivist thinkers of experiential L&T advocate that learners' knowledge is the outcome of acquiring and experiencing practical knowledge by reflecting on the real-life environment. Hence, throughout this textbook you will engage with theoretical discussions which are supported by numerous real-life sustainable marketing/business examples and case studies, which will help you acquire the theoretical knowledge, while also experiencing its practical application.

Our Focus on Society (People)

As a social science discipline, marketing thought and practice have evolved with societal issues at the core of this progression. Marketers deal with different socioeconomic challenges. For example, marketers offer different versions of the same product in different price ranges, so that customers across different income levels can afford the product.

For sustainable marketing, marketers' societal orientation is more important to contribute to people in local and global society, while underpinning the brands' profit-making ventures. Expectations have changed; today's society demands marketers help people to adapt the changing socioeconomic environments, which are affected by technological progress, climate change, economic instability, geopolitical uncertainty, and inequity. Realizing the societal challenges is therefore instrumental for sustainable marketers to play their role to give back to society. Throughout this textbook, you will see the application of sustainable marketing and will be encouraged to reflect on the challenges related to changing societal demands.

Our Focus on Environment (Planet)

Climate change is perhaps the most pressing challenge that our global community is dealing with today. If we don't focus on protecting the ecological setting of our planet, we cannot contribute to a sustainable future. The green forest area, which is considered the lung of our planet, is shrinking across the world. Excessive environmental pollution and ongoing CO_2 emissions are harming life on land and life below water. Animals in the oceans are dying because of plastic pollution. The excessive and irresponsible use of plastics/plastic products resulted in the existence of microplastics in abundance, even in human blood! The challenge lies in how we reverse this cycle of pollution. It is not easy, and there is no overnight solution. Therefore, we need to work together across different sectors to address these environmental challenges. Similarly, sustainable marketers need to play their role on this journey toward sustainability. Across different sections and subsections of each of the chapters of this core textbook, you will see a number of practical cases related to our environmental challenges, which are discussed to demonstrate the applications of marketing principles that contribute to the global pursuit of protecting our environment. The discussions will be useful to prepare you as a marketer to play your role in protecting our global environment.

Our Focus on Sustainable Development Goals (SDGs) and the Triple Bottom Line - Namely People, Planet, and Profit

The United Nations has unveiled the Sustainable Development Goals (SDGs) for the global community, aiming to tackle the grand challenges linked to socioeconomic and environmental issues. The process began with discussions at the United Nations Conference on Sustainable Development, also known as Rio+20, held in Rio de Janeiro in 2012. Following Rio+20, an inclusive and transparent process was initiated, involving extensive consultations with a wide range of stakeholders, including governments, civil society, the private sector, and academia. In September 2015, after much negotiation and consideration, the UN General Assembly adopted the 2030 Agenda for Sustainable Development, which includes the 17 SDGs at the historic UN Summit at the UN Headquarters in New York. In adopting the SDGs, the then UN Secretary-General Ban Ki-moon described "we have a big, bold agenda before us—now we must work to make it real in people's lives everywhere." The 17 UN SDGs are: No Poverty (SDG 1); Zero Hunger (SDG 2); Good Health and Well-being (SDG 3); Quality Education (SDG 4); Gender Equality (SDG 5); Clean Water and Sanitation (SDG 6); Affordable and Clean Energy (SDG 7); Decent Work and Economic Growth (SDG 8); Industry, Innovation, and Infrastructure (SDG 9); Reduced Inequalities (SDG 10); Sustainable Cities and Communities (SDG 11); Responsible Consumption and Production (SDG 12);

Climate Action (SDG 13); Life Below Water (SDG 14); Life on Land (SDG 15); Peace, Justice, and Strong Institutions (SDG 16); and Partnerships for the Goals (SDG 17).

On the one hand, marketing is responsible for stimulating excessive production and consumption in our global society, which is in contradiction to the theme of SDG 12 (Responsible Consumption and Production). Conversely, SDG 17 (Partnerships for the Goals) essentially represents a form of strategic marketing. The United Nations encourages establishing partnerships with stakeholders at local and global levels to advance the other 16 SDGs. This aligns closely with the definition of marketing, which involves developing, advancing, and evaluating relationships with key stakeholders. Thus, marketing principles are fundamentally embedded in the pursuit of these global goals. From this context, and based on different marketing principles, sustainable marketers can and should contribute to the global pursuit of these 17 UN SDGs. Practical marketing cases are discussed throughout all chapters to demonstrate how sustainable marketers can successfully pursue the 17 UN SDGs. The last section of each chapter also summarizes these chapter-specific practical discussions and examples of how sustainable marketing principles can help marketers contribute to society (people) and the environment (planet), and pursue the 17 SDGs in real-life marketing dynamics, while underpinning their brands' profit-making ventures.

Our Research-Led Teaching Approach

The research-led teaching approach is a well-recognized L&T method to underpin students' learning experience. Discussions throughout this textbook have evolved based on the findings from conceptual and empirical research studies published in leading journals. Alongside adopting and adapting the findings from our own published research studies, we also have cited numerous research works by leading researchers in our field to support the discussion.

Our Cross-disciplinary and Cross-functional Focus

Marketing as an academic discipline has been evolving based on inputs from many other cross-disciplinary and cross-functional fields of study and practice. For example, economics, statistics, sociology, mathematics, psychology, strategy, leadership, etc. In this context, marketing is both art and science (which is discussed in ► Sect. 1.2.6, the Art and Science of Marketing, in ► Chap. 1 of this text). Across different chapters, we have discussed various cross-disciplinary and cross-functional theories and clarified how the knowledge from different disciplinary and functional areas can help sustainable marketers to empower and influence their marketing dynamics further.

Our Global Perspective

Thirty to forty years ago, people had limited knowledge about how children were growing up in different parts of the world. The experiences and upbringing of children in various global regions were largely unknown to those outside their immediate locales. Today's children know about the world much more than the older generations, as just a fingertip away can connect them with the other regions of the world through the internet. On the one hand, this global outlook brings opportunities for marketers to expand their business globally. On the other hand, today's supermarkets in cities across the world sell more cross-cultural products than they used to sell 20–30 years prior. We have more cross-cultural customers today in our local markets than ever before. Our textbook discusses practical case examples from different global regions so that cases from different socioeconomic and cultural settings can help enrich your understanding of the applications of sustainable marketing principles from diverse perspectives.

Our Focus on Technology

Following the inception of the World Wide Web (the internet), the contemporary and embryonic aspects of these digital technologies, tools and platforms, and the availability of many of these technological inventions to mass people in recent decades, introduced the era of the digital age in our global village. Such digital and mobile technologies include but are not limited to social media, artificial intelligence (AI), robotics, augmented reality, virtual reality, big data, database marketing, advanced analytics, cloud computing, blockchain, internet-of-things, search engine optimization, etc., influence our marketing dynamics significantly. But, it does not mean, we overlook our traditional marketing principles. Instead, in support of the traditional, contemporary, and embryonic marketing principles, marketers attempt to plan for strategies that would be win-win for their customers and other stakeholders in these ever-evolving marketing dynamics. We have discussed these digital technologies and their applications for sustainable marketing in this textbook to enrich students' learning experience. In addition, each chapter of this textbook contains web exercises.

Electronic and in Print Versions

As part of our focus on technology, *Sustainable Marketing: Strategic Marketing for People, Planet and Profit* is available online in contemporary electronic version and in traditional print version. Accordingly, you can choose to use the version that best suits your needs and preferences; you may prefer to use both the online and the print versions together. It provides the flexibility to choose a combination of resources that works best for you.

Our Policy Focus

As a social science discipline, marketing has policy implications for its stakeholders in society and the environment. On the one hand, if a brand is to be successful as a sustainable brand in the long run the ethical, societal, environmental, and legal policy dimensions of sustainability principles should be assimilated into sustainable marketing strategies. Conversely, using their experience, marketers can influence the formation of policy for collective well-being. Our policy focus helps us to incorporate the pragmatic perspectives of policy implications of sustainable marketing.

Our Target Audience and Our User-Friendliness

Undergraduate and postgraduate students are the target audience of this textbook. We have embraced a balanced approach to demystify complex business and management theories, as well as marketing principles, by presenting them in a straightforward and accessible manner. Our efforts will only be successful if you find the text engaging and enjoyable to read. We have therefore worked hard at composing interesting and thought-provoking discussions to give you an enjoyable experience. Our goal is to stimulate your curiosity, to look forward to exploring the next section, case study, or exercise. Drawing on our finest writing and superior design, we have strived to make this textbook both accessible and enjoyable, providing you with an engaging exploration of sustainable marketing dynamics.

We appreciate the value of dialogue. Should you have any questions, comments, or suggestions, please do not hesitate to get in touch with our respective co-authors of each chapter and receive feedback from us:

- Dr Riad Shams (riad.shams@northumbria.ac.uk): ► Chaps. 1, 6 and 7.
- Dr David M Brown (david.brown@hw.ac.uk): ► Chaps. 3 and 5.
- Dr Kimberley Hardcastle (kimberley.hardcastle@northumbria.ac.uk): ► Chaps. 2 and 4.

Last but not least, we conclude this preface with a philosophical thought:

> Your sickness is from you, but you do not perceive it and your remedy is within you, but you do not sense it. You presume you are a small entity, but within you is enfolded the entire Universe. You are indeed the evident book, by whose alphabet the hidden becomes manifest. Therefore, you have no need to look beyond yourself. What you seek is within you, if only you reflect. (Ali Ibn Abi Talib R.A., as cited in Think Full Circle,[1] 2015)

1 Think Full Circle. (2015). Eloquent variations of time and space. Retrieved May 18, 2015, from ► https://thinkfullcircle.wordpress.com/

We hope that this textbook will help you explore some contexts of your universe. Explore your own universe and write your own book that will have a sustainable optimistic impact on your loved ones, society, the environment, the economy, your organization, and yourself to make a difference. You can choose to be the same as ordinary marketers who have gone before you and have promoted consumption at all costs or you can be the change leader—you can choose a different path, a more sustainable path to a better world, by promoting "responsible consumption and production."

S. M. Riad Shams
Newcastle upon Tyne, UK

David M Brown
Edinburgh, UK

Kimberley Hardcastle
Newcastle upon Tyne, UK

Contents

Supplementary Information

About the Authors

S. M. Riad Shams

Riad is an Assistant Professor in Marketing and PhD Program Leader at Newcastle Business School, Northumbria University. He has extensive academic and industry experience. He taught in Australia, Bangladesh, Russia, the Netherlands, and the UK. He is a Board Member of the Northern Advanced Research Training Initiative, a Fellow of the Higher Education Academy, and a Fellow of the EuroMed Academy of Business. Riad has published twenty-seven books, including two core textbooks, *Marketing Management: Past, Present and Future*, and this *Sustainable Marketing* text. He led Business with Marketing program and leads on Strategic Marketing, Sustainable Marketing, and Marketing Management for Competitive Advantage modules on the MSc Marketing and Business programs.

Riad's research attempts to analyze the "cause and consequence of stakeholder relationships and interactions as a stakeholder causal scope (SCS)" to recognize their value anticipation, in order to develop strategic frameworks for win-win value co-creation. Following this cross-disciplinary approach, his research addresses environmental, societal, and managerial issues, mainly in the areas of Sustainable Marketing, Social Business, Brand Management, Digital Marketing, and International Marketing. Riad has a sustained record of papers published in top-tier journals. He delivered a keynote at the SIDREA International Conference in January 2020 at the Sapienza University of Rome. He is an associate editor of the *Journal of Social Entrepreneurship,* and (guest) edited a number of high impact journals, i.e., *Industrial Marketing Management*, *Regional Studies*, *International Marketing Review*, *Journal of Business Research*, *Journal of International Management*, *Technological Forecasting and Social Change*, *Asia Pacific Journal of Management* and the *European Business Review*. He is the founding editor of the *Annals of Business Research* and editor-in-chief of the *International Journal of Big Data Management*. He is co-editor of the *Palgrave Studies in Cross-Disciplinary Business Research.* He chaired the 17th NARTI Annual Conference 2024 and led academic tracks in several international conferences. He has been chairing the Marketing Research Committee at the EuroMed Academy of Business.

Riad was nominated for the Student-Led Teaching Award—2024 and the Northumbria Shine Award—2023. He is the author of the most cited paper in the *Journal of International Management*. Riad obtained the second place in the Cambridge-Kent-Czinkota Competition for Excellence in International Business Research—2019 and entered into the Kent Business School's Honours Board. He received the Emerald Literati Award (Outstanding Paper) in 2021 and 2019, the Emerald Literati Award (Outstanding Reviewer) in 2018, and the EuroMed Research Award in 2014. As per Google Scholar, he is the fourth most cited author in Social Business research and eighth most cited author in Sustainable Marketing research.

David M Brown

David is an Associate Professor in Marketing at Heriot-Watt University, Edinburgh. A practice-oriented scholar with 18 years of industry experience before joining academia in 2012, he has authored two previous books. His research addresses marketing's contribution to social, economic, and environmental sustainability, and particularly the needs of diverse, often marginalized, stakeholders. He believes passionately that universities should be "anchor institutions," collaborating with regional partners to enrich communities. He works with charities and public sector organizations to explore social sustainability issues such as the potential societal benefits of no/lo-alcohol beverage adoption. His research has appeared in publications including the British Journal of Management, *Journal of World Business*, *Journal of Business Research*, *Industrial Marketing Management*, *Journal of Marketing Management*, *Strategic Change*, *Leisure Studies*, *International Journal of Management Education*, and *the International Journal of Entrepreneurship and Innovation*, for which he is an Editorial Board Member. He has designed and facilitated courses at the University of Exeter, and for German Hochschulen and other institutions internationally.

Kimberley Hardcastle

Kimberley is an Assistant Professor in Marketing at Newcastle Business School (AACSB), Northumbria University. Prior to entering academia, she has worked in various marketing and business management roles in the F&B service sector. She has a PhD in Marketing and is a researcher, writer, speaker, and advisor in branding, consumer habits, AI, algorithms, and platforms. She is keen to understand how brands can use successfully proven influential buyer strategies to help consumers create good sustainable habits; her research also extends to examining the societal implications of emerging technologies. Her work has been published at several international conferences and in peer-reviewed journals. Her current projects include The Responsible Development of AI and the societal implications of this in the North-East of England. Kimberley is working with NorthEast partners to understand how the region can adopt a citizen well-being approach to safeguarding and upskilling the region to ensure the NorthEast is not left behind with emerging technology. Kimberley's Business and Marketing expertise resonates beyond academia as she is also a contributor at the World Advertising Research Centre (WARC).

Marketing Dynamics and Stakeholder Value Creation for People, Planet, and Profit

Contents

S. M. R. Shams et al., *Sustainable Marketing*, Classroom Companion: Business,
https://doi.org/10.1007/978-3-031-79130-7_1

Preliminary Vignette: The Value Chain of a For-Profit Firm for Its People, Planet, and Profit

Source: ▶ pexels.com/royalty-free-images/

Aspuna, a British small- and medium-sized profit-focused firm headquartered in London, process raw agricultural produce of farmers and sell processed food products across different global regions. When they were trying to expand their value network in Africa to source African farmers' produces, they found that farmers in Africa do not receive the price that they deserve in return of their hard labor for cultivation. As a result, the farmers cannot take care of their family properly.

In response to this socioeconomic challenge in the African value network, Aspuna helped the farmers to develop a farmers' cooperative, provided technology to the farmers, and established crops processing plants in Africa. Under the cooperative, the farmers can apply technologies to increase their production and bring their raw crops to the processing plants to process and package it, to make it as branded finished agricultural products. In this way, Aspuna established a socially-caring value network to support the farmers in Africa, created more jobs at the crops processing plants for local community, and ensured that they can source their supplies (i.e., raw agricultural produces) in competitive price, to produce their branded finished agricultural products. This value network helps Aspuna to offer a competitive price to their customers in different global regions.

To market their branded finished agricultural products, Aspuna partnered with local micro entrepreneurs to sell those finished agricultural products in local African markets, while exporting it to other global regions. It helped Aspuna create further value for those local micro entrepreneurs, beside the farmers. To describe the core focus of their business, Aspuna state on their website (▶ https://www.aspuna.com/about/) that "our social impact commodities asset management business is supercharging a profitable and sustainable African agricultural sector by investing in processing capacity, implementing innovation and positively impacting *people, place and profitability*, where we operate."

To establish such a socially-caring value network, it is imperative for marketers to establish and pursue a vision that would have social and/or ecological, and economic implications for the key stakeholders, who are directly associated with that value network. For example, the internal and external stakeholders who help marketers to create value to serve their customers. For the case of Aspuna, it is the farmers, micro entrepreneurs and workers in their crops processing plants in Africa, who help them to create social and economic value in a competitive way that helps Aspuna to offer their branded finished products at a competitive price to their global customers. In return, Aspuna help to improve the life of those key stakeholders, i.e., the African farmers, workers and micro entrepreneurs, who were not able to receive the right value for their labor. Reflecting on this socially-caring value network of Aspuna, we see that they pursue a business vision for "a world where all globally traded commodities add social and economic value to the communities from where they originate" (Aspuna, n.d., np).

1.1 Introduction

Sustainable marketing is not a philanthropy. The purpose of a sustainable marketing strategy is to contribute back to the society and/or the natural environment of a market where you serve as a marketer, so that the societal and/or environmental contribution of your brand can underpin the brand's commercial competitive advantage. In this context, sustainable marketing aims to pursue the triple bottom line, i.e., people (society), planet (environment) and profit (commercial success), instead of pursuing profit only.

Sustainable marketing strategies are applicable to any organization, such as profit-only business firms and government or non-government organizations across different sectors. However, it is important that an organization that intends to pursue the triple bottom line (i.e., people, planet, and profit) upon applying sustainable marketing strategies, they need to find out a proper fit of a particular social or ecological cause that they may want to align with their brand to promote it. For example, if a tobacco manufacturing company wants to support cancer research, they would attract criticism in some cases. Yes, supporting cancer research is highly admirable always; however, critics would argue that you (the tobacco manufacturing company) are causing cancer in the global society, and now you are trying to cure it! Some critics would go further criticizing that it is like offering us a pair of leather shoes after killing our cows. Therefore, firms need to carefully select a particular social or environmental cause that would align better with their brand to promote the brand based on a well-aligned, sustainable marketing strategy.

Source: ► pexels.com/royalty-free-images/

A good example of such a brand and cause-related alignments for a sustainable marketing strategy is Jeep and its promotion based on supporting the wounded members (and the families of the soldiers) in the United States military forces who were injured or died overseas to protect their national interests. For example, Jeep runs several cause-related promotional campaigns, which include the campaigns such as "Jeep saluting US military and veterans with $1000 cash-back deal" (Gastelu, 2022, np), "Jeep adds all-new freedom trim in honor of the U.S. military" (Julian, 2021, np), or Drive for troops—Help support our soldiers, veterans and their families (Jeep, 2014). Such a marriage between the brand Jeep and the cause is well-aligned because the nature of the product or service that is offered by the brand (i.e., Jeep) and the type of vehicles that are used by those wounded soldiers are the same or similar. Jeep has been promoting them successfully based on such social cause-related marketing campaigns. Similarly, irrespective of the type of organization from the public, private, or third sector you will be working for, it is important to ensure such a good synchronization between your organizational brand (and its market offerings) with the social or environmental cause, based on what you may want to promote your organization.

In general, the marketing dynamics as it is known today have been evolving based on the mutual contribution of marketing academics and practitioners. The diverse customer and non-customer stakeholder groups of a brand collectively consist of its overall marketing dynamics. Therefore, it is imperative for marketers to understand the current and latent needs of their stakeholders to prolifically and responsively influence their overall marketing dynamics. The ongoing emphasis on socioeconomic and ecological concerns of a market that is served by a brand needs to be considered in the brand's marketing policy and strategy, so that the market-specific socioeconomic and ecological value contribution of the brand could underpin the brand's commercial profit-making ventures in its target market, as well.

This chapter will focus on the views of reflective marketing academics and practitioners from historical and embryonic contexts to sequentially clarify the evolution and ongoing transformation in our past, present, and future (embryonic) marketing dynamics. The discussion here will adopt a strategic viewpoint to help marketers develop their marketing orientation on different government and non-government stakeholders (including customers), with an aim to understand stakeholder-specific current and latent needs. The discussions of such needs include the economic, social, and environmental interests and expectations of the involved stakeholders, who have an influence on a marketer's overall value or stakeholder network. Consequently, the reflection on stakeholders' economic, social, and environmental needs will be instrumental for marketers to create stakeholder value to contribute to the people (society) and planet (ecology) of a market that is served by marketers while making a profit.

The discussion in this chapter will be based on the historical and transformational considerations in marketing dynamics, market structure, and business environment from diverse socioeconomic and ecological contexts, stakeholders' needs and distress, competitive factors, and application of cross-functional stakeholder-focused business theories (including marketing principles) and value-creating strategies to influence the market-specific competitive factors. Eventually, the social, ecological, and economic value of a brand that could be created, disseminated, and evaluated based on a sustainable marketing strategy would be acceptable by the diverse customer and non-customer stakeholder groups while enabling the brand to contribute to its people and planet and underpin its commercial sustainability, simultaneously.

It should be noted that when we use the terms in this textbook, such as "sustainability" or "sustainable," we refer not only to societal or environmental sustainability, but also to commercial sustainability (i.e., creating a business that is commercially sustainable and simultaneously contributes to its society and/or environment for societal and/or environmental sustainability, so that societal and/or environmental contribution of the business can underpin its profit-making ventures further).

Web Exercise 1.1

Brand and cause-related alignment is a key decision-making issue for a brand to promote it. For marketing decision making, selecting a social or environmental cause is vital that may not attract criticisms, but should have a well-fit between the brand's overall market offerings, value propositions and the cause, based on what, the brand would be promoted. Let us have a look at different cause-related marketing campaigns to realize the extent of brand and cause alignment. For example, the Dove's "Reverse Selfie" commercial (▶ https://www.youtube.com/watch?v=z2T-Rh838GA), where they are positioning the brand, based on promoting a message that "the pressure of social media is hurting our girls' self-esteem." IKEA, as a furniture manufacturing and self-assembling brand promote them, based on the message "Climate Action Starts at Home" (▶ https://www.youtube.com/watch?v=qA4GQDUgQBc). Coca-Cola promote them emphasizing on the exigency of Recycling and Reusing existing materials (▶ https://www.youtube.com/watch?v=0Crpab4WWic). If these web links do not work, you can find these commercials, based on Googling or YouTubing the titles or information about these commercials.

Have a look at these and different other TV or web commercials of different brands, and think of:

- How are the brands promoting them based on a specific societal or ecological cause?
- What is the core societal or ecological message the brands are promoting?
- Do the cause and the brand fit well for the brand's overall market offering to promote them?
- What would be the impact of such socially or environmentally caring brand message on the brand's customer and non-customer stakeholders?

1.2 Definitions and Scope

1.2.1 Marketing Definition and Purpose of Marketing

◘ Table 1.1 presents the definitions of marketing by the American Marketing Association (AMA) in different eras. If we look at the AMA's 1935 definition, it focuses purely on operational activities, which are centered on profit-making initiatives only, to make sure that produced goods and services are delivered at the right

◘ **Table 1.1** Definitions of marketing by the American Marketing Association (AMA) in different eras

Year	AMA definition of marketing
1935	"Marketing is the performance of business activities that direct the flow of goods and services from producers to consumers" (In Keefe, 2004; as cited in Ringold & Weitz, 2007, p. 255)
1985	"Marketing is the process of planning and executing the conception, pricing, promotion, and distribution of goods, ideas, and services to create exchanges that satisfy individual and organizational goals" (In Thomas, 2020, p. 958)
2004	"Marketing is an organizational function and a set of processes for creating, communicating, and delivering value to customers and for managing customer relationships in ways that benefit the organization and its stakeholders" (In Gundlach, 2007, p. 243)
2007	"Marketing is the activity, set of institutions, and processes for creating, communicating, delivering, and exchanging offerings that have value for customers, clients, partners, and society at large" (In Gundlach & Wilkie, 2009, p. 260; AMA, 2017)

time to its customers. It was because, during the late nineteenth century and early decades of the twentieth century, the world experienced the industrial revolution, such as the automated process of production, transportation, and communication. It changes the lifestyle of mass people (i.e., the consumers), especially in the more industrialized countries. Factories started to produce more than before because of the newly invented automated production technologies. As a result, businesses needed to adopt a more pragmatic process of business performance management to ensure that the produced goods and services do not remain unsold. However, the gradual expansion of the Industrial Revolution across different nations by the mid to late decades of the twentieth century ensured an extremely prominent level of production across the global market. As a result, many produced goods during this period remained unsold. To sell these unsold goods, marketers needed to think of a more structured way related to the varied needs of customers. The focus on customers' diverse needs derives from the AMA's 1985 definition of marketing, where we see marketers are focused on customers' diverse needs to promote their market offerings, centered on product or service design, competitive pricing, accessible distribution channels (place), and appealing promotional strategies. This 4P-based (product, price, place, and promotion) marketing thought and practice was initially coined by Neil Borden (1964), the Harvard scholar, in his AMA presidential note in 1953 and later in 1964. In the global marketing management dynamics, the marketing mix concept was popularized further by McCarthy (1960) and McCarthy and Brogowicz (1981). It helped marketers to plan in a more systematic manner, focusing on customers' needs related to product, price, place/distribution (channel), and promotional relevance.

Looking at AMA's 2004 definition of marketing lets us understand that the 4Ps marketing mix concept was not sufficient alone to win the target market, because of increasingly intensified market competition. Consequently, the words "value," "customer value," "customer relationships," and "stakeholder" appeared in the marketing definition. It introduced different value-based loyalty schemes across different industries to promote a brand in response to the intensified market competition among the competitive brands. However, creating value for customers and commercial stakeholders (e.g., investors, suppliers) only, are becoming not enough anymore to win a target market, considering that the global population (who are also customer and non-customer stakeholders of a commercial brand) have been becoming more socially- and environmentally-concerned to protect community and ecology. As a result, we see the inclusion of the word "society" in AMA's 2007 marketing definition. The 2007 definition emphasizes the necessity of a brand to delight its customers and non-customer stakeholders based on contributing back to a social or ecological cause so that the social or ecological contribution of a brand could underpin its commercial competitive advantage, as well.

A basic definition of marketing is recognizing and meeting the target market's needs based on competitive value propositions. It is about clarifying to the target audiences of a brand why doing business with us is more prolific for you in comparison to doing business with our competitive brands. The target audience or target market would include customers and all non-customer stakeholders (e.g., suppliers, investors, employees, government, and media among others). Recognizing the target market's needs is a complex task for marketers. This is because an inaccurate or partial understanding of the target market's needs could lead to a marketing myopia. Marketing myopia is a short-sightedness in measuring the target market's needs, which could be addressed based on marketing research. More discussions on how to avoid marketing myopia are shared in ► Sect. 1.4.1 (Ongoing Market Orientation) of this current chapter. The core definition of marketing is simple (i.e., recognizing and meeting the target market's needs). However, we emphasize that recognizing and meeting the needs of the target markets is a complex task, considering that even a slight misinterpretation of the target markets' needs that marketers aim to satisfy can lead to marketing myopia, bring an enormous adverse impact on their brand and its mage, and can offer significant competitive disadvantage to the brand, and advantage to the competitive brands. Therefore, marketers need to be careful while anticipating the target market's needs, and those needs to be reflected through product/service design, pricing strategy, distribution, and promotion plans to pursue a competitive advantage for the brand. In this context, we define *marketing as a dynamic of designing, delivering, and evaluating a utility or value in the form of a product or service to meet the target markets' needs to establish a competitive advantage for the product or service.*

Marketing is a dynamic of designing, delivering and evaluating a utility or value in a form of product or service to meet the target markets' needs, to establish a competitive advantage for the product or service

Designing, delivering, and evaluating a particular utility or value as a solution to a target market's one specific problem in relation to a certain need of the target market is, therefore, a primary task of marketers. For example, thirst or dehydration is a problem for human beings or any living creature. In relation to this problem, our need is to drink liquids that are good for the human body. As a marketer of a drink, let us think about how many alternative competitive value propositions currently are available in the market to provide a solution to our customers' problem of dehydration. We have mineral water, cola, coffee, and other competitive drinks in general in supermarkets to purchase to satisfy our need to rehydrate. As a marketer of a new drink, if we market fresh green coconut water or fresh sugarcane juice without preservatives through supermarkets or, if our resources permit, through our own distribution channel, don't we think that there would be a considerable segment in our global market, where customers would prefer to purchase more of such healthy drinks?

◘ Source: ▶ pexels.com/royalty-free-images/

Indeed, reflecting on the target market's (i.e., customer and non-customer stakeholders) real-life problems and relevant needs, marketers design, deliver, and evaluate their marketing mix, namely product, price, place (the distribution channel), and promotion. How such reflection can ensue is discussed in ▶ Chaps. 2 (Product and Service Development), ▶ 3 (Pricing Strategy), ▶ 4 (Brand Management, Marketing Communication, and Promotion Management), and ▶ 5 (Distribution Management). More discussion about the target market is shared in ▶ Sects. 1.2.1 (Marketing Definition and Purpose of Marketing), about needs in ▶ 1.2.2 (Strategic Marketing and Sustainable Marketing), and in ▶ 1.2.3 (Customer's Problems, and Expressed vs. Latent Need: Marketing a Solution, Not a Product Itself).

The purpose of marketing activities is primarily threefold: Need management, differentiation management, and relationship management to establish a brand's competitive advantage. In addition to the physical marketing dynamics (e.g., traditional or online distribution channels including wholesaling or retailing shops, promotional campaigns, developing products at a lab, and so on), the actors involved in the marketing dynamics (e.g., marketers/sellers, buyers, distributors, investors, competitors, and other stakeholders), the marketing dynamics itself belong in the minds of marketers and their target market, which includes all customer and non-customer stakeholders. To influence these marketing dynamics, marketers first need to orient with their target markets' needs, to design, deliver, and evaluate marketing offerings (value, based on a product, service, or idea) reflecting on those market needs. We have more discussion on need in ▶ Sect. 1.2.3 (Customer's Problems and Expressed vs. Latent Need: Marketing a Solution, Not a Product Itself).

Purpose of marketing activities is primarily threefold: Need management, differentiation management and relationship management to establish a brand's competitive advantage.

Marketers also need to differentiate their market offerings in comparison to the value of competitive market offerings to clarify how uniquely our product or service satisfies the needs of our customers, who are our and our competitors' common target audience. Branding and re-branding or brand positioning and re-positioning help us to differentiate the distinctiveness of the utility of our market offering through the lens of marketing mix. For example, often, we see customer complaints as customers share their dissatisfactory remarks related to a product, saying that it is not user-friendly. Adding features to our product that would make it more user-friendly would be an option to enrich the uniqueness of our product. Similarly, creating and promoting a unique design of a product, which comes with features that can demonstrate the distinctiveness of our product in comparison to the competitive products, is important. For example, the unique design of the Apple iPhone and its security features differentiate it from other competitive smartphone brands.

■ Source: ▶ pexels.com/royalty-free-images/

Another core purpose of all marketing activities is relationship management with all customer and non-customer stakeholders. In fact, one of the definitions of marketing is developing and sustaining relationships (Brodie et al., 1997). Developing and maintaining a relationship with whom? Developing and maintaining relationships with all customers and non-customer stakeholders, such as suppliers, investors, employees, government, media, retailers, and others, who help a brand to create value and critique the brand's value proposition to improve its market offering. Building and sustaining relationship with all these stakeholder groups, as well as the customers, is imperative for marketers to survive and prosper. This is why, the terms such as "customer lifetime value" or "share of wallet" are more important for marketers, in comparison to "market share." For example, if a customer spends $100.00 in their entire lifetime and spends $20.00 to purchase a particular toothpaste brand, in this case, this $20.00 is this customer's lifetime value or share of wallet for that toothpaste brand. If marketers can increase their customers' share of wallet based on loyalty programs to influence customers to re-purchase and remain loyal to the brand (as we see varied loyalty programs across different industries), the brand's market share will eventually be increased.

Share of wallet can be defined as the percentage of a particular customer's total spent to purchase a specific brand in their entire lifetime, in comparison to the total spending to purchase other competitive brands in the lifetime.

1.2.2 Strategic Marketing and Sustainable Marketing

Strategic marketing is a dual task of marketers. We have seen the alternative definitions of marketing in ► Sect. 1.2.1 (Marketing Definition and Purpose of Marketing). Strategic marketing is the same as marketing; however, the only difference is marketers need to act simultaneously twice as per the marketing definition to exploit the current competitive advantage of a brand, while working for the brand's future competitive advantage. For example, reflecting on Hitt et al.'s (2011) arguments on strategic entrepreneurship to create value for individuals, organizations, and society at large, strategic marketing could be defined as developing insights into key stakeholders' needs, with an aim to managing a firm's resources to simultaneously developing and exploiting the firm's current competitive advantage, while exploring ongoing business opportunities to establish future competitive advantages as a path of value creation.

Strategic marketing is exploiting the current competitive advantage of a brand, while exploring value further creation opportunities to pursue its future competitive advantage.

Let us have a look at a simulated example from the higher education industry to shed more light on the definition of strategic marketing. In general, universities that have full-fledged medical science faculty to offer medical degrees and undertake medical research, these universities usually have higher rankings in the global university league table. This is because the outputs of medical research have a wider impact on society and the global population in comparison to the research outputs of other faculties, such as business. For example, the University of Oxford has invented the Coronavirus vaccine. Imagine how many people around the world have been benefited by the Oxford vaccine since the spread of Coronavirus in early 2020. Now, let us imagine that a university that does not have an existing full-fledged medical faculty but is now planning to launch a complete medical faculty in the next 5-year period. Launching a full-fledged medical faculty is a strategic decision for the university, as they need to think about whether there would be sufficient demand for the faculty, whether they will establish their own hospital or partner with an existing hospital in their city, who would be investors and many other relevant decisions they need to undertake, implement and monitor to launch their medical faculty over the next 5 year period, while the university also need to exploit their current competitive advantage, based on their existing operations at other faculties, e.g., Business, Law, Arts and Humanities, Engineering, etc. It means that the university cannot neglect the current operations of its existing faculties to exploit its existing competitive advantage across those existing faculties while working for the medical faculty to launch in the near future to pursue its future competitive advantage.

Web Exercise 1.2

What is and is not a strategy? Professor Michal Porter is a prominent scholar in strategy. He has a number of seminal contributions to strategic management and strategic marketing research and practice. For example, you may know Porter's competitive forces that help marketers to analyze their competitive market environment. Let's have a look at a keynote of Porter on Strategy here ("Keynote on Strategy by Michael Porter, Professor Harvard Business School"): ► https://www.youtube.com/watch?v=EvvnoNAUPS0, which is just little more than an hour. But, listening at this keynote would be worthwhile, as many students and colleagues (who have watched this keynote) told us that after watching the video of this keynote speech, their perception about what is and is not a strategy has been changed. Therefore, it is highly recommended to watch it at your own pace to think of what is and is not a strategy.

In ► Sect. 1.1 (Introduction), we have discussed that sustainable marketing is not a philanthrophy. It is because a 2018 study on consumption reveals that "if your brand isn't helping your consumers improve their environmental and social footprint, then you're in danger of disappointing 88% of them" (Townsend, 2018, np). In comparison to 2018, it is likely that the number of such socially and environmentally caring customers has increased over the period. In general, we see that a brand, in comparison to its competitive brands, contributes more to a societal or environmental cause, the brand attracts more loyal customers, in comparison to its competitors. For example, two coffee shops that are located next to each other and offer the same price, the same taste of coffee, and the same customer service; overall, both coffee shops offer the same extent of materialistic benefits to their customers. However, one of these two coffee shops, for example, advertises a sign at their front door saying that a percentage of the price that you pay for each cup of coffee, we donate to a particular social or ecological cause. In this case, from where, out of these two coffee shops, would you be purchasing your coffee; however, both offer the same materialistic benefits? Asking this question to a number of post-graduate marketing students across different student cohorts (in between mid-2019 and mid-2024), they answer that all of them would be purchasing their coffee from the shop that contributes to a social or ecological cause. It indicates a latent need or aspiration of contemporary customers and non-customer stakeholders that they prefer that organizations/brands should help them avoid harming society and the environment, but protect it. This is why sustainable marketing is not a philanthropy but a necessity for brands to survive and prosper in the contemporary marketplace.

Sustainable marketing can be defined as contributing back to society and/or the natural environment of a market where a brand operates its business, so that the societal and/or environmental contribution of the brand could underpin its commercial competitive advantage. Sustainable marketing is more challenging for marketers, in comparison to traditional profit-only marketing. It is because, for the former case, marketers need to satisfy the needs of three different groups of stakeholders, such as societal stakeholders (e.g., community groups), environmental stakeholders (e.g., environmentalists), and commercial stakeholders (e.g., business partners, customers), whereas, for the latter case, marketers need to satisfy only the commercial stakeholders. In this context, sustainable marketers should develop insights on the needs of their triple Ps' (i.e., people, planet, and profit, or societal, environmental, and commercial) stakeholders in relation to a particular marketing strategy to design, implement and evaluate. Upon developing an orientation on all three groups of stakeholders' needs and value anticipation, sustainable marketers need to introduce (unlock) a progressive transformation in the value delivery process in order to ensure all three groups of stakeholders' satisfaction based on innovative market offerings that could bring shared growth for marketers, their brands, and their three groups of stakeholders. Such sustainable marketing dynamics are illustrated in ◘ Fig. 1.1, and an example of such sustainable marketing dynamics is clarified in ► Marketing in Action 1.1.

◘ **Fig. 1.1** Visualizing sustainable marketing

Sustainable marketing is contributing back to the society or the natural environment of a market where a brand operates its business, so that the societal and/or environmental contribution of the brand could underpin its commercial competitive advantage.

Following the discussions of this current section (► Sect. 1.2.2), we posit that the aim of sustainable marketing is to pursue the triple bottom line, i.e., people (society), planet (environment), and profit (commercial success), instead of pursuing only profit. It is not necessarily for a philanthropy purpose, but sustainable marketing is a necessity for brands today to adopt to survive and prosper. For example, have you ever imagined how you produce carbon dioxide while using the services of Meta (Facebook) and how you negatively impact the environment while using Facebook? Since each Facebook user produces carbon dioxide while using this service, there was pressure on Facebook from governments and activist groups to address this environmental challenge. Meta invested in improving their technology and the improvement was reported in 2021: "Every one of Facebook's 2.85 million users produces 12 grams of carbon dioxide by using the company's services every year. That is 25 times better than in 2016 when each user produced a little less than 300 g of CO_2 annually" (Collins, 2021, np). It means whether you are a corporate giant, a mid-sized firm, a small and medium-sized business, or a micro entrepreneur, you as a marketer need to invest time and resources to pursue the triple bottom line (people, planet, and profit) to remain relevant with your market offering with the needs and distresses of your target audiences in the contemporary and embryonic contexts of global marketing dynamics.

Marketing in Action 1.1: A Sustainable Marketing Dynamic to Deliver Value to Customers, Partners and Society at Large

Source: ► pexels.com/royalty-free-images/

1

Throughout the world, we have huge food waste every day. In contrary, we cannot feed our people properly in our society, irrespective to the developing and developed economic countries. United Nations (np) reported on 29 September 2023 that "globally, around 13 percent of food produced is lost between harvest and retail, while an estimated 17 percent of total global food production is wasted in households, in the food service and in retail altogether… (which is accounted) for 38 percent of total energy usage in the global food system" (np). In fact, "we are wasting on average 74 kilograms of food at home per person per year" (UN, 2022). ◘ Figure 1.2 shows the percentage of annual average global food waste in meat and poultry, dairy products, cereals, fish and seafood, and fruits and vegetables. In contrary to this massive annual food waste, "between 691 and 783 million people faced hunger in 2022" (United Nations, 2023) across the world. In response to this significant issue across the globe, entrepreneurial marketers have launched an app in 2016, namely "Look, Smell, Taste," under the Too Good To Go (2023) sustainable marketing campaign. Chefs at at commercial kitchens need to chuck away some food by end of each day. It is because that those foods may be good to consume in next 5–6 h but may not be good for human body to consume after 12–20 h once the restaurants reopen on next day. To avoid this operational loss at commercial kitchens, innovative marketers helped to design and promote the "Look, Smell, Taste" app to the commercial kitchens. Since those leftover foods are yet "too good to go," chefs at restaurants closer to the end of a day, e.g., at 8.30 pm upload those leftover foods on to the app and offer them on a reduced price, e.g., a $10.00 food at $5.10. Subscribers from local areas receive instant notification of these offers to purchase this food within next 2–3 h before the restaurant closes for the day. As a result, restaurants are able to not only reduce their operational loss, as before they needed to chuck away those foods, but also they are able to make some profit, e.g., instead of making $5.00 profit by selling a food at $10.00, they still make $0.10 profit by selling it at $5.10 instead of chucking it away. This sustainable marketing campaign helped saving 78,873,299 bags of meals in 2022 (Impact Report, 2022). In this way, the innovative sustainable

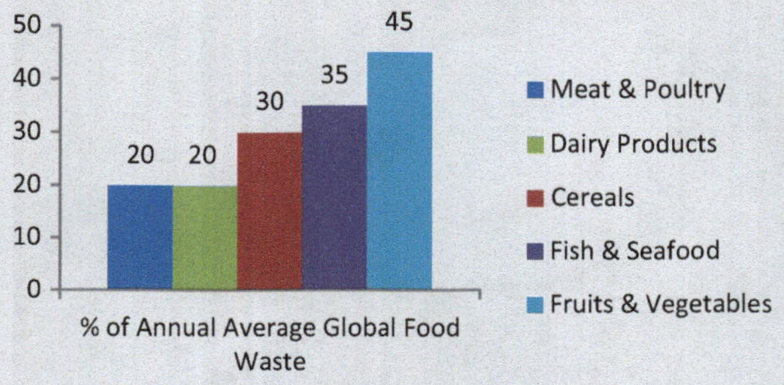

◘ **Fig. 1.2** Percentage of annual average global food waste (IFCO Systems, 2023)

marketers are able to contribute the following utilities (value) to their business partners, customers, and society at large:
Feed more people in our wider society.
Help restaurants (the customers of "Look, Smell, Taste" app) to reduce their operational loss, while enabling them to make some profit.
Share profit with the app developers as business partner.
And contribute to environment by reducing energy waste, considering that the amount of our annual global food waste is accounted for 38% total energy usage in the global food system (United Nations, 2023).

Like the "Too Good to Go" sustainable marketing campaign, as a sustainable marketer, think about different global socioeconomic and ecological challenges to integrate additional value to the market offerings across different industries, to pursue the triple bottom line, namely People, Planet, and Profit.

1.2.3 Customers' Problems and Expressed Need vs. Latent Need: Marketing a Solution, Not a Product Itself

In ► Sect. 1.2.1 (Marketing Definition and Purpose of Marketing), we have discussed dehydration as a problem for customers. In relation to the dehydration problem, customers need to drink more liquid. In response to this customer need, we have also discussed that alternative solutions, such as mineral water, cola, or coffee, are currently available at supermarkets to satisfy the rehydration need to address the dehydration problem. We also have discussed that if a supermarket or just a micro entrepreneur starts selling 100% preservative-free fresh green coconut water or sugarcane juice, some health-conscious customers would be delighted to purchase preservative-free, fresh green coconut water or sugarcane juice instead of purchasing cola or coffee. In this essence, the solutions to customers' problems that are already available in the market (e.g., cola or coffee, in this case) can be considered as already expressed solutions for customers to purchase one to satisfy their rehydration need to address their dehydration problem. In this context, offering preservative-free, fresh green coconut water or sugarcane juice could be considered as customers' latent aspiration or latent need to satisfy the same dehydration problem, considering that a considerable segment, i.e., the health-conscious customer group may be delighted to purchase and consume more healthy drinks; however, those preservative-free healthy drinks, e.g., fresh green coconut water or fresh sugarcane juice may not be available currently at supermarkets at least in many developed economic countries. Therefore, we define latent need as accepting a new utility or value as the solution to a customer problem with delightful satisfaction and without any predetermined plan to consume the utility. It means a health-conscious customer went to a supermarket to purchase a cola to rehydrate. But suddenly, the customer discovered that the supermarket also sells a new drink that is preservative-free and better for the human body. In this case, the health-conscious customer would delightfully purchase that new drink instead of previously available substitute products, e.g., cola or coffee, and such consumption behaviors are not an outcome of a predetermined plan to purchase and consume that new drink because the customer did not know about the new healthy drink, prior to their visit to the supermarket.

Latent need could be defined as accepting a new utility or value as solution of a customer problem with delightful satisfaction, and without any predetermined plan to consume the utility.

A latent need can also be explained further based on the presents/gifts that we receive from time to time from our family, friends, and colleagues. It happens sometimes that we have received a present, but we never thought about the utility of that present (i.e., a specific product or service), before receiving it. But, as soon as we receive the present, we feel a "wow" feeling, like this is exactly what I needed to satisfy one of my consumption needs, in relation to addressing a particular problem. Like this case, many alternative solutions currently remain latent in the minds of customers and non-customer stakeholders. As sustainable marketers, we need to explore and explain those latent needs of our triple Ps' stakeholders, i.e., People, Planet, and Profit (more precisely, societal, environmental, and commercial) stakeholders, to delight them, to create that "wow" feeling in the mind of our triple Ps' stakeholders.

Human psychology is a wonder of all time, considering that, in general, it is difficult to satisfy human beings. There is an ancient saying that "a mountain full of gold, double that will not be enough to satisfy the desire for single soul!" (as cited in Sterling, 2021). But for marketers, the good thing is we do not need to satisfy everyone; that is not possible either. So, the question is, how could we strive to bring that "wow" feeling into the minds of our triple Ps' stakeholders? To answer this question, marketers need to dive deep into the mind of their target audience (i.e., the triple Ps' stakeholders), considering that marketing (influencing our target audience to let them trust why doing business with us is more profitable for them, instead of making business with our competitors) occurs in between the mind of marketers and their target audience. The solution or value you sell as a product or service to satisfy a particular need of your target audience, to sell that product or service, put yourself first in the position of your target audience, and think of the utilities (related to its marketing mix) that you would have preferred, if you were the customer of that product or service that you are selling. Try to explore what additional utility of the product would be cherished by your target audience but is not currently offered by your brand and its competitors.

Such additional utilities may not need to be that expensive to delight your target audience. For example, someone needs to walk for 20 min on an early breezy winter morning to catch the metro to reach the office, and the person feels cold every morning while walking for these 20 min. Let's assume that the person was presented by their loved one with a jacket that comes with an extra-high collar, and that extra-high collar now protects that person's shoulder, neck, and outer ear areas from the chilly wind during early winter morning. As a result, that person started feeling thankful for this present because of its extra warmth, as they never thought before about a jacket with an extra-high collar to protect from the chilly winter breeze. If you are a marketer of garments products, adding that extra warmth (extra utility) in your product could bring a "wow" feeling in the mind of your customers, as similar to this simulative case example, there are another five hundred thousand customers who may want that extra warmth, but they did not think about it before you started offering it. If marketers could dive deep into the mind of their target audience to understand their latent aspiration precisely, marketers would realize that there are numerous aspirations (needs) currently latent in the minds of their customer and non-customer stakeholders. If we compare the currently available expressed needs in the market, i.e., the existing products and their utilities, which are currently offered by competitive brands, we may realize that the needs (i.e., the diverse small contexts of extra utility for the same product) that are currently latent in the mind of our customers, we may find that the currently available expressed needs are actually the tip of an iceberg on the ocean of human psychology, which is represented symbolically in ◘ Fig. 1.3. Therefore, to delight our target audience, as marketers, we need to dive deep into the mind of our target audience to:

- Understand our target audience's needs related to the problem that they want to address.
- Analyze currently available expressed needs in the market and compare them with prospective latent needs, to add those extra features to our product to address our target audience's problem in a way that would bring that "wow" feeling in the mind of our target audience.

While designing our new market offerings centered on a latent need, marketers should consider the marketing mix (i.e., decision on product, price, place, and promotion), to align that latent need with the product design, pricing option, delivery strategy and promotional plan. In terms of promotional plan, we posit that marketers do not sell a particular product or service itself, but the utility of their product or service as a solution of customers' problem. Let us have a look at an example

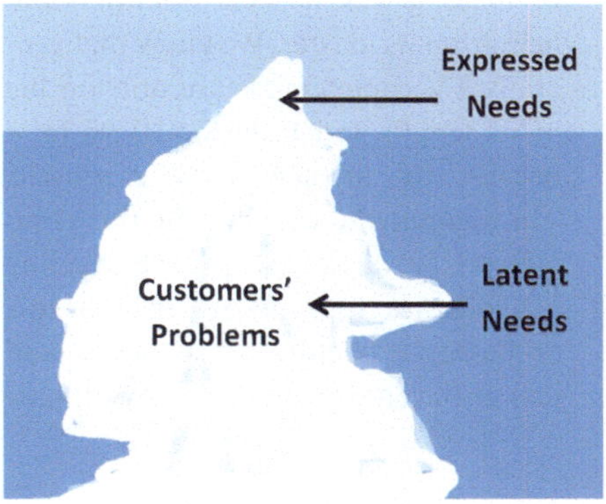

◘ **Fig. 1.3** Expressed needs vs. latent needs

in ► Marketing in Action 1.2 where we see how Novartis, the pharmaceutical brand sells a solution (i.e., the benefit of having a healthy pet) to address a customer problem (i.e., a poorly pet), instead of selling the pet medicine, itself.

Marketing in Action 1.2: Novartis Sell a Solution of Their Customers' Problem, But Not Their Product Itself

■ Source: ► pexels.com/royalty-free-images/

In a TV commercial of Novartis' pet medicine product category, it was shown that few people at home spending casual time at living room, and an unhealthy-looking dog of that home was staring from the living room's door to the people at home. The expression of the dog was, it needs attention and help from its master. We know that generally TV commercials are not too big in its length and lasts for 30–90 min, because the airtime is too expensive. Throughout the duration of this commercial, Novartis was mostly reminding its audience (promoting) the benefits of having a healthy pet at home, and just at the end of the commercial, how Novartis can help their customers to keep their pets healthy was added very briefly. It is because that Novartis' customers are concerned about the health problem of their pets. To promote one of their pet medicines, Cosequin (2023) via ► https://www.vetscriptions.co.uk/collections/dog/novartis, Novartis clarify that "Cosequin is used in dogs to aid in maintaining optimal joint function, cartilage production and to protect existing cartilage from breakdown" (np). It is because that optimizing the function of cartilage and protecting it from breakdown in dogs' body is the main concern of Novartis' customers, in relation to the pet medicines offered by Novartis. Look at Novartis' pet medicines/products at ▸ https://www.vetscriptions.co.uk/collections/dog/novartis, and at the website of other pharmaceutical retailers, and think of the solution Novartis (and other retailers) are promoting to address the problem of their customers, the pet owners, instead of selling their product itself at first instance.

Information source: Czinkota et al. (2021)

1.2.4 Identifying and Nurturing a Latent Need

Source: ▸ pexels.com/royalty-free-images/

Precisely identifying a latent need of customers that would be expected and accepted by them is a challenge; however, nurturing or creating the market demand for a latent need is further challenging. In general, we do not need rocket science to design a product or service based on latent needs, or the new or modified product that has been developed or modified for the first time does not necessarily need to be expensive to satisfy its customers and to establish a competitive edge. For example, when we make espresso coffee at home or at a commercial kitchen, we need to put the coffee powder in a small container, i.e., the coffee tamper, to make a very good blend of the powder. It is because the better we can make the blend of coffee powder, the better the taste of coffee will be. Imagine on a busy weekend night at a restaurant, the restaurant owner's priority would be making as many cups of coffee as each cup of coffee would taste as best as possible to serve more customers in less time while employing fewer casual staff members. However, making a good blend of coffee powder using a traditional tamper would take some time. Even after spending some considerable time (e.g., 20–40 s) to make a good blend, it would be difficult to ascertain how good the blend can be made using the traditional tamper. Recognizing this latent need of restaurant owners (i.e., serving more customers with the best possible coffee within less possible time and employing a less possible number of casual restaurant staff members), entrepreneurial marketers launched the push tamper using one of the simplest technology that you can see on this video clip of push tamper ("PUSH coffee tamper by Rinaldo's"): ► https://www.youtube.com/watch?v=YLUVAHB16w8. As you see, the technology that is used to develop that new product, the push tamper, to replace the traditional tamper, is not only a quite simple technology but also the push tamper itself as a new product is not that expensive, in comparison to the traditional tamper. Nevertheless, the push tamper satisfies the restaurant owners' latent need to serve more customers in less time with better-tasting coffee, as the push tamper helps make a much better blend of coffee powder in comparison to the traditional tamper, while saving time for coffee makers.

◘ Source: ► pexels.com/royalty-free-images/

Let us have a look at Facebook's example to see how they nurtured the target market or created a market for a new product or service that is developed to serve a latent need. We know that for a long time, there was only the "like" button to press on Facebook's wall for its users to share their expression using Facebook's service. On the one hand, Facebook wanted to increase the interactions of their users on Facebook wall because more interactions from more users would bring more business opportunities for Facebook. On the other hand, Facebook anticipated that if they introduce a few other buttons for users besides the traditional "like" button, such as "love," "wow," "yay," or "dislike" buttons, Facebook users, as their latent need to communicate their expressions more expressively, would delightfully accept all or most of those new buttons to use and express their emotions more precisely on Facebook wall. However, Facebook did not introduce those additional buttons straightaway, without understanding their customers' reactions to the prospective new buttons. To understand their customers' reactions, Facebook conducted market research, inviting a number of sample groups across several types of Facebook users around the world. Facebook requested these sample groups to use those new buttons and share their experience with Facebook. Following the customers' insights derived from this market research, Facebook has not introduced the "yay" and the "dislike" buttons. To clarify the reason for not introducing the "yay" and "dislike" buttons, Chris Cox, Facebook's Chief Product Officer, described to Bloomberg that "a 'dislike' button had been 'rejected' on the grounds that it would sow too much negativity," and "the expression 'yay' was also shot down because it's not universally understood (among the Facebook's users across the world)" (Eadicicco, 2016, np). Therefore, it is important to understand the customers' reactions first through market research, before launching a new or modified product based on a latent need or before investing too much in it.

■ Source: ► pexels.com/royalty-free-images/

We have seen in the cases of push tamper and Facebook, marketers did need neither rocket science, nor big investment to introduce a new or modified product to serve a latent need, which could be true in many cases. However, in some cases, we may need some considerable investment to serve a latent need, especially depending on the extent of societal, environmental, and commercial impacts we may want to bring in, based on serving a particular latent need. For example, when we go shopping for our groceries at supermarkets, we as customers' problem is that we cannot carry the bulk amount of groceries by hand without any bags. In relation to this customer problem, customers need a bag or number of bags to carry their purchases from supermarkets to home. Unfortunately, we are yet to use environmentally damaging plastic bags throughout the world to carry our groceries and other shopping items. Therefore, "every day, the equivalent of 2,000 garbage trucks full of plastic are dumped into the world's oceans, rivers, and lakes. Plastic pollution is a global problem. Every year, 19–23 million tonnes of plastic waste leaks into aquatic ecosystems, polluting lakes, rivers, and seas" (UNDP, 2023a, np). Because of this wide and non-responsible use of plastic and plastic products across the globe (Plastic Oceans, 2022; Surfers Against Sewage, 2023):

- One million marine creatures are murdered each year because of plastic pollution!
- 100% of the mussels that we consume today contain microplastic!
- One in three fish caught today for human consumption contains plastic!
- 40 pounds of plastic are consumed by human beings today in their lifetime!
- 90% of plastic is made from fossil fuel!
- 2.5 tonnes of CO_2 is originated by producing 1 tonne of plastic!
- A warning is that "there will be more plastic in our oceans than fish by 2050!" (np).

▪ Jute trees, processed jute, and 100% biodegradable jute-made products. Source: ▸ pexels.com/royalty-free-images/

To address this global societal and ecological challenge related to plastic and plastic products, we have seen Thailand start using banana leaves for packaging at supermarkets, replacing plastic bags (Forbes, 2019). "In a bid to decrease plastic waste, Bangladeshi scientist Mubarak Ahmad Khan invented a plastic bag alternative with the help of a local natural fiber" (CNN, 2023, np). The local natural fiber used to invent this 100% biodegradable plastic bag is jute, a natural fiber produced in Bangladesh. This decomposable jute-made plastic bag is branded as "Sonali Bag." The meaning of Sonali in Bangla is golden. Since the processed jute looks golden, the jute-made fully biodegradable plastic bag is branded as "Sonali Bag," which has been adopted by UNDP (2023b) to combat global plastic pollution.

▪ Source: ▸ pexels.com/royalty-free-images/

The 100% biodegradable "Sonali Bag" as an alternative to non-biodegradable plastic bags, or the "Look, Smell, Taste" app that is discussed in ▸ Marketing in Action 1.1 are, however, invented by scientists or programmers; sustainable marketers have options to introduce such ideas and partner with scientists and investors to address the local, regional and global societal and environmental challenges. In this way, responsible marketers can introduce positive changes in our global society for a better and sustainable tomorrow for us and our future generations. In this current section, we are discussing about identifying and nurturing latent needs. Let us have a look at how tea was made popular first in India by Europeans. Tea was an unknown new drink to Indians before the local Indian tea market was nurtured by British companies to market tea in India. During the medieval era, once Europeans first went to India, they found that India was a mountainous country, which was particularly good for cultivating tea commercially. However, cultivating tea commercially in India was not sensible enough unless marketers could create a local market for this new drink in India. It was because, during the medieval era, cultivating tea commercially in India solely for the purpose of exporting it to other global regions was not feasible from the context of cost-effectiveness. The European marketers envisaged the latent need for tea as a new drink for Indians to plan a marketing campaign around India to create its local market. There is a myth that the European marketers went to various places around India, where people hang around with their family and friends, such as marketplaces and other social gathering places, and started offering the new hot drink for free. First week, the second week, and again during the third week, the new drink was offered for free around India. On the same day and at the same time during the fourth week, more people (through word-of-mouth) gathered together at those venues to have that new drink again or for the first time. But, during the fourth week, the local potential customers did not find the Europeans there offering that new drink. Once the local market was nurtured, i.e., once the demand for that new drink started to increase over those four weeks around India, the marketers then started offering tea subsequently with a price that would be affordable by their local customers in India, as well as good for retrieving their return on investment. In support of this myth, the Tea Lovers (2021) describe that "initially, free samples of tea were offered from horse-drawn carts belonging to various companies…Brooke Bond, an English tea company, started experimenting with a fleet of horse-drawn vans for distributing teas. The British tradition of taking tea with a little milk and sugar was introduced along with the samples" (np).

1.2.5 Value, the Core Concept of Marketing

Source: ► pexels.com/royalty-free-images/

The utility of a product or service to satisfy customers' needs is the core value of marketing. For example, airlines offer a core utility, which is helping its customers to travel a longer distance within a shorter period of time, in comparison to other modes of public transportation, such as bus, train, self-driving, and boat. To obtain this core utility, airlines' customers need to sacrifice certain things. Time to search for the best possible deal, the amount to pay for a flight, and travel time to reach the airport are a few of such things that customers need to sacrifice in order to obtain that core utility of airliners' service. Therefore, value is simply what we, as customers, sacrifice and what we receive in return for that sacrifice.

To obtain that utility or value of a product or service to satisfy our needs, as customers, we need to plan and organize certain activities based on our budget, tastes, and preferences in order to compare the alternative value propositions among the competitive products or services to finally select the most relevant value that would satisfy our needs most meaningfully. For example, you have decided to invite some of your friends for dinner. You have a certain budget, preferences for one or a few diverse types of cuisines, and other utilities, such as live music at the restaurant for your dinner. Based on these preferences, you have shortlisted four restaurants and finally have selected the restaurant from this shortlist where you are inviting your friends. As soon as you have selected the restaurant out of the alternative competitive restaurants, you have considered a plan in an organized way, that is, you have now anticipated what you will sacrifice and, in return, what utility you will obtain from the restaurant, which can be called as your anticipated value as your expectation before consuming the dinner. So, customers' anticipated value could be defined as customers' own interpretation of a particular product's utility before they consume it and how they think of what they may need to sacrifice and what utility they may obtain in return for their sacrifice after they consume it.

Anticipated value is customers' own interpretation about a particular product's utility before they purchase and consume it, and how they think of what they may need to sacrifice and what utility they may obtain in return of their sacrifice after they consume it.

Once the dinner commences, you have started comparing consciously or subconsciously what the restaurant promised to you before you made the booking and what they are serving during the dinner. Such comparisons may relate to the taste of food, the friendliness of restaurant staff members, whether the live music is as good as you thought, and so on. By the time the dinner is completed, you have developed your actual perceived value based on your overall dinner (consumption) experience related to your initially anticipated value. Thus, perceived value can be defined as the difference between a brand's promises to its customers and what it delivers to the customers. If customers' perceived value meets or exceeds their initially anticipated value, i.e., you are more satisfied after the dinner than you expected before the dinner, it will be a win-win for both you and the restaurant owner. It is because the restaurant would be able to attract more customers through your satisfied word-of-mouth, and you may receive discounts in the future because of your loyalty.

Perceived value is the difference between a brand's promises to its customers and what it delivers to the customers.

On the contrary, after the dinner, if your perceived value is that the restaurant delivered less in comparison to what they promised before the dinner, you may not repeat purchases from them. Also, the restaurant will not receive additional customers from your satisfactory word-of-mouth, but there would be a chance that they would lose some potential customers from your family, friends, and colleagues because of your dissatisfactory word-of-mouth about them. It means that customers' actual perceived value depends on the extent of their satisfaction and dissatisfaction after they consume a particular product, which can be positive or negative, depending on what a brand promises to its customers and what they deliver. Persistency in keeping promises is also important to ensure an uninterrupted value-delivery process for a flawless customer experience to establish a crowd of loyal customer base.

Following this dinner example related to initially anticipated value (before the consumption) and actual perceived value (after the consumption), let us summarize a few things that you needed to confirm for this dinner. You needed to plan some activities in an organized way, such as shortlisting and selecting the restaurants related to your preferences, letting your friends know the time and duration of the dinner, address of the restaurant, transportation options to and from the restaurant, and so on. You needed to sacrifice your monetary resources (e.g., the price for the dinner), psychic resources (e.g., deciding the best possible restaurant among the alternative competitive ones), and physical resources (e.g., outbound and inbound walking or driving to and from the restaurant) to organize the dinner. Also, whether you will become a repeat customer of that restaurant or not will depend on the difference between your initially anticipated value and your actual perceived value before and after the dinner. Therefore, value can be defined as "an anticipated outcome of any sort of planned and organized activity. The activity could be derived from monetary, psychic, or physical resources. The more the outcome meets initial anticipation, the more the possibility of win-win outcomes or value optimization for all involved stakeholders" (Shams, 2013, p. 244). ► Section 1.5.4 (Value Proposition for People, Planet, and Profit) and ► Manager's Corner 1.3 (IKEA's Core and Added Value for Customers, Partners, and Society at Large) discuss more about value and value proposition for sustainable marketing, i.e., targeting to the triple Ps' (People, Planet and Profit or societal, environmental and commercial) stakeholder groups.

Value is "an anticipated outcome of any sort of planned and organized activity. The activity could be derived from monetary, psychic, or physical resources. The more the outcome meets initial anticipation, the more the possibility of win-win outcomes or value optimization for all involved stakeholders" (Shams, 2013, p. 244).

1.2.6 The Art and Science of Marketing

What comes to your mind first when you compare the core value offered by ALDI, the German multinational supermarket chain, with its competitive supermarkets across different global regions, such as Woolworths or Coles in Australia; Tesco,

Sainsbury's, or Asda in the UK; Kroger or Target in the USA? In response to this question, a number of postgraduate marketing students across different cohorts answered between mid-2019 and mid-2024 that ALDI is more affordable or slightly cheaper than its competitors, which is exactly what ALDI's marketers wanted their customers to think and feel about them. Indeed, influencing or forming the perception of our target audience about the value of our brand is the art part of marketing. It is not easy to influence or form that perception of a brand's target audience. It needs a consistent, integrated marketing communication or branding strategy to convey the brand message to its target audiences (customers and non-customer stakeholders) about the brand's value and promises to its customers. In such efforts to influence or form the target audiences' perception, the most important consideration is ensuring consistency in what a brand promises to its customers and non-customer stakeholders and what it delivers over an extended period of time.

Art of marketing is influencing or forming the perception of a brand's target audience about the brand's core value and its promises.

As part of ALDI's (n.d.) promise as an affordable supermarket and their "everyday low prices" commitment, they assert that "we do our best to keep prices stable through all the world events and unexpected challenges that arise, but occasionally there will be fluctuations. As part of our commitment to everyday low prices, if prices do go up, we make it a priority to bring them down again as soon as possible" (np). As ALDI is quite consistently keeping their "everyday low prices," and many of their customers may trust it; ALDI is becoming America's fastest-growing grocer (Dean, 2023a), and they are "buying 400 grocery stores in Florida and the southeast as it continues its rapid US expansion" (Dean, 2023b, np).

It, however, could vary from customer to customer based on their personal shopping experience to ascertain whether ALDI is more affordable than most of its competitive supermarket brands. One of the authors' (of this core textbook's) firsthand experience is that ALDI was cheaper in the pre-COVID pandemic era, but in the post-pandemic era, ALDI is struggling to keep its promise of "everyday low prices" in comparison to its competitive supermarkets. Again, it is just one individual customer's perceived experience about ALDI's value, which can contradict the perceptions of many other grocery shoppers. For example, many postgraduate marketing students, even in the post-pandemic era, confirmed that they found ALDI is still relatively cheaper than its competitors. So, the question is, how many ALDI customers in their total market think about ALDI in what way? Perhaps some ALDI's customers perceive that they deliver 100%, some may perceive 50%, and some other customers may perceive that they deliver 0% of their promises. As part of marketing research, collecting customer data to understand different customer segments' perceptions about a brand's value propositions and analyzing that data based on business statistics or business mathematics in order to

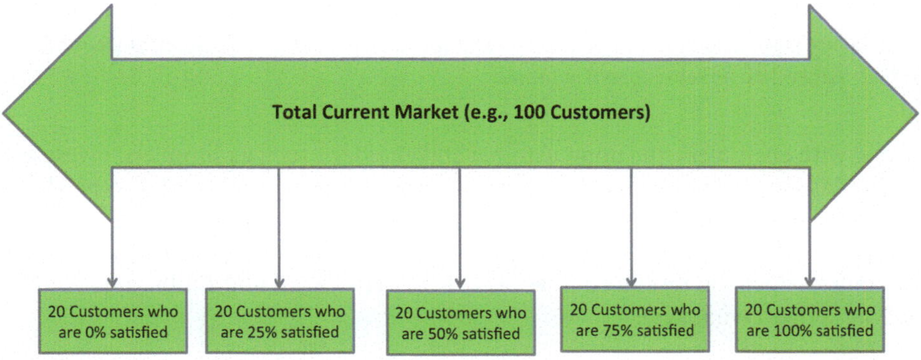

Fig. 1.4 Symbolic representation of the application of marketing science to understand our customer and non-customer stakeholders

re-strategize the marketing plan reflecting on customer segment-specific perceptions is, therefore, the science part of marketing. Figure 1.4 represents such applications of marketing science to understand our customers and non-customer stakeholders' perceptions.

> **Science of marketing** is collecting market data to analyze it in order to re-strategize marketing plan reflecting on customer segment-specific perceived experience about a brand.

1.3 Marketing Dynamics

> Marketing management and the allied dynamics—as it is acknowledged today have been evolving persistently based on the mutual endeavors of academics and reflective industry-based practitioners. Marketing academics and practitioners proactively deal with various market (socioeconomic and ecological) issues, competitive forces, and challenges, as well as flourish opportunities and create value to nurture contemporary and latent market needs. (Shams, 2018, p. 151).

In economics, the term "market" is traditionally referred to as a place (either a physical place or an online space) where two or more parties meet to exchange value among themselves. However, in marketing, the "market" is the people who are current and potential customers and non-customer stakeholders of a brand. Figure 1.8 represents the key non-customer stakeholders of a brand.

1

Market is the combination of people, who are current and potential customers and non-customer stakeholders of a brand.

Along with customers, the non-customer stakeholders, such as suppliers, investors, or employees, are considered as a market, considering that a brand also needs to satisfy these key stakeholders, as they help the brand to create value for its customers. Different factors influence the ongoing evolvement of our global marketing dynamics, which are discussed in ► Sect. 1.3.1 (Historical and Embryonic Nature of Marketing Dynamics).

1.3.1 Historical and Embryonic Nature of Marketing Dynamics

Marketing (i.e., influencing others to act in a way that is preferred by the influencer) as a field of practice has been in existence since the ancient era of human civilization. Let us clarify it based on the words of Professor Philip Kotler. As per the sacred values, according to Kotler, the snake who influenced Adam and Eve to act in a way that was preferred by the snake was the first creature who practiced marketing for the first time to influence a target audience. A keynote by Professor Kotler ("Marketing with Philip Kotler") can be accessed here: ► https://www.youtube.com/watch?v=sR-qL7QdVZQ, where Professor Kotler (2012) talked about the evolvement of marketing, including the snake and Aristotle, the Greek philosopher.

◘ Source: ► pexels.com/royalty-free-images/

"Rhetoric is the faculty of observing in any given case the available means of persuasion" (Kotler, 2012, np). In rhetoric faculty, a speaker's ability, i.e., crafting words to make his speech credible, stimulating the emotions of the target audience, and ensuring the authenticity of the message delivered, is important to influence the target audience (as discussed by Kotler, 2012). Persuasion or influencing or forming public or individuals' opinions about a brand's promise, credibility, or authenticity is still a particularly essential element to planning for marketing mix strategy for the contemporary marketplace. We see that marketing as a field of practice has been in existence since the ancient era; however, marketing as a field of academic discipline has started evolving since the late nineteenth century. ◘ Table 1.2 presents a brief history of marketing as a field of academic discipline and the evolving aspects of marketing dynamics in practice since the first decade of the twentieth century, summarizing the different eras and relevant markets, market influences, and marketing's response to the market influences over the period.

◘ Source: ► pexels.com/royalty-free-images/

Following the Industrial Revolution, which started in the mid-late nineteenth century, the production of goods increased radically. As a result, investors and marketers needed to think of a more structured plan to sell those increased amounts of produced goods. Thus, the transactional sales concept was considered in the marketing school of thought in a more structured way to analyze business performance based on the "flow of goods and services from producers to consumers" (Keefe, 2004; as cited in Ringold & Weitz, 2007, p. 255). This early era of the marketing school of thought remained until the late 1940s, and we saw that during this era, marketing as defined its performance was measured based on ensuring a smooth flow or transaction of goods between sellers and buyers. This era of marketing dynamics belongs to the age of the early modernist market. The ongoing improvements in technology and the waves of industrialization propelled human civilization to produce increased goods. However, during the next decade (the 1950s), a massive amount of increasingly produced goods in factories remained unsold. To sell these unsold goods, marketers needed to think of more structured ways of offering utilities to customers centered on customers' needs, wants, and expectations related to product design, pricing preference, distribution ease, and

Table 1.2 The evolvement of marketing dynamics: Different eras, the market influences, and marketing's response to the changes over time. (Reprinted a revised version by permission of the copyright owner, © Springer)

Era	The market	Market influences	The role of marketing in response to the market influencers
Early 1900s to late 1940s	Early modernist market	– Industrialization and mass-production (Hassard, 1993; Thomas, 1997)	– Transactional sales concept
Early 1950s to late 1960s	Late modernist market	– Unsold mass-produced goods	– Introduction of added utility based on the 4Ps marketing mix
Early 1970s to late 1980s	Early postmodernist market	– Knowledge democracy (Graham, 2003) – Ease of access to information – Value optimization (Graham, 2003) – Symbolic (form and style) representation of value – Service preference (Gummesson, 2008)	– Value-based marketing – Niche and customized value proposition (Stavros, 2005)
Early 1990s to pre-2000	Mid postmodernist market		– Stakeholder relationship management for value optimization within a value network (Gummesson, 2008)
Post 2000	The millennial postmodernist market		– Service dominance logic: Integration of resources among the key stakeholders and value co-creation (Gummesson, 2008); – Corporate social responsibilities, environmental issues and social well-being (Gummesson, 2014)

Era	The market	Market influences	The role of marketing in response to the market influencers
Post 2010	The triple Ps' stakeholders focused market	– Increased concerns for climate change and intensified societal, environmental and geopolitical concerns	– Social business – Pursuing the Sustainable Development Goals through business activities – Sustainable marketing – Circular economy
Post 2020	The mixed-zone market	– Metaphorical space – Synchronized space – Traditional space	– Pursuing the Sustainable Development Goals through business activities – Value proposition for triple Ps' stakeholders – Extended marketing mix: Product, price, place, promotion, public opinion formation, political power, people, participants, physical evidence, process, public relations, probe, partition, priority, position, profit, plan, performance, positive implementation, planet, preference, proving advantage (Baumgartner, 1991; Booms & Bitner, 1981; Borden, 1964; Judd, 1987; Kotler, 1984; Kurtz, 2009; Looney & Seiler, 2003; McCarthy, 1960; McCarthy & Brogowicz, 1981)

promotional appeal. As a result, the 4 Ps' (Product, Price, Place, and Promotion) Marketing Mix concept (Borden, 1964; McCarthy, 1960) was introduced to offer more utilities to customers while competing among the competitive brands based on offering superior added utility (or value) centered on the 4Ps value proposition. This era (the early 1950s to late 1960s) was known as the late modernist market.

◘ Source: ► pexels.com/royalty-free-images/

Since the following decades, the late 1960s and onward, civilization witnessed an information revolution. During this time, man first landed on the moon. Because of the information revolution, in the next three decades, until the year 2000, the concepts of knowledge democracy (Graham, 2003) and ease of access to information were introduced, and subsequently, the launch of the internet ensured the enormous flow of information between countries, and buyers and sellers across different industries. In the following decade, the first decade of the twenty-first century witnessed the launch of social media. The enormous flow of information about competitive brands propelled value optimization (Graham, 2003) battle among the competitive brands through additional service propositions (Gummesson, 2008). During these four to five decades (late 1960s to late 2010s), a kind of hyper or symbolic form of style as an extended representation of value was introduced that intensified the mass consumerism culture across the world. Because of such a symbolic form of value optimization, we have seen numerous prestigious brands were born during this period. For example, a basic Toyota or an expensive Lamborghini can both transport its drivers and passengers from one place to another place within the same time, or a simple Casio wristwatch or an expensive Rolex wristwatch both offer the same core utility, i.e., informing time; however, the symbolic representation of value propelled the waves of mass consumerism across the global market, and has created a huge number of brand-savvy customers.

During this period, stakeholder relationship management-focused marketing or relationship marketing was introduced to offer more individualized value to customers. As a result, the concepts of niche marketing and customized value proposition were applied to influence the target market.

We have seen the two forms of individualization or customization. One is 100% individualized customer experience. For example, haircut and beauty parlors, making your own pizza, or designing your own car (as we see on the websites of different pizza or car manufacturing brands) attempt to offer a high-extent customer experience based on individualized preference of value. Another form of customization we have seen is called "mass customization." For example, Unilever may offer ten to fifteen (or more) different types of shampoos to its customers. One shampoo is better for silky hair, one is better for reducing dandruff, one is good for fluffy hair, and the other one may be good for glossy hair. There may be three million customers in one national market who all want silky hair. Unilever's shampoo, which is better for silky hair, is mass-customized for all those three million customers, who all want their hair to look silkier. During this period, the service-dominant logic (Gummesson, 2008) received marketers' interest in optimizing service-based value propositions. This period could be divided into three eras, based on the gradual evolvements in market influences, customers' preferences, and marketers' responses, such as the early postmodernist market (early 1970s to late 1980s), mid postmodernist market (early 1990s to pre-2000), and the millennial postmodernist market (post-2000, the first decade of the twenty-first century). By the turn of the century, from the twentieth to twenty-first century, marketers also started introducing more socially and environmentally caring factors to compete among the competitors, such as corporate social responsibility, environmental concern, and social well-being (Gummesson, 2014), considering that the global society started becoming more conscious about the impact of climate change, and societal and geopolitical issues.

Web Exercise 1.3

Visit the websites of different car manufacturing brands, such as ► ford.com or ► hyundai.com and think of the extent of individualized experience the brands are offering to their customers.

Replicate the exercise with different pizza brands, such as ► pizzahut.com or ► dominos.com.

Remember that you may need to visit the local websites of such brands, based on your current country of residence.

■ Source: ▶ pexels.com/royalty-free-images/

The increased concerns for Climate change and intensified societal, environmental, and geopolitical concerns propelled human civilization more than ever before in the second decade of the twenty-first century, i.e., the post-2010 era. As a result, the "Sustainable Development Goals (SDGs) were adopted in 2015 by the 193 United Nations (UN) member states. There are 17 goals which address economic, environmental, and social impacts, and are designed to form a blueprint for good growth, nationally and internationally" (PricewaterhouseCoopers, 2023, np). We have also seen that a 2018 study reveals that "if your brand isn't helping your consumers improve their environmental and social footprint, then you're in danger of disappointing 88% of them" (Townsend, 2018, np), and this number of responsible consumers may have increased since 2018, especially in the post-COVID pandemic era. Therefore, sustainable marketing is not philanthropy, but marketers have adopted the sustainable marketing concept to satisfy the increasing number of responsible consumers around the world while contributing to the global society's pursuit of the SDGs.

Source: ▶ pexels.com/royalty-free-images/

Also, the social business concept has received global interest since the year 2006, when Professor Muhammad Yunus and his social business organization, the Grameen Bank, received a Nobel Prize in 2006 because of changing people's lives through the practice of social business. Today, there are different forms of social businesses. However, the "microcredit" form of social business was initially practiced in the early postmodernist market in 1973 by Professor Muhammad Yunus to help underprivileged communities. For example, in the traditional banking system, a core eligibility to obtain a bank loan to start a business is that the loan receiver needs to mortgage valuable property to the bank in return for the bank loan. Unfortunately, not too many people around us have the ability to mortgage a valuable property to obtain a bank loan to launch even a small business. However, every day, millions of novel ideas to contribute to human society around the world are becoming futile, considering that many individuals do not have the required support and empowerment to work for their ideas, with an aim to change their own lives while contributing back to society. To address this problem, the "microcredit" form of social business lends a small amount of money to a small group of underprivileged people who work together for their own small venture in order to create and multiply value based on that microcredit, with an aim to gradually pay back the loan, while progressing toward self-reliant to improve their socioeconomic condition. A 2018 report shows that there are Yunus Social Business Centers at ninety-seven universities across the world (YSBC, 2018). Most universities in the world today teach social business to Business Faculty, and social business is widely practiced today across the world, considering that the concept helps marketers pursue our global SDGs.

The urgency to pursue global SDGs propels business researchers and executives to think of more sustainable means of resource usage to optimize the value of limited resources. In this context, the concept of circular economy received popularity across the world during this post-2010 era. The application of the circular economy concept helps reduce material waste while refurbishing the waste to integrate value into the waste to reuse it so that the lifespan of a single material can be extended. "Circular economy (, 2023, np) is therefore to make the most of the material resources available to us by applying three basic principles: reduce, reuse and recycle," which is also referred to as the Rs of circular economy. Subsequently, the Rs in the circular economy concept is expanded to transform waste into a new resource. ◘ Figure 1.5 describes the Rs of the circular economy.

The circular economy model is defined as a responsible process of production and consumption that focuses on sharing, recycling, refurbishing, and reusing limited current resources (e.g., a product) as long as possible, so that a product's lifecycle can be prolonged and more people could be benefited by that product (Adapted from European Parliament, 2023). Examples are discussed in this current section while clarifying the Rs of the circular economy (◘ Fig. 1.5). The first three Rs (Refuse, Rethink, and Reduce) of ◘ Fig. 1.5 are focused on increasing manufacturing efficiency to optimize more responsible and endurable production of products. For example, Refuse is for abandoning the production of unnecessary and socially or environmentally harmful products, like producing more eco-friendly cars and biodegradable plastic bags, and gradually discontinuing manufacturing fossil-fuel-based cars, plastics, and other products that are damaging to the environment. Rethink at the manufacturing stage refers to increasing the functionality of a sole product so that the same product can be used for multiple purposes. For example, the utilities that we see in contemporary smartphones, that along with making a phone call, a smartphone also offers the utilities of the camera, clock, mailing service, global positioning system-centered navigation process (replacing the traditional compass and the paper-based street directory), the social media network, the traditional paper-based filing system, which are just a few to name. At the designing stage of a product, reducing the use of raw materials to produce a product or utility that would need fewer materials to deliver and consume than

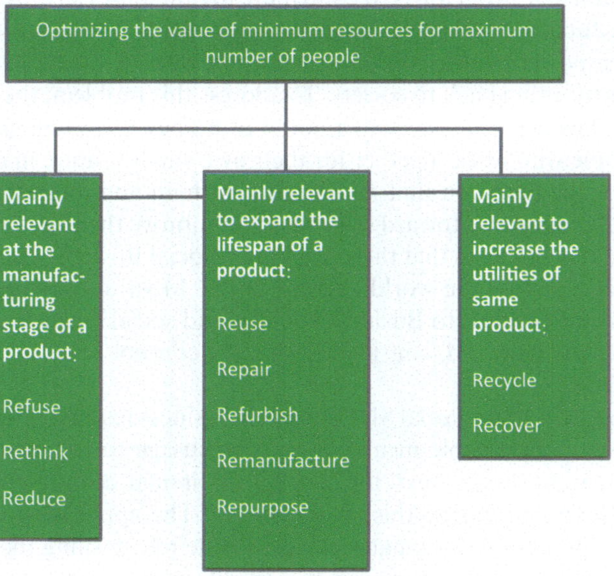

◘ **Fig. 1.5** The Rs of circular economy (Information source: Morseletto, 2020)

before. For example, using fewer packaging materials for a product or designing, delivering, and evaluating a paperless service delivery process.

Circular economy is a responsible process of production and consumption that focuses on sharing, recycling, refurbishing and reusing of limited current resources (e.g., a product) as much as possible, so that a product's lifecycle can be prolonged, and more people could be benefited by that product.

The next five Rs (Reuse, Repair, Refurbish, Remanufacture, and Repurpose) in ◘ Fig. 1.5 aim to expand the lifespan of a product so that the same product can be used for a longer period instead of manufacturing more new products. "Every year, we increase the goods we own by 25 billion tonnes—the equivalent of 93,000 Empire State Buildings." (Rethink Sustainability, 2023, np). To reduce waste of these goods, considering these five Rs can help marketers to design, manufacture, deliver, and evaluate a product in a way that can offer longer utility to its users, instead of influencing the users to purchase a new product replacing the existing one for the same utility. For example, "each year four billion toothbrushes are thrown away; enough to circle the earth twelve times," and a further environmental challenge is that "every toothbrush you've ever owned still exists" (SURI, 2023a, np), which has a detrimental impact on our ecology. To address this environmental challenge, ► suri.com designed a toothbrush that can be reusable by replacing a small portion of the original toothbrush. This is why, targeting environmentally caring customers, SURI (2023b, np) promotes them with a branding maxim: "Meet SURI. The last toothbrush you will ever buy." ► Section 1.2.4 (Identifying and Nurturing a Latent Need) discusses customers' latent need, a need that customers are not thinking now to satisfy, but if marketers design and offer a product to serve such latent needs, customers would be delighted to use that product. SURI's reusable toothbrush is designed to satisfy the latent needs of environmentally conscious customers, as it helps customers improve their environmental footprint.

Web Exercise 1.4

Reusable-product beauty-product ► zerostore.co.uk, ► thegreenturtle.co.uk, and similar other stores showcase a number of reusable products from water bottle to diaper or from beauty products to microfiber mop pads. Visit these and similar other websites, and:

- Consider how such reusable products help us to improve our social and environmental footprint.
- Think of a product that you are marketing now or would like to market in future, and envisage alternative features of the product to consider how you as its marketer could help designing a reusable version of this product, replacing its one time use version.

The next two Rs in ◘ Fig. 1.5 focus on expanding the lifespan of a specific product are Repair and Refurbish, which may be similar, but the refurbishment of a product is more than repairing it. Repairing a product is fixing the problem of an old product so that the product's lifetime can be increased to use it for a longer period of time. On the contrary, refurbishment of an old product may include repairing its damaged parts while upgrading its overall functionality to its initial state of performance, cleaning it, making a cosmetic improvement, and repackaging for resale (Ecotech Services, n.d.). For example, the Apple store offers refurbished smartphones to replace their customers' faulty phones instead of just repairing the faulty ones so that the refurbished version can serve the customer for a longer period, in comparison to just the repaired version. The next two Rs (Remanufacture and Repurpose) for expanding the lifespan of a product are also similar but not the same. For example, remanufacture is manufacturing a new car by using some of the refurbished parts of a written-off car, whereas repurpose would be using a refurbished chip of a completely scrapped laptop to produce another type of electronic product that offers a different utility, in comparison to the original product.

The next two Rs of circular economy presented in ◘ Fig. 1.5 are Recycle and Recover. The purpose here is to increase the valuable utilities of the same material that is used to manufacture a product. For example, every day, we print massive amounts of paper at office or at home office, and we throw away a considerable percentage of such printed papers in bins by the end of the day. Instead of throwing away such used papers, we collect and return them to our supplier, and the supplier can return them to its manufacturer so that the used papers can be recycled to reduce the production cost of new papers. As a result, the supplier (e.g., the paper wholesaler) can obtain a better price from the manufacturer, and we can obtain a better price from our supplier when we purchase printing papers next time while collectively reducing our environmental footprint. The "resource recovery model is emphasizing the conversion of waste into secondary raw materials. The goal of this model is to obtain additional uses from resources and to extract more value from them by avoiding final disposal for as long as possible" (np). And examples of recover include recovering heat from product destruction, reusing food and agricultural waste, and recovering oil and wastewater for reuse (Veolia, 2020). The applications of these Rs of circular economy help marketers (World Economic Forum, 2023):

- To design, deliver, and evaluate more attractive products by serving socially and environmentally conscious customers' latent needs.
- To increase customer intimacy by escalating customer dialogue, in comparison to the traditional linear economy.
- To lower the costs, recycling and reusing materials can help us reduce manufacturing costs.
- To lower the impact on the climate and environment, the circular economy helps us improve our environmental footprint.
- For more extensive use of resources, i.e., products and its components.
- To reduce production times, refurbishing an old product to give it a completely new or near-new look would take less time in most cases in comparison to manufacturing an entirely new product.

- To improve the resilience of materials.
- To explore recurrent income sources by repurposing the existing resources.

"Unlike the traditional linear economic model based on a "take-make-consume-throw away" pattern, a circular economy is based on sharing, leasing, reuse, repair, refurbishment and recycling, in an (almost) closed loop, where products and the materials they contain are highly valued" (European Parliament, 2016, p. 1), so that the eventual waste of material would be as minimum as possible. The "social business" and "circular economy" concepts are linked with the "sustainable marketing" concept, considering that all these three concepts have a common focus on satisfying the needs of the triple bottom line stakeholders, i.e., people, planet, and profit. Pursuing the triple bottom line would have two interdependent outcomes. For example, pursuing a brand's commercial affluence so that the brand can contribute back to people (society) and the planet (ecology), and subsequently, the societal and ecological contribution of the brand can underpin its commercial competitive advantage, as well. Like marketing, the sustainable marketing concept is applicable to any organization across public, private, and the third sector to design, deliver, promote, and evaluate their products and relevant value propositions in a way that would help marketers contribute to the global pursuit of SDGs. Because of the high emphasis on the triple bottom line or triple Ps (i.e., People, Planet and Profit, or Societal, Environmental and Commercial) (i.e., People, Planet and Profit, or Societal, Environmental and Commercial) stakeholders, the post-2010 era could be referred to as the triple Ps' stakeholders-focused market.

The turn of the decade from the 2010s to 2020s experienced the global COVID-19 pandemic, which further increased concerns for climate change and intensified societal, environmental, and geopolitical concerns (Singh et al., 2022). The beginning of this post-2020 era also has witnessed the expansion of a metaphorical space in our global marketplace, following the fuzzy technological advancements, such as the next generation of cloud computing, artificial intelligence (AI), and metaspace, which is a space where two or more people interact from a distance but they would have a feeling that they are interacting in-person.

Such next-generation technological advances introduce an additional space in our traditional marketplace, which can be referred to as a metaphorical space, which is symbolic of marketers' and customers' interactions from a distance while they will have a feeling of in-person interactions. The AI technology, investments in cryptocurrency, and the evolvement of metaspace are signifying the development of this metaphorical space as a newer form of marketspace (i.e., the internet-enhanced digital marketplace) that will make the already closer global market further close. Until the introduction of the internet and the subsequent World Wide Web in the early 1990s for the usage of mass people, a child in one global region did not know much about how a child in another global region used to grow up. Similarly, customers across different global regions used to know much less about each other's needs, wants, and expectations. The internet-enhanced marketspace reduced this gap in customers' knowledge and contributed to the evolvement of a uniform global customer culture to nurture the global market for products and services. The metaphorical space will ensure livelier marketers' and customers' interaction experience from a distance; however, the more tech-savvy customers in our global market will benefit more from the metaphorical space.

Generation A or Alpha (who were born between 2013 and 2025) and Generation Z (who were born between 1997 and 2012) are more tech-savvy in our global market, considering that they were born in the internet age. In addition to these two tech-savvy generations, we also have other generations in our global market, such as Generation Y, who were born between 1981 and 1996; Generation X, who were born between 1965 and 1980, and Generation Baby Boomers who were born in between 1946 and 1964 (Kasasa, 2021). Among these older generations, the baby boomers and Generation X are the most traditional customers, who prefer the traditional means of transactions more than the transaction patterns introduced by the internet-enhanced marketspace or the metaphorical space. Among these generations of customers in our global market, Generation Y (who were born between 1981 and 1996) would have characteristics related to both the traditional customers and the more tech-savvy customers, considering that they were born in the middle of the first two traditional customer groups and the two more recent tech-savvy customer groups. In this context, marketers need to adapt their value propositions for the same product or service, focusing on the needs and preferences of three different customer groups such as traditional customers who are comfortable with the traditional marketspace, the tech-savvy customers groups who are comfortable with the metaphorical marketspace, and the customer groups from Generation Y who have synchronized customer characteristics (some from the older generations and some from the younger generations) considering that Generation Y is the middle generation out of these five generations of global customer groups.

Furthermore, out of these five generations of customers in global customer groups, some traditional customers (e.g., from Generation X), some tech-savvy customers (e.g., from Generation Z), and some Generation Y customers may have common characteristics in their consumer behavior. This group of customers across different generations who share some common characteristics constitute together a mixed-zone space in the contemporary global market. Focusing on the customer behavior of this mixed-zone customer group would be useful if marketers prefer to offer mass-customized products, considering that these customers share characteristics of customers across different generations. In this context, our global marketspace can be divided into four categories, namely the traditional space, metaphorical space, synchronized space, and mixed-zone space, which is symbolized in ◘ Fig. 1.6.

The market influences that we have seen since the early 1990s to post-2020 (which are listed in ◘ Table 1.2), such as industrialization and mass production, unsold goods, knowledge democracy, ease of information, symbolic representation of value, service preference, increased concerns for climate change and intensified societal, environmental, and geopolitical concerns; all of them still exist in our traditional, synchronized, and the evolving metaphorical marketspaces. It means marketers cannot ignore the traditional transactional marketing concept or the 1960s' 4P-based marketing mix concept. Over the progress in time, different new complexities (such as unsold goods, ease of information that have advantages and challenges, climate change, and geopolitical concerns) are adding newer influences to our global marketing dynamics. Therefore, in support of the traditional methods of marketing (i.e., transactional marketing and the 4Ps marketing concepts), marketers have been adapting newer approaches, such as value-based marketing, niche marketing, stakeholder relationship marketing, service-dominant logic, corporate social responsibility, social business, circular economy concept and designing triple bottom line stakeholders' value proposition based on sustainable marketing, in order to address the evolving complexities in our global marketing management dynamics. As a result, none of these marketing concepts can be overthrown, as they complement each other to address the existing and evolving concerns in our post-2020 marketing dynamics. In the same vein, the 4Ps marketing mix concept has been extended to the 22 Ps marketing mix to satisfy the needs of different contemporary customer groups across the traditional marketspace, metaphoric space, synchronized space and mixed-zone marketspace, which is specified in ◘ Fig. 1.6.

◘ Source: ► pexels.com/royalty-free-images/

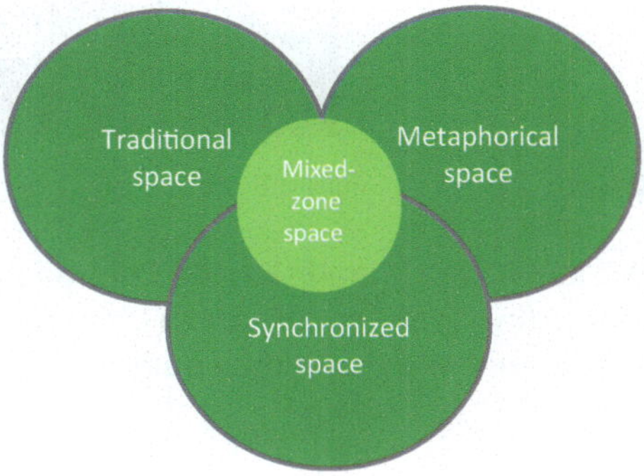

◘ **Fig. 1.6** The symbolic representation of mixed-zone marketspace

1

The 22Ps marketing mix is discussed in ▶ Sect. 1.3.2 (Marketing Mix and the Extended Marketing Mix). Marketing dynamics are always futuristic dynamics. In the post-2020 market, in support of the traditional and embryonic marketing concepts, marketers are designing value propositions for their triple bottom line stakeholders to help the global community to pursue SDGs, while the marketers are also working to influence the future market. For example, the ongoing progress of technology will help you (who are reading this textbook today as an undergraduate or postgraduate student) to market space tourism or market a passenger flight that will take passengers from London to Sydney in 4 h, which are merely very few examples of how rapidly and how radically the global marketing dynamics could be progressed over time.

1.3.2 Marketing Mix and the Extended Marketing Mix

▫ Source: ▶ pexels.com/royalty-free-images/

We know that the 4P-based traditional marketing mix was first introduced by McCarthy (1960) and Borden (1964) to help marketers plan in a more structured way to influence their target audience, such as customers and non-customer stakeholders. For example, some consumers may want to purchase just one 150 ML Coke at 2.00 o'clock in the morning. So, Coke's consideration, related to their distribution channel (i.e., place in the 4Ps marketing mix) is whether they have placed a self-service vending machine next to these group of consumers, e.g., at a town center so that the consumer could walk for 10 min to purchase that 150 ML Coke at 2.00 am! These types of consumers in total may be very few in comparison to Coke's total target market, yet they are a considerable niche because, if Coke does not extend its distribution channel for this group, Pepsi may have a competitive edge over Coke, if Pepsi does it. Similarly, customers' needs, wants, and preferences should be considered in product design and pricing strategy to confirm that the product is user-friendly and the price is affordable and competitive. The promotional strategies and branding campaigns also should be focused on pursuing two goals. One is creating as much awareness as possible or as much awareness as marketers want to create about their value proposition among their target market, clarifying why doing business with us is more profitable for you instead of doing business with our competitive brands. The second goal of a promotional strategy or a branding campaign is to influence or form the perception of the target audience of a brand, e.g., "everyday low prices" of ALDI. To successfully pursue any of these two promotional goals, marketers need to understand their target audience's lifestyle and preferences and align their promotional campaigns reflecting on that lifestyle and preferences. For example, Pepsi promotes them as the choice of the new generation (Findlay, 1988), and therefore, they include some activities from the lifestyle of young people, e.g., concert or mountain tracking in their commercials, in order to reflect on their core target market's lifestyle.

To support the traditional 4Ps, marketers have added additional Ps in the marketing mix over time to address the embryonic nature of our ever-evolving marketing dynamics. ◘ Figure 1.7 lists 22Ps in our expanded marketing mix. In 1981, Booms and Bitner added three new Ps (Participants, Physical evidence, and Process) to the marketing mix. Participants include the internal and external stakeholders who help a brand create value for its customers (Brown et al., 2024). For example, employees, suppliers, and investors help create value for a brand. Even customers are considered important participants in a value creation process, as marketers understand customers' perceptions through marketing research and reflect on those customer perceptions to design a new product. In this context, customers are also involved as participants in such a co-creation process of new product development. From the 1970s to the 1990s, when the symbolic (form and style of) representation of value became prominent as a market influencer, marketers needed to differentiate their products or services even with the physical evidence associated with their value proposition. For example, someone is visiting a McDonald's or a KFC, where neither the McDonald's nor the KFC has a signboard of their brand name or brand logo displayed outside or inside of their outlet. Nevertheless, the way McDonald's or KFC's interior and exterior are designed, its color combination, arrangement of furniture, etc., would help the visitor to recognize whether it is a McDonald's, a KFC, or another competitive fast-food brand. Process in mar-

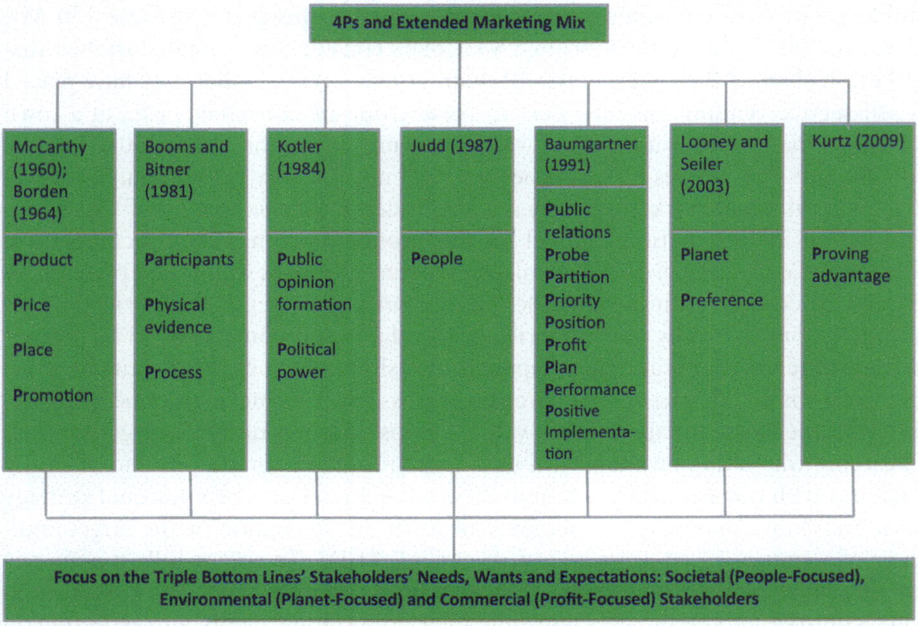

Fig. 1.7 The 22Ps' extended marketing mix for triple Ps' stakeholders

keting mix refers to the entire process of value delivery dynamics, including product design, delivery, and evaluation. For example, competitive universities or competitive banks offer the same degrees (e.g., master's in marketing management) or the same financial service (e.g., car loan); however, the design, delivery, and evaluation process of such same services could be different to a different extent among the competitive brands. This is why, when even a very experienced professor joins a new university, the professor usually is not given a high workload during the first semester, as during this initial period at a new workplace, the new professor needs to adapt to the different process of the new employer, in order to deliver even the same teaching, research or academic leadership services. Like product, price, place, and promotion, considering participants, differentiation in physical evidence, and value delivery process is important for marketers to align from the contexts of their triple Ps' societal, environmental, and commercial stakeholders.

Public opinion formation and political power as two new Ps were added to the marketing mix by Kotler in 1984. Irrespective of a purely profit-oriented market offering or a triple Ps focused socially and environmentally caring commercial brand, celebrity endorsement (e.g., brand ambassador role) is a popular strategy among marketers to influence public opinion, considering that celebrities across different sectors (e.g., sports, music, cinema, etc.) have a huge fan base. These fans follow their celebrities and consciously or subconsciously maintain a relationship with their preferred celebrities from a distance, even without ever meeting them. As a result, the opinion of celebrities who have a decent image among their followers has an impact on influencing public opinion about the value proposition of a brand. Another example of public opinion formation is that example of ALDI, when we compare them with other competitive supermarket brands, it comes first to our mind that ALDI is more affordable than other supermarkets. Following the art of marketing as discussed in ► Sect. 1.2.6 (The Art and Science of Marketing), ALDI's marketers form the public opinion as everyday low prices. Promoting a brand's value proposition related to its triple Ps' stakeholders and influencing public opinion about the brand's contribution to its triple Ps' stakeholders can help the brand establish a socially and environmentally caring brand image. Similarly, negotiating and working with government policymakers to contribute to a country's society or environment can help a brand establish a good relationship with political parties. Such political power or proximity to power can help a brand establish a positive brand image if the brand is able to consistently deliver what it promises for societal, environmental, and commercial progress. Based on Curtin Malaysia, an example of such People, Planet, and Profit-focused proximity to power is discussed in ► Sect. 1.3.4 (Stakeholders' Needs and Distress: An Opportunity to Delight Them Based on Pursuing the Triple Bottom Line).

Judd added People as another P in the marketing mix in 1987. People are referred to as both those who develop and deliver a product and those who consume it. Considering the influence of a product on people, including those who consume it in a very remote market across our global society, is important. This is why when Oxford University and other producers produced their Corona vaccine, the vaccines needed to be examined based on a thorough process of quality assurance at its different development stages. Before the vaccines were permitted to be applied publicly to mass consumers, they were tested on diverse sample groups among different races, age groups, and genders so that the vaccine could be improved to have a consistent performance and avoid side effects for people across the world. Baumgartner, in 1991, introduced nine more Ps in the marketing mix, namely Public relations, Probe, Partition, Priority, Position, Profit, Plan, Performance, and Positive implementation. Public relations (e.g., press conference, press release) and priority in between different objects (e.g., timing, target segment, pricing, online or in-person delivery, etc.) in a value delivery process, help to position a brand in its target audience's mind; positive implementation of plans; and communicating the brand's overall performance related to its triple bottom line including profit, which are important considerations for marketers.

Kover et al. (2002) described partition as the difference between the two versions of the same product, based on a generic product and an expanded or expected product. Generic product offers the core utility where the expected product offers a better quality version of the same product, e.g., with value-added features. For example, a textbook at its optimum quality (e.g., with better quality papers, color print, hardbound) that is priced at $140 for customers in developed economic countries cannot be sold at the same price to the customers in developing economic countries. For customers in developing economic countries, publishers offer the same textbook in its generic form (e.g., with black and white print, low-quality paper, not hard bound), which helps the publishers to reduce its production cost to offer it at a lower price that would be affordable for the customers in developing economic countries. Like product design, the partition concept applies to other contexts of marketing strategies. Partitioned price means an overall product or service could come with a basic price and an added price for the value-added components of the product or service (Bambauer & Gierl, 2008). For example, to purchase a flight, customers need to pay a basic price. On top of the basic price, different prices are also added for value-added components, such as health insurance, hotel, and taxi service, on-flight meals, checked-in luggage, etc. Probe in the marketing mix is another essential element, which is a thorough investigation of customers' perceptions based on market research for marketing orientation to discover the customers' needs, wants, and expectations more closely. Probing is instrumental in recognizing customers' latent needs. Ongoing market orientation is helpful for probing. More discussion on market orientation is shared in ► Sect. 1.4.1 (Ongoing Market Orientation).

Planet and Preference were added to the marketing mix by Looney and Seiler in 2003. The increased concerns about climate change propelled marketers to add the planet to the marketing mix, to design their value propositions in a way that would be more eco-friendly. Preference in the marketing mix is an easily comprehendible concept for marketers, as without properly understanding the target customers' and non-customer stakeholders' preferences and reflecting on those preferences in product design and its pricing, promotion, and delivery strategies, it would not be possible for marketers to win their target market. Proving advantage through customers' experience (Kurtz, 2009) could be considered as another marketing mix element, considering that it is important to ensure customers have a positive brand experience so that they can understand why doing business with our brand is more profitable for them, instead of doing business with our competitive brands. Another aspect of proving advantage through customers' experience is promoting that positive customer experience to attract new customers. The extended marketing mix (22 Ps as represented by ◘ Fig. 1.7) can help marketers to design, deliver and evaluate a market offering (a product, service or idea) to prolifically pursue the triple bottom line, if all societal (people-focused), environmental (planet-focused), and commercial (profit-focused) stakeholders' needs, wants, and expectations are considered closely in the product's overall value proposition.

1.3.3 Key Stakeholder Value Network: Whom You Must Need to Satisfy

Source: ► pexels.com/royalty-free-images/

Let us assume that you are marketing a branded flour (flour, based on what we make our bread or cake). The key stakeholder value network of your flour brand starts from the individual or firm that produces the seeds of your flour. If the individual or firm who produces the seeds cannot offer the seeds at a competitive price to farmers, the farmers who produce the raw flour will not be able to supply you with the raw flour at a competitive price. Therefore, as a marketer of that branded flour, you need to consider your relationships and interactions with those suppliers (the farmers and seed producers) who supply you with the raw materials of your finished product. Once the farmers produce the flour, you may need an outsourced transportation company to bring the raw flour from farmers' fields to your factory. At your factory, your employees process the raw flour, weigh them in small units (e.g., 1 KG, 2 KG), put it in bags, and print the name, logo, and promotional message on the bags to transform the raw flour into branded flour. Again, you may need the service of your outsourced transportation partner to transport your branded flour as a finished product from your factory to different wholesalers and retailers across the country, who help you to bring your finished product to your consumers. Consumers purchase your product from retailers, they consume it, and at the end of their consumption, they dump the packaging of your branded flour in the bin. The council people come by the weekend to pick up those used product packaging. Now, if you can develop a good business relationship with the council, you can collect back those used plastic bags (e.g., the packaging) to refurbish and reuse them, which can help you to reduce your production costs while enabling you to reduce your brand's and consumers' environmental footprint.

All these internal and external stakeholders (such as the seed producer, farmer, inbound and outbound logistics partner, employees, wholesalers, retailers, and council) who help you to produce value for your customers are included in the key stakeholder value network of your branded flour. Of course, the consumers are also key stakeholders, and you need to understand your customers. Simultaneously, you also need to understand your key stakeholders in your value network to satisfy them, as without these key stakeholders' contributions, you are unable to create value for your customers. This key stakeholder value network is represented in ◘ Fig. 1.8. The key stakeholder value network also includes the business partners who invest into your value proposition to create value for your triple Ps bottom line's societal, environmental, and commercial stakeholders. You also need to be concerned about the societal and environmental activist groups, as you need to satisfy their needs, if you really want to contribute back to society and environment. The local, regional, national, and international chamber of commerce and trade associations are also included in your key stakeholder value network, considering that you need to synchronize your value proposition complying with the policy of these industry-based stakeholders. Similarly, your value proposition

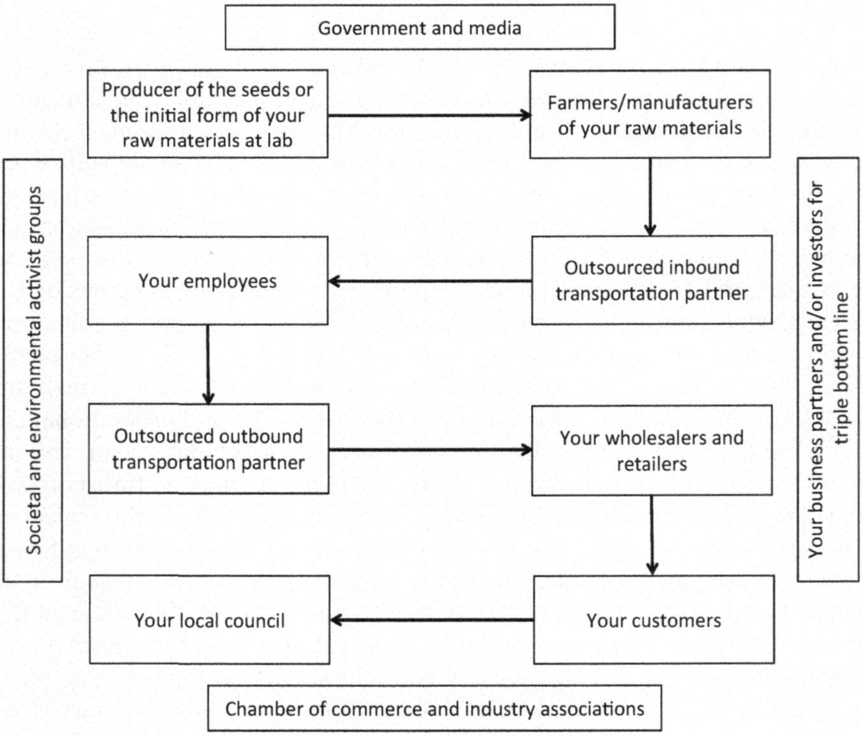

◘ **Fig. 1.8** Key stakeholder value network (Adapted from European Environment Agency, 2016)

needs to be complied with the business law imposed by your national or state government. Media is another key stakeholder, considering that it takes a long time to establish an authentic socially or environmentally caring brand image, but a single media report can bring permanent harm to the image of a brand.

1.3.4 Stakeholders' Needs and Distress: An Opportunity to Delight Them Based on Pursuing the Triple Bottom Line

Source: ► pexels.com/royalty-free-images/

People, Planet, and Profit are the triple bottom line of sustainable marketers; however, purely profit-oriented firms have only a single bottom line, i.e., making more profit at any cost. Since the profit-only marketers need to satisfy only their commercial stakeholders, and sustainable marketers need to satisfy their three stakeholder groups, namely societal, environmental, and commercial stakeholders, in general, sustainable marketers' mission is more challenging than the profit-only marketers. However, looking at different societal and environmental stakeholders' distress can be instrumental even for profit-focused businesses to align their value proposition centered on societal, environmental, and commercial stakeholders to make more profit. Let us have a look at an example of Curtin Malaysia, which is a profit-focused higher education business. Curtin is an Australian University. It has its main campus in Perth, Australia. Curtin has a transnational campus in Malaysia, which has been branded/promoted as Curtin Malaysia.

A core focus of such overseas transnational education (TNE) campuses is expanding business and making a profit. Many TNE businesses are successful; however, many other TNE operations struggle to survive. The key revenue source of such transnational campuses is the tuition fees that they receive from their transnational students at their overseas locations. However, TNE campuses, from time to time, struggle to attract enough full-fee-paying students to run their overseas operations. Curtin Malaysia (n.d.) was opened in 1999 on the island of Borneo in Malaysia. Considering the challenge of attracting sufficient full-fee-paying transnational students, Curtin Malaysia started to explore its options to expand its alternative revenue sources in its Malaysian transnational market. In this effort, on the one hand, Curtin Malaysia recognized that in Malaysian society and ecology,

1

their societal and environmental stakeholders have some significant distress in their socioeconomic and ecological dynamics. For example, in Malaysia, they have some endangered animals, such as Borneo Sun Bear. Curtin Malaysia also found that when Malaysian fishermen go for fishing, they often have confrontations with crocodiles in local rivers, which is neither good for the fishermen nor the crocodiles, as they injure each other. On the other hand, Curtin Malaysia knew that they had a well-established zoological research culture at their home campus in Perth. Therefore, in response to such social and ecological distresses of their wider societal and environmental stakeholders in Malaysia, and with an aim to expand their revenue sources, Curtin Malaysia considered how they could apply their zoological research strength to address their transnational stakeholders' distress to satisfy them. From this context, Curtin Malaysia applied for the Malaysian Government's research fund.

Source: ▶ pexels.com/royalty-free-images/

As the first foreign university, Curtin Malaysia received Malaysian Ringgit 1000,000,000 (approximately US$231,192,570) as research grants from the Malaysian Government (Research and Development, n.d.). Alongside the Central Government's funds, Curtin also received funds from the state government and different non-government partners in Malaysia (Shams, 2016). Following their research and knowledge exchange initiatives, Curtin has contributed to sustainably managing the local wildlife, the Malaysian Sun Bear, and other endangered animals in Malaysia, and has reduced the Malaysian fishermen and crocodile confrontations (The Road Ahead, 2015). For example, they studied the lifestyle of Malaysian crocodiles and educated the fishermen, e.g., when it is better to go for fishing and when it is better to let the crocodiles have more freedom. In this way, Curtin Malaysia was able to contribute back to the society and environment in their Malaysian market while expanding their revenue sources. It is because such research grants from different government and non-government funding bodies are also one of the substantial revenue sources for higher education institutions.

To discuss their societal and environmental contribution, Professor Jim Mienczakoniski (the Curtin Malaysia CEO) explains that:

> Eric (a Borneo sun bear) is a part of our research and the environment that we are concerned for and looking after. Eric lives here at the crocodile firm and nature park near Curtin University in Miri, Sarawak, Borneo. The nature park (offers) a chance to… (work closer to) …the entire range of creatures; most of them are unique to this region. (Shams, 2016, p. 148)

To expand their revenue sources and pursue commercial success in the Malaysian market, Curtin Malaysia identified and addressed the distress of the Malaysian government and their societal and environmental stakeholders in Malaysia. For example, the fishermen want to protect themselves from crocodiles, and the Malaysian Government wants to protect the fishermen, their crocodiles, sun bears, and other endangered animals. Upon satisfying these latent needs of the Malaysian triple bottom line stakeholders and the associated stakeholder-specific distresses, Curtin Malaysia was also able to ensure their commercial success to survive and prosper with their Malaysian operation. Furthermore, Curtin Malaysia established a brand image in the Malaysian market that they are a socially and environmentally caring higher education brand that can make a difference. This brand image can help them pursue future commercial successes in the Malaysian market, as well. For example, the fishermen who benefited from Curtin Malaysia based on their education on crocodile lifestyle; those fishermen would consider Curtin and influence the next generation to obtain their higher education from Curtin. Furthermore, as a change-maker, by addressing some of the Malaysian societal and environmental challenges, Curtin would have a favorable image in the mind of the Malaysian Government. Such proximity to political power would also be favorable for Curtin to run their operation in the Malaysian market, with an aim to expand their commercial success while contributing back to their societal and environmental stakeholders in Malaysia. Following this case example, it can be proclaimed that Curtin Malaysia's contribution to their societal and environmental stakeholders also has reinforced the commercial competitive advantage of the brand Curtin in Malaysia.

1.4 Understanding Your Business Environment

1.4.1 Ongoing Market Orientation

The purpose of market orientation is to design, deliver, and evaluate a product, closely reflecting on the needs, wants, and expectations of a brand's target audiences to provide them with the most competitive value (Kumar et al., 2011). Market orientation has been studied both in strategic marketing and strategic management literature. In traditional marketing literature, market orientation is defined as an "organization-wide generation of market intelligence that pertains to current and future customer needs, dissemination of intelligence across departments, and organization-wide responsiveness" (Kohli & Jaworski, 1990); as cited in Hurley & Hult, 1998, p. 43). "Such a generation of market intelligence requires continual monitoring of market conditions, relevant to customers, competitors, and market environments to develop and offer products and services (in a way that would be accepted) by customers" (Kohli & Jaworski, 1990; Narver & Slater, 1990, 1994; as cited in Shams, 2016, p. 145). Collecting data from current and prospective customers about their needs and maintaining a learning orientation about customers' needs are characteristics of market-oriented brands (Kumar et al., 2011). Since marketing dynamics involve all customers and non-customer stakeholders, we define market orientation as closely recognizing and analyzing the needs, wants, and expectations of a brand's target audiences, i.e., customers' and key non-customer stakeholders, to create competitive value reflecting on the target audiences' needs.

Market orientation is closely recognizing and analyzing the needs, wants and expectations of a brand's target audiences, i.e., customers' and key non-customer stakeholders, to create competitive value reflecting on the target audiences' needs.

Homburg et al. (2011) discussed about five steps in market orientation:
- Problem and need recognition.
- Integrating solution.
- Dealing with feedback.
- Further consideration of customers' interest.
- Proffering a competitive value proposition based on informative promotion.

Based on marketing research, recognizing customers' problems and relevant needs is the first step of market orientation. Like the example of Curtin Malaysia as discussed in ► Sect. 1.3.4 (Stakeholders' Needs and Distress: An Opportunity to Delight Them Based on Pursuing the Triple Bottom Line), sustainable marketers should recognize the problem, distress, and relevant needs of their societal, environmental, and commercial stakeholders. Upon recognizing stakeholders' problems and relevant needs, marketers need to analyze their available resources and capabilities and the identified stakeholders' problems and need to consider how a superior solution can be provided in comparison to the competitive brands' solutions to address the stakeholders' problems and satisfy their needs. Once a prototype solution as a market offering (e.g., product or service) is developed, the next step in ongoing market orientation is to share the prototype solution with representative sample groups across different customer segments to understand their views about the prototype product in order to deal with any customer feedback, with an aim to improve the prototype product further. The next step is sharing the improved version of the product with customers to understand their views about the modified product features to confirm whether the product could offer a superior solution to customers' problems or not. Once it is confirmed that customers are happy, the last step of ongoing market orientation is promoting the finished product with an informative demonstration to clarify how the product can most competitively address the customers' problems and satisfy their needs.

We are referring to it as an ongoing market orientation, considering that once a new product is launched, marketers' need to collect and analyze market data on an ongoing basis to reflect on the product performance in relation to the customers and non-customer stakeholders' perceptions about the value of the product. Some core focus of market data collection and analysis would include:

- Customers' perceptions across different customer segments.
- Competitors' reactions to the performance of our value proposition.
- Other societal and environmental stakeholders' reactions.
- Reactions of government, media, and industry-based stakeholders, e.g., trade associations.
- Reactions from key stakeholders, such as suppliers, investors, and others, help us to create value for our customers.

Since focusing on all such customer and non-customer stakeholders' reactions among all societal, environmental, and commercial stakeholders is imperative to continually improve a brand's value proposition, market data collection, analysis, and synthesis should focus on such inclusive contexts of diverse stakeholders' reactions and interests to flourish ongoing market intelligence. Alongside developing a product or service based on the needs of customers, ongoing market orientation helps marketers to retain their customers. For example:

> When marketers understand how much a given customer might be worth to the organization over time, they can tailor the product/service offering according to that customer's changing needs and requirements and still ensure an adequate lifetime return on investment (Berger et al. 2002). Therefore, market-oriented organizations are capable of better customer retention (Narver et al., 1999). (As cited in Kumar et al., 2011, p. 20)

In this context, it can be argued that ongoing market orientation is imperative for the commercial sustainability of a brand, and once a brand is commercially sustainable, its opportunities to contribute back to society and the environment increase further. In fact, "market orientation is an important basis of business sustainability, thus, "the company's market-oriented approach to sustainability has resulted in the receipt of numerous awards as a leader in sustainability (Crittenden et al., 2011, p. 81)" (as cited in Shams, 2016, p. 145).

1.4.2 Competitive Forces in Your Market

While discussing ongoing market orientation in ► Sect. 1.4.1, we have discussed that a focus of market data collection and analysis for market orientation is recognizing the reactions and interests of a brand's customers and non-customer stakeholders about the brand's value proposition. In this current section, let us have a further look at these customers and non-customer stakeholders, who, in general, collectively constitute the competitive environment for a brand in any given market. In the seminal paper on strategic management (Industry Structure and Competitive Strategy: Keys to Profitability), Professor Michael Porter (1980) discussed five competitive forces in a market. The remainder of this current ► Sect. 1.4.2 discusses about these five competitive forces coined by Porter.

Source: ► pexels.com/royalty-free-images/

Prospective New Entrants In each industry in any given market, there would be a few key competitors (e.g., two to five, more or less) who individually or jointly own most of the market share in that market. There would be a few other smaller competitors, who together also own a percentage of the total market share. In addition to these existing competitors of a brand in a particular market, there would be a scope for entrepreneurs to launch a new business in the industry in that market. How easy or difficult would it be for such prospective new entrants in competition to enter the market? Marketers need to consider the relevant socioeconomic conditions, e.g., government support for entrepreneurs, customers' buying power, the extent of economic and political stability, and other relevant issues related to a particular market, to understand the potential competition that would arise in the future from such potential new entrants.

Threat of Substitute Products Brands encounter severe competition from the competitive brands within their own industry, such as Coca-Cola vs. Pepsi-Cola or iPhone vs. Galaxy. In addition, brands can also encounter competition from outside of their own industry. For example, during the 1960s, colas were synonyms for soft drinks in the minds of mass consumers; however, during the 1980s and 1990s, both Coca-Cola and Pepsi-Cola encountered competition from the coffee industry, as customers had a need for something (e.g., caffeine) that would keep them active (Coca Cola and Pepsi, 2018). Here, cola's caffeine had been substituted by coffee's caffeine to satisfy the same needs of common customers of cola brands and coffee brands; however, both coffee and cola generally do not belong to a single industry. Similarly, some supermarkets, such as Sainsbury's (2022), or other brands, such as IKEA (2023), offer food and drink products in the format of a café or restaurant. As a result, regular fast-food brands and restaurant chains also encounter competition from the supermarket and furniture industries, as their offerings substitute consumers' needs that can be satisfied by McDonalds, KFCs, Nando's, or other fast-food or restaurant brands. Therefore, marketers need to be concerned about the competition that they may encounter with substitute products or services even from the outside of their own industry.

Competition Between Existing Competitors In the rent-a-car industry, a 2021 Bloomberg (2023) report shows that the Avis Group owns 31% and the Hertz Group owns 26% market share. In the past, once Avis was trailing Hertz, the former market leader, in their head-to-head competition, Avis was promoting them through their new advertising campaign appealing that "When you're only No. 2, you try harder" (Stevenson, 2013, np). It means that since Avis was the market challenger in terms of market share during that period of their rivalry with Hertz, Avis was promoting them, saying that we are hungrier to serve you, considering that we have more appetite as number 2. In such direct head-to-head competitions among the major competitors, there is one interesting thing to look at. See, Avis was not saying that our fleets are better than the fleets of Hertz, but they are appealing to the customers based on the "value" that is appreciated by customers, which is "more customer service" in the rent-a-car industry.

Bargaining Power of Buyers Buyers' (customers' and consumers') buying power, ability, and propensity to switch between competitive brands, changes in buyers' preferences, and relevant other factors in buyers' buying and consumption decision-making process need to be considered by marketers, to plan, implement, and monitor a competitive value proposition. There is a distinction between customers and consumers. Customers are resellers, such as wholesalers or retailers who purchase your product in bulk units to resell to the end users of your product. For example, Unilever's products are purchased first by different supermarkets as retailers to sell those products to its ultimate end users. Therefore, buyers who purchase Unilever's products from supermarkets to consume it are the consumers of Unilever. Unilever, as a marketer needs to consider the bargaining power of both their customers and consumers, considering that a supermarket may prefer to resell the competitive brands more than Unilever's brands if the supermarket receives higher incentives from the competitive brands. Similarly, end users may prefer to switch to a competitive brand if they perceive more value from the competitive brands in comparison to the existing value proposition of Unilever.

Bargaining Power of Suppliers Let us assume that you have seven alternative suppliers to source a core raw material to manufacture your product. Each of these seven suppliers has bargaining power, to different extents, to do business with you. Analyzing each supplier's bargaining power is important for marketers, considering that it would not be viable to dissatisfy a supplier who offers the best possible raw material at the possible competitive prices. ▶ Section 1.4.3 (Stakeholders' Bargaining Power) discusses some key contexts to analyze stakeholders' bargaining power.

1.4.3 Stakeholders' Bargaining Power

Let us assume that you are preparing for a client meeting with a current or prospective investor, supplier, buyer, or any other stakeholder who can influence your business, or you want to influence them. Before such client meetings, it would be good to determine who, among you and your client, has more bargaining power against each other. Following Porter's five competitive forces, Short (2015) described four factors to analyze and determine who has more bargaining power among two stakeholders in a given situation. The four factors are:

- Number and size of buyers and suppliers.
- Importance of a specific buyer or supplier.
- Is the item (the product or service based on what two parties are negotiating) vital?
- And the ability to enter each other's industry.

◘ Source: ▶ pexels.com/royalty-free-images/

Number and Size of Buyers and Suppliers Considering whether there are more buyers than available suppliers, and whether among two buyer and supplier who is comparatively bigger in terms of the size of their businesses and resources.

Importance of Specific Buyer and Supplier It is to determine whether a buyer can survive without the service of a specific supplier among the alternative suppliers, and vice-versa. For example, if there are seven suppliers of a core raw material of your product, the question is whether you can run your production without sourcing your raw material from one of these seven suppliers, and vice versa, that is, whether a particular supplier can survive without supplying the raw materials to you (as there may or may not be many other buyers to purchase that raw material from that supplier, even if you do not purchase it from them).

Is the Item Vital? Here, an item is referred to as a product or service based on what two parties are negotiating among themselves. The question is whether, in negotiation, one party can survive, e.g., run their business without any interruption, without purchasing or selling the product or service, based on what they are negotiating. For example, a supermarket could survive without selling the shampoo of a particular brand because there are many other alternative competitive shampoo brands available in the market. But think of another situation. Often, consumers at supermarkets ask customer service attendants whether they have the shampoo (or any other product) of a particular brand. If this is the case for the shampoo brand that we are referring to here, in this case, it would be difficult for the supermarket not to sell that shampoo brand because the brand has established a huge loyal customer base, and customers ask about that shampoo brand often at the supermarket.

Ability to Enter Each Other's Industry Supermarkets sell a number of food products and could survive if they do not source cake from an individual cake maker. In this case, the individual cake maker may be able to open their own physical or online outlet to sell the cake directly to its consumers instead of selling it via the supermarket.

Look at this video clip: ▶ https://www.youtube.com/watch?v=fIpJq_rYgmU (or search it on YouTube based on "bargaining power buyers and sellers" of Jeff Short). In this video, these four factors to determine bargaining power among two stakeholders are discussed further. After looking at this video clip, let us have an exercise on determining bargaining power, based on the exercise specified in ◘ Table 1.3.

Table 1.3 Determining who among two stakeholders in a given situation has more bargaining power

Issues of analysis	Industry	The item to negotiate	Buyer	Supplier	Who has more bargaining power?
First exercise					
The number and size of a buyer and supplier	Mobile phone service providers	SIM card	Your country's largest mobile phone service provider	A Chinese SIM manufacturer	
Importance of a specific buyer and supplier	Pharmaceutical industry	Liquid rehydration saline	National health service, who have several hospitals around the country	Country's largest pharmaceutical company	
Is the item vital?	Financial industry	ATM installation and maintenance around the country	Country's largest bank	An ATM machine installation and maintenance company	
Ability to enter each other's industry	Financial industry	ATM installation and maintenance around the country	Country's largest bank	An ATM machine installation and maintenance company	
Second exercise					
The number and size of a buyer and supplier	Coffee	Coffee beans	A corner café	An individual coffee bean supplier	
Importance of a specific buyer and supplier	Garments industry	Cheap labor	Walmart	Garments (e.g., shirts, shorts etc.) supplier from a developing economic country	
Is the item vital?	Information technology (IT)	National security	National government	A high-profile IT expert, specialized in digital security	
Ability to enter each other's industry	Fast food	Chicken	KFC	A local medium-sized poultry firm	

1.5 Sustainable Marketing Strategy

1.5.1 Sustainable Vision and Mission

■ Source: ► pexels.com/royalty-free-images/

The ultimate target of sustainable marketers is to ensure equity, diversity, and inclusion (EDI) of all customers in our local, regional, national, and global societies in a consumption culture that ensures fairness in the consumption of at least the basic human needs (i.e., food, cloth, home, medical service, the opportunity to work and education) across all customer groups. For example, without sacrificing the profit-motive of a brand, how the brand can ensure EDI of low-income consumer groups during times of recession is a fundamental responsibility of the brand if it wants to establish a brand image as a sustainable brand, i.e., a socially- and/or environmentally-caring brand. We understand that pursuing this EDI target of ensuring fairness in our global consumption culture is not easy, and it takes a lot of time, planning, and endeavors to pursue this target. Therefore, sustainable marketers need to play their part in this global effort of sustainable consumption, based on their available minimal resources, even on a small scale, starting from their local market (Brown, 2023), if their currently available resources do not permit them to focus initially on the national or global context. Consumption culture could be defined as the extent of availability of basic and luxurious products and services among different consumer groups in a society. Sustainable consumption can be defined as consuming a product or service (e.g., at least related to basic human needs) in a way that ensures equity, diversity, and inclusion of all current and future consumer groups in a society to fulfill consumers' basic needs to survive and prosper. To ensure a sustainable consumption practice in our wider society or market, sustainable marketers design their value proposition to implement and evaluate the proposition in a way that would be instrumental in contributing value to their societal (people-focused), environmental (planet-focused), and commercial (profit-focused) stakeholders. Examples of such triple bottom line focused value propositions are discussed in ► Manager's Corner 1.1, 1.2, and ► 1.3.

- **Consumption culture** is the extent of availability of basic and luxurious products and services among different consumer groups in a society.
- **Sustainable consumption** is consuming a product or service (e.g., at least related to the basic human needs) in a way that ensures equity, diversity, and inclusion of all current and future consumer groups in a society to fulfil consumers' basic needs to survive and prosper.

Since it takes lots of planning and effort to ensure EDI in our local or global market, sustainable marketers set a long-term target to achieve their goal, which is known as vision. A vision is establishing a target as a lighthouse that can help us navigate toward our destination where we want to reach within a targeted period. For example, the United Nations set 17 global Sustainable Development Goals (SDGs) as a target in 2015 to achieve by 2030 (PricewaterhouseCoopers, 2023, np) as a vision to address global social, environmental, and economic concerns. We know that the SDGs cannot be accomplished by 2030, but setting the 2030 target in 2015 can help us start working and gradually addressing the concerns related to the SDGs. The Preliminary Vignette of this chapter (the value chain of a for-profit firm for its people, planet, and profit) describes the sustainable vision of Aspuna, which is working for "a world where all globally traded commodities add social and economic value to the communities from where they originate" (Aspuna, n.d., np). In the Preliminary Vignette, we see how Aspuna, as a profit-focused business firm, is working to improve the life of farmers and micro entrepreneurs in their African value network to pursue their vision while making more profit by exporting processed food products from Africa to other global regions. Establishing such a triple Ps' stakeholder-focused (People, Planet, and Profit-focused) vision helps Aspuna to reflect on the distress of their stakeholders in their socially caring value network (i.e., ensuring that the farmers and micro entrepreneurs can improve their lives), as well as helps them to source their processed food products in the more competitive price from Africa to distribute to different global markets, with an aim to make a profit.

Vision is establishing a long-term target as a lighthouse that can help us accurately navigating toward our destination, where we want to reach within a targeted period.

To pursue a sustainable vision, i.e., a vision that helps sustainable marketers to contribute value to their triple bottom line's stakeholders, marketers need to plan, implement, and evaluate a number of activities over a targeted period, which is known as a mission. A mission can be defined as the combination of planning for activities and actual activities that we strategize, implement, and evaluate to pursue our vision. For example, Microsoft's initial mission in 1992 was to confirm "a computer on every desk and in every home" (Bae, 2015). By the turn of the century (from the twentieth to the twenty-first century), we have seen that Microsoft has achieved its initial mission (i.e., "a computer on every desk and in every home"), at least at their home market in the USA. In order to accomplish this initial mission, Microsoft needed to plan, implement, and monitor a set of systematic activities. These activities are the elaborated form of the mission statement of Microsoft to convey the message of how they were trying to contribute value to their stakeholders. In this context, a mission statement could be defined as summarizing the core focus of a brand's overall activities to convey the message of how the brand's activities collectively help the brand to contribute value to its different stakeholder groups in order to pursue the brand vision.

- **Mission** is the combination of plan of activities and actual activities that we strategize, implement and evaluate to pursue our vision.
- **Mission statement** is summarizing the core focus of a brand's all activities to convey the message how the brand's overall activities collectively help the brand to contribute value to its different stakeholder groups, in order pursue the brand vision.

Conveying the message of how a brand's overall activities collectively help the brand to contribute value to its different stakeholder groups in order to pursue the brand vision through its mission statement is important for commercial success. It is because different non-customer stakeholder groups, such as investors, government, media, social, and environmental activist groups, are more concerned today to see how distinct brands across the world can minimize their detrimental impact on society and/or environment and can enhance their societal and environmental contribution for a sustainable future. Once Microsoft felt that it had accomplished its initial mission (i.e., "a computer on every desk and in every home"), it redefined its mission statement in 2015. ► Manager's Corner 1.1 shares more discussion on Microsoft's redefined mission statement to clarify how they convey their triple bottom line focused value proposition, to clarify their mission to contribute to their societal, environmental, and commercial stakeholders.

1

Managers' Corner 1.1: Microsoft's Redefined Mission Statement to Empower the Triple Bottom Line's Stakeholders

Source: ► pexels.com/royalty-free-images/

Microsoft have redefined their mission statement in 2015, to re-align with the changing need of their target market. Since 1992, for a quite long time their mission statement was to put "a computer on every desk and in every home." Microsoft CEO, Satya Nadella described that "when I joined the company in 1992, we used to talk about our mission as putting a PC in every home, and by the end of the decade we have done that, at least in the developed world." Over time, once Microsoft has accomplished this initial mission at least in the developed economies, and once the United Nations' Sustainable Development Goals (SDGs) were announced in 2015, Microsoft needed to re-focus on the contemporary need of their expanded target market beyond the developed countries. This re-alignment plan with the expanded markets' need enables them to re-define their corporate mission, with an aim to ensure their global expansion.

Following this drive to adapt with the change in the market environment, Microsoft took "forward and force it (their corporate mission) to innovate beyond a vague statement in 2015 for *Microsoft* to 'empower every person and every organization on the planet to achieve more'. In recent years, this has seen Microsoft switch its focus from *Windows* and PC toward cloud computing." "Once Microsoft has accomplished their initial mission statement the question was largely what's next?" At that stage, Microsoft needed to expand their global reach, while aligning with the global SDGs. One of the lucrative options to expand their global reach is exploiting the contemporary mobile technology, which is considered as "a turning point in its evolution as a mobile-first, cloud-first software development company."

In this context, Microsoft has shifted their focus on to the mobile technology and cloud-based software from their traditional focus on operating system software. This shift in company's mission, enables Microsoft to reach to their global target markets beyond the developed economies, in support of the cloud computing system. As a result, "these changes are leading the company down a different and potentially more lucrative path that focuses on productivity, emphasizes cloud computing, and embraces an agnostic attitude toward the operating systems." This redefined mission statement focusing on empowering people and organizations across the globe clarifies that the business is all about empowering their customer and non-customer stakeholders in different international markets, which could be accomplished through the "utility of the company's (cloud) computing products."

To explain the redefined corporate mission statement, the Microsoft spokesperson described that "at Microsoft, our mission is to enable people and businesses throughout the world to realize their full potential. We consider our mission statement a commitment to our customers." This re-alignment initiative to cope up with the changing market environment (e.g., global expansion and promotion of the mobile technology, and the introduction of SDGs), and redefining their mission statement in line with this change enables Microsoft to operate "under a 'customer (or people) first' philosophy." According to the market analysts, this redefined mission statement focusing on the "customer philosophy" "frees Microsoft from the self-imposed and unnecessary burden of always trying to be better than a particular competitor," as well as help them to deal with the changing market environment after accomplishing their initial corporate mission. Furthermore, this redefined mission statement enables Microsoft to freely engage in R&D activities to pursue excellence in innovation to contribute value to their different stakeholder groups, as "making decisions solely on the basis of beating a competitor like Apple or Oracle, instead of making decisions based on what is best for your stakeholders, stifles innovation and often leads companies down the slow spiral of decay and eventually death."

"Empower every person and every organization on the planet to achieve more." Reflecting on this redefined mission statement of Microsoft, let's first, think of what the meaning of empowering someone is. It is to ensure resources, training and other relevant support provisions for someone to achieve her or his full potential as a human being to survive and prosper, so that the empowered person also can contribute back to her or his society and environment. Secondly, analyze and summarize your views on:

- How authentic the Microsoft's redefined mission statement is to empower their commercial, societal and environmental stakeholders in their global market?
- Is the Microsoft's redefined mission statement feasible to accomplish?

Information Sources Miller, C. (2008). Bill Gates: how a geek changed the world. BBC. Retrieved June 15, 2019, from ► http://news.bbc.co.uk/2/hi/business/7461783.stm; Kaelin, M. (2017). Microsoft has set itself up for success-now it's time to make something happen. TechRepublic. Retrieved June 15, 2019, from ► https://www.techrepublic.com/article/microsoft-has-set-itself-up-for-success-but-now-its-time-to-make-something-happen/; Weinberger, M. (2017). Microsoft CEO Satya Nadella says Bill Gates' original mission 'always bothered me.' Business Insider. Retrieved June 15, 2019, from ► https://www.businessinsider.com/microsoft-ceo-satya-nadella-bothered-by-bill-gates-mission-2017-2; Media Marketing (2017). Satya Nadella says Microsoft's old mission statement is obsolete. Media Marketing. Retrieved June 15, 2019, from ► https://www.media-marketing.com/en/news/satya-nadella-says-microsofts-old-mission-statement-is-obsolete/; Gregory, L. (2019). Microsoft's mission statement & vision statement (An analysis). Panmore Institute. Retrieved June 15, 2019, from ► http://panmore.com/microsoft-corporation-vision-statement-mission-statement-analysis. (Reprinted an expanded version by permission of the copyright owner, © Springer))

1.5.2 Feasible Vision and Mission Statement

Strategic marketing scholars describe six characteristics to define a feasible mission statement so that a brand's mission statement would be authentic and feasible to implement in the minds of its target audiences. These six characteristics are (Drummond & Ensor, 2001):

- Market-oriented
- Motivating
- Realistic
- Distinctive competencies
- Specific environment
- Fit market environment

◘ Figure 1.9 adapted these six characteristics of a feasible mission statement from the contexts of a sustainable brand that aims to satisfy the needs of triple Ps' stakeholders (People's, Planet's, and Profit's stakeholders) while addressing stakeholder-specific distress. For example, to satisfy the triple Ps' stakeholders, sustainable marketers need to develop an accurate orientation about the needs, wants, and expectations of those stakeholders, and their distresses to satisfy them. ► Section 1.4.1 (Ongoing Market Orientation) discusses how marketers can develop orientation about their market, which includes all customers and non-customer stakeholders. Proper market orientation can help marketers to devise a mission statement that could be relevant to the marketers' societal, environmental, and commercial

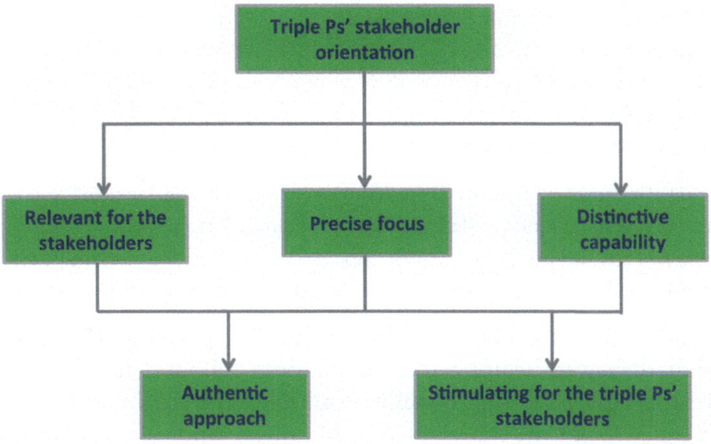

Fig. 1.9 Key considerations to define a feasible mission statement targeting the triple Ps' stakeholders. (Adapted from Drummond & Ensor, 2001)

stakeholders' needs. The mission statement then should specify a precise focus on the triple Ps' stakeholder needs that the brand aims to satisfy. It is because that a particular stakeholder would have diversified needs and distresses; however, it is not possible for a brand to satisfy someone from all contexts. Therefore, the brand's mission statement should precisely indicate the specific needs of each stakeholder that it aims to satisfy. A brand's mission statement should also provide an impression about the distinctive capability of the brand to elucidate how distinctively the brand is capable of delivering value in comparison to the competitive brands' capabilities to satisfy stakeholders' needs. It helps marketers differentiate their brand's value proposition in comparison to the competitive value propositions. Upon accurately understanding the triple Ps' stakeholders' needs and analyzing a brand's available resources to satisfy those needs, sustainable marketers could consider how their specific capabilities could help them to focus on the precise needs of their stakeholders. Then, marketers need to demonstrate an authentic approach through their overall activities to implement and evaluate their mission, in a way that would be stimulating for their triple Ps' stakeholders to establish trust on the brand's value proposition.

Sustainable marketers are like visionary statesmen. There is a distinct difference between a statesperson and a leader. Focusing on the countrymen's needs and distresses, a visionary statesperson influences her or his compatriots, clarifying how working together can help them accomplish their mission, to pursue their vision. Therefore, the compatriots willingly respond to the statesman's call to work together, as the compatriots can see that their endeavors are for collective development. On the contrary, an ordinary leader tries to influence her or his followers quite often based on the regulatory bindings to perform a particular task. Similarly, the sustainable marketers are visionary statesperson who aims to influence their target audiences (i.e., the triple Ps' or triple bottom line stakeholders), based on conveying a message and action plan that will be mutually beneficial for societal, environmental, and economic progress. In this context, Nelson Mandela, as a visionary statesperson, was a (sustainable) marketer too, considering that he was able to influence his people based on conveying a message and action plan for collective progress. The visionary statesperson is like the flute player of Hamelin. In medieval literature, the author has narrated a fantasy that:

> In the mid14th Century, a monk from Minden, Heinrich von Herford, puts together a collection of holy legends called the "Catena Aurea." It speaks of a "miracle" that took place in 1284 in Hamlyn. A youth appeared and played on a strange silver flute. Every child that heard the flute, followed the stranger. (Museum Hameln, n.d., np)

Since the children found a tune in the flute of the flute player of Hamelin that they felt was what they were waiting for, the children followed the flute player and migrated with the player from Hamelin to another place. Similarly, the visionary statesmen's call, such as the call of Nelson Mandela, made his compatriots believe that this was the call that they were waiting to respond to, as the compatriots believe the call is for collective development. In the same vein, sustainable marketers as visionary statesmen need to identify the latent needs of their triple bottom line's stakeholders in order to convey a message and indicate an action plan through their sustainable mission statement so that their stakeholders can have trust in them that their value proposition is for collective progress for their society, environment, and economy. Based on the Grameen-Danone partnership and their joint value proposition, ▶ Manager's Corner 1.2 shares an example of how Grameen-Danone adopts the six characteristics to define their mission statement, focusing on the needs and distresses of their triple bottom line's stakeholders, in order to pursue their vision.

Like the Grameen Danone Foods' (GDF's) mission statement as discussed in ▶ Manager's Corner 1.2, GDF also has established a vision (Grameen-Danone, 2023) relevant to the needs of wider society, i.e., "enabling Bangladesh to maximize untapped human potentials" (np). Through accomplishing their mission statement, i.e., improving kids' health, creating jobs for the vulnerable, and creating impactful partnerships at upstream and downstream value networks, GDF also justifies how truthful they are to pursue their vision, considering that their target audiences (e.g., the micro farmers and salespeople from vulnerable groups, or families who are worried about their kids' undernutrition) can realize how accomplishing their mission statement, GDF is working to maximize the untapped human potentials, in order to pursue their vision. Like a statesman (in the case of GDF, it is Professor Yunus, who is a sustainable marketer too, as he is able to influence his target audience), sustainable marketers should set their own career vision to contribute back to society and the environment, so that their societal and environmental contribution could help them underpinning their brands' commercial competitive advantage. In the same context, the career vision of one of the co-authors of this core sustainable marketing textbook is "inspiring our students, colleagues and stakeholders at government, non-government and third sector, eyeing on reinforcing equity, diversity, and inclusion of our global community for a shared sustainable development, based on sustainable marketing research and education."

Manager's Corner 1.2: Vision and Mission of Grameen-Danone to Deliver Value to Customers, Partners and Society at Large

Source: ▶ pexels.com/royalty-free-images/

"Severe malnutrition affects 45 million children across the globe each year" (UN World Food Program, 2023, np). To address this malnutrition and/or undernutrition challenge in Bangladesh, the Grameen-Danone Foods (GDF) Limited was launched in 2006

» Considering the high rates of micronutrient deficiencies in Bangladeshi children, Franck Riboud (Chairman and CEO of Danone-2005) and Professor Dr. Muhammad Yunus (founder of Grameen microcredit bank and 2006 Nobel Peace Prize Laureate), announced the creation of Grameen Danone Foods Limited by combining the expertise of Danone in dairy and the experience of Grameen in microcredits for deprived populations. (Grameen-Danone, 2023, np)

The objective of GDF is to offer "daily healthy nutrition (at competitive price) to low income nutritionally deprived populations in Bangladesh and alleviate poverty through the implementation of a community-based business model" (Safi, 2023, np). The GDF products include a range of yogurts, biscuits, milk and other drinks that provide required vitamins and other food value for human body, especially for kids during their growth stage. The vision of GDF is "enabling Bangladesh to maximize untapped human potentials" (np), and their mission statement includes (Grameen-Danone, 2023):

- "Providing health through nutrition.
- Creating employment opportunities among vulnerable communities.
- Scaling-up ecosystems development with impactful partnerships" (np).

Since its inception in 2006, GDF have offered 33 fortified food products to address the undernutrition challenge in Bangladesh and have impacted on more 300,000 lives to help them to fulfil their everyday healthy nutrition need (Grameen-Danone, 2023). These figures demonstrate that GDF are on-track in pursuing their mission to "providing health through nutrition," which helps them to work for their vision, i.e., maximizing the untapped human potentials, considering that empowering people with basic nutrition requirements is a prerequisite to maximize the human potentials, especially from the underprivileged communities. Following the partnership between Grameen and Danone, GDF as a Social Business (2013):

> Affects people's lives not only by improving their health. Benefits exist along the whole value chain. The milk for the yoghurt is purchased from micro farmers. The production is designed in such a way as to give as many people as possible a job. Sales ladies distribute the yoghurt door-to-door and receive a 10% commission. Unsold yoghurts are taken back. In total, Grameen Danone Foods is responsible for the creation of about 1600 jobs within a 30 km radius around the plant. (np)

Therefore, the GDF can expand their partnership with local micro entrepreneurs (e.g., the micro farmers), while creating more jobs for people from disadvantaged communities to accomplish their mission of addressing the (1) undernutrition challenge in wider society, (2) creating jobs for vulnerable communities, (3) expanding impactful partnership to improve people' life, while (4) making profit for both Grameen and Danone. In this context, GDF deliver value to their customers, partners, and society at large.

Let's have a look at the GDF's mission statement from the lens of the characteristics of a feasible mission statement, which is discussed in ◘ Fig. 1.9. Orientation on triple Ps' stakeholders' needs and distress, precise focus and distinct capability to satisfy the relevant needs of stakeholders through an authentic approach that can stimulate the triple Ps' stakeholders to associate with a brand's value proposition, are the key characteristics of a feasible mission statement. In the case of GDF's mission statement, it can be confirmed that the statement reflects on the undernutrition need and distress of people from wider society who are their customers too. Their mission statement is focused on three specific socioeconomic issues, which are improving kids' health, creating employments for underprivileged communities, and creating impactful partnership, which are also relevant to the needs of their target audiences. GDF also have adopted an authentic approach to deliver and evaluate their value proposition, e.g., ensuring fair price for micro farmers or creating jobs for jobless in society, which stimulates their target audiences to partner with GDF to co-create value for GDF's customers, partners and wider society.

1.5.3 Sustainable Marketing Strategy: Four Essentials to Pursue the Triple Bottom Line (People, Planet, and Profit)

Why some organizations are successful in pursuing the triple bottom line? Why do many other organizations struggle to successfully pursue them? How the pursuit of the triple bottom line could be more meaningful for all concerned stakeholders? This section presents arguments based on four essentials for the successful pursuit of the triple bottom line of a business, i.e., people, planet, and profit.

There are few successful businesses in the world that have been effectively addressing the different societal (people-oriented) and ecological (planet-related) challenges in the long run while making a profit. On the contrary, there are many other businesses, and their sustainable marketing projects also have been struggling to retrieve their investment back. It is because of this that sustainable marketing projects need to satisfy three distinct types of stakeholder groups, namely, commercial, societal, and/or environmental stakeholders, simultaneously, which is more challenging in comparison to satisfying only the commercial stakeholders for profit-only marketing projects. Sustainable marketing, i.e., contributing to people and/or the planet through marketing strategies, is not a charity, but it is today an unprecedented need for any commercial brand to survive and prosper. Stakeholders of a brand are now more concerned about the brand's societal and ecological performance than ever before. Therefore, marketers have been striving more in recent decades to contribute back to the society and ecology in their markets so that such societal and ecological contributions could be instrumental in sustaining or prolonging their commercial competitive advantage, as well.

Sustainable marketing is a well-recognized field of research and practice today; however, we lack conceptual developments and empirical replications in this field to help more sustainable marketers to contribute to people and the planet while making more profit. Four essentials for the successful pursuit of the triple bottom line (people, planet, and profit) are discussed here, which are:

- Equity in orientation
- Triple Ps' stakeholder-led co-creation
- Co-evaluation
- And visible change for the triple Ps' stakeholders

These four essentials for sustainable marketing strategy are depicted in ◘ Fig. 1.10. These essentials for planning, implementing, and evaluating sustainable marketing strategies will be imperative for the success of sustainable marketing strategies.

Equity in Orientation Stakeholder orientation is a fundamental concept to understand the customer and non-customer stakeholders' expectations so that a brand's value proposition can be designed to meet those stakeholder-specific expectations. On the one hand, understanding different stakeholders' expectations is key to business success. On the other hand, an equal or balanced emphasis on different groups

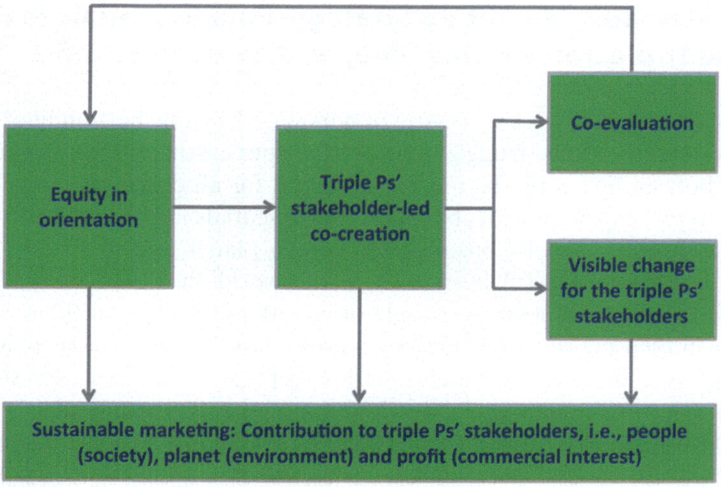

Fig. 1.10 Four essentials for sustainable marketing strategy

of stakeholders' expectations at all stages of a business cycle is instrumental for ongoing success and avoiding criticisms from stakeholders in the long run. In terms of equity in orientation, a brand, therefore, should consider two things. One is equally emphasizing all key societal, environmental, and commercial stakeholders' expectations, and the other one is an ongoing focus on such diverse stakeholders' expectations in varied business settings. For example, in the case of a successful partnership between Grameen and Danone, we have seen in ► Manager's Corner 1.2 that Grameen and Danone aim to satisfy their societal stakeholders' needs and distress (i.e., addressing the undernutrition of kids) while satisfying the needs of their commercial stakeholders (i.e., Grameen, Danone, the micro farmers as their suppliers, and creating jobs for disadvantaged communities). In terms of satisfying the environmental need to improve their stakeholders' environmental footprint in their value network, Grameen-Danone ensures that:

> A minimum number of automatic devices and complex machinery were included to increase the need for labor and thus create more local jobs, all while reducing initial investment costs and limiting future maintenance costs. This also helped keep the carbon footprint and environmental impact of the plant low. (Renouard, 2012, np)

Inequity in orientation from any of these three contexts (commercial, societal, and/or environmental contexts) of multi-stakeholder orientation can lead to the failure of a brand to meaningfully pursue the triple bottom line. Another example is related to Facebook, as they address the climate change issue to a decent extent by reducing *the* CO_2 emissions of each Facebook user from 2016 to 2020 (Collins, 2021). It demonstrates Facebook's focus on environmental needs and contribution to the planet. However, Facebook's concurrent failure to sufficiently orient with people's privacy while they use Facebook is a key reason for them not being able to establish a truly socially caring global brand image, as it (the failure to protect users' privacy) can be identical to an impression that Facebook emphasizes profit over their users' privacy. Consequently, such people-oriented, privacy-related scandals, over-emphasis on profits, and the subsequent damaging brand image obscure Facebook's current position as a responsible Meta for bringing meaningful changes related to the most unprecedented societal and ethical challenges in today's world, irrespective of their minimal socio-ecological contribution from time to time, such as reducing CO_2 emissions per user. It implies that equal emphasis on all key societal, environmental, and commercial stakeholders' expectations and the ongoing focus on such diverse stakeholders' expectations in varied business settings and the different life cycles of a business is vital for meaningful pursuit of the triple bottom line in the long run.

Triple Ps' Stakeholder-Led Co-creation Similar to equity in orientation, another essential consideration for the successful pursuit of the triple bottom line is ensuring all three groups of stakeholders' contributions, i.e., the stakeholders who are concerned for people, planet, and profit (triple Ps stakeholders) co-create a solution together to address the commercial, societal and environmental challenges based on a mutually-beneficial venture. For example, Grameen-Danone Foods (as discussed in ► Manager's Corner 1.2) co-creates nutritious food products based on the supplies of milk from micro farmers to not only make a profit but also to address the undernutrition challenges of the underprivileged people in the market, where they operate their business. In this effort, they have contributed to the healthy growth of 300,000 underprivileged children (Grameen-Danone, 2023), ensured sustainable revenue (fair price) for 500 farmers who are the suppliers of Grameen-Danone, and empowered 1500 deprived women from rural areas by creating jobs for them at each mini-factory (Kaplan, 2006). Such co-creation of production and in-bound and out-bound supply chain management of Grameen, Danone, and their common stakeholders in their target market (i.e., the underprivileged women and micro farmers from rural areas) enabled them to promote their food products together, to reach their huge customer-base in rural areas. Consequently, they can address the societal challenges of nutrition and a better life for the deprived communities in their market while making a profit.

In general, without equity in orientation and ongoing focus on different commercial, societal, and/or environmental stakeholders' expectations, it is not possible to understand the pragmatic co-creation scopes of such diverse stakeholders. Since Grameen can equitably orient with the needs of their commercial stakeholders (i.e., Danone and their suppliers and employees) and the needs of their societal stakeholders (i.e., the disadvantaged kids, women, and micro farmers), the Grameen-Danone venture is able to successfully pursue their commercial and societal goals, simultaneously. Another example of a successful sustainable marketing case study similar to Grameen-Danone is described in the Preliminary Vignette of this chapter based on Aspuna, as Aspuna is able to orient proactively with the distress of their societal stakeholders (i.e., disadvantaged farmers) in the African value network of their proceed-food business.

▪ Visible Change for the Triple Ps' Stakeholders

Grameen-Danone case demonstrates that their venture brings real change in the life of their societal and commercial stakeholders, as they can increase the living quality of the underprivileged communities in their market while making profit. In fact, more consumers want that commercial brands should help their consumers more to make a real difference to contribute to our society and protect our environment, e.g., through more eco-friendly packaging, recyclable materials, and ethical and more socially- and/or environmentally caring market offerings (Townsend, 2018). The same study by Townsend in 2018 reached the conclusion that "if your brand isn't helping your consumers improve their environmental and social footprint, then you're in danger of disappointing 88% of them." In the case of Grameen-Danone's sustainable marketing-centered commercial venture, they are also promoting the facts, i.e., already helping 300,000 kids to meet their nutrition needs, improving the lives of 500 micro farmers and women in underprivileged communities (Kaplan, 2006). It helps them to establish trust among their triple Ps' stakeholders that they deliver their promises. Therefore, making such visible changes to address the societal and/or ecological challenges not only enables the commercial brands to contribute back to the people and/or planet but also increases the potential to make a large crowd of loyal customer and non-customer stakeholder groups, who prefer more responsible production and consumption.

Co-evaluation Like equitable orientation, co-creation, and practical change for diverse commercial, societal, and/or environmental stakeholders, co-evaluation of the ongoing progress of a sustainable marketing venture by its diverse groups of stakeholders is imperative for the venture's long-term success and contribution to people and planet, while making a profit (Apostolidis et al., 2021; Brown et al., 2022). Co-evaluation of a business venture by its different stakeholders can be an extended form of initial co-creation of the business processes and functions, which are co-created by the involved stakeholders. Like the initial stage of co-creation, the associated commercial, societal, and/or environmental stakeholders should be involved in the co-evaluation of the processes and functions at different life stages of a sustainable marketing-centered commercial venture. Since rationally protecting the interests of all commercial, societal, and environmental stakeholders is the key to the successful pursuit of the triple bottom line, establishing a process of co-evaluation by all these stakeholder groups at every stage of the venture and its value chain can underpin its potential for sustainable operation.

1.5.4 Value Proposition for People, Planet, and Profit

Source: ► pexels.com/royalty-free-images/

► Section 1.2.5 (Value, the Core Concept of Marketing) discusses about the concept of value, which (i.e., value) is the core concept of marketing. It is because the competitive brands compete among themselves based on planning, delivering, and evaluating the most competitive utility or value of their market offerings to satisfy the same needs of their common customers. In a simplistic sense, value is what we sacrifice as consumers and, in return, what we receive from competitive brands to consume. Consumers' sacrifice to consume the value of a product or service would include the money they spent to purchase the product or service, the associated search, and the time cost to find the most competitive value among competitive brands. Therefore, marketers' need to plan how they could integrate value into the basic 4Ps marketing mix (Product, Price, Place, Promotion) and the extended marketing mix, where relevant, to make their product or service most competitive with an aim to offer superior value, in comparison to the competitive brands. For example, in terms of integrating the most superior value, marketers need to consider how the product features would be most user-friendly, how the pricing strategy would be most affordable, how place or distribution channel (online and/or traditional physical outlets) would be most convenient for customers to reach out and obtain the product or service, and how the promotional message would be most appealing to the target audiences related to their needs, wants and expectations.

1

Value proposition is a comprehensive plan of designing, delivering and evaluating value to satisfy the customers' and non-customer stakeholders' needs, wants and expectations.

Such marketing-mix focused plan for value should not only be targeted to meet the customers' needs, wants, and expectations but also be targeted to meet the needs of non-customer stakeholders, such as investors, suppliers, employees, business partners, and others, who help a brand to create value to satisfy the brand's customers. It is because if a brand cannot satisfy its suppliers to source raw materials at the most competitive price, the brand will not be able to offer the most competitive price for its product or service to its customers. In this context, the **value proposition** is a comprehensive plan of designing, delivering, and evaluating value to satisfy the customers' and non-customer stakeholders' needs, wants, and expectations. For example, ◘ Fig. 1.8 (Key stakeholder value network) portrays the key stakeholders of a brand that the brand must satisfy, considering that these key stakeholders help the brand create value for its customers.

For sustainable marketers, these non-customer stakeholders include not only the profit-focused commercial stakeholders but also the people- and/or planet-focused stakeholders, as well. For example, the socially- and environmentally-conscious customers and activist groups, media, and different local communities (e.g., in ► Sect. 1.3.2, Marketing Mix and the Extended Marketing Mix), we have seen that Curtin Malaysia as an educational institution is also satisfying the needs of Malaysian fishermen (i.e., reducing the collision between fishermen and crocodile) to expand their revenue sources in the Malaysian market. To design, deliver, and evaluate customer and non-customer stakeholder-specific value propositions, sustainable marketers should:

- Recognize first their target customer segment/s and customer segment-specific needs, wants, and expectations.
- Identify the core commercial stakeholders who would help marketers to create value to satisfy their customers and such commercial stakeholder-specific (e.g., supplier, employee, partner, etc.) needs, wants, and expectations.
- Identify the relevant societal and/or environmental stakeholders who would be concerned about the marketers' commercial value proposition (e.g., whether the factory of a brand pollutes air or local river or not), and such stakeholder-specific needs and distresses.
- Develop an integrated mapping of the key stakeholder network, as represented in ◘ Fig. 1.8, to figure out the short-, mid-, and long-term strategies to address the needs and distresses of their key commercial, societal, and/or environmental stakeholders.

Such integrated customer and non-customer stakeholder-focused value propositions could be devised based on only core utility or value or based on the combination of core and added value of a product or service. For example, the core utility offered by passenger airline brands is helping passengers to travel to a far-off destination in less time, in comparison to other means of transportation, such as train or coach service. In addition to this core utility, we see that the competitive airlines plan to differentiate their value proposition in comparison to the competitive value propositions based on added value. For example, Emirates customizes its in-flight meal offerings based on the culture-specific food habits of its passengers. One of the co-authors of this textbook experienced that Emirates offers fish curry and rice instead of the traditional sandwich as an in-flight meal. It indicates that Emirates monitors the passports of their passengers and customizes their in-flight meals based on the flight destination and their passengers' food culture- or food habit-specific in-flight meals; however, they may not need to spend more money to offer such customized meals instead of generalized (e.g., sandwich) in-flight meals. On the contrary, budget airways, such as EasyJet, Ryanair, and others, differentiate themselves by offering more affordable pricing options as added value based on their passenger-specific needs. For example, in this case, if a passenger does not want to consume the in-flight meal, the passenger will pay less, or if another passenger does not want to carry check-in luggage, the passenger will pay less. In this way, competitive airlines design their value propositions based on both core values (i.e., helping passengers to travel to a far-off destination in less time) and added value (e.g., customized in-flight meals or pricing options). ► Manager's Corner 1.3 discusses IKEA's value proposition based on both core value and added value, targeting their commercial, societal, and environmental stakeholders.

Web Exercise 1.5

Browse the websites of different brands from five or more industries, such as supermarket (e.g., ASDA), fast-food (e.g., McDonalds), airlines (e.g., United Airlines), information technology (e.g., IBM), sports industry (e.g., Newcastle United), and Google for more brand-specific information to understand:

- The core and added value offered by these brands.
- Who are included in these brands commercial, societal and/or environmental stakeholder groups.
- How these brands plan for their value propositions to satisfy the needs of their commercial, societal, and/or environmental stakeholder groups.
- How the triple Ps' (People, Planet, and Profit) value propositions differ in between these brands across different industries.
- How these brands are targeting their industry-specific triple Ps' stakeholders.

Manager's Corner 1.3: IKEA's Core and Added Value for Customers, Partners and Society at Large

Source: ► pexels.com/royalty-free-images/

IKEA, the furniture brand offers customized added value, based on the pricing strategy of their furniture's price, in addition to offering the furniture itself as core value. For example, IKEA offer non-assembled furniture. If customers purchase non-assembled furniture, instead of fully-assembled furniture and assemble it by them, in this case, customers will pay comparatively less price. In this way, IKEA offer lower price in comparison to their competitive furniture brands to offer additional value to their customers. In addition to satisfying the commercial stakeholders, i.e., the customers with core and added value, under their Social Business initiative, IKEA also partner with small entrepreneurs, with an aim to create opportunities for those small entrepreneurs. For example:

> IKEA Social Entrepreneurship is a business that started with the vision of honouring unique skills and competencies and gives them a global platform to stand on—resulting in better lives for those who need it most. This has been accomplished by partnering with enterprises that produce products and services that in turn empower people who struggle to provide for themselves and their loved ones. The result is financial independence and life-changing opportunities for families and communities, with improved health care, education and gender empowerment. (People & Planet, 2023, np)

In this way, IKEA have been not only contributing to the personal and professional life of their business partners, i.e., creating business opportunity while improving health care, education and gender empowerment, perhaps, but also able to source their supplies, e.g., non-assembled furniture in a better competitive price from the community of their business partners, i.e., those small entrepreneurs. It also helps IKEA to offer competitive price to their customers in comparison to their competitive brands.

Think about this people, planet, and profit oriented value proposition of IKEA, and if you were IKEA's Marketing Director, think about how you would diversify IKEA's partnership with their existing and potentially new business partners, reflecting on their different areas of operations, e.g., sourcing furniture, showroom management, inbound and outbound logistics and so on, to offer more win-win value propositions to their triple Ps' stakeholders.

1.6 Sharing Dependency, Risk, and Uncertainty Among the Triple Bottom Line's Stakeholders

In general, even during times of peace or without natural calamity, there are business risks and uncertainties in our local, regional, and global business dynamics. In addition, the complex value-delivery network, ongoing technological progress, climate change, geo-political uncertainty, and inequity pose unprecedented challenges for business leaders seeking to achieve the triple-bottom line by delivering on People, Planet, and Profit, which bring more business risks and uncertainties. Imagine you have been planning to launch a business for the last eighteen months in your local or an overseas international market. In these eighteen months, you have spent time and resources to develop a new product, set up an inbound and outbound logistics network, spend on promotional campaigns, and so on, and now you are about to launch your new business in the next month. Just before launching the business, you have found that the government tariff on your product has been increased in your local or international market. Because of this sudden change in your market environment, you now need to re-design your pricing strategy, which may impact your overall plans for your relationships with your inbound and outbound logistics partners, your ad agency, and so on. Government tariff is just one example related to such business risks and uncertainties; however, business risks and uncertainties could derive from many other contexts of a marketers' overall business dynamics. For sustainable marketers, predicting and addressing such business risks and uncertainties are more complex. It is because the traditional profit-only marketers need to rethink only the commercial stakeholders, whereas the sustainable marketers need to rethink their business relationships not only with their commercial stakeholders but also the societal and/or environmental ones. Therefore, sharing dependency, risk, and uncertainty among the stakeholders, e.g., business partners, from the beginning of a new venture is a prerequisite to nurturing stakeholder relationships while sharing business profit or loss among them based on a predetermined partnership contract.

Source: ► pexels.com/royalty-free-images/

Before we have a look at a real-life example, let us first have a look at a simulative example of sharing dependency, business risk, and uncertainty among the key stakeholders based on a simulative real-life business scenario. Let us assume that you are planning to launch a new sports league, like the National Basketball League in the USA or the English Premier League for football in the UK, or the Indian Premier League in India. Now, you are planning to secure a broadcasting partner for your sports league. Based on the negotiation, it has been decided that your new broadcasting partner will give you, let us say, $1000.00 over the period of the first 10 years of the league. However, the broadcasting partner will not give you an equal amount each year, which will be determined centered on the success of the broadcaster's television rating point (TRP) each year that they would have each year, based on the extent of popularity of your sports league to attract television viewers. For example, during the first season of your league, the broadcaster will give you $100.00. If the league can meet the first year's TRP expectation, you will receive $100.00 as well in the second season of the league. If the league can exceed the first year's TRP expectation, you will receive $110.00, for example, in the second season of the league. On the contrary, if the league is not able to meet the first year's TRP expectation, you will receive $90.00, for example, in the second season of the league. In this way, the sports league and its broadcasting partner can appreciate the dependency among themselves while sharing business risk and uncertainty over the period of the first ten seasons of the league.

A practical example of such sharing dependency, risk, and uncertainty is noted in the Indian Premier League (IPL) for cricket. IPL was first launched in 2008. Prior to the inauguration season, IPL signed a contract with their broadcasting partners, Sony and the World Sports Group (WSG). Based on this initial contract, Sony, WSG, and IPL share dependency, risk, and uncertainty in the same way that is discussed in the simulative example of sharing dependency, risk, and uncertainty in this current section, based on the fictional sports league of yours and your broadcasting partner. IPL rehabilitated their initial 2008's US$1.026 billion 10-year broadcasting contract with Sony and WSG on US$1.64 billion for the rest of the 9 years of their initial contract, where the next year's annual income of IPL from Sony and WSG will rely on the preceding years TRP success of Sony and WSG that they would derive based on the extent of IPL's popularity (Patil, 2009; Shams, 2013). Sharing dependency, business risks, and uncertainties in this method among the business partners is like following a variable rate of interest instead of a fixed rate over a period. Since variable rates over different years would differ between each year depending on the changes in economic conditions in our overall market environment, this method of sharing dependency, risk, and uncertainty among business partners helps each other to avoid a huge economic loss. For example, if IPL cannot ensure sufficient TRP for Sony and WSG, Sony and WSG do not need to pay IPL an excess amount of broadcasting sponsorship fee, considering that the business situation of Sony and WSG (i.e., not meeting the TRP expectation) in the current year does not allow them to pay IPL the excess sponsorship amount in the subsequent year. In the same vein, sustainable marketers need to consider how they could share business profit or loss synchronizing with the rate of economic inflation, recession, business success, or failure to ensure equity among their societal, environmental, and commercial stakeholders.

1.7 Marketing Management and Value Creation for the Sustainable Development Goals

◘ Source: ► pexels.com/royalty-free-images/

A number of practical case examples from across different local and global market regions and from different industries are discussed throughout this current chapter to demonstrate how the discussed marketing and aligned business principles can be applied to contribute back to society and the environment while making more profit. For example, the Preliminary Vignette of this chapter (The value chain of a for-profit firm for its people, planet, and profit) describes how Aspuna, the British food-processing SME, established a commercial competitive advantage in its African value network. To source raw food grains at a more competitive price and to expand their business in Africa, Aspuna observed that the African farmers do not receive the correct price for their produce, which has an adverse impact on the farmer community's life, health, and well-being. To pursue their commercial goal, Aspuna realized that investing in infrastructure (i.e., small food-processing plants) and establishing a win-win value network (i.e., the farmers' cooperative and partnership with local micro entrepreneurs) in Africa would be instrumental in improving the life of the African farmers' and micro entrepreneurs' communities while sourcing food grains in more competitive price. Following this initiative, African farmers sell their food grains to Aspuna and process those raw grains in Aspuna's food processing plants in order to help Aspuna sell the processed food products in the local African markets based on their partnership with local micro entrepreneurs, as well as to export it to other global regions. In this way, Aspuna creates value for Sustainable Development Goal (SDG) 1 (no poverty), SDG 2 (zero hunger), SDG 3 (good health and well-being), SDG 8 (decent work and economic growth) for the African farmers' community, ensuring infrastructure (i.e., food processing plants) and innovation (i.e., win-win value network) for SDG 9 (industry, innovation and infrastructure). Eventually, this win-win value network of Aspuna helps them reduce inequalities (SDG 10) in the African communities and contribute to sustainable cities and communities (SDG 11) in Africa, based on their partnerships with the farmers' and micro entrepreneurs' communities to pursue the SDGs (SDG 17).

► Section 1.1 (Introduction) describes that it is important to select a particular societal or environmental cause relevant to the type of product or service (or its core value) that is offered by a brand to promote the brand based on a well-aligned cause. For example, Jeep promotes them, based on their socially caring brand message and value proposition, focusing on the American soldiers (and soldiers' families) who have wounded or died to protect the American interest overseas. Such a marriage between the brand Jeep and the cause is well-aligned because the nature of the product or service that is offered by the brand (i.e., Jeep) and the type of vehicles that are used by those soldiers are the same or similar. In such marketing campaigns, Jeep helps those soldiers and their families, the overall soldiers' community who suffered casualty for their country, revitalizing again to normal life. Eventually, the marketing campaigns help Jeep create value for SDG 11 (sustainable cities and communities). ► Web Exercise 1.1 shares the "Reverse Selfie" commercial of Dove, where Dove promotes how excessive and improper use of social media would be depriving for young girls. The brand communication message here, "the pressure of social media is hurting our girls' self-esteem," promotes responsible consumption behavior while using social media, which helps Dove create value (i.e., public awareness) for good mental health and well-being (SDG 3) of young girls, and SDG 12 (responsible consumption and production). ▶ Web Exercise 1.1 also shares IKEA's marketing communication message, "Climate Action Starts at Home," and Coca-Cola's promotion emphasizing the exigency of Recycling and Reusing existing materials. It helps both brands to create public awareness, which is important for SDG 12 (responsible consumption and production) and SDG 13 (climate action).

To address the criticisms of environmentalist groups, Facebook invested a lot of money to improve its system to reduce the CO_2 emission produced by each Facebook user. ▶ Section 1.2.2 (Strategic Marketing and Sustainable Marketing) presents more discussion about this environmental action of Facebook. This climate action of Facebook indicates that the wider stakeholder groups drove them to create value for SDG 12 (responsible consumption and production), and SDG 13 (climate action). It again reminds us that sustainable marketing is not a philanthrophy but a necessity for brands to survive and prosper, considering that increased number of customers and non-customer stakeholders are concerned today about society and environment than ever before. To discuss about latent need, ▶ Sect. 1.2.3 (Customers' Problems, and Expressed Need vs. Latent Need: Marketing a Solution, Not a Product Itself) describes about healthy alternative drinks (e.g., 100% preservative-free fresh coconut water or sugarcane juice) to address the dehydration problem of customers, considering that health-conscious customers would prefer more healthy alternatives, in comparison to the drinks with added preservatives, which are widely sold by supermarkets. In this context, if a supermarket starts promoting and selling more healthy drinks, they will not only have a commercial competitive advantage among the segment of health-conscious consumers, but also be able to create value for SDG 3 (good health and well-being) and SDG 12 (responsible consumption and production).

To address our enormous food waste across the world and to feed more people in our society, ▶ Marketing in Action 1.1 (A Sustainable Marketing Dynamic to Deliver Value to Customers, Partners, and Society at Large) presents the sustainable marketers' "Too Good to Go" campaign. Under this campaign, we see that using the "Look, Smell, Taste" app, marketers are helping commercial kitchens not only to reduce their operational loss but also helping to feed more people in our society. The "Too Good To Go" campaign eventually creates value for SDG 2 (zero hunger), SDG 3 (good health and wealth-being), SDG 10 (reduced inequalities), SDG 11 (sustainable cities and communities), SDG 12 (responsible consumption and production), SDG 13 (climate action) while reducing the environmental waste of food grains, and SDG 17 (partnerships for the goals) by partnering with different commercial kitchens and app developers. ▶ Section 1.2.4 (Identifying and Nurturing a Latent Need) shares an example of a 100% biodegradable plastic bag to replace the existing plastic bags that we use every day across the world. Since customers and non-customer stakeholders are concerned about the environment, investing in developing more environmentally-friendly products based on 100% biodegradable materials could help marketers to create value for SDG 3 (good health and well-being), as the Guardian (2019, np) reported that "people eat at least 50,000 plastic particle a year," because of our excessive use of plastic. Also, developing environmentally-friendly products can help marketers contribute to SDG 12 (responsible consumption and production), SDG 13 (climate action), SDG 14 (life below water) and SDG 15 (life on land) as animals on the ocean and on land have also been suffering because of the pollution caused by non-biodegradable product materials (Earth.org, 2022).

Differentiating market offerings in comparison to competitive offerings is a key aspect of marketing communication and branding. ► Section 1.2.6 (The Art and Science of Marketing) discusses that ALDI, as a supermarket, differentiates its value proposition based on its "Everyday Low Prices" campaign. It helps ALDI to offer more affordable food products to low-income consumers while creating value for SDG 2 (zero hunger) and SDG 3 (good health and well-being). ► Section 1.3.4 (Stakeholders' Needs and Distress: An Opportunity to Delight Them Based on Pursuing the Triple Bottom Line) discusses Curtin University's efforts to diversify their revenue sources in their transnational market in Malaysia. In addition to attracting more fee-paying local students in Malaysia to offer quality education, based on their Malaysian Government partnership and their biodiversity research, Curtin can reduce the collision between Malaysian fishermen and crocodiles and improve the life of the local endangered animals (i.e., the Malaysian sun bears). Such sustainable efforts to diversify their revenue sources enable Curtin Malaysia not only to meet their commercial target but also to contribute back to the local community and ecology in Malaysia while creating value for SDG 3 (good health and well-being), SDG 4 (quality education), SDG 11 (sustainable cities and communities), SDG 14 (life below water), SDG 15 (life on land), and SDG 17 (partnerships for the goals).

To address the children's malnutrition challenge, the Grameen-Danone partnership (SDG 17, partnership for the goals) has created a sustainable value network, where they create employment and businesses for micro dairy farmers and women in rural areas (SDG1, no poverty; SDG 5, gender equality; SDG 8, decent work and economic growth; SDG 10, reduced inequalities; SDG 11, sustainable cities and communities), and produce healthy dairy-products and sell it in competitive price (SDG 3, good health and well-being). ► Manager's Corner 1.3 (IKEA's Core and Added Value for Customers, Partners, and Society at Large) presents the case of IKEA and clarifies how IKEA creates partnerships (SDG 17) with their micro suppliers to source the furniture at a more competitive price while creating business for their micro suppliers (SDG 1, no poverty; and SDG 8, decent work and economic growth), and providing health care and education services (SDG 3, good health and well-being; SDG 4, quality education; and SDG 11, sustainable cities and communities) for their supplier community, while ensuring gender empowerment in the community (SDG 5, gender equality).

Similar to these examples of sustainable marketing campaigns in ◘ Table 1.4, the Rs of the circular economy concept (Refuse, Rethink, Reduce, Reuse, Repair, Refurbish, Remanufacture, Repurpose, Recycle, and Recover), which are illustrated in ◘ Fig. 1.5, can help marketers to integrate value in their marketing mix, centered on the needs and expectations of their customers and non-customer stakeholders, with an aim to contribute to the pursuit of our global SDGs. Since the circular economy is a responsible process of production and consumption that

Table 1.4 Sustainable marketing for creating value for the Sustainable Development Goals

Marketing management projects	Section in this chapter, where it is discussed	Value creation for the Sustainable Development Goals (SDGs)
Aspuna's expansion in the African market	Preliminary Vignette	SDG 1 (no poverty), SDG 2 (zero hunger), SDG 3 (good health and well-being), SDG 8 (decent work and economic growth), SDG 9 (industry, innovation and infrastructure), SDG 10 (reduced inequalities), SDG 11 (sustainable cities and communities), SDG 17 (partnerships for the goals)
Jeep's cause-related promotional campaigns	► Sect. 1.1	SDG 11 (sustainable cities and communities)
Dove's "Reverse Selfie" commercial	► Web Exercise 1.1	SDG 3 (good health and well-being), SDG 12 (responsible consumption and production)
IKEA's and Coke's campaigns for climate action	► Web Exercise 1.1	SDG 12 (responsible consumption and production), SDG 13 (climate action)
Facebook's reduction of CO_2 emission and environmental care	► Sect. 1.2.2	SDG 12 (responsible consumption and production), SDG 13 (climate action)
Promoting healthy drink	► Sect. 1.2.3	SDG 3 (good health and well-being), SDG 12 (responsible consumption and production)
Too Good to Go campaign	► Marketing in Action 1.1	SDG 2 (zero hunger), SDG 3 (good health and wealth-being), SDG 10 (reduced inequalities), SDG 11 (sustainable cities and communities), SDG 12 (responsible consumption and production), SDG 13 (climate action), 17 (partnerships for the goals)
Biodegradable packaging	► Sect. 1.2.4	SDG 3 (good health and well-being), SDG 12 (responsible consumption and production), SDG 13 (climate action), SDG 14 (life below water), SDG 15 (life on land)
ALDI's Everyday Low-Price campaign	► Sect. 1.2.6	SDG 2 (zero hunger), SDG 3 (good health and well-being)

(continued)

Table 1.4 (continued)

Marketing management projects	Section in this chapter, where it is discussed	Value creation for the Sustainable Development Goals (SDGs)
Diversification of Curtin Malaysia's revenue sources	► Sect. 1.3.4	SDG 3 (good health and well-being), SDG 4 (quality education), SDG 11 (sustainable cities and communities), SDG 14 (life below water), SDG 15 (life on land), SDG 17 (partnerships for the goals)
Grameen-Danone partnership	► Manager's Corner 1.2	SDG 1 (no poverty), SDG 3 (good health and well-being), SDG 5 (gender equality), SDG 8 (decent work and economic growth), SDG 10 (reduced inequalities), 11 (sustainable cities and communities), SDG 12 (responsible consumption and production), SDG 17 (partnerships for the goals)
IKEA's partnership with its suppliers	► Manager's Corner 1.3	SDG 1 (no poverty), SDG 3 (good health and well-being), SDG 4 (quality education), SDG 5 (gender equality), SDG 8 (decent work and economic growth), SDG 11 (sustainable cities and communities), SDG 17 (partnerships for the goals)

focuses on sharing, recycling, refurbishing, and reusing limited current resources (e.g., a product or materials) as much as possible so that a product's lifecycle can be prolonged, and more people could be benefited by that product. ► Section 1.3.1 (Historical and Embryonic Nature of Marketing Dynamics) presents more discussion on the concept of circular economy.

The last but most important consideration for sustainable marketers would be to remember their professional vision "to inspire their customers and non-customer stakeholders, eying on reinforcing equity, diversity and inclusion of our global community for a shared sustainable development, based on sustainable marketing." Working to pursue such a professional vision of sustainable marketing could also be instrumental for marketers to differentiate themselves in order to make their contribution standalone from the crowds of ordinary marketers, as an effort of personal branding to accelerate in their marketing career.

Summary

Marketing dynamics, as it is known today, have been evolving based on the mutual contribution of marketing academics and practitioners. The diverse customer and non-customer stakeholder groups of a brand collectively consist of its overall marketing dynamics. Therefore, it is imperative for marketers to understand the current and latent needs of their customers and non-customer stakeholders to prolifically and responsively influence their overall marketing dynamics.

The customers and other stakeholders of a brand are now more concerned about a brand's societal and environmental performance than ever before. Thus, sustainable marketing (i.e., contributing back to society and/or the environment through a profit-making venture) is not philanthropy but a necessity for brands to survive and prosper in contemporary markets. The socioeconomic and ecological concerns in a market that could be addressed by a brand need to be considered in the brand's marketing policy and strategy so that the market-specific socioeconomic and ecological value contribution of the brand could underpin the brand's commercial competitive advantage, as well as make a profit.

In these contexts, this chapter focuses on the views of reflective marketing academics and practitioners from a historical context to sequentially clarify the evolution and ongoing transformation in our past, present, and future (embryonic) marketing dynamics. The discussion here adopts a strategic viewpoint to help marketers develop their marketing orientation on different government and non-government stakeholders (including customers), with an aim to understand stakeholder-specific current and latent needs. The analysis of such needs includes the economic, social, and environmental interests, expectations, and distresses of the involved stakeholders, who have an influence on marketers' overall value network. The chapter has a comprehensive discussion on how sustainable marketers can identify and analyze their customer and non-customer stakeholders' economic, social, and environmental needs to create stakeholder value to contribute to people (society) and planet (ecology) of a market that is served by them while making a profit.

The chapter clarifies the application of established and embryonic marketing principles in a simplified manner and shares practical case examples that demonstrate how sustainable marketers create value for their societal, environmental, and commercial stakeholders. Finally, the chapter clarifies how sustainable marketing principles can help marketers to create value for the global sustainable development goals, based on a profit-making venture.

Review Questions

1. Let us assume that McDonald's chair wants to create a more socially and environmentally caring value proposition for McDonald's. If you were McDonald's Marketing Director, think about the key stakeholder network as specified in ◘ Fig. 1.8, in order to determine the specific needs, wants, and distresses of Mac's commercial, societal, and environmental stakeholders that you would be focusing on to create a more socially and environmentally caring value proposition for them.
2. As a sustainable marketer, let us assume that you are launching a new business in the hospitality industry. To play your role in addressing global food waste, at least to some extent, and to feed more people in our society, how would you be planning to set the vision and mission of this new hospitality business? How would you be applying the key considerations to make a feasible mission statement (as specified in ◘ Fig. 1.9) targeting the triple Ps' stakeholders of this business?
3. Let us assume that you have joined the Danish sea-liner, Maersk, as its Chief Marketing Officer. During the interview, you promised the Maersk Chair that you would apply the circular economy principles to make Maersk's marketing, promotion, and supply chain operations more cost-effective and socially and environmentally friendly. How would you be applying the Rs of circular economy (see ◘ Fig. 1.5) to keep your promise as Chief Marketing Officer?
4. Let us assume that you are the Head of Marketing of a globally top tobacco manufacturing brand, e.g., British American Tobacco (BTC). Tobacco products are harmful to the human body, and in general, the promotion of tobacco products is prohibited in all or most countries. Also, tobacco products attract numerous criticisms from different stakeholder groups. However, as Head of Marketing, you want to create a socially or environmentally valuable brand image of BTC based on cause-related marketing. List a number of possible societal and environmental causes that you think would be a good fit to align with BTC's products. See the discussion on cause-related marketing and the good fit of a brand with a particular societal or ecological cause in ► Sect. 1.1 (Introduction). Justify: why do you think the causes you have listed would be a good fit with the BTC's products?
5. Let us assume that you have secured a 1-year placement (internship) at United Airlines. By the end of 1 year, you have developed a good understanding of the overall marketing dynamics of United Airlines. In your internship end report to the management, in addition to the core 4Ps, what additional Ps from the extended marketing mix (the 22 Ps), as specified in ◘ Fig. 1.7, would you be

suggesting to the management to adapt to make United's value proposition more focused on its triple line stakeholders? Justify your selection of the additional Ps.

6. ALDI has a marketing message, "Everyday Low Prices," which could be considered a socially caring marketing message, considering that if ALDI is able to keep this promise, it can be helpful, especially for the low-income consumer groups. Over the decades of ALDI's global operation, in general, when grocery consumers compare the market offerings of ALDI in comparison to the offerings of other competitive supermarket brands, the consumers assume that ALDI is more affordable. Let us assume that you have commenced a new role as Marketing Manager at the Australian supermarket brand Woolworths. How would you define or redefine Woolworths' marketing message in the national market to encounter your international counterpart, ALDI?
7. ► Section 1.3.4 clarifies how Curtin Malaysia focused on their triple bottom line stakeholders' needs and distresses to delight them. Let us do quick secondary research by just Googling a few of the University of Oxford's societal and environmental contributions and consider the University's contributions to our global population and environment that you think enabled the University to make more profit, as well.

Discussion Questions

1. Facebook attracted a lot of criticism because of the data breach scandal, especially in the second decade of this current century. In addition to the data breach scandal, there are other criticisms that they receive from their global market. For example, the 2021 Nobel Peace Laureate, Maria Ressa, accused "Facebook is 'biased against facts'" (The Guardian, 2021a, np), which has been featured not only in activist groups' protests but also in mainstream newspapers. For example, "Facebook and Fear" (The Guardian, 2021b, np), "Facebook broke democracy in many countries around the world" (The Centre for International Governance Innovation, 2019, np). The marketing team needs to respond to such criticisms with pragmatic actions and messages that could calm down the protesters. Read these news stories (the full URLs are available in the Reference Section) and Google for more information related to such criticisms of Facebook. Discuss the message and a plan for a marketing campaign that you would implement for Facebook to address such criticisms.
2. Google for Barbie's "Imagine the Possibilities" marketing campaign, or look at this 1.55-min web commercial of Barbie on YouTube (► https://www.youtube.com/watch?v=l1vnsqbnAkk). In this commercial, Barbie shows that some 4–6-year-old girls are playing roles as professionals, such as football coaches, professors, doctors, businesspersons, and others. The last few seconds of the commercial show that a little girl is playing with a Barbie doll while thinking about her fantasies. Discuss (1) who is the target audience of this commercial? (2) Share your views of how Barbie attempts to empower young girls, e.g., by helping them to think about their future lives through this commercial. (3) If you market a product or service (any product or service) targeting the kids'

segment, discuss your plan for a socially caring marketing campaign to enlighten kids about life, society, or the environment.

3. Following the transfer of ownership of Newcastle United in 2021, the football club received criticisms from activist groups. For example, the Guardian (2021c) reported criticisms of "English football will sell itself to anyone" (np). Google for more information about Newcastle's criticisms following this transfer of ownership, and think how, as Chief Marketing Officer, you would have been responding to such criticisms to clarify the position of your brand.
4. Marketing ethics has sometimes been called an oxymoron. An example of this can be found in corporate sponsorship of Girl Scout badges, such as the Fashion Adventurer program offered by the retailer Limited Too. Combining fashion and sewing education with in-store visits and coupons redeemable after completion of the program, such initiatives are considered by some to be little more than artful sales pitches cloaked in a sash full of contemporary patches. What is your assessment of corporate sponsorship of Girl Scout badges? Can you, by using marketing concepts, develop a rationale for ethics to be fully incorporated into marketing thinking? (Czinkota et al., 2021, p. 38; (Reprinted by permission of the copyright owner, © Springer).
5. As you are a marketing expert, you have been invited by a local Charity, for example, to help them diversify their revenue sources and increase their operational surpluses so that they can pay their employees well and reinvest the additional surpluses to support their charitable social or environmental cause further. Evaluate the Charity's current operation and existing key stakeholders. In relation to the charitable cause, the Charity's current operation, and the stakeholders, think about how you, as a marketing expert, could identify the latent needs of this Charity's customers and non-customer stakeholders, with an aim to satisfy those needs, to diversify the Charity's revenue sources.
6. In 2022, Lidl (score 74%), Waitrose (score 74%), Asda (score 71%), Sainsbury's (score 71%), and Tesco (score 69%) were rated as the top five eco-friendly supermarkets in the UK (Smiley Movement, 2023). Reducing greenhouse gas emissions, becoming electrified, tackling plastics, and cutting food waste are considered key indicators of becoming a sustainable supermarket (The Independent, 2023). Let us assume that you are the Marketing Director of a large supermarket brand, but your supermarket is not scoring well against those sustainability indicators. Along with these four sustainability indicators, discuss other indicators that you think would be relevant for supermarkets' operations to make them more eco-friendly. Discuss the issues that you, as Marketing Director, may need to consider implementing a plan for your supermarket brand to score higher against your selected sustainability indicators.
7. Dolce & Gabbana, the Italian fashion brand, had a marketing myopia when they were trying to win the Chinese market in 2018. They hired a Chinese model to launch a commercial in China. The commercial showed that the model wore a Dolce & Gabbana dress and was trying to eat pizza using a Chinese chopstick. However, the commercial was not well received by Chinese customers, as naturally, they found it a criticism of their culture, considering that someone is struggling with chopsticks to eat. However, Dolce & Gabbana's

initial thought was it would be well received by their Chinese target audience, as they hired a Chinese model and designed the commercial relating to Chinese culture, i.e., the chopstick. The commercial has attracted a lot of controversies and was featured in mainstream media. For example, the Independent (2018) reported that "Dolce & Gabbana accused of racism in the new campaign" (np). BBC reported in 2019 (np) that the "Racist D&G ad: Chinese model says (the) campaign almost ruined (her) career." In 2021, CNN (np) reported that "3 years after ad controversy, D&G is still struggling to win back China." Google for more information about this ad controversy and discuss how you would have been planning for a fresh marketing campaign focusing on Chinese norms and values to win back the Chinese market.

Practice Quiz

1. For a brand, sustainable marketing is
   (a) A notional strategy
   (b) A charity
   (c) An optional strategy
   (d) A necessity
2. The word "society" was included in the American Marketing Association's definition of marketing in
   (a) 1985
   (b) 1935
   (c) 2007
   (d) 2004
3. Target market of a brand includes
   (a) All customers and shareholders
   (b) All customers and non-customer stakeholders
   (c) All customers
   (d) All shareholders and suppliers
4. The core purpose of overall marketing activities include
   (a) Need management, industry management, and relationship management
   (b) Need management, differentiation management, and relationship management
   (c) Industry management, relationship management, and target audience management
   (d) Need management, differentiation management, and industry management
5. For marketers, strategic marketing is a
   (a) Dual task
   (b) Three-way task
   (c) Single task
   (d) Four-way task
6. The triple bottom line (three Ps) of sustainable marketers include
   (a) People, planet, and promise keeping
   (b) People, planet, and profit

(c) Price, payment, and profit
(d) Price, profit, and people

7. The simplest definition of value is
   (a) What we produce in return of what we receive
   (b) What we receive in return of what we sacrifice
   (c) What we donate in return of what we receive
   (d) What our memory forgets in return of what we sacrifice
8. In general, customers can develop an anticipated value
   (a) Before they consume a product or service
   (b) Instantly after they complete consumption of a product or service
   (c) Both before and after they consume a product or service
   (d) After a long time, they complete consumption of a product or service
9. In marketing, the market includes only
   (a) Current customers
   (b) Both current and potential non-customer stakeholders
   (c) Both current and potential customers and non-customer stakeholders
   (d) Potential customers and non-customer stakeholders
10. Symbolic form of value started to influence customers from
    (a) Early 2010s
    (b) Late 1960s
    (c) Mid-1940s
    (d) Early 1990s
11. Marketing is
    (a) Science and technology
    (b) Science of market analysis
    (c) Art of promise
    (d) Art and science
12. The United Nations introduced the Sustainable Development Goals in
    (a) 2014
    (b) 2013
    (c) 2012
    (d) 2015
13. The Rs of the circular economy include
    (a) Refuse, rethink, reduce, reuse, refurbish, remanufacture, repurpose, recycle, recover
    (b) Refuse, rethink, reduce, redirect, return, refurbish, remanufacture, repurpose, recycle, recover
    (c) Refuse, rethink, reduce, redirect, repair, refurbish, remanufacture, repurpose, recycle, recover
    (d) Refuse, rethink, reduce, reuse, repair, refurbish, remanufacture, repurpose, recycle, recover
14. How many Ps are discussed in the expanded marketing mix of this chapter?
    (a) 25
    (b) 16
    (c) 22
    (d) 21

15. Homburg et al. (2011) discussed how many steps in market orientation?
    (a) 16
    (b) 7
    (c) 5
    (d) 11
16. Porter's five competitive market forces include
    (a) New customers, substitute products, future competitors, buyers, and suppliers
    (b) New entrants, non-substitute products, future competitors, buyers, and suppliers
    (c) New entrants, substitute products, future competitors, buyers, and suppliers
    (d) New entrants, substitute products, current competitors, buyers, and suppliers
17. Stakeholders' bargaining power analysis include
    (a) Number and size of buyers and suppliers, importance of specific buyer and supplier, is the item vital, and ability to snatch each other's employees
    (b) Number and size of government and non-government stakeholders, importance of specific government and non-government stakeholder, is the item vital, and ability to enter each other's industry
    (c) Number and size of buyers and suppliers, importance of specific buyer and supplier, is the item vital, and ability to enter each other's industry
    (d) Number and size of government and non-government stakeholders, importance of specific buyer and supplier, is the item vital, and ability to enter each other's industry
18. Drummond and Ensor (2001) describe how many characteristics of a feasible mission statement?
    (a) 4
    (b) 10
    (c) 6
    (d) 8
19. Four essentials to pursue the triple bottom line (people, planet, and profit) include
    (a) Equity in orientation, triple Ps' stakeholder-led co-creation, visible change for the society, and co-evaluation
    (b) Equity in orientation, triple Ps' stakeholder-led co-creation, visible change for the triple Ps' stakeholders, and co-evaluation
    (c) Equity in orientation, triple Ps' stakeholder-led co-creation, visible change for the environment, and co-evaluation
    (d) Equity in orientation, triple Ps' stakeholder-led co-creation, visible change for the triple Ps' stakeholders, and co-destruction

(You can find the correct answers to these questions by taking the quiz and then comparing your answers with the answers that are provided at the end of the book, under the 'Practice Quiz Answers' part of the book.)

References

ALDI. (n.d.). Saving is easier with permanently low prices. Retrieved November 21, 2023, from https://www.aldi.com.au/en/about-aldi/aldis-low-prices/

AMA. (2017). Definitions of marketing. Retrieved November 19, 2023, from https://www.ama.org/the-definition-of-marketing-what-is-marketing/

Apostolidis, C., Brown, D. M., Wijetunga, D., & Kathriarachchi, E. (2021). Sustainable value co-creation at the bottom of the pyramid: Using mobile applications to reduce food waste and improve food security. *Journal of Marketing Management, 37*(9–10), 856–886.

Aspuna. (n.d.). Who we are. Retrieved December 3, 2023, from https://www.aspuna.com/about/

Bae, H. (2015). Bill Gates' 40th anniversary email: Goal was 'a computer on every desk'. CNN. Retrieved December 2, 2023, from https://money.cnn.com/2015/04/05/technology/bill-gates-email-microsoft-40-anniversary/index.html

Bambauer, S., & Gierl, H. (2008). Should marketeers use price partitioning or total prices? In Lee, A. Y., Soman, D., and Duluth, M. N. (Eds). *Advances in Consumer Research, 35*, 262–268.

Baumgartner, J. (1991). Nonmarketing professionals need more than 4Ps. *Marketing News, 22*, 28.

BBC. (2019). Chinese model says (the) campaign almost ruined (her) career. Retrieved December 25, 2023, from https://www.bbc.co.uk/news/world-asia-china-46968750

Berger P. D., Bolton R. N., Bowman D., Briggs E., Kumar V., Parasuraman A., and Terry C. (2002), "Marketing Actions and the Value of Customer Assets: A Framework for Customer Asset Management," *Journal of Service Research, 5* (August), 39–55.

Bloomberg (2023). Rental car sales accelerate amid nationwide shortage. Retrieved November 27, 2023, from https://secondmeasure.com/datapoints/top-car-rental-companies-competitors-turo-getaround-hertz-avis-enterprise/#:~:text=Avis%20Budget%20Group%2C%20which%20includes,of%20sales%20in%20June%202021

Booms, B. H., & Bitner, M. J. (1981). Marketing strategies and organization structures for service firms. In J. H. Donnelly & W. R. George (Eds.), *Marketing of services* (pp. 47–51). American Marketing Association.

Borden, N. H. (1964). The concept of the marketing mix. *Journal of Advertising Research, 1964*, 2–7.

Brodie, R. J., Coviello, N. E., Brookes, R. W., & Little, V. (1997). Towards a paradigm shift in marketing? An examination of current marketing practices. *Journal of Marketing Management, 13*, 383–406.

Brown, D. M. (2023). Managing neolocalism outside the locale in real ale and craft beer entrepreneurship: The case of Ben Lui Brewery. *International Journal of Entrepreneurship and Innovation.* https://doi.org/10.1177/14657503231194001

Brown, D. M., Apostolidis, C., Dey, B. L., Singh, P., Thrassou, A., Kretsos, L., & Babu, M. M. (2024). Sustainability starts from within: A critical analysis of internal marketing in supporting sustainable value co-creation in B2B organizations. *Industrial Marketing Management, 117*, 14–27.

Brown, D. M., Apostolidis, C., Singh, P., Dey, B. L., & Chelekis, J. (2022). Understanding multi-stakeholder co-creation as an entrepreneurial approach to survival and sustainability: The case of Pennine Pubs. *International Journal of Entrepreneurship and Innovation.* https://doi.org/10.1177/14657503221145101

Circular Economy. (2023). *REPSOL global.* Retrieved November 25, 2023, from https://www.repsol.com/en/sustainability/sustainability-pillars/environment/circular-economy/index.cshtml#:~:text=The%20aim%20of%20the%20circular,model%20is%20established%20over%20time

CNN. (2021). Three years after ad controversy, D&G is still struggling to win back China. Retrieved December 25, 2023, from https://edition.cnn.com/style/article/dolce-gabbana-karen-mok-china/index.html

CNN. (2023). Fighting plastic pollution with a biodegradable bag. Retrieved November 20, 2023, from https://edition.cnn.com/videos/tv/2022/04/01/sonali-bag-spc-intl.cnn

Coca Cola and Pepsi. (2018). Retrieved November 27, 2023, from https://www.ukessays.com/essays/marketing/the-competition-between-coca-cola-and-pepsi-marketing-essay.php#citethis

Collins, K. (2021). How much CO2 do you produce using Facebook? Less than in 2016, report shows. CNET. Retrieved November 19, 2023, from https://www.cnet.com/tech/services-and-software/how-much-co2-do-you-produce-by-using-facebook-less-than-in-2016-report-shows/
Cosequin. (2023). Retrieved November 20, 2023, from https://www.vetscriptions.co.uk/products/cosequin-ds
Crittenden, V., Crittenden, W. F., Ferrell, L. K., Ferrell, O. C., & Pinney, C. C. (2011). Market-oriented sustainability: A conceptual framework and propositions. *Journal of the Academy of Marketing Science, 39*, 71–85.
Curtin Malaysia. (n.d.). Curtin University Malaysia. Retrieved November 26, 2025, from https://www.curtin.edu.au/about/campus-locations/malaysia/#:~:text=Situated%20on%20the%20north%2Dwest,excellence%20in%20South%2DEast%20Asia
Czinkota, M., Kotabe, M., Vrontis, D., & Shams, S. M. R. (2021). *Marketing management: Past, present and future* (4th ed.). Springer International Publishing.
Dean, G. (2023a). ALDI is America's fastest-growing grocer. Here's how a small German convenience store run by 2 brothers turned into one of the world's biggest retail chains. Business Insider. Retrieved November 21, 2023, from https://www.businessinsider.com/aldi-history-albrecht-family-karl-theo-grocery-store-shop-retail-2022-12?r=US&IR=T
Dean, G. (2023b). ALDI is buying 400 grocery stores in Florida and the southeast as it continues its rapid US expansion. Retrieved November 21, 2023, from https://www.businessinsider.com/aldi--buying-winn-dixie-harveys-supermarket-grocery-stores-florida-southeast-2023-8?r=US&IR=T
Drummond, G., & Ensor, J. (2001). *Strategic marketing planning and control*. ButterworthHeinemann.
Eadicicco, L. (2016). Why Facebook doesn't have a dislike button. The time. Retrieved November 20, 2023, from https://time.com/4235311/facebook-dislike-button-responses/
Earth.org. (2022). The detrimental impacts of plastic pollution on animals. Retrieved December 23, 2023, from https://earth.org/plastic-pollution-animals/
Ecotech Services. (n.d.). Repair versus refurbishment. Retrieved November 25, 2023, from https://www.ecotechservices.co.nz/resources/knowledge-base/repair-versus-refurbishment/
European Environment Agency. (2016). Retrieved from How does the food we buy, eat and don't eat impact the environment? Retrieved November 26, 2023, from https://www.eea.europa.eu/media/infographics/how-does-the-food-we/view
European Parliament. (2016). Briefing—Closing the loop: New circular economy package. Retrieved November 25, 2023, from https://www.europarl.europa.eu/RegData/etudes/BRIE/2016/573899/EPRS_BRI%282016%29573899_EN.pdf
European Parliament. (2023). Circular economy: Definition, importance and benefits. Retrieved November 25, 2023, from https://www.europarl.europa.eu/news/en/headlines/economy/20151201STO05603/circular-economy-definition-importance-and-benefits#:~:text=What%20is%20the%20circular%20economy,cycle%20of%20products%20is%20extended
Findlay, T. (1988). PEPSI—The choice of a new generation. *Media International Australia, 48*(1). https://doi.org/10.1177/1329878X8804800107
Forbes. (2019). Thailand supermarket ditches plastic packaging for banana leaves. Retrieved November 20, 2023, from https://www.forbes.com/sites/trevornace/2019/03/25/thailand-supermarket-uses-banana-leaves-instead-of-plastic-packaging/?sh=45988cd67102
Gastelu, G. (2022). Jeep saluting US military and veterans with $1,000 cash-back deal. Retrieved January 10, 2023, from https://www.foxnews.com/auto/jeep-saluting-u-s-military-veterans-cash-back-deal
Graham, C. (2003). The transition from mode-1 to mode-2 society: The implications for education. Retrieved August 11, 2009, from http://www.futures1.net/docs/Mode-2%20Paper.doc.pdf
Grameen-Danone. (2023). About Grameen Danone. Retrieved December 3, 2023, from https://www.grameendanone.net/who-we-are/#:~:text=Founded%20in%20March%202006%2C%20Grameen,employment%20opportunities%20among%20vulnerable%20communities
Gummesson, E. (2008). *Total relationship marketing* (3rd ed.). Elsevier/Butterworth-Heinemann.
Gummesson, E. (2014). Productivity, quality and relationship marketing in service operations: A revisit in a new service paradigm. *International Journal of Contemporary Hospitality Management, 26*(5), 656–662.

1

Gundlach, G. T. (2007). The American marketing association's 2004 definition of marketing: Perspectives on its implications for scholarship and the role and responsibility of marketing in society. *Journal of Public Policy & Marketing, 26*(2), 259–264.

Gundlach, G. T., & Wilkie, W. L. (2009). The American marketing Association's new definition of marketing: Perspective and commentary on the 2007 revision. *Journal of Public Policy & Marketing, 28*(2), 243–250.

Hassard, J. (1993). *Sociology and organizational theory: Positivsm, paradigms, and postmodernity*. Cambridge University Press.

Hitt, M. A., Ireland, D., Sirmon, D. G., & Trahms, C. A. (2011). Strategic entrepreneurship: Creating value for individuals, organizations, and society. *Academy of Management Perspectives., 25*(2), 57–75.

Homburg, C., Müller, M., & Klarmann, M. (2011). When should the customer really be king? On the optimum level of salesperson customer orientation in sales encounters. *Journal of Marketing, 75*(2), 55–74.

Hurley, R. F., & Hult, G. T. M. (1998). Innovation, market orintation, and organizational learning: An integration and empirical examination. *Journal of Marketing, 62*, 42–54.

IFCO Systems. (2023). Global food waste by country: Who's the biggest waster? Retrieved November 19, 2023, from https://www.ifco.com/countries-with-the-least-and-most-food-waste/#:~:text=Global%20food%20waste%20in%202023&text=At%20the%20time%2C%20the%20FAO,almost%20one%20trillion%20US%20dollars

IKEA. (2023). IKEA food. Retrieved November 27, 2023, from https://www.ikea.com/gb/en/stores/restaurant/

Impact Report. (2022). Fighting food waste together. Retrieved November 19, 2023, from https://tgtg-mkt-cms-prod.s3.eu-west-1.amazonaws.com/40187/ImpactReport2022_ENG.pdf

Jeep. (2014). Drive for troops—Help support our soldiers. Retrieved January 10, 2023, from https://www.autoparkchryslerjeep.com/blogs/2638/lifestyle/drive-troops-help-support-soldiers/

Judd, V. C. (1987). Differentiate with the 5th P: People. *Industrial Marketing Management, 16*(4), 241–247.

Julian. (2021). Jeep adds all-new freedom trim in honor of the U.S. military. Retrieved January 10, 2023, from https://www.southernchryslerjeepgreenbrier.com/jeep-adds-all-new-freedom-trim-in-honor-of-the-u-s-military/

Kaplan, A. (2006). Case study—Grameen Danone: A case for 'hybrid' models. Alliance Magazine. Retrieved December 3, 2023, from https://www.alliancemagazine.org/feature/case-study-grameen-danone-a-case-for-hybrid-models/

Kasasa. (2021). Boomers, Gen X, Gen Y, Gen Z, and Gen A explained. Retrieved November 25, 2023, from https://offer.kasasa.com/exchange/articles/generations/gen-x-gen-y-gen-z#:~:text=Gen%20X%3A%20Gen%20X%20was,72.1%20million%20in%20the%20U.S

Keefe, L. M. (2004). What is the meaning of marketing? *Marketing News* (September 15), 17–18.

Kohli, A., & Jaworski, B. (1990). Market orientation: The construct, research propositions and managerial implications. *Journal of Marketing, 54*(2), 1–18.

Kotler, P. (1984). *Marketing management: Analysis, planning and control* (5th ed.). Prentice-Hall.

Kotler, P. (2012). Keynote: Marketing with Philip Kotler. Retrieved November 22, 2023, from https://www.youtube.com/watch?v=sR-qL7QdVZQ

Kover, G., Szakaly, Z., & Kovach, G. (2002). The product on the breeding pig market in the point of view of marketing. *Acta Agraria Kaposvariensis, 6*(1), 22–34.

Kumar, V., Jones, E., Venkatesan, R., & Leone, R. P. (2011). Is market orientation a source of sustainable competitive advantage or simply the cost of competing? *Journal of Marketing, 75*(1), 16–30.

Kurtz, D. L. (2009). *Contemporary marketing* (2009th ed.). South-Western Cengage Learning.

Looney, H. F., & Seiler, M. (2003). Invitation marketing: Using customer preferences to overcome ad avoidance. Accenture. Retrieved May 26, 2010, from http://www.accenture.com/Global/Research_and_Insights/Outlook/By_Alphabet/InvitationAvoidance.htm

McCarthy, E. J. (1960). *Basic marketing: A managerial approach*. Irwin.

McCarthy, E. J., & Brogowicz, A. A. (1981). *Basic marketing: A managerial approach*. R.D. Irwin.

Morseletto, P. (2020). Targets for a circular economy. *Resources, Conservation and Recycling, 153*, 1–12.

Museum Hameln. (n.d.). The pied piper in literature and arts: The children's exodus. Retrieved December 3, 2023, from https://museumhameln.de/dauerausstellung_engl_rf09/

Narver, J., Jacobson, R. L., & Slater, S. (1999). Market orientation and business performance: An analysis of panel data. In R. Deshpandé (Ed.), *Developing a market orientation* (pp. 195–216). Sage Publications.

Narver, J., & Slater, S. (1990). The effect of a market orientation on business profitability. *Journal of Marketing, 54*, 20–36.

Narver, J., & Slater, S. (1994). Does competitive environment moderate the market orientation-performance relationship? *Journal of Marketing, 58*, 46–55.

Patil, A. (2009). Re: Anurag Patil's blog. Anurag-patil.com. Retrieved April 12, 2010, from www.aunragpatil.com/wordpress/?p=74

People & Planet. (2023). The business of the future? It's fair and inclusive. IKEA. Retrieved December 16, 2023, from https://www.ikea.com/global/en/our-business/people-planet/social-entrepreneurship/#:~:text=The%20partnerships%20between%20IKEA%20and,people%2C%20with%20products%20available%20globally

Plastic Oceans. (2022). The facts. Retrieved November 20, 2023, from https://plasticoceans.org/the--facts/

Porter, M. (1980). Industry structure and competitive strategy: Keys to profitability. *Financial Analysts Journal, 36*(4), 30–41.

PricewaterhouseCoopers. (2023). The sustainable development goals. Retrieved November 22, 2023, from https://www.pwc.co.uk/who-we-are/our-purpose/strategy/sustainable-development-goals.html#:~:text=The%20Sustainable%20Development%20Goals%20(SDGs,nationally%20and%20internationally%2C%20by%202030

Renouard, C. (2012). A social business success story: Grameen Donone in Bangladesh. Essec Business School. Retrieved December 3, 2023, from https://knowledge.essec.edu/en/sustainability/a-social-business-success-story.html

Research and Development. (n.d.). Curtin University, Sarawak, Malaysia. Retrieved October 10, 2015, from http://www.curtin.edu.my/R&D/Rob

Rethink Sustainability. (2023). The 10 principles of a circular economy. Retrieved November 25, 2023, from https://www.lombardodier.com/contents/corporate-news/responsible-capital/2020/september/the-10-steps-to-a-circular-econo.html

Ringold, D. J., & Weitz, B. (2007). The American marketing association definition of marketing: Moving from lagging to leading indicator. *Journal of Public Policy & Marketing, 26*(2), 251–260.

Safi, S. A. (2023). Brand management. Retrieved December 3, 2023, from https://bd.linkedin.com/in/abedinsafi#:~:text=Grameen%20Danone%20Foods%20%E2%80%93%20a%20social,a%20community%20based%20business%20model

Sainsbury. (2022). Sainsbury's Café. Retrieved November 27, 2023, from https://help.sainsburys.co.uk/help/terms-and-conditions/sainsburyscafe

Shams, S. M. R. (2013). Implications of relationship marketing indicators to enable organisational growth: A stakeholder causal scope analysis. In H. Kaufmann & M. F. A. K. Panni (Eds.), *Customer-centric marketing strategies: Tools for building organizational performance* (pp. 214–244). Business Science Reference.

Shams, S. M. R. (2016). Sustainability issues in transnational education service: A conceptual framework and empirical insights. *Journal of Global Marketing, 29*(3), 139–155.

Shams, S. M. R. (2018). The evolution of marketing as an innovative knowledge stream: The evolving role of stakeholder causal scope. In D. Vrontis, Y. Weber, A. Thrassou, S. M. R. Shams, & E. Tsoukatos (Eds.), *Innovation and capacity building: Cross-disciplinary management theories for practical applications* (pp. 151–166). Palgrave Macmillan.

Short, J. (2015). Bargaining power buyers and sellers. Retrieved November 27, 2023, from https://www.youtube.com/watch?v=fIpJq_rYgmU

Singh, P., Brown, D. M., Chelekis, J., Apostolidis, C., & Dey, B. L. (2022). Sustainability in the beer and pub industry during the Covid-19 period: An emerging new normal. *Journal of Business Research, 141*, 656–672.

Smiley Movement. (2023). The top 5 eco-friendly supermarkets in 2022. Retrieved December 25, 2023, from https://smileymovement.org/news/top-5-most-eco-friendly-supermarkets

Social Business. (2013). Grameen Danone Foods Ltd—Nutrition. Retrieved December 3, 2023, from https://www.yunussb.com/articles/grameen-danone-foods-ltd-nutrition#:~:text=The%20production%20is%20designed%20in,30km%20radius%20around%20the%20plant

Stavros, C. (2005). *Relationship marketing in Australian professional sport: An organizational perspective* (doctoral dissertation). Brisbane, Australia: Griffith University.

Sterling, O. (2021). Same earth different worlds. *Amazon.* Retrieved November 20, 2023, from https://www.amazon.com/dp/B097PFJJSR/ref=sr_1_1?crid=1L88PY6WLI8Z4&keywords=a+mountain+full+of+gold&marketplaceId=ATVPDKIKX0DER&musicTerritory=US&qid=1700483200&sprefix=a+mountain+full+of+gold%2Caps%2C268&sr=8-1&trackAsin=B097PFJJSR

Stevenson, S. (2013). Rivalries: We're no. 2! We're no. 2! How a mad men-era ad firm discoved the perks of being an underdog. Retrieved November 27, 2023, from https://slate.com/business/2013/08/hertz-vs-avis-advertising-wars-how-an-ad-firm-made-a-virtue-out-of-second-place.html

Surfers Against Sewage. (2023). Plastic pollution: Facts and figures. Retrieved November 20, 2023, from https://www.sas.org.uk/plastic-pollution/plastic-pollution-facts-figures/

SURI. (2023a). Every toothbrush you've ever owned still exists. Retrieved November 25, 2023, from https://www.trysuri.com/

SURI. (2023b). Retrieved November 25, 2023, from https://www.trysuri.com/pages/sustainable--electric-toothbrush-v12?tw_source=google&tw_adid=681310358390&tw_campaign=20779-383060&gclid=EAIaIQobChMIwJD33uzdggMVhI9oCR0ijQnUEAEYAiAAEgIYpfD_BwE

The Centre for International Governance Innovation. (2019). Maria Ressa: "Facebook broke democracy in many countries around the world, including in mine". Retrieved December 25, 2023, from https://www.cigionline.org/articles/maria-ressa-facebook-broke-democracy-many-countries-around-world-including-mine/

The Guardian. (2019). People eat at least 50,000 plastic particles a year, study finds. Retrieved December 23, 2023, from https://www.theguardian.com/environment/2019/jun/05/people-eat-at-least-50000-plastic-particles-a-year-study-finds#:~:text=The%20average%20person%20eats%20at,human%20ingestion%20of%20plastic%20pollution

The Guardian. (2021a). Facebook is 'biased against facts', says Nobel prize winner. Retrieved December 25, 2023, from https://www.theguardian.com/technology/2021/oct/09/facebook-biased-against-facts-nobel-peace-prize-winner-philippines-maria-ressa-misinformation

The Guardian. (2021b). Facebook and fear in Manila: Maria Ressa's fight for facts. Retrieved December 25, 2023, from https://www.theguardian.com/world/2021/apr/18/facebook-and-fear-in-manila-maria-ressas-fight-for-facts

The Guardian. (2021c). English football will sell itself to anyone: Human rights groups condemn Saudi-Newcastle deal. Retrieved December 25, 2023, from https://www.theguardian.com/football/2021/oct/07/english-football-will-sell-itself-to-anyone-human-rights-groups-on-saudi-newcastle-deal

The Independent. (2018). Dolce & Gabbana accused of racism in new campaign. Retrieved December 25, 2023, from https://www.independent.co.uk/life-style/fashion/dolce-and-gabbana-campaign-racism-china-dg-loves-china-fashion-designer-a8643761.html

The Independent. (2023). Which supermarket is the most sustainable and how can you tell? Retrieved December 25, 2023, from https://www.independent.co.uk/life-style/food-and-drink/which-supermarket-sustainable-tesco-morrisons-b2321119.html

The Road Ahead. (2015). Curtin University, Sarawak, Malaysia. Retrieved October 10, 2015, from https://www.youtube.com/watch? v=Sv1alyzv_s4

The Tea Lovers. (2021). Retrieved November 20, 2023, from https://www.facebook.com/groups/1108434145852746/posts/4560839503945509/

Thomas, J. (2020). Marketing concept: Examining AMA definitions, evolution, influences. Paper present at the 17th International Conference on Management (Jan 2-4).

Thomas, M. (1997). Consumer market research: Does it have validity? Some postmodern thoughts. *Marketing Intelligence & Planning, 15*(2), 54–59.

Too Good To Go. (2023). Best before labels are causing food waste. Retrieved November 19, 2023, from https://www.toogoodtogo.com/en-gb/initiative/look-smell-taste/look-smell-taste

Townsend, S. (2018). 88% of consumers want you to help them make a difference. Forbes. Retrieved April 10, 2022, from https://www.forbes.com/sites/solitairetownsend/2018/11/21/consumers-want-you-to-help-them-make-a-difference/?sh=3dfef6866954

UN. (2022). Why the global fight to tackle food waste has only just begun. Retrieved November 19, 2023, from https://www.unep.org/news-and-stories/story/why-global-fight-tackle-food-waste-has-only-just-begun

UN World Food Program. (2023). Retrieved December 3, 2023, from https://www.wfpusa.org/articles/what-you-need-to-know-about-child-malnutrition/

UNDP. (2023a). Plastic pollution. Retrieved November 20, 2023, from https://www.unep.org/plastic-pollution#:~:text=Every%20day%2C%20the%20equivalent%20of,polluting%20lakes%2C%20rivers%20and%20seas

UNDP. (2023b). UNDP to raise awareness on plastic pollution by introducing jute-polymer bags. Retrieved November 20, 2023, from https://www.undp.org/bangladesh/press-releases/undp-raise-awareness-plastic-pollution-introducing-jute-polymer-bags

United Nations. (2023). International day of awareness on food loss and waste reduction, 29 September. Retrieved November 19, 2023, from https://www.un.org/en/observances/end-food-waste-day#:~:text=the%20consumption%20level.-,An%20estimated%2017%20percent%20of%20total%20global%20food%20production%20is,in%20the%20global%20food%20system

Veolia. (2020). The circular economy: What is a resource recovery model? Retrieved November 25, 2023, from https://blog.veolianorthamerica.com/circular-economy-what-is-resource-recovery-model

World Economic Forum. (2023). 8 ways the circular economy will transform how business is done. Retrieved November 25, 2023, from https://www.weforum.org/agenda/2023/03/8-ways-the--circular-economy-outperforms-traditional-business-models/

YSBC. (2018). List of Yunus Social Business Centre. Retrieved November 22, 2023, from https://www.muhammadyunus.org/post/1712/list-of-yunus-social-business-centre-ysbc

Further Readings

Bernstein, R. S., Bulger, M., Salipante, P., Weisinger, J., & y. (2020). From diversity to inclusion to equity: A theory of generative interactions. *Journal of Business Ethics., 167*, 395–410.

Bertrand, M., Mullainathan, S., & Shafir, E. (2006). Behavioral economics and marketing in aid of decision making among the poor. *Journal of Public Policy & Marketing, 25*(1), 8–23.

Bhattacharya, C. B., & Sen, S. (2004). Doing better at doing good: When, why, and how consumers respond to corporate social initiatives. *California Management Review, 47*(1), 9–24.

Blasi, S., Caporin, M., & Fontini, F. (2018). A multidimensional analysis of the relationship between corporate social responsibility and firms' economic performance. *Ecological Economics, 147*, 218–229.

Brunk, K. H. (2010). Exploring origins of ethical company/brand perceptions—A consumer perspective of corporate ethics. *Journal of Business Research, 63*(3), 255–262.

Bryant, A., & Hill, R. P. (2019). Poverty, consumption, and counterintuitive behaviour. *Marketing Letters, 30*(3), 233–243.

Chandy, R. K., Johar, G. V., Moorman, C., & Roberts, J. H. (2021). Better marketing for a better world. *Journal of Marketing, 85*(3), 1–9.

Crockett, D., & Grier, S. A. (2021). Race in the marketplace and COVID-19. *Journal of Public Policy & Marketing, 40*(1), 89–91.

Gonzalez-Arcos, C., Joubert, A. M., Scaraboto, D., Guesalaga, R., & Sandberg, J. (2021). How do I carry all this now? Understanding consumer resistance to sustainability interventions. *Journal of Marketing, 85*(3), 44–61.

Gosliner, W., Brown, D. M., Sun, B. C., Woodward-Lopez, G., & Crawford, P. B. (2019). Availability, quality and Price of produce in low-income neighbourhood food Stores in California raise equity issues. *Public Health Nutrition, 21*(9), 1639–1648.

Habel, J., Alavi, S., & Linsenmayer, K. (2021). Variable compensation and salesperson health. *Journal of Marketing, 85*(3), 130–149.

Harris, A.-M. G., Henderson, G. R., & Williams, J. D. (2005). Courting customers: Assessing consumer racial profiling and other marketplace discrimination. *Journal of Public Policy & Marketing, 24*(1), 163–171.

Hein, W., Steinfield, L., Ourahmoune, N., Coleman, C. A., Zayer, L. T., & Littlefield, J. (2016). Gender justice and the market: A transformative consumer research perspective. *Journal of Public Policy & Marketing, 35*(2), 223–236.

Henderson, G. R., & Williams, J. D. (2013). From exclusion to inclusion: An introduction to the special issue on marketplace diversity and inclusion. *Journal of Public Policy & Marketing, 32*(1), 1–5.

Jacob, J., Vieites, Y., Goldszmidt, R., & Andrade, E. B. (2022). Expected socioeconomic-status-based discrimination reduces price sensitivity among the poor. *Journal of Marketing Research, 59*(6), 1083–1100.

Kim, S., Gupta, S., & Lee, C. (2021). Managing members, donors, and member-donors for effective nonprofit fundraising. *Journal of Marketing, 85*(3), 220–239.

Kipnis, E., Demangeot, C., Pullig, C., Cross, S. N. N., Cui, C. C., Galalae, C., Kearney, S., Licsandru, T. C., Mari, C., Ruiz, V. M., Swanepoel, S., Licsandru, T. C., Mari, C., Ruiz, V. M., Swanepoel, S., Vorster, L., & Williams, J. D. (2021). Institutionalizing diversity-and-inclusion-engaged marketing for multicultural marketplace well-being. *Journal of Public Policy & Marketing, 40*(2), 143–164.

Laczniak, G. R., & Murphy, P. E. (2012). Stakeholder theory and marketing: Moving from a firm-centric to a societal perspective. *Journal of Public Policy & Marketing, 31*(2), 284–292.

Lee, J., & Lee, Y. (2015). The interactions of CSR, self-congruity and purchase intention among Chinese consumers. *Australasian Marketing Journal, 23*(1), 19–26.

Luo, X., & Bhattacharya, C. B. (2006). Corporate social responsibility, customer satisfaction, and market value. *Journal of Marketing, 70*(4), 1–18.

Maignan, I., & Ferrell, O. C. (2004). Corporate social responsibility and marketing: An integrative framework. *Journal of the Academy of Marketing Science., 32*(1), 3–19.

Mart'ınez, P., & del Bosque, I. R. (2013). CSR and customer loyalty: The roles of trust, customer identification with the company and satisfaction. *International Journal of Hospitality Management, 35*, 89–99.

Minton, E. A., Cabano, F., Gardner, M., Mathras, D., Elliot, E., & Mandel, N. (2017). LGBTQ and religious identity conflict in service settings. *Journal of Services Marketing, 31*(4/5), 351–361.

Mishra, S., & Modi, S. B. (2016). Corporate social responsibility and shareholder wealth: The role of marketing capability. *Journal of Marketing, 80*(1), 26–46.

Mulki, J. P., & Jaramillo, F. (2011). Ethical reputation and value received: Customer perceptions. *International Journal of Bank Marketing, 29*(5), 358–372.

Park, Y. W., Voss, G. B., & Voss, Z. G. (2023). Advancing customer diversity, equity, and inclusion: Measurement, stakeholder influence, and the role of marketing. *Journal of the Academy of Marketing Science, 51*, 174–197.

Patel, P. C., & Feng, C. (2021). LGBT workplace equality policy and customer satisfaction: The roles of marketing capability and demand instability. *Journal of Public Policy & Marketing, 40*(1), 7–26.

Piff, K. P., Kraus, M. W., Côté, S., Cheng, B. H., & Keltner, D. (2010). Having less, giving more: The influence of social class on prosocial behavior. *Journal of Personality and Social Psychology, 99*(5), 771–784.

Richard, O. C., Steward, M. M., McKay, P. F., & Sackett, T. W. (2017). The impact of store-unit-community racial diversity congruence on store-unit sales performance. *Journal of Management, 43*(7), 2386–2403.

Rifkin, J. R., Du, K. M., & Berger, J. (2021). Penny for your preferences: Leveraging self-expression to encourage small prosocial gifts. *Journal of Marketing, 85*(3), 204–219.

Rust, R. T. (2020). The future of marketing. *International Journal of Research in Marketing, 37*(1), 15–26.

Seitanidi, M. M., & Crane, A. (2009). Implementing CSR through partnerships: Understanding the selection, design and institutionalisation of non-profit-business partnerships. *Journal of Business Ethics, 85*(2), 413–429.

Sen, S., & Bhattacharya, C. B. (2001). Does doing good always lead to doing better? Consumer reactions to corporate social responsibility. *Journal of Marketing Research, 38*(2), 225–243.

Servaes, H., & Tamayo, A. (2013). The impact of corporate social responsibility on firm value: The role of customer awareness. *Management Science, 59*(5), 1045–1061.

Sun, J. J., Bellezza, S., & Paharia, N. (2021). Buy less, buy luxury: Understanding and overcoming product durability neglect for sustainable consumption. *Journal of Marketing, 85*(3), 28–43.

Varadarajan, P. R., & Menon, A. (1988). Cause-related marketing: A coalignment of marketing strategy and corporate philanthropy. *Journal of Marketing, 52*(3), 58–74.

Viswanathan, M., Umashankar, N., Sreekumar, A., & Goreczny, A. (2021). Marketplace literacy as a pathway to a better world: Evidence from field experiments in low-access subsistence marketplaces. *Journal of Marketing, 85*(3), 113–129.

Waddock, S. A., & Graves, S. B. (1997). The corporate social performance–financial performance link. *Strategic Management Journal, 18*(4), 303–319.

Wang, Q., Dou, J., & Jia, S. (2016). A meta-analytic review of corporate social responsibility and corporate financial performance: The moderating effect of contextual factors. *Business & Society, 55*(8), 1083–1121.

Wang, Y., Lewis, M., & Singh, V. (2021). Investigating the effects of excise taxes, public usage restrictions, and anti-smoking ads across cigarette brands. *Journal of Marketing, 85*(3), 150–167.

Weihrauch, A., & Huang, S.-C. (2021). Portraying humans as machines to promote health: Unintended risks, mechanisms, and solutions. *Journal of Marketing, 85*(3), 184–203.

White, K., Habib, R., & Hardisty, D. J. (2019). How to SHIFT consumer behaviors to be more sustainable: A literature review and guiding framework. *Journal of Marketing, 83*(3), 22–49.

Wilkie, W. L., & Moore, E. S. (1999). Marketing's contributions to society. *Journal of Marketing, 63*(4), 198–218.

Yim, S., Bae, Y. H., Lim, H., & Kwon, J. (2019). The role of marketing capability in linking CSR to corporate financial performance: When CSR gives positive signals to stakeholders. *European Journal of Marketing, 53*(7), 1333–1354.

Zhang, K., Cai, F., & Shi, Z. (2021). Do promotions make consumers more generous? The impact of price promotions on consumers' donation behaviour. *Journal of Marketing, 85*(3), 240–255.

Zhang, W., Chintagunta, P., & Kalwani, M. (2021). Social media, influencers, and adoption of an eco-friendly product: Field experiment evidence from rural China. *Journal of Marketing, 85*(3), 10–27.

Product and Service Development for People, Planet, and Profit

Contents

S. M. R. Shams et al., *Sustainable Marketing*, Classroom Companion: Business,
https://doi.org/10.1007/978-3-031-79130-7_2

Preliminary Vignette—The Interconnected Themes of Production, Consumption and Sustainability

The factory purred with relentless efficiency, a symphony of gears and whirring machinery that echoed through the vast, eery space. Workers moved with expert precision, assembling products at a pace dictated by the relentless march of the clock. Amidst the pulsing dance of production, Maria, a seasoned worker, gazed out of a high window. Beyond the factory walls, she could see the sprawling landscape transformed by the relentless demand for more. Smokestacks reached for the sky, spewing forth clouds that mingled with the grayish tones of a setting sun. In the heart of this industrial orchestra, Maria felt the pulse of a society caught in the throes of insatiable consumption. The products they churned out, once shiny and new, now seemed like artefacts of a disposable culture. The assembly line moved perpetually, a mechanical beast that demanded sacrifice in the form of finite resources and ecological balance. As Maria tightened the last screw on yet another product, she could not shake the weight of the realization that each twist and turn of the screwdriver was a stitch in the fabric of an unsustainable narrative. The products that left the factory floor carried with them the burden of a predatory appetite for more, driving a cycle that seemed oblivious to the toll it took on the environment. The discarded remnants of yesterday's gadgets littered the outskirts of the factory like a graveyard of neglected desires. Mountains of plastic, obsolete electronics, and discarded packaging bore witness to a society that had embraced convenience at the cost of its habitat. Maria sighed, feeling the weight of responsibility settle on her shoulders. In the shadows of the assembly line, she pondered the consequences of a world addicted to constant production and consumption. The once vibrant landscapes had been replaced by a landscape scarred by excess, a reflection of a society on the brink of reckoning. As the factory continued its unyielding march, Maria could not help but wonder if the products they produced were the architects of their demise—a realization that would soon echo through the corridors of the upcoming chapter.

Assembly line is a manufacturing process where products are sequentially assembled by workers along a conveyor belt, often associated with mass production, volume and linear manner.

Disposable culture is a societal trend characterized by the frequent disposal of products, contributing to environmental concerns and waste.

2.1 Introduction

The purpose of the marketing mix is to develop interwoven strategies to encourage as many consumers as possible to purchase while the brand attempts to expend minimal costs (Borden, 1964). Part of the marketer's role when building the marketing mix is to develop an attractive product or service that creates value for a defined group of consumers (Kotler, 2016), who have identical needs for a particular product or service. Irrespective of how efficiently production costs are controlled, a product or service can only generate profits through sales. Consequently, a proficient marketer should formulate products and services that cater to the diverse wants, needs, and desires of customers. Since each customer is distinct and seeks specific benefits from products and services, businesses should ideally customize their offerings to fulfil the unique wants, needs, and desires of each customer. However, for many businesses, this is not always achievable, so firms need a way of classifying products in a system that is aligned to customer segments, as characterized by their needs and wants. The more flexibility a business has the more it must configure products to different customer segments at minimal cost, the more segments it can target with the core product. Economies of scale become important here; this refers to the cost advantages reaped by companies when production becomes efficient. Economies of scale refer to the cost advantages that a business gains as its production or scale of operation increases, leading to lower average costs per unit of output. Because of this, it is vital to develop new products with flexibility as a key feature. Once this is achieved, a brand is then evaluated by consumers on whether the brand has developed a better product/service than other competitors in the market. Marketers have typically responded to these challenges by developing several tools for stimulating the consumers' desire to buy. For many businesses, product and service development is the most crucial factor driving a business's success or failure.

Cost reduction is the process where companies reduce their costs (e.g., labor) and increase their profits.

Beyond the challenge of crafting a valuable product or service that addresses diverse customer wants and needs while concurrently minimizing costs, marketers in the contemporary milieu must now consider the impact of such activity on the environment and society. Perpetual consumption of products and services has been damaging the planet for an exceptionally long time (See BBC article—how the world embraced consumerism), and questions have emerged regarding the marketer's role in this. It is frequently noted that marketing, specifically product development, can be wasteful. There is also a real risk of insincere actions from a marketer's perspective; this is because consumers now want to purchase products and services from brands that are doing the right thing concerning the sourcing and sustainability of products. In response to this, when making products, brands often pledge to be more ethical, socially responsible, and environmentally friendly by embracing sustainable practices across their business (Du et al., 2022). However, for many brands, this does not happen, leading to claims from consumers that businesses often do not practice what they preach.

Materialism refers to how consumers use brands as props to create a desired identity with an emphasis on acquiring and consuming material goods as a measure of success or happiness. The concept of consumerism is linked to the concept of materialism.

Web Exercise 2.1

How the world embraced consumerism.

This BBC Future article describes that "over the course of the twentieth century, capitalism moulded the ordinary person into a consumer. Higgs (2021) traces the historical roots of the world's unquenchable thirst for more stuff" (np).

Google this article and Google "consumerism: past, present and future" to read and explore more about the origin, developmental trends and future of consumerism.

In this chapter, we build upon "classic" strategic product and service development approaches by exploring how companies have traditionally developed products and how that has contributed to the situation we find ourselves in the contemporary environment. We also consider how brands can effectively cultivate a sustainability focus that is aligned with customer expectations and offers value propositions that contribute to the triple bottom line, i.e., people, planet, and profit (Elkington, 1994). A people-oriented focus includes product and service development strategies that assist consumers to make better choices. This is because brands that help customers create better habits using behavior design techniques and the power of action prompts will have a huge advantage over those that do not. In addition, customers often find themselves overwhelmed with the process of choosing products or services, therefore, to improve the living conditions and culture of people, brands need to explore the idea of cutting down their product ranges, therefore reducing choice and helping customers to make easier sustainable decisions. To help protect the planet, products and services, development strategies must be developed to assist in discouraging the perpetual/endless cycle of consumption in which brands find themselves at the center of debate. Many big brands release a product only to launch a newer, better product weeks or months later; this results in appalling amounts of waste and harms the planet. This chapter will carefully consider product and service development strategies according to customer perceptions of worth, dynamic competitor activities and business needs to better understand product and service development for sustainable frameworks and people, planet, and profit.

Triple bottom line—The triple bottom line framework posits that businesses should also focus on social and environmental concerns rather than purely profit. Instead of one bottom line, there should be three—People, Planet, Profit.

2

2.2 What Are Products?

Organizations need to develop and implement activities that satisfy customer needs. Marketers use a mix of tactics to adapt products to meet the needs and wants of their target customers. Product is one of the elements of the marketing mix and is usually the starting point for most marketers in their quest to create value for customers. A brand creates value for customers by (1) developing an attractive product or service, (2) pricing it right, (3) placing it based on a distribution channel that would be convenient for consumers so that it is easy to find and acquire, and (4) promoting it so that the intended customer/s knows about the product's existence and develops a desire for it (Borden, 1964; Kotler, 2016). We call this the 4Ps of the Marketing Mix: Product, Price, Place and Promotion, it is called a mix because each element needs to work together to be successful.

A product can be defined as a collection of physical, psychological and symbolic attributes that can yield satisfaction or benefits to a buyer or user. Essentially, "*a product is anything that can be offered to a market for attention, acquisition, use or consumption that might satisfy a want or need*" (Kotler et al., 2013, p. 276). Developing a product is about understanding the benefits on offer to a customer; therefore, the product can be known as a bundle of advantages which a marketer presents to the customer for a price. Products tend to be made up of features with benefits; features are specific attributes that enable a product or service to perform its function, and benefits are need-satisfaction outcomes, these both result in value for the consumer.

Single-use products are products designed for one-time use, often contributing to disposable culture and environmental issues.

Satisfaction is the extent to which customers' expectations and needs are met or exceeded by a product or service.

2.3 Product Features and Benefits

Creating customer value is fundamental to both profit and non-profit organizations. In fact, creating superior customer value is a necessary condition for a company to secure a niche in a competitive environment, not to mention a leadership position in the market (Day, 1990). From a product production perspective, customer value is frequently created by adding specific benefits to a product. This is expressed in the following "value equation":

> Customer's Perceived Benefits—Customer's Perceived Costs = Perceived Customer Value (Day, 1990, p. 142)

People achieve personal values (or an identity) through actions or activities, such as social interaction, economic exchange, possession, and consumption (Sheth et al., 1991). To help individuals attain their values, customers can accomplish certain identity goals through consumption. Belk (1988) explains that part of a person's self-identity is defined by their possessions (such as the products they buy), this means the extended self consists of the person plus their possessions. He suggests the self provides a "sense of who and what we are," and possessions help support the sense of self because we are mostly what we have and possess (Belk, 1988). Therefore, relative to personal values, consumption values are instrumental. For example, owning a stylish house or buying a prestigious car are, for some people, valuable ways of creating a desired identity and a way of achieving self-satisfaction. Consumption activities usually include a range of goods and services. For example, owning a stylish house requires house owners to acquire many goods and services in addition to the house itself, just as taking a vacation involves many other related purchases that are not just the hotel or flight. Because of this, value is influenced by the psychology of the customer and the culture the customer is part of. As a result, value can change during the use of a product on the customer journey, and brands often respond to this by adding, removing or improving product features to enhance the customer journey experience.

Conscious consumer identity The conscious consumer identity reflects an individual's intentional and informed choices in purchasing and consumption, prioritizing ethical, sustainable, and socially responsible products and practices.

In summary, customer value refers to the important personal goals that customers are seeking. Customers buy products not solely for the sake of their "transactional value" but for the product's benefits that will satisfy their needs or personal values.

When developing these product benefits, brands have typically achieved value for customers by (1) Adding new product features, (2) Improving packaging or design, (3) Continuous innovation, and (4) Increased the speed of product cycle and delivery (in response to customers' need for newness) (Keller, 1993). Product development helps build value for individual customers or market segments. This is often accomplished by gathering data from multiple customers in that segment. Creating customer value in this way increases customer satisfaction and customer experience, for the brand, this can also lead to increased loyalty, market share, price and increased efficiency, which in turn leads to increased profits. This is why companies dedicate a vast number of resources to product development and often why it is considered the most important task of the marketing mix. The next section (▶ Sect. 2.4) explores the classic ways in which companies have habitually developed products to assist in maximizing profit. ▶ Manager's Corner 2.1 describes a case of sustainable apparel (product) development.

Product development is the process of conceiving, designing, prototyping, and bringing a new product or service to market, involving various stages such as ideation, research, design, testing, and refinement.

Manager's Corner 2.1

Patagonia Leading the Way in Sustainable Apparel.

Background Patagonia, a renowned outdoor clothing company, has been a trailblazer in adopting sustainable and ethical practices in its production processes. On their website, you will find a wealth of information detailing the challenges associated with the "costs" (both to the humans and the planet). According to Patagonia, the clothing industry contributes up to 10% of the pollution driving the climate crisis. Apparel workers are among the lowest-paid laborers in the world. Patagonia also highlights that the world's largest clothing brands hide dirty, irresponsible practices and misuse words like "sustainable," "green," and "conscious." Leading to cut-rate prices, overconsumption and a culture of convenience as well as pollution, labor abuses, and waste.

Changes Made Patagonia shifted some time ago to using recycled materials for its clothing lines, reducing the demand for added resources and minimizing environmental impact. They have also embraced fair labor practices, ensuring that workers throughout their supply chain are treated ethically. Patagonia believes that "Together, We Can Change How Clothes Are Made." For Spring 2024, 98% of Patagonia styles use recycled materials. Switching to recycled allowed Patagonia to avoid 8500 metric tons of CO_2e, enough to power more than 1070 homes for 1 year in the United States, according to the EPA. Caring for their workers is paramount—"We are dedicated to improving conditions for our workers. Over 85% of our product line was made in a Fair Trade Certified™ factory, impacting more than 75,000 workers." They are continuously looking for ways to do better, according to Patagonia, the best thing we can do for the planet is to stop buying new clothes and get more use out of the stuff we already own, cutting down on consumption. Patagonia's Worn Wear® program does just that. Patagonia passionately believes that informed buyers will force the clothing industry to drop their dirty practices. Demand better practices—what you buy is what the industry will become.

Impact By prioritizing sustainability, Patagonia not only appeals to environmentally conscious consumers but also sets an industry standard. The company's commitment to transparency and environmental responsibility has influenced other apparel brands to reconsider their production methods.

Reflecting on the changes Patagonia has made over the years and their form stance about demanding better practices, analyze, and summarize your views on:

- Assess the feasibility of other companies adopting a similar strategy to Patagonia's shift toward using recycled materials in its production processes.
- How practical is this approach for other businesses in different industries, and what challenges might other companies face in ensuring the continued success of these sustainability initiatives?
- How realistic is the statement: informed buyers will force the clothing industry to drop their dirty practices?

Information sources:

- https://www.unep.org/news-and-stories/press-release/us-outdoor-clothing-brand-patagonia-wins-un-champions-earth-award
- https://www.patagonia.com/our-footprint/

Fair trade is designed to help producers in developing countries achieve sustainable and equitable trade relationships.

Feasibility analysis involves assessing the practicality and possibility of the suggested concept, considering factors such as technical, economic, and operational feasibility.

2.4 Classic Models of Product Development

2.4.1 Kotler's Five Levels

When developing product features to create customer benefits, marketers have used classic models to assist with answering some core questions that determine the likelihood of success for a new product. Kotler (1967) suggested there are three key questions which need answering to determine this success they include (1) Do people need the product? (2) Is it different and better than competitors' offerings? (3) Would people be willing to pay the proposed price? Kotler (1967) devised a model that recognizes customers have five levels of need, ranging from functional/core needs to emotional needs. The model also acknowledges that products are merely a means to satisfy customers' varying needs or wants. He distinguished three drivers of how customers attach value to a product:

1. Need: a lack of a basic requirement.
2. Want: a specific requirement of products to satisfy a need.
3. Demand: a set of wants plus the desire and ability to pay for the product.

Customers will choose a product based on their perceived value of it. Satisfaction is the degree to which the actual use of a product matches the perceived value at the time of the purchase. A customer is only satisfied if the actual value is the same or exceeds the perceived value. Kotler attributed five levels to products to match needs:

In greater detail, using the carbonated drinks beverage Coca-Cola, the five product levels are:

1. Core benefit:
   The basic need or want that consumers satisfy by consuming the product or service. For example, the core benefit of Coca-Cola is to quench a thirst.
2. Generic product:
   An edition of the product containing only those attributes or characteristics absolutely necessary for it to function. For example, the generic product is a burnt vanilla-smelling, black, carbonated, and sweetened fizzy drink.
3. Expected product:
   The set of attributes or characteristics that customers typically expect and subscribe to when they purchase a product. For example, the expected product is that the customer's Coca-Cola is cold. If this is not the case, then expectations will not be met, and the drink will not taste its best in the mind of the customer.
4. Augmented product:
   The inclusion of additional features, benefits, attributes or related services that serve to differentiate the product from its competitors. For example, Coca-Cola's augmented product is that it offers Coke Zero. This exceeds customers' expectations by offering all the great taste of Coca-Cola, but with no sugar and zero calories, therefore achieving other goals for the consumer.

5. Potential product:

   This includes all the extensions and transformations a product might undertake in the future. To ensure future customer loyalty, a business must aim to surprise and delight customers in the future by continuing to enhance products. For example, one way in which Coca-Cola delights customers is by running competitions. The prizes in these competitions are often things that "money can't buy," such as celebrity experiences. To continue to delight customers over time, the competition prizes change frequently. Another way Coca-Cola does this is by using innovative promotional campaigns that excite and engage customers, such as putting customers' names on the bottle and creating a hunter-gatherer/scarcity situation which excites the customers' brains.

Kotler's (1967) five-product-level model provides businesses with a method for structuring their product portfolio to target various customer segments. This enables them to analyze product and customer profitability (sales and costs) in a structured way. By organizing products according to this model, a business's sales processes can be aligned to its customer needs and help focus other operational processes around its customers—such as research, design and engineering, procurement, production planning, costing and pricing, logistics, and sales and marketing. Grouping products into product groups that align with customer segments helps the shaping and planning of sales, as well as the production and new product planning across the product lifecycle. The image below demonstrates how this works in action using the automobile industry.

Product lifecycle management (PLM) is a holistic approach that involves the strategic management of a product or service from its initial concept and design through manufacturing/production, distribution, and use, to its eventual disposal or recycling.

2.4.2 The Product Lifecycle

Life-cycle assessment (LCA) is a process of evaluating the effects that a product has on the environment over the entire period of its life thereby increasing resource-use efficiency and reducing impact on the planet.

The life story of most successful products is a chronicle of the products passing through specific identifiable stages called the product lifecycle (PLC).

The lifecycle of products varies depending on the type of product. It has been argued that products have finite lives, that they pass through stages from birth, through growth to maturity and eventually through decline to death. Implemented with strategic rigor, the PLC not only predicts the end of all products but also makes suggestions about appropriate marketing strategies at each stage:

Stage 1. Market Development

This is when a new product is first brought to market before there is proven demand for it, and often before it has been fully established technically in all respects. Sales are low and creep along slowly.

Stage 2. Market Growth

Demand begins to accelerate, and the size of the total market expands rapidly. It might also be called the "Take-off Stage."

Stage 3. Market Maturity

Demand levels off and grows, for the most part, only at the replacement and new family-formation rate.

Stage 4. Market Decline

The product begins to lose consumer appeal, and sales drift downward.

(Source: Theodore Levitt, Harvard Business Review)

Traditionally, the lifecycle starts when raw materials are extracted from the ground. The raw materials are processed/refined so that they can be used in a factory to manufacture the product. The product is sold to a customer who uses the product until it cannot be used anymore, for example, it might break or it is no longer useful. The product is then disposed of or recycled, and product development teams begin to think about making changes/adaptations to the product or develop a new product entirely.

2.4.3 The Boston Matrix

New product development covers the complete process of bringing a new product to market, renewing an existing product or introducing a product in a new market. No business activity is heralded more for its promise and approach with more justified optimism than the development and manufacture of new products. Whether in mature businesses like automobiles and electrical appliances or more dynamic sectors like the tech industry, managers correctly view new products as a chance to get ahead of the competition.

Ideally, a successful new product can set industry standards, standards that become another company's barrier to entry or open crucial new markets. New product campaigns offer brands opportunities to reorganize and to get more out of a sales force, factory, or field service network, for example. New products capitalize on old investments. Newness is a subjective issue; some see it as "new to the world," others as "any improvement" or "product changes/adaptations." Newness is important because, according to Henderson, who worked for the Boston Consulting Group (BCG), a company should have a portfolio of products with different growth rates and different market shares. The portfolio composition is a function of the balance between cash flows. High-growth products require cash inputs to grow. Low-growth products should generate excess cash. Both forms are needed simultaneously (BCG, 1968).

Standardization is the process of establishing and adhering to uniform specifications, or criteria in manufacturing and service delivery process of a product.

According to BCG (1968), four rules determine the cash flow of a product.

- Margins and cash generated are a function of market share. High margins and high market share go together. This is a matter of common observation, explained by the experience curve effect.
- Growth requires cash input to finance added assets. The added cash required to hold shares is a function of growth rates.
- High market share must be earned or bought. Buying market share requires an additional increment of investment.
- No product market can grow indefinitely. The payoff from growth must come when the growth slows, or it never will. The payoff is cash that cannot be reinvested in that product.

This can be diagrammatically presented as the Boston Matrix.

This matrix helps brands to analyze their portfolio of products. Some tensions can arise when a business needs to assess where to allocate investment across the portfolio, e.g., in product development or promotion.

The four categories can be described as follows:

- Stars are high-growth products competing in markets where they are strong compared with the competition. Often, Stars need heavy investment to sustain growth. Eventually, growth will slow down and assuming they keep their market share, Stars will become Cash Cows.
- Cash Cows are low-growth products with a high market share. These are mature, successful products with little need for investment. They need to be managed for continued profit—so that they continue to generate the strong cash flows that the company needs for its Stars.
- Question marks are products with low market share operating in high-growth markets. This suggests that they have potential but may need substantial investment to grow market share at the expense of larger competitors. Management must think hard about "Question Marks"—which ones should they invest in? Which ones should they allow to fail or shrink?
- Unsurprisingly, the term "dogs" refers to products that have a low market share in unattractive, low-growth markets. Dogs may generate enough cash to break even, but they are rarely, if ever, worth investing in. Dogs are usually sold or closed.

The Boston Matrix makes a series of key assumptions:

1. Market share can be gained by investment in marketing.
2. Market share gains will always generate cash surpluses.

3. Cash surpluses will be generated when the product is in the maturity stage of the PLC.
4. The best opportunity to build a dominant market position is during the growth phase.

Ideally, a business would prefer products in all categories (apart from dogs!) to give it a balanced portfolio of products. Brands with a "balanced" product portfolio have products at various stages of the PLC, not just at the product development stage.

2.4.4 New Product Development Process

Of course, not all products can be new, especially if it is a successful brand that has been established for years therefore, businesses are faced with balancing their portfolio:

Returning to the term "new," what is meant by new when we are referring to the product development process? Innovation is often discussed when a new product is being developed. Innovation helps us to understand what is meant by the term new within the new product development (NPD) process. New refers to products that are, in fact, new to the world, whereas innovation indicates changes or any adaptations made to existing products. Before a business can launch a new product, it needs to go through several stages before it appears in the marketplace:

(Section source: Kotler et al., 2013)

Idea Generation

> **Idea generation** is the process of generating a wide range of potential concepts or solutions to address a particular problem of customers or an opportunity for them.

The first step of ideation occurs long before a brand's product comes to fruition. It begins with identifying market gaps for new products or discovering opportunities to improve your existing offerings and processes. The ideas can be collected from various sources: marketing research aimed at discovering customer needs and shifts in their preferences, internal brainstorming sessions or analyzing new successful technologies offered by competitors.

Airbnb idea generation: The idea of peer-to-peer accommodation services was born out of vital necessity. Back in 2007, jobless roommates Brian Chesky and Joe Gebbia ran out of cash and could not pay rent for their San Francisco apartment. Luckily for them, the city was about to host a huge design conference, and all hotel rooms were fully booked, with many conference attendees still needing accommodation. So, the duo decided "to make a few bucks" by offering visitors a bed and breakfast in their loft. They bought three air mattresses, made a simple website, and even provided Pop-Tarts for breakfast; this is how it all started (Source: Business Insider).

Market research refers to the gathering of data, insight and skills used to understand a market and the consumers within that market. Often used to roll out a new product development or a new service.

Idea Evaluation/Screening

After ideas are generated, a business should decide which of them will work best and which are not worth investing in. This is the very first stage of screening points that new products go through. The main goal of this stage is to reject poor ideas or those not aligned with the brand strategy or company's objectives and thus prevent wasting money on unprofitable projects.

Idea screening involves assessing and filtering the generated ideas to identify those with the most potential for further development.

Concept Development and Testing

The ultimate step of the ideation phase helps you understand whether customers really need your innovation. To prove its viability, you should develop a new product concept or a detailed description of the idea from the customer's point of view. The concept reflects the core benefits of a new solution in terms of price, convenience, usability, performance, quality, and functionality.

Conceptualization phase is the initial stage where ideas are generated and explored to form the basis for a new product or service.

When the detailed description is ready, this should be taken to targeted customers, customer surveys or focus group discussions are typically used to gather a mixed-method approach to gaining feedback. The outcome of this will help the business to make improvements to increase the chances of achieving product/market fit.

Concept development and market testing—At these stages ideas are refined into more detailed concepts, followed by testing and confirmation with target consumers to ensure market acceptance and feasibility.

Development Phase

At this phase, a strong idea becomes a genuine phenomenon targeting a real audience and promising tangible revenue. To add more value to the innovation, a product development team works alongside the product management team and marketing department. This should be a collaborative effort where all parties align their efforts with a clear product vision established across the business.

Product Development is the systematic process of creating, designing, and bringing a new product or service to market often using product lifecycle management.

Business Analysis and Marketing Strategy

A concept that survives an ideation phase should be studied from the business point of view. Will the new product be profitable? How much will it earn? What will R&D, production, and distribution promotional expenditures be? These and other relevant questions are addressed by a business analyst whose responsibilities include eliminating wasteful ideas and preventing unnecessary costs.

Proof of concept is the stage where a prototype is developed to confirm and demonstrate the technical feasibility and potential success of the concept.

Product Engineering/Development

Product engineering is the process of designing, developing and testing a new product or feature. This step involves a product development team supervised by a project manager who tracks progress, keeps the process within budget, and, in the ideal world, guarantees on-time delivery.

Post-Development/Introduction Phase

Once the initial working version of a forthcoming new product is ready, it enters post-development, the last phase, and this has three steps: (1) Test marketing, (2) Commercialization (launch), and (3) Post-launch review.

Test Marketing

First, the new product or its prototype is introduced to a small set of potential consumers from the target audience to collect feedback and understand what needs to be improved before a full-fledged launch. In this step, the company can conduct user acceptance testing—a type of quality assurance activity. After test marketing, the brand may need to modify some features or design elements to make the product more useful and attractive to the targeted users.

Commercialization

After test marketing, the company can finally concentrate on a large-scale launch of the new product in the mass market. The introduction of the business's unique solution needs to attract the attention of as many people as possible. Three aspects are critical for launch success:

- Customer awareness: Support the introduction with advertising, direct social media or email marketing campaigns and other promotional activities that are suitable for the target audience. Build excitement around your offering.
- Competitive pricing: Determine the price based on the covering cost and what the market will bear.
- Appropriate timing: To time the launch of your product, consider your schedule, releases by your competitors, and user readiness to adopt your solution.

Commercialization refers to the process of introducing a new product, service, or technology into the market and ultimately, make available for consumption to generate revenue and achieve market success.

Post-Launch Review

Considerations in the post-launch review stage include approximately 30,000 new products being introduced each year, and 95% of them fail, according to Clayton Christensen, a professor at Harvard Business School. Therefore, businesses need to question why product development fails. The risks of failure can be caused by: inferior quality, non-competitive pricing, a small target market, poor product-market fit, premature or delayed market entry, and/or weak promotional support and low brand awareness.

Section source: ► https://hbr.org/2011/04/why-most-product-launches-fail#:~:text=(See%20the%20sidebar%20%E2%80%9C40%20Ways,too%20late%20in%20the%20game.

2.4.5 Summary and Critical Sustainability Appraisal of Classic Models of Product Development

So far in this chapter, we have discussed some of the classic ways in which product development has been defined and theorized to create value for a target user and generate profit for a business. Historically, models have been employed to support marketing strategy development, particularly in relation to product development, growth and profit maximization. Academics and practitioners have developed models to provide a practical framework to assess how businesses are performing compared with competitors, and an outcome of the assessment might be to adjust planned growth strategies. As with all models, the information presented aims to assist clear and simple practical application, but most models bring with them a fair amount of critique. Critique is not always rooted in the reductionist nature of the information provided. Take the product lifecycle, for example, in recent years, criticism has been thrown at this model regarding its impact on the environment, particularly relating to the beginning of the life stage of the product in which the product is designed and developed. At this stage, decisions about materials and processes required to create the product can have a significant impact on the product's environmental footprint. But this is not always emphasized, and it is important that such classic models start to adapt and include such emphasis. For example, the measure of sustainability is determined during the beginning of life, but for a sustainable product life cycle, it is important to reduce the product's environmental impact to a minimum to zero at every stage of its life cycle, from design to end of life. Product lifecycle sustainability should be an approach to managing the stages of a product's existence so that any negative impact on the triple bottom line is minimized. If materials are chosen with an eye to reuse, that will increase the sustainability of the practices at the end of the product's usable life. The goal of sustainability in this context is a closed-loop product lifecycle, in which all elements that go into the creation of a product are reused, recycled or remanufactured rather than discarded. For example, car manufacturers such as Volvo use carbon-neutral energy sources, such as hydroelectricity and wind power, for their electrical needs. Modern, efficient equipment and industrial processes are used throughout the entire manufacturing process. Energy-efficient engines have been designed and developed to reduce pollution and CO_2 emissions. Efficient engines require less fuel, reducing negative effects on the environment. Approximately 85% of Volvo cars' parts and materials can undergo recycling, with many components repurposed for reuse, including materials like steel, rubber, plastics, and glass. Incorporating these sustainable considerations is essential during the new product development process (Volvo). However, often, this has not been the case, and although things are changing, the manufacturing of cars in the past has been done differently. The use of raw materials, perhaps some recycled materials and components, enters the factory at one end of the enormous manufacturing facility. An enormous amount of electrical energy is required to manufacture each car. Producing electrical energy often means burning fossil fuels such as coal and oil, producing pollution. Completed cars are driven out of the opposite end of the factory and transported to their owners. The car is driven throughout its useful lifetime, using fuel and requiring fluids such as oil. These cause environmental pollution. Tires, parts/components will also need replacing, using more raw materials and using energy during their manufacture. Eventually, the car breaks down for the last time, and scrapping is necessary.

Section source: ► https://www.secorvolvocars.com/volvo-recycles.htm

There are other criticisms directed at the product growth models, mostly surrounding the focus on market share and market growth and overlooking issues such as developing a sustainable competitive advantage. It has also been stated that market growth is an inadequate measure of a market's attractiveness, and market share is an adequate measure of a product's ability to generate cash. A Harvard Business Review article entitled "Why Most Product Launches Fail" describes a component of the difficulty in new product development. Often cited as reasons for product failure include the product falling short of claims, failing to glean a true and unmet customer need, the product defining a new category and requiring substantial consumer education but not getting it, the product is revolutionary, but there is not a market for it, the business cannot support fast growth. Among all these failures are two key themes: (1) Customer well-being and (2) The planet's resources.

One thing is clear: with emphasis from all models on the best way to achieve growth and profit, over time, businesses have interpreted this as encouraging buyers to consume as much as possible. The product developer's role in producing the greatest volume and diversity of goods and services by developing several tools (using growth models, for example) to stimulate consumers' desire to buy (Kotler, 2016) has played a huge role in this present-day perpetual cycle of consumption. Classic approaches have focused on the financial rewards of product development yet have overlooked environmental and social objectives. ► Marketing in Action 2.1 presents a case study on sustainable new product development.

Volume production is the manufacturing and production of large quantities of products to meet high demand.

Marketing in Action 2.1

Scenario Planning Case Study Sustainable New Product Development for Tony's Chocolonely

Background The "confectionery" industry, which encompasses the production and sale of various chocolate-based products, including chocolate bars, sweets/candies, truffles, and other confectionery items, has had a difficult few years regarding claims of unethical practices. Imagine you are a seasoned marketer tasked with fronting the new product development process for Tony's Chocolonely. Tony's Chocolonely has built a reputation for its fair trade practices, with its aim to stop slavery, empower cocoa farmers, and produce high-quality chocolates. The company is now looking to expand its product line with a new affordable, innovative and sustainable offering.

Objective Your objective is to guide Tony's Chocolonely through the new product development process to introduce a groundbreaking product that aligns with the company's commitment to sustainability and ethical sourcing. The goal is to create a product that not only resonates with existing consumers but also attracts a new audience interested in eco-friendly and socially responsible products at an accessible price.

Scenario Tony's Chocolonely recognizes the growing demand for plant-based alternatives and wants to introduce a line of sustainably sourced and ethically produced vegan chocolate bars. The challenge is to develop a product that maintains the company's high-quality standards while ensuring it meets the expectations of health-conscious and environmentally-conscious consumers. While Tony's Chocolonely is renowned for its commitment to sustainability and ethical sourcing, it is important to note that its products typically carry a higher price point compared to mainstream brands like Cadbury or Galaxy. The primary challenge in this scenario is to navigate the new product development process in a way that aligns with Tony's Chocolonely values while achieving an affordable and accessible price range for a new targeted consumer segment.

Ethical sourcing involves the procurement of goods and services in a socially and environmentally responsible manner, considering factors such as fair labor practices, environmental sustainability, and ethical treatment of workers throughout the supply chain.

2.5 A Brief History of How We Arrived at the Perpetual Consumption of Products

Product development is a core business function. The process has been considered critical for brands competing in growing national and international markets for decades. Product development is complex, as it includes innovating through identifying the needs and expectations of current and future markets. It is a process that involves all production activities ranging from research and design to the disposal of the product at the end of its life. Product development is one of the most important business activities for driving a brand's competitive advantage. Because of this, business and marketing activity has typically centered on the promotion of a product for decades. The product orientation holds that consumers will favor products that offer the most quality, superior performance, or innovative features. Therefore, the emphasis has been on making outstanding products and improving them over time, and attention has concentrated on product orientation rather than customer orientation.

The drive toward a product orientation was propelled in the manufacturing and mass production era. The production of goods played a significant role in the economy; hence, this era witnessed a transition from handmade products and goods to machine-produced goods. The output of products increased exponentially; this was coupled with an exceptional rise in the population, particularly in Great Britain, Europe and the United States. When the factory emerged, consumption of mass-produced goods was widespread, and this included not only entirely new products but old products in new forms. Businesses addressed the scalability issue brought about by enlightenment thinking, which led to new scientific and industrial discoveries. The introduction of interchangeability and the moving assembly lines, this enabled the development of "mass" production, which provided low-cost products through large-scale manufacturing. Initially, the number of varieties offered by such production was limited, as evidenced by the famous quote from Henry Ford, "Any customer can have a car painted any color that he wants, so long as it's black." Here, it is useful to consider the economic theory of supply and demand concerning the famous economist Adam Smith, who said if the desire for goods increases while their availability decreases, prices rise, on the other hand, if the availability of goods increases and the desire for it decreases the price comes down. Mass production provided the gateway to a new era; it revealed that economic value was hidden in new and poorly understood mass consumers who wanted to buy goods but only at a price that they could afford. Therefore, purportedly, Henry Ford reduced the price of an automobile by 60% to increase demand using a revolutionary industrial logic that combined high volume and low unit cost, he called it mass production (summarized in his famous quote). Henry Ford was among the first to strike gold by tapping into the new mass consumption logic with the Model T Ford.

Mass manufacturing or mass production is the production of goods on a large scale using standardized processes and machinery.

Scalability is the ability of a business to handle increased demands or growth without compromising performance or efficiency.

Source: ► https://unsplash.com/

The Model T factory had to adapt to a sudden surge in demand, yet the principles of mass production it employed proved applicable to a wide array of products, fundamentally altering the landscape of production. This transformation reverberated across the global economy, solidifying the ascendancy of a new era of mass-production capitalism as the primary driver of wealth creation in the twentieth century (Zuboff, 2015). Breakthroughs in low-cost goods and services unleashed expansions of production, manufacturers designed products and forced them onto consumers with limited input from consumers. As expected, the quality of products deteriorated, and businesses failed to include their consumers in the production process. It took quite a bit of time before mass production gave way to mass customization (Hu, 2013).

Capitalism is an economic system used for producing the greatest volume and diversity of goods and services. For it to succeed, however, there must be enough buyers for all the goods and services that capitalism is capable of producing.

The traditional model of producing standardized, large quantities of goods yielded way to a dynamic approach that prioritized flexible, individualized production tailored to the unique preferences and needs of each customer. This led to a paradigm change and resulted in mass customization in the 1980s and 1990s as brands started to include consumers in the production process. Demand for product variety increased; therefore, the number of varieties offered by consumer product manufacturers increased significantly. Product variation has dramatically increased since, for example, the car industry in the USA increased from 44 distinct types of vehicle models in 1969 to 165 in 2006. The shift from mass production to mass customization resulted from the convergence of technological advancements, changing consumer preferences, and economic considerations. Innovations in automation and digital manufacturing enabled flexible production processes, while the Information Age heightened consumer expectations for personalized products. Businesses, driven by globalization and market competition, sought to differentiate themselves through customization to meet diverse consumer demands. Supply chain efficiencies, economic benefits, and the rise of e-commerce further contributed to the appeal of this transformative shift, allowing companies to offer unique, tailored experiences in response to evolving market dynamics. The 1980s and 1990s witnessed significant developments in technology, manufacturing practices, and consumer behavior that contributed to the emergence of mass customization as a prominent trend in various industries.

Automation is the use of machinery and technology to perform tasks without direct human involvement, often a time-saving resource.

Sustainable supply chain A sustainable supply chain integrates environmentally and socially responsible practices across sourcing, production, distribution, and consumption to minimize adverse impacts and promote ethical standards throughout the entire supply chain.

In the contemporary environment, Apple is a notable example that illustrates the underlying objectives of capitalism at the core of these transitional eras. Apple launched onto the music scene amid a war between supply and demand. Young consumers were enthusiastic about sharing music, but as time went on, music started to be shared differently, consumers with evolving tastes and identities wanted different ways to share music with peers. A great deal of this change happened in the late 1990s and 2000s. In June 1996, MP3 was founded, MP3 used connections inside record companies and enabled music consumers to rip pre-released music CDs and make them available for downloads. In 1997, Windows Media Player emerged, and MP3 evolved even more. In January 2000, users were able to stream their own files, but just before 2000, music consumers started to rely heavily on the audio streaming service provider Napster, which let users search across all users who shared the platform and provided a centralized server that indexed files for music consumers and conducted searches on users' behalf. However, individual files remained on the user's computer and could be transferred directly from peer to peer with no other third-party involvement.

The 2000s marked an era in terms of computer science for modern file sharing, several file-sharing protocols and far more formats were introduced along with a near decade in product protocol experimentation. In February 2001, Napster peaked at around 25 million users (and was eventually shut down due to copyright infringement), and Apple released their first iPod, which would eventually become the world's most popular portable MP3 player. The music-sharing industry was disrupted, and a new era of music on demand emerged, prompting Apple iPod users to claim, "*I can get what I want, when I want it and where I want it.*" Apple recognized the changing needs of music consumers, and things developed fast. The next step was to make Microsoft Windows compatible with the iTunes platform, prompting Steve Jobs to announce less than 1 week after launching the software, "*We've broken every record and become the largest online music company in the world.*" Within a month, there were 5,000,000 downloads, then ten million in 3 months and later, in 2008, Apple went on to mark the world's largest music retailer when iPod sales spectacularly exploded from 1,000,000 units per month after the music stores launched 100 million less than 4 years later. But the iPod iTunes innovation flipped the century-old logic that we talked about when we were discussing Henry Ford and the Model T consumption. The iPod iTunes innovation that leverages the new capabilities of technical digital technologies to invert the consumption experience enabled Apple to rewrite the relationship between listeners and the music with a distinct commercial logic. While familiar to us now, when the new era of music listening was first introduced, this innovation depended on a few key characteristics; digitalization made it possible to rescue valued assets (in this case songs from institutional spaces which were trapped by the costly procedures for producing CDs) and in the case of the CD apple bypassed the physical production, the product along with its packaging, inventory, storage, marketing, transportation, distribution and physical retailing made it possible for listeners to continuously reconfigure their songs at will. So, just as Ford back in the day tapped into a new mass consumption model, Apple was among the first to experience explosive commercial success by exploiting a new society of individuals and their demand for individual music consumption.

(Section sources: Medium, The Conversation, Geurts & Cepa, 2023)

▶ https://medium.com/@alisaalexander/digital-music-streaming-disruption-or-innovation-73ef02aefffd

▶ https://theconversation.com/apples-ipod-came-out-two-decades-ago-and-changed-how-we-listen-to-music-where-are-we-headed-now-169272

Source: ▶ https://unsplash.com/

Product innovation is at the core of Apple's business model. Apple is well known for its innovations in hardware, software, retail services, physical stores, diligence and luxury brand persona. After a below-average start in the business world (some attribute this to the Mackintosh products), Apple expanded fast. According to the Harvard Business Review, Apple expanded from some 8000 employees and $7 billion in revenue in 1997 (when Steve Jobs returned) to 237,000 employees and $260 billion in revenue in 2019. When Steve Jobs came back to Apple, the brand had a conventional structure for a company of its size and scope. The activities outlined in the product development models were executed sequentially, leading to significant challenges in conforming to specifications between the interfaces of functional departments. Apple was divided into business units, each with its own profit and loss responsibilities. For example, general managers ran the Macintosh products group, other managers ran the information appliances division on the server products division, and there were many more groups operationalized in this way. As is often the case with decentralized business units, managers were inclined to disagree with one another over transfer prices. Believing that this conventional management had stifled innovation, Steve Jobs, in his first year of returning as CEO, laid off multiple general managers of all business units and put the entire company under one profit and loss account. This is an example of how the wider business practices and foundations of an organization, such as the underpinning business model, shape the entire operation; this includes products and product innovation.

In the same Harvard Business Review paper, it was reported that Apple's main purpose is to create products that enrich people's daily lives. If we dig a little bit deeper into the brand and what it stands for (which we will cover in ▶ Chap. 4, Brand Management, Marketing Communication, and Promotion Management for People, Planet, and Profit), you will see that this forms a huge part of Apple's identity. Product innovation was at the core of this identity, and this involves not only developing entirely new product categories, such as the iPhone, the iPad, and the Apple Watch, but also continually innovating within those categories. No product feature better reflects Apple's commitment to continuous innovations than the iPhone camera when the iPhone was introduced in 2007. Steve Jobs devoted only 6 s to its camera in the annual keynote event for unveiling the new product. Since then, the iPhone camera technology has contributed to the photography industry with a stream of innovations. Hi-dynamic range imaging in 2010, panorama photos in 2012, true tone flash in 2013, optical image stabilization in 2015, the dual-lens camera in 2016, portrait mode in 2016, portrait lighting in 2017, and a night mode in 2019 are but a few of the improvements made to the entire product of the iPhone. Apple has become fixated on detail, and to create such innovation, Apple relies on a business structure that centers on functional expertise. The fundamental belief is that those with the most expertise and experience in a domain should make decisions for that domain. Part of driving this product innovation is Apple competes in a market where the rates of technological change and disruption are extremely high, so it must rely on the judgement and intuition of people with deep knowledge of the technology required for disruption. But interestingly and evidently, where Apple comes under inspection for the sustainability of its business activities is that long before it can market a product or get any market feedback and solid market forecasts, Apple must make bets about which technologies are likely to succeed. Significantly, the bonuses of senior research and design executives responsible for Apple products are based on a companywide performance rather than the costs of/or revenue from products therefore, those product decisions are insulated from short-term financial pressure.

2.6 Evolution of Perpetual Consumption of Products

2.6.1 Planned Obsolescence

Apple's focus on continuously inventing new products allows us to explore a business strategy that emerged around the time of mass production, influenced by Ford's Model T. This approach, known as planned obsolescence, is a way companies encourage consumers to upgrade to newer versions of their products. The strategy deliberately shortens the lifespan of a product and forces people to repeatedly purchase a newer version of that product, a problematic process for the environment and society. The impetus for companies to adopt this business strategy lies in the pursuit of profit maximization. Corporations, characterized by industry leaders like Apple, must adapt by extending annual manufacturing of the same product with incremental modifications to sustain and expand their customer base. Apple has faced ongoing scrutiny for its adoption of such practices. Take, for instance, the tangible properties of computing devices—PCs, laptops, mobile smartphones, printers, tablets—which inherently possess a lifespan of 10 years or more. Nevertheless, the prevalent trend of significantly shorter usage periods is a direct consequence of planned obsolescence. According to a 2023 BBC article, in the USA, people replace their phones every 18 months on average as new models with upgraded features are released.

Section source: ▶ https://www.bbc.com/future/article/20231121-what-does-a-sustainable-smartphone-look-like

Apple is not the only brand guilty of engaging in these practices; the situation is similar for many electronic devices such as desktop computers, tablets, wearables and other smartphone producers (Android). However, the adoption of these practices has caused a problem in producing goods with uneconomically and unsustainable short useful lives, which encourages customers to make repeat purchases. Let us stick with Apple to demonstrate the sheer ubiquitous adoption of this business strategy. Let us consider every time a new iPhone is released, Apple also issues new software, if you do not upgrade, it makes the user experience extremely miserable. Many have engaged in numerous conversations where, upon the release of the latest iPhone, the immediate inclination has been, "I should get the new iPhone." The counterargument to this is because technology is constantly improving, tech companies want to continue to offer the latest, greatest, most efficient software, and therefore, sometimes the older software just does not work as well anymore. Undoubtedly, this has been an ongoing phenomenon for years. However, the issue arises when a brand, such as Apple, deliberately impedes the performance of older devices to actively incentivize consumers to purchase new iterations of their products. In fact, Apple has faced legal action for indulging in such practices. A BBC News article published in 2018 (Apple investigated by France for "planned obsolescence") details how French prosecutors launched an investigation over allegations of planned obsolescence. Consumer advocates claim a deliberate slowing down of all devices and imply it was designed to drive sales of new models. During the same year (December), Apple admitted that older iPhone models were deliberately slowed down through software updates, but Apple insisted it was because the phone's battery performance diminished over time. Apple issued a public apology and said the update was intended to extend the life of all the phones because lithium batteries become less capable of supplying peak current demands as they age over time, this can result in the device unexpectedly shutting down to protect its electronic components. Apple said its updates with software on the newer phones were made to smooth out the instantaneous peaks, and only when needed has it slashed the price of replacing an aged battery by 2/3, which Apple said would restore devices to the former speed and pledged to be more transparent in the future. A further example illustrating such practices was a lawsuit from Chile. Back in 2017, Apple agreed to pay $3.4 million to settle a litigation in Chile that accused the company of iPhone's planned obsolescence. The lawsuit focused on users of the iPhone 6, iPhone 6S, iPhone 7 and iPhone SE product ranges. The complaint accused Apple of planning obsolescence into its devices to get users to upgrade to new iPhone models (see Apple Inside article for more information).

Apple's alleged practice of intentionally slowing down older iPhones under the guise of battery optimization has been criticized for appearing more as a strategy to incentivize consumers to upgrade to newer models rather than a genuine concern for device performance and longevity. Planned obsolescence has played a huge role in encouraging consumers to perpetually consume. Many cases involving planned obsolescence have occurred in a technologically advanced era, but that is not to suggest that encouraging consumers to consume more is a new marketing tactic. We have seen from the manufacturing era there are many traditional selling techniques which have been used for decades by big companies wanting to maximize profits. The next section discusses some of the typical ways companies have accomplished more sales through product purchases.

2.6.2 Upsizing/Upselling/Cross-Selling/Add-Ons

Upsizing, upselling, cross-selling or add-ons are well-established selling techniques used to increase consumer buying of more products. There are, of course, some slight differences between these techniques, which we will discuss, but they are minor, and the point of this section is to highlight in the technological era, planned obsolescence obscures some of these traditional techniques yet at the same time has increased usage.

Upsizing is the process of encouraging a customer to buy a little more product or to upgrade on the final purchase. This selling technique is often used in competitive environments such as the travel industry or hospitality and is even used in the computer hardware and software industry. The upselling technique is used to generate more profit—whether the marketer encourages the customer to purchase more expensive items or promotes other add-ons that usually involve the customers spending more money on consuming more of the product. Sales employees play a vital role in upselling products, techniques such as providing the customer with a time-bound offer are a way of upselling as it creates fear over the scarcity of the purchase. Typically, for it to be most effective, employees using upselling are usually aware of customers' backgrounds and budgets, allowing them to understand what the consumer values for the upselling to become successful. Examples can include encouraging consumers to purchase a higher specification or a new desktop or tablet and selling luxury items such as more expensive trainers from a unique collection. An example of both upsizing and upselling working in conjunction with each other to maximize sales is often found in the fast-food industry—persuading the consumer to choose a larger meal size at a fast-food restaurant:

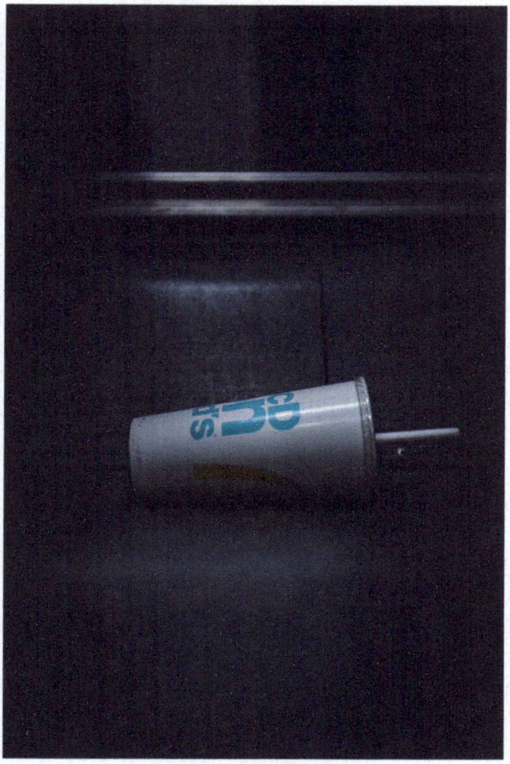

■ Source: ► https://unsplash.com/

An add-on is simply the technique of persuading a consumer to purchase an additional product, an example of this may include a retail scenario where a female consumer is buying a dress to attend a wedding as a guest, this consumer may also need a fascinator, shoes and a handbag to match the dress. A common cross-selling technique involves the seller enhancing the value of a purchase by recommending a complementary product. This strategy is frequently employed in the hospitality industry. To illustrate, consider the scenario at a McDonald's fast-food restaurant, where an employee might suggest adding fries to a customer's order along with a burger. Over time, these specific selling techniques have been successful for profit margins however. However, there are clear ethical issues pushing products onto consumers when the consumers do not want them or in a more covert tactic when the consumer is unaware these sales tactics are happening.

Source: Wiesman (2006).

2.6.3 Premiumization

To emphasize superior quality and exclusivity, brands often engage in premiumization to make a product more appealing. Premiumization is used to overcome competitive pressures from the market; it is simply a process of making customers pay higher prices for a product. We once again return to classic economic theory and supply and demand. Brands typically create a demand for a customer to pay higher prices within a certain category, such as the trainer category, and the aim is to stretch the category norms upwards via this demand. When referring to premiumization, we are not talking about luxury but rather reasons why customers will be willing to pay more for products (see pricing and willingness to pay). At its core, luxury extends beyond fashion items, premium goods, or mere trading up. The concept of premium products encompasses numerous factors, including the functional experience, product value, performance-driven pricing, accessibility, comparative value and overall impact. Luxury goods are characterized by associations such as fulfilling dreams, providing emotional experiences, personal value, a sense of privilege, exceptional value, magnification, and aesthetic appeal. For example, if the average category price for a bottle of water is 0.47 pence, premiumising bottled water is getting customers to pay 0.60 pence per bottle. Premiumization can be understood in a similar vein to creating brand equity. Brand equity is the commercial value that is derived from consumers' perceptions (Keller, 1993); therefore, premiumization is the vehicle from which that value is achieved. It is important to understand how customers perceive an assigned value; we will cover this in ▶ Chap. 4 when we discuss creating brand meaning. But for the purpose of this example, we need to understand how choices are not just about rational product attributes or functional benefits but also include societal, cultural, psychological and emotional motivations for consumers paying a premium price for products. According to Statista, water is the most consumed beverage on earth, and bottled water is taking an increasingly larger share of that consumption. Bottled water accounts for 50% of sales in the non-alcoholic beverages market worldwide, the UK market for bottled water was worth 1.6 billion British pounds in 2021. Yet, water is tasteless, odorless and colorless, it is also scarcer in some places but still in much of the world water is delivered directly to people's homes at a low cost and is easily available in great quantity so how then can one command a premium price for water?

Source: ► https://unsplash.com/

The concept of premiumization encompasses factors that motivate customers to pay a premium, including product attributes that comprise the following considerations: trusted, safe, high awareness, popular, unique, novel, hedonic, signals image, status, aesthetics, design, scarcity, and limited production. According to Statista, consumption of bottled water is going to continue to grow, so what is driving this premiumization? An obvious reason is changes in consumer trends and habits and an eagerness to access higher quality products or beverages. But we cannot ignore the impact that COVID-19 has had on consumers either. Consumption trends were impacted, and people were at home more, therefore, the only tangible way consumers could indulge themselves was to buy premium products, fueled by months of imposed savings for some, the appetite for premium products may stay; however, this will be dependent upon the state of economies worldwide going forward in the next few years. Given the explicit shift in consumer behavior prompted by the pandemic and evolving trends over the years, there is a growing emphasis on health consciousness. This paradigm shift is evident in the food and beverage market, where consumers are increasingly prioritizing health in their purchasing decisions. Beverages, being no exception, present opportunities to produce superior-quality, health-conscious products.

Section source: ▸ https://www.statista.com/topics/7151/bottled-water-market-in-the-uk/#topicOverview

Planned obsolescence is not the only strategy used as a sales tactic for profit-making; however, it is the most devious in contributing to the marketer's role in non-stop consumerism. Brands do their best to get consumers to buy more things from them because that is how most successful companies have made a profit. Tech companies clearly benefit from encouraging consumers to upgrade regularly, but other businesses also do their best to make things obsolete so that consumers upgrade and buy more products from them. It is evident that planned obsolescence of a product or the premature end of life of a product is wasteful, unethical, bad for the environment and costly for consumers but nonetheless has been a lucrative part of investors' and shareholders' business model. The next section in this chapter discusses the foundational cause of this strategy and, more widely, the perpetual cycle of consumption.

2.7 Why Capitalism and the Shareholder Business Model Underpinning Production Matter

Hopefully, by now, you will have realized the inextricable links between core business activities, such as the business models that underpin production and the continuous production of products which promote the relentless levels of consumerism we witness today. The previous section highlighted the scale in which businesses had adopted planned obsolescence as a business strategy to sell more products; it is evident that this strategy is not beneficial for the environment or consumers. Strategies that emerged out of mass production clearly advantage shareholders/investors of businesses. This is why it is important to understand the foundational underpinnings of where we have ended up today with mass consumption. This includes not only the economic models that are driving the political landscape but also the business models that are linked to profit maximization. Kotler, with a background in economics as well as an expert marketer, has claimed for quite some time that capitalism is dependent on constant consumption, he even points to the fact that during the Industrial Revolution in the eighteenth–nineteenth century, technology improved and increased the production capacity which in turn led to the rise in consumption levels. For example, people evolved from consuming as a "need" to adopting consumption as a culture and a lifestyle. Capitalism is an economic system characterized by private ownership of the means of production, profit-driven businesses and market-based allocation of resources. Returning to economic theory, the market-based allocation of resources includes goods/services being distributed and prices determined by supply and demand in a competitive marketplace; this means that products are bought and sold, and their prices are decided by how much people want them and how many are available. Many economists claim that capitalism has evolved over time and has developed into something different today. Some economists go as far as to say that different countries are adopting different variations of capitalism. Though the extent of government intervention in markets can vary significantly from one country to another, capitalism is the predominant economic model globally, and it has been widely adopted in various forms in many countries. Why has capitalism survived for so long as the preferred economic system? Capitalism encourages businesses and individuals to innovate and create innovative ideas/products to stay competitive, the profit motive associated with this serves as a strong incentive for businesses to innovate and take risks in the modern marketplace. Although heavily debated, some argue that capitalism has created economic growth, job opportunities, and higher standards of living, which, over time, have contributed to reducing poverty levels. Furthermore, with various businesses competing, consumers have the freedom to choose from assorted products, this has led to efficiency in production and resource allocation as businesses strive to minimize costs and maximize output.

Business models refer to the way a company creates, delivers, and captures value usually through revenue streams. Business models provide strategic direction for the key components and relationships between customers, products or services, channels, revenue streams, and cost structures to achieve sustainable and profitable operations.

Production efficiency is the effectiveness and optimization of the manufacturing process to minimize waste and maximize output.

While the advantages of embracing such a system on a worldwide scale are apparent, it is imperative to scrutinize and confront potential criticisms and shortcomings, the relentless pursuit of economic growth by shareholders is trashing the environment and leading to a growing gap between rich and poor and therefore raises questions about whether capitalism does increase the quality of consumers lives. Shareholder value capitalism began around the 1970s, its main premise being that the leading purpose of every business should be to maximize shareholders' wealth. If we are being kind, some would say that the underlying assertion of this was if businesses pursue this objective, both shareholders and society will benefit; however, we have come to see that this is a flawed idea. One of the main advocates of this approach was economist Milton Friedman, Friedman argued that the primary responsibility of a business is to its shareholders and that attempts to address social issues should be subordinate to the goal of maximizing profits. This idea gained traction in the following decades: CEOs of the late 1970s, early 1980s and 1990s swore allegiance to maximizing shareholder value and two of the biggest companies around that time, Coca-Cola and General Electric, became central to the movement for a profits-first focus. Interestingly, the CEO of Coca-Cola was the first American manager to become a billionaire on the basis of stock holdings in a company that he neither founded nor taken public, thus the first boss in America to become a billionaire just by owning a lot of Coca-Cola's shares, even though he didn't start the company or make it public (see Harvard Business Review article—

The Age of Customer Capitalism). Nevertheless, our attention turns to much more recent times to a company that has shed light on multiple ethical, irresponsible and unsustainable issues, that company is Amazon. In 1997, Jeff Bezos, the founder and CEO of ▶ Amazon.com, wrote a letter to shareholders printed from the annual report of that year:

To our shareholders:

» ▶ Amazon.com *passed many milestones in 1997: by year-end, we had served more than 1.5 million customers, yielding 838% revenue growth to $147.8 million, and extended our market leadership despite aggressive competitive entry.*

But this is Day 1 for the Internet and, if we execute well, for ▶ Amazon.com. *Today, online commerce saves customers money and precious time. Tomorrow, through personalization, online commerce will accelerate the very process of discovery.* ▶ Amazon.com *uses the Internet to create real value for its customers and, by doing so, hopes to create an enduring franchise, even in established and large markets.*

We have a window of opportunity as larger players marshal the resources to pursue the online opportunity and as customers new to purchasing online are receptive to forming new relationships. The competitive landscape has continued to evolve at a fast pace. Many large players have moved online with credible offerings and have devoted substantial energy and resources to building awareness, traffic, and sales. Our goal is to move quickly to solidify and extend our current position while we begin to pursue the online commerce opportunities in other areas. We see substantial opportunity in the large markets we are targeting. This strategy is not without risk: it requires serious investment and crisp execution against established franchise leaders.

Shareholder value refers to the financial worth or returns that a company generates for its shareholders, where the focus has often been on maximizing value for its investors.

As a result of this considered direction, Amazon implemented a distinctive strategy that sets it apart in the corporate landscape. The company, under the leadership of Jeff Bezos, pursued a groundbreaking timeline of capital allocation that defied the conventional focus on short-term returns. Unlike many companies of Amazon's maturity, which often prioritize immediate shareholder gains, Amazon prioritized long-term goals over quick profits. This approach involved expansive investments in various sectors, from building a vast infrastructure. Amazon's disregard for short-term needs in favor of long-term objectives has allowed the company to dominate e-commerce and become a technology giant. Amazon's revolutionary timeline of capital allocation had a total disregard for the short-term needs of investments in pursuit of long-term goals, and this was rare for a company of Amazon's maturity.

Typically, big companies have more assets to innovate, but they rarely take big risks or innovate at the cost of cannibalization of the brand or any current business, neither would they chance to alienate suppliers or investors. They engage in a game of not losing, and shareholders commend them for their approach. Most shareholders ask management how they can build the greatest advantage with the least amount of capital investment. Amazon reversed the question—What can we do that gives us an advantage that is usually expensive and that no one else can afford? Amazon's business strategy has established dominance as an online platform thanks to two elements of its strategy: 1. A willingness to sustain losses and 2. Investing aggressively at the expense of profits and integration across multiple business lines. Jeff Bezos wrote the following to shareholders (in the Amazon 1997 shareholder letter) regarding these two objectives:

> *We believe that a fundamental measure of our success will be the shareholder value we create over the long term. This value will be a direct result of our ability to extend and solidify our current market leadership position. We first measure ourselves in terms of the metrics most indicative of our market leadership: customer and revenue growth, the degree to which our customers continue to purchase from us on a repeat basis, and the strength of our brand. We have invested and will continue to invest aggressively to expand and leverage our customer base, brand, and infrastructure as we move to establish an enduring franchise.*

Put simply, the premise of Amazon's business model was to establish scale. To achieve scale, the company prioritized growth. Under this approach, aggressive investing would be key, even if that involved slashing prices or spending billions on expanding capacity to become consumers' one-stop-shop and the continuous consumption of products was at the core of this strategy (Source: Khan, 2019).

Amazon serves as a notable case study, displaying a range of concerning ethical practices and shortcomings across the triple bottom line. Amazon's business strategy, structure and conduct defy antitrust regulations relating to consumer welfare (Khan, 2019). As a leader in e-commerce and a company that was light years ahead of every other company on the Internet during the late 1990s, they have been central to the new market creations which have emerged online; Amazon has changed the economics of markets. At the core of Amazon's success is its aspiration to function as a crucial infrastructure for numerous businesses, a concept known as vertical integration (see ▶ Chap. 5 on distribution). Amazon prioritized facilitating a direct path for products from the factory to the consumer, aiming to streamline distribution channels by eliminating unnecessary layers. This has resulted in many businesses relying on Amazon to control the distribution of their products, thus placing Amazon in complete control of the customer journey and creating a situation where competition (particularly SMEs) depends on Amazon.

Customer journey is the overall experience and interactions a customer has with a company, from initial awareness through to purchase and post-sales support. The customer can entail many touchpoints with a business and a business needs to manage a seamless experience between these touchpoints.

Amazon's market dominance continues despite criticisms regarding labor practices, monopolistic tendencies, and environmental impact.

Source: ► https://unsplash.com/

An article written by the Drum in December 2021 (See article—Public votes Amazon worst offender for using too much packaging) highlighted that for a second year in a row, Amazon was the worst offender when it comes to using too much packaging. A study performed by the Chartered Institute of Marketing (CIM) conducted a poll of U.K. consumers, and Amazon received 40% of the vote, the CIM claims that companies who refuse to address their product packaging impact damaging not only the planet but their reputation as well, as overpackaging is affecting how consumers shop online with over 30% saying it is off-putting. A different article in Marketing Week (see article—Amazon proves consumers are choosing easy over ethical) highlights that Amazon proves consumers are choosing easy over ethical and suggests that consumers are complicit in this waste. The Marketing Week article goes on to say that consumers are ultimately and happily paying for Amazon's absolute level of profiteering. In an era of supposedly "woke" consumerism in which corporate social responsibility, brand purpose and brand activism (see branding, ► Chap. 4) are high on consumers' agenda, most consumers are joyfully (albeit covertly) one-clicking their casual shopping via Amazon and accepting unnecessarily over packaged parcels every week, if not every day.

Social responsibility is an ethical framework in which a person or a business must work and cooperate with other people and organizations for the benefit of the community.

Corporate social responsibility is a management concept whereby companies integrate social and environmental concerns in their business operations.

In the realm of consumer choices, Amazon's undeniable success emphasizes an established trend where convenience often takes precedence over ethical considerations. The ease of accessibility and swift delivery of products, however, come with their own set of ethical and environmental implications, raising questions about the collective responsibility of consumers in the discussion of excessive consumption. The relationship between consumers and a convenience-driven marketplace serves as a transition to the exploration of perpetual consumption, capitalism, and the triple bottom line; it becomes imperative to scrutinize the ramifications of product choices in a world where convenience frequently comes at the expense of ethical and sustainable practices.

2.7.1 Perpetual Consumption of Products, Capitalism and the Triple Bottom Line

According to Kotler (2016), marketing is the engine of capitalism; this is not to say he does not believe in capitalism he states that capitalism is the best economic system for producing the greatest volume and diversity of goods and services, which in turn benefits innovation and quality delivered to the consumer. But he recognizes that for capitalism to succeed, there must be enough buyers for all goods and services which capitalism can produce. We arrive at the marketer's role within the ongoing mass production crisis, a crisis stemming from unsustainable consumption and production patterns identified as a fundamental driver of the triple planetary crisis, as outlined by the United Nations Sustainable Development Goals. Marketers developed many tactics for stimulating consumers' desire to buy products, these were often price-related, the marketer plays a role in promoting continuous consumption through a range of tactics, including discounts, zero-interest financing, gift certificates, coupons, guarantees, warranties, credit cards, subscriptions, loyalty programs and branding. Regardless of the many discounts marketers used to stimulate increased purchasing of products, consumers couldn't keep up with the expense; therefore, Kotler identified an additional responsibility of marketers: persuading consumers to take loans to make purchases, paving the way for the marketing of credit cards, mortgages, and other credit instruments that played a pivotal role in enabling mass consumption of products and more widely capitalism. Interestingly, Kotler suggests that to understand the current capitalist economy, we need to replace the term "consumers" with the term "borrowers." This is a recent problem because if we study past culture groups and consumers' working trends, we know that by working 4–5 h a day, most people can sufficiently provide for their basic needs, but of course, this is currently not enough for many groups in society. In the contemporary environment, consumers are constantly in need of a bigger house, a flashier car, a more luxurious holiday, and a more advanced smartphone. Marketers often employ the strategy of fostering dissatisfaction among buyers. At the peak of a trend, a brand would introduce a product only to unveil a newer, improved version shortly after, a phenomenon prominently observed in the case of the Apple iPhone discussed earlier. This approach frequently leaves consumers feeling as if they lack the latest and best product and prompts them to continuously seek out novel purchases.

Sustainable development is a complete approach that seeks to meet the needs of the present without compromising the ability of future generations to meet their own needs, incorporating economic, social, and environmental considerations.

Addiction to novelty is problematic for the modern human brain. In his book, *Sapiens, A Brief History of Humankind* (Harari, 2014), Harari proposed that the human brain's pursuit of novelty conferred evolutionary advantages. Seeking out new information, environments and experiences not only kept humans safe but stimulated cognitive growth and complexity, contributing to the development of larger and more sophisticated brains over time. This adaptive behavior facilitated problem-solving, learning, and innovation, enhancing the survival chances of early humans. However, in the contemporary context, Harari notes that this same inclination for novelty can lead to challenges, such as the modern phenomenon of addiction to constant stimulation and the pursuit of new things, which may not always be beneficial for individuals or societies. Humans now work longer days/weeks to buy more new products and material items to fill the emptiness and dissatisfaction in their lives. Remember that capitalism flourishes if marketers can stimulate buying interest; therefore, marketers are acutely aware that the modern brain is attracted to novelty and often use strategies that capture and maintain consumer attention in a landscape saturated with information and constant stimuli.

The link between marketers of products, the capitalist economic system and perpetual consumption is clear: get people to want and buy all the things that are made and get consumers to accept debt to obtain these products. People have adopted consumption as a cultural lifestyle; it is deeply embedded in society, and therefore, the task of balancing traditional economic targets both for countries and businesses with social and environmental concerns is vast. Capitalism's commitment to profit generation has, at times, led to significant waste and with an abundance of available products, contemporary consumers often experience feeling overwhelmed.

The abundance of choices can lead to "choice paralysis," causing consumers to feel an overwhelming sense of anxiety as the sheer volume of information makes any decision seem potentially unsuitable, prompting many to refrain from taking any action. In marketing, it is often assumed that giving consumers "more" has beneficial outcomes for product choice, but providing customers with more options can be paralyzing. Too many choices overwhelm the brain, and this not only affects the well-being of the consumer but also sales.

A well-cited study conducted at Columbia University by Iyengar & Lepper (2000) researched the phenomena of choice overload. The researchers set up a booth of jam samples, every few hours, they would switch from a selection of 24 jams to a group of six jams, when the 24 jams were on display, 60% of customers would stop to get a sample, and 3% of these customers would buy a jar. When there were six jams on display, only 40% stopped, but here is the critical point—30% of these consumers bought jam. According to those researchers, the moral of the story—the trap of providing lots of options, which marketers often fall into does attract customers to browse; however, fewer choices convert into a sale.

2.7.2 Summary of Product Development and the Impact on People, Planet, and Profit

So far, this chapter has discussed one of the elements of the marketing mix: product. We have covered how production has evolved and developed in the modern economy. Some of the traditional and well-established models that have been used to enhance this growth have been presented, and some of the challenges we now face because of this continued growth have been discussed. As the needs and expectations of customers have changed since the emergence of manufacturing and mass production, customers have become more demanding, and marketing thought has become more complex. Although an excellent road map for marketing decision-making has stood the test of time and still warrants a central position in any marketing strategy for guiding tactical execution, the classic 4Ps of the marketing mix (Product, Price, Place, and Promotion) did not leave much room to consider the consumer. The explicit focus of the mix on internal processes undermines the elements of customer feedback and interaction as a basis for building up relationships and retention. As the evolution took place from mass production to mass customization, this marked a shift from uniform, large-scale manufacturing to tailored and personalized production to meet individual customer preferences. Therefore, to meet the evolving needs of modern-day consumers and industry, an extension to the 4Ps was required. This added a different focus to the marketing mix, and within the product element, services were considered to enhance the customer experience, this led to three additional Ps. Understanding how the extended 3Ps impact the triple bottom line—people, process and physical evidence is crucial to this chapter, the next section discusses the growth and development of services, service characteristics and the service economy before going on to discuss the additional 3Ps and developing sustainable products and services.

2.8 The Service Economy

According to the Office of National Statistics (ONS), the services sector makes up approximately 80% of the UK gross domestic product (GDP), the services sector has been in a period of consistent, steady growth since 2010 (ONS).

Source: Office for National Statistics

Many factors have contributed to this increase, notably, consumer lifestyles and increased deregulation of numerous industries; meanwhile, the public sector is gradually being privatized, this includes health and education. The process of globalization has facilitated the integration of economies, and market-oriented policies have been adopted by countries in various regions, providing opportunities for increased profit in the service sector. Services differ fundamentally from products. Services consist of activities, benefits or satisfaction offered for sale that are intangible and do not result in the ownership of goods (Kotler & Armstrong, 2000). A key factor distinguishing services marketing from marketing of physical products is the human element, often included as a new parameter in the services marketing mix (Melewar & Saunders, 2000). The human factor underlines the personal nature of service marketing; service providers play a double role in the marketing process as service delivering factors the personnel is a powerful element tool of customer persuasion and a major parameter affecting the customer's perception of the delivered service quality.

2.8.1 Service Characteristics

The product-services continuum visually presents some key differences between products and services; it also represents the spectrum of offerings that businesses provide, ranging from tangible goods to intangible services. At one end of the continuum are pure products, such as physical goods with no associated services, while at the opposite end are pure services, which involve intangible experiences or expertise without a physical product. Between these extremes lie hybrid offerings, where products and services are interconnected. This continuum reflects the evolving nature of business models, with brands increasingly integrating services to enhance the value proposition of their products and differentiate themselves in the competitive market.

There are four key characteristics which differentiate services from products (Dall'Olmo Riley & De Chernatony, 2000), these include:

1. Intangibility,
2. Inseparability,
3. Heterogeneity,
4. Perishability.

2.8.2 Intangibility

Intangibility illustrates the uniqueness of services in comparison with products; this is because services cannot be tasted, seen, heard, smelled or stamped with a product quality guarantee. Therefore, the value of the service is more difficult to communicate compared to products, and it is also harder for customers to evaluate before purchasing and, therefore, differentiate between competitors. Thus, it has been argued that branding of services is even more important than branding of products since the customer has no tangible characteristics to use when assessing the brand for the first time. Consequently, some services have a high perceived risk inherent in the purchase decision. The perception of a high-quality service plays a significant role in influencing the consumers' decision to select it. As such, customers tend to buy a service only when they trust a brand, as it is difficult for consumers to assess its quality before experiencing it. A service can also complement a product even in the most product-dominant markets, in fact, in the contemporary environment, consumers are unlikely to receive a product without some form of contact with the provider. Therefore, the intangible aspect of the product, such as the experience a customer has with a car dealer salesperson when buying a car, is just as important as stamping that product quality.

2.8.3 Inseparability

Inseparability refers to the role of both the consumer and the company within the production process. In services, the consumers are integrated into the production process, where the service provider and consumer mutually create value. Therefore, since services are connected to their consumers, production is not viewed as a separate activity, which has been the case in the past, where tangible products tend to be produced in a factory, then sold in a store and often consumed at home. In contrast, services tend to be sold and only then produced and consumed and therefore, the production and consumption of the services take place simultaneously (Bateson, 1992). As a result, the seller directly affects consumer loyalty, hence managing the interpersonal interactions between the service provider and the consumer results in better service quality and enhances consumer satisfaction.

> **Service quality** refers to the level of excellence or satisfaction experienced by customers about the services provided by a business.

2.8.4 Heterogeneity

Since most services are labor-intensive and rely on people, the performance quality of the service varies depending on the performance of the seller/servicer/deliverer, who is typically in contact with different buyers. As Gross et al., comment, "*every time a business service is performed, the process and the customer experience will be different*" (p. 377). With traditional products, more established companies develop a process of quality-checking products, this often involves a quality checklist and includes checking random samples for inspection, verifying package requirements, reporting quality defects and checking against a standardized specification. This process is normally conducted on-site (in a factory or warehouse, perhaps) prior to any involvement from consumers. As such, consumer involvement (apart from being involved in some initial feedback on products in the form of focus groups) is minimal. In services, customer involvement within the service delivery directly affects the quality of the service provided. In addition to this, service providers must also consider how employees impact the delivery. A key difference here is that businesses cannot quality-check service delivery in the same way they check product quality. There is a significant human element involved in service delivery, and humans are inherently subjective, which makes it difficult to standardize employee behavior. Therefore, heterogeneity of experiences plays a significant role in distinguishing the differences between products and services.

Sustainability reporting involves the systematic communication of a company's economic, environmental, and social performance, providing stakeholders with insights into its sustainability efforts and impacts.

2.8.5 Perishability

Perishability further distinguishes services from goods (Bateson, 1992). Services are produced and consumed simultaneously. They are considered non-inventoriable commodities because they cannot be stored. This means services are time-bound and lack tangible ownership. Therefore, service brands face the challenge of developing an image and reputation to attract consumers before the service brand has been experienced (De Chernatony & Dall'Olmo Riley, 1998; Dall'Olmo Riley & De Chernatony, 2000). Perishability also refers to the time-sensitive nature of the service. Take, for example, a plane ticket or a music concert ticket. Once the departure time for a flight or the start time for a music concert has passed, the opportunity to use or experience these services is lost. Plane tickets are perishable in the sense that the specific departure time is fixed, and if a passenger misses the flight, the ticket loses its value. Similarly, music concert tickets are perishable because the event occurs at a specific time, and once the concert concludes, the ticket becomes obsolete.

Let us look at an example from sport—aspects of sport are intangible for example, you cannot touch the actual competition that takes place between two teams. The emotions and experiences of participants and spectators represent it. The heterogeneous nature of sports makes sports inconsistent and unpredictable; and sports marketers have little or no control over these factors. Sport is subjective because the impressions, experiences and interpretations about sport vary from person to person. In a team sport, there are usually two teams competing in the match; therefore, the marketer must consider how they will enhance the customer experience even if a sizeable proportion of fans leaving the stadium are dissatisfied with the outcome of the game because their team lost. Sport is inseparable because the sporting experience is simultaneously produced and consumed. For example, as the athletes are competing (producing), the spectators are watching (consuming). Sport is perishable because as soon as the game is finished, the experience and the emotions felt by consumers cannot be stored for sale at a future date or resold.

Let us now put this into practice using an example from English football. Liverpool Football Club (LFC), established in 1892, stands as one of the (if not thee) most successful and celebrated sports teams in the world. With a rich history, an enthusiastic fan base, and an illustrious record of achievements, LFC is a powerhouse in the world of football, synonymous with excellence and a commitment to delivering unparalleled experiences to its supporters.

Source: ► https://unsplash.com/

Let us explore how the aspects of intangibility, inseparability, heterogeneity, and perishability apply to service delivery in the context of LFC:

Intangibility In the context of a football club like LFC, intangibility is evident in fan experience. The emotional connection, the thrill of a match, and the sense of community are intangible elements that contribute to the overall service experience.

Inseparability is reflected in the immediate interaction between the football club and its fans during matches, events, or through various communication channels. The experiences provided are often directly tied to the involvement and engagement of both the club and its supporters.

Stakeholder engagement is the systematic identification, analysis, planning and implementation of actions designed to influence stakeholders, an organization involves people who may be affected by the decisions it makes.

Heterogeneity is evident in the diverse range of services and experiences offered by LFC. This includes matchday experiences, merchandise, digital content, and community engagement initiatives, catering to the varied preferences and interests of the fan base.

> **Community engagement** encompasses the active involvement of businesses with local communities, considering their needs, fostering positive relationships, and contributing to social development.

Perishability is apparent in the time-sensitive nature of services offered by LFC, particularly in the context of matchday events. Once a match is played, the specific experience of that matchday is gone, and fans cannot revisit it in the same way. Additionally, the relevance and impact of promotional activities, community events, and communication strategies have a shelf life.

2.9 Services and the Traditional 4Ps

Services have increasingly become a part of physical products as an element of the augmented product dimension—see Kotler's levels model (Kotler & Armstrong, 2000). As such, services became a significant parameter of product differentiation and an important basis of competitive advantages and, therefore, a major generator of economic activity and a substantial source of corporate revenue in Western post-industrial economies (Hu, 2013).

2.10 The Additional 3Ps

In line with changes in consumer lifestyles, industry shifts and globalization, the service economy developed. Criticism began to emerge relating to the shortcomings of the 4Ps framework. These focused on the internal orientation of the marketing mix with product and production at the forefront. The 4Ps were transactional in approach, and there was a need to incorporate some form of interactivity with consumers to meet the increasing role played by the service element of a product, this led to developing a relational orientation. Thus, where once there were the 4Ps to explain the marketing mix, today it is more commonly accepted that a more developed 7Ps adds a much-needed layer of depth to the marketing mix to include People, Process and Physical evidence.

- People—This term is often confused with the consumer, but the people element of the 3Ps is about the individuals from the organization who encounter the customers and, therefore, have a direct impact on the perceived quality of the service, for example, the employees who produce or deliver the service. Services encompass personal interactions between employees and customers, which may be a source of service quality perceptions as well as a source of competitive advantage. The actions of employees influence the success of action and function of a business with training, communication, learning, and skills, and help achieve optimum value for a company. Implementation might consider the following questions: What are the expectations of customer service? Does the organization have any? How is this implemented? Furthermore, the main challenge in services lies in the consistent delivery of exceptional service, even when staff change.

- Processes include the systems and processes that deliver a product to a customer. It involves all the tasks, schedules, activities and routines that enable service delivery. Knowing the process means that companies can build in key activities to enhance the customer experience. The pace of the process, as well as the skill set of the service provider, is revealed to the customer, and it forms the basis of customer satisfaction. In the face of simultaneous consumption and production, managing the process of service demand with service supply is extremely difficult. The design and the implementation of product elements are crucial to the creation and delivery of products.
- Physical Evidence/Environment—Customers need tangible cues to deduce quality. This makes up the elements of the physical environment the customer experiences. Encompassing the surroundings where the service takes place can aid in the success of the service process. This can also be a differentiator between organizations. It holds significant importance because the customer often judges the quality of the service provided through physical evidence. Similarly, noticeable environments can influence impressions perceived by customers about quality service. The physical environment may comprise ambience, the comfort of the seating, background music, the physical layout of the service facility, and the appearance of the staff, which in turn can affect a customer's satisfaction with a service experience.

2.10.1 Additional 3Ps in Action

In recent times, The Body Shop has made substantial attempts to renew its brand purpose as a response to the current economic climate. Since the brand's inception in the 1970s in the UK, The Body Shop has been driving a purpose to fight for a fairer, more beautiful world, with a focus on believing in the beauty of the planet and the good in people. Activism was and still is central to its brand purpose. Founded in 1976 by the late British environmental and human rights campaigner Dame Anita Roddick, the brand is a notable example of how using the three additional Ps not only works effectively for a business with a service element but also does an incredible job of placing people and planet first. Let us look:

- Physical Environment—The Body Shop stores are fitted with sustainable fixtures, these are made from repossessed wood and recycled plastics, the purpose is to help minimize the brand's environmental footprint by using 100% recycled material destined for landfill. Since 2019, new stores have featured an activism hub, where people can discover the brand's activist roots as well as campaign for social change around issues ranging from gender equality to fighting cosmetic testing on animals. The Body Shop attempts to communicate the brand purpose and specific campaigns through every brand touchpoint; this is not just limited to in-store, the brand has worked hard at reflecting its environment goals online too, to bring product and activism closer together. Furthermore, in-store visuals often highlight the company's dedication to cruelty-free and environmentally conscious products. The layout of the physical environment encourages customers to explore different product ranges, enhancing the overall shopping experience. Additionally, knowledgeable and approachable staff members are strategically placed throughout the store to provide personalized assistance and product recommendations, contributing to excellent customer service and reinforcing The Body Shop's commitment to a positive and ethical retail environment.

- Processes—According to Marketing Week, a core part of ensuring activism is at the heart of The Body Shop stores involves up-skilling teams to run local activism projects. This goes beyond the typical retail staff training. Education is at the center of this, and therefore, processes have involved inviting grassroots activist groups into stores to provide workshops. Another notable example of the processes The Body Shop has in place is the inclusion of informational displays and product labels that highlight the brand's commitment to cruelty-free and ethically sourced ingredients. The labeling system provides transparency about the sourcing and production methods of their products, educating consumers about the positive impact of their purchases. In addition, The Body Shop often employs interactive elements within the store, such as video displays or leaflets, to educate customers on the company's initiatives related to environmental sustainability and community empowerment. The systems and processes that The Body Shop have in place have a clear goal of bringing the product and brand purpose closer together and educating the consumer.
- People—The Body Shop does not merely recruit sales assistants; their goal is to create a team of activists who are enthusiastic about the brand, the consumers and creating a beautiful, fairer world. The Body Shop strives to create employees who are friendly and knowledgeable and assist shoppers in experiencing and having fun with products. As well as educating consumers, the activist team are trained to offer expert advice, personalized makeovers and body care, skincare and haircare experiential demonstrations. Training and empowerment of store staff are key to The Body Shop's brand values and delivering service experience; they are trained not only to assist customers with product choices but also to share meaningful stories and information about The Body Shop's commitment to social and environmental responsibility. This strategic use of employees as brand ambassadors contributes to both customer education and the alignment of the product with the brand's overarching purpose.

Customer empowerment is when brands provide customers with the information and the tools that they need to make a decision, while also attempting to avoid manipulation/exploitation.

(Source for the section: ▶ https://www.marketingweek.com/the-body-shop-stores-activist-hubs/)

A paradigm shift in marketing has been notably shaped by the emphasis on the additional 3Ps—People, Process, and Physical evidence. This evolution reflects a departure from traditional profit-centric approaches to a more holistic perspective that prioritizes the well-being of consumers. Brands, recognizing the significance of societal and environmental impact, have embraced a well-being maximization model. By integrating people-centric values, environmental sustainability, and a clear sense of purpose, businesses are not only seeking financial success but are also striving to create a positive impact on individuals and the planet.

2.11 Co-creating Value with the Consumer—From Value Maximization to Well-being Maximization

Due to the heightened emphasis on brand services, the additional 3Ps have directed attention toward generating value for customers who have historically been overlooked in company decision-making processes. As brands actively involve customers in decision-making processes by fostering a collaborative approach to enhance overall product and service experiences, co-creating value with the consumer has become integral to the services paradigm. In marketing theory, value co-creation is rooted in the service-dominant logic (SD-Logic). The SD Logic (Vargo & Lusch, 2004, 2008) suggests that all products are valued for the services they provide and that the value derived from a physical good is not the good itself but the value created only by value-in-use (Lovelock & Patterson, 2015). It is only when customers engage or participate during the production, delivery, and consumption of the service that customer value is co-created. Value co-creation is integral to one of the fundamental premises of the S-D logic, which considers the exchange of service central to the interactions among organizations, markets and society (Singh et al., 2022). In the contemporary environment, the terms value and identity have evolved to play a significant role in consumer's lives. Value co-creation can also integrate economic, social and environmental dimensions, leading to the creation of "sustainable value."

From the mid-nineteenth century toward the onset of the Millennium, value co-creation has played a vital role in transforming a growth orientation into a well-being orientation. The concept of value has been integral to this. The changing notion of value from value maximization (the mass production era) to well-being maximization in the current environmentally concerned era has transformed marketing thought. The idea of value has been challenged by adverse developments and concerns regarding social and economic crises. Value transformations have started to redirect marketing focus toward a more human-centric future that will redesign people's lifestyles. Collaboratively generating value with consumers has propelled the shift from a transactional marketing orientation that detrimentally impacted both people and the planet for decades to a human-centric concept of value. This approach aims to introduce innovative well-being solutions to the market for products, prioritizing inclusivity for both individuals and the planet (Kelleci & Yildiz, 2021). ► Marketing in Action 2.2 presents a case of Airbnb to demonstrate how Airbnb leverage its service characteristics for sustainable competitive advantage.

Co-creation is the process of collaborating with all stakeholders involved to produce value.

Marketing in Action 2.2

Scenario Planning Case Study Leveraging Service Characteristics for Sustainable Competitive Advantage—Airbnb

Background Imagine you are a seasoned marketer tasked with guiding a leading hospitality company, Airbnb, through a scenario planning process. In a highly competitive market where continual product innovation is challenging, Airbnb is exploring alternative sustainable strategies to maintain and enhance its competitive edge. The company wants to shift its focus from constantly producing new offerings to leveraging service characteristics as a unique selling proposition.

Objective Your objective is to explore how Airbnb can strategically use service characteristics to differentiate itself in the competitive marketplace. The goal is to provide an exceptional and distinctive service experience that not only retains existing customers but also attracts new users. This approach aims to go beyond the traditional product-centric competition, focusing on responsibility and sustainability in the travel and accommodation industry.

Scenario Airbnb has observed a shift in consumer preferences, where travelers are not just looking for a place to stay but are seeking unique and personalized experiences. The company wants to position itself as a service-oriented platform that offers more than just accommodation and wants to consider the environmental impact of provision. Instead of focusing solely on the introduction of new property listings, Airbnb aims to differentiate itself by enhancing service characteristics that cater to the evolving needs and desires of its users. You will need to think beyond traditional product-centric competition and explore how a platform like Airbnb can leverage service characteristics to create a unique and compelling value proposition for users in the highly competitive travel and accommodation industry. Please also consider how the brand can respond to the SDGs in this scenario.

2.12 Potential Ways Forward—Developing Sustainable Products and Services and Influencing Behavior

Competitive branding of products became essential in the contemporary machine age (which is yet at its embryonic stage), as manufacturing sameness ran rife throughout production, mostly starting in the USA (then spreading to Europe and beyond). The market was swamped with uniform mass-produced products that were almost indistinguishable from one another (Source: Klein, 2000). There were clear benefits to this: mass production was efficient and established huge profits for the best businesses therefore, significant economic progress was attained, and as a result, the living standards of people increased tremendously. Yet, some aspects were neglected, as the primary duty of a business was seen as using its resources to solely boost profits and thrive. Consideration of people and planet was omitted from business processes. Although economic prosperity is crucially important in the contemporary environment, the world is now far more interested in ***how*** we achieve profit, not just profit at all costs. Further analysis of the emergence and development of branding and the triple bottom line (People, Planet, and Profit) will be discussed in ▸ Chap. 4. We finish this chapter by considering some of the ways forward for products and services.

Sustainable, ethical and responsible consumerism of products are huge priorities (see a recent report from Deloitte on The Sustainable Consumer 2023) for the modern consumer. Marketers are perfectly positioned to initiate immediate change! But how do marketers achieve this after experiencing decades in a growth-oriented system with little attention devoted to people or the environment? The last section of this chapter will dedicate some time to considering some of the opportunities businesses, and marketers have in the current environment to assist in transforming product production processes into a sustainable and well-being-oriented system. Our discussion begins at the start of the product journey to consider how some of the practices surrounding the design of new product development can be more innovative in terms of their responsible sustainability.

2.12.1 Sustainable Innovation for New Product Development (NPD) and Ethical Co-Design

Increasingly, consumers are looking for brands that are sustainable and produce sustainable products (▸ https://www.statista.com/statistics/1377869/global-shift-to-buying-sustainable-products/). Many ideas are emerging that focus on design strategies. For example, even during the early discussions around product development with the research and design team, ideas might include reducing the environmental impact of a product on the planet through its entire life cycle. However, as we have highlighted in this chapter, this is only one part of the problem, less attention has been devoted to influencing the behavior of the users of products: the consumers. While the research and design team of a new product might seek consumer feedback, often, the process of designing and bringing the product to market takes place behind closed doors with limited co-creative input. Therefore, an ethical design system for sustainable innovation of a new product must consider how the consumer uses that product throughout its life cycle because gains in operational efficiency throughout the supply chain must not be counteracted by consumer's unsustainable levels of consumption.

Sustainable innovation refers to the development and implementation of new ideas, processes, or products/services that address environmental, social, and economic challenges while minimizing negative impacts and promoting long-term ecological and societal well-being.

It is useful to revisit our discussion on co-creation at this point to enhance discussion. We know that co-creation is a process that brings together multiple parties to jointly produce mutually valued outcomes (Prahalad & Ramaswamy, 2004), there are vast amounts of opportunities to be gained via a co-creative approach for the production process. In addition, co-creative thought can provide opportunities for both businesses and consumers to become more sustainable when producing and consuming products going forward. Companies have long known that design decisions determine most of a product's manufacturing, operating, and maintenance costs. The same logic applies to sustainability where an ethical co-design approach includes consumers and all stakeholders who are then integrated into the design thinking process of the product's lifecycle.

Ethical co-design is a collaborative and inclusive design process that prioritizes fairness, respect, and consideration for diverse perspectives, ensuring the ethical treatment of stakeholders and fostering equitable outcomes in the creation of products, services, or solutions.

Inclusive Design is based on the simple principle that designing for the widest range of people creates better designs and benefits everyone.

Often, it is assumed that the research and design team work unaccompanied on design thinking, and therefore, design ideas are individualistic by nature. Most assume that the research and design team, who usually possess formal training, know what is best and what is needed for the user. Yet, when developing products, embracing a co-creative design thinking approach involves actively engaging consumers and key stakeholders from within and outside the marketing team. This collaborative strategy has yielded innovative ideas, promoting the development of solutions that are effective while keeping a keen focus on both environmental sustainability and meeting the consumers' needs. For instance, companies like Lego involve users in co-designing new products through crowd-sourcing ideas, ensuring that the final creations resonate with their customer base. LEGO IDEAS serves as a web-based platform that allows customers to actively participate in the product ideation process. This innovative approach invites LEGO enthusiasts and creative individuals to submit their unique product ideas for consideration. Users can display their designs, and fellow community members can vote on the submissions. Ideas that garner enough support have the potential to be developed into actual LEGO sets, creating a direct link between the brand and its passionate consumer base. While LEGO IDEAS primarily focuses on creativity and innovation, it operates within the broader framework of LEGO Group's commitment to ethical and sustainable business practices, ensuring that the products and ideas generated through the platform align with the company's values and standards (Source for section: ► https://ideas.lego.com/).

◘ Source: ► https://unsplash.com/

Let us now look at a case that focuses on creating shared value for sustainability. Co-creation has been at the heart of Nestlé's strategy around research and design for some time. Nestlé, a multinational food and beverage company, has faced criticism and ethical challenges over the years. One notable issue pertained to controversies surrounding the company's supply chain practices, such as child labor in cocoa production and deforestation linked to palm oil sourcing. In response to these challenges, Nestlé has taken significant steps to address ethical concerns and enhance sustainability around their products. The company has embraced the concept of Creating Shared Value (CSV), focusing on simultaneously creating value for both shareholders and society. The company has committed to eradicating child labor from its cocoa supply chain, enhancing traceability in palm oil sourcing to ensure responsible production practices, and promoting sustainability across its agricultural supply chains. Nestlé's efforts include engaging with stakeholders, collaborating with industry partners, and investing in programs to improve the livelihoods of farmers and address environmental impacts. For further information, please access the sustainability report from the brand's website, which serves as a platform for Nestlé to communicate its progress and actions in these critical areas.

We have highlighted some success stories, but with such large communities, it is not always possible to include all stakeholders. Trends change, consumers change—they evolve. Therefore, some companies and academic researchers have suggested that an open approach to both innovation and design is the way forward. This is why a co-creative process of product design requires open and active participation from people who are affected by the design and the product's outcome. Both academia and industry have been working hard to understand and implement better approaches to new product development, and this begins with design. Co-creative design, open design and open innovation are all processes with their limitations that have been and continue to be developed in response to more responsible ways of producing products. Striving for sustainability through innovation involves coordinating the interests and welfare of diverse stakeholders, encompassing shareholders, consumers, employees, partners, communities, and the planet. (Du et al., 2022). A commitment to sustainability in product development entails the deliberate incorporation of sustainability criteria throughout a brand's new product development process, aiming to generate not just economic but also social and environmental value (Varadarajan, 2017). This comprises (1) new concept development, (2) selection of low-impact materials, (3) reduction of material usage, (4) optimization of production techniques, (5) optimization of the distribution system, (6) reduction of impact during use, (7) optimization of initial lifetime, and (8) optimization of end-of-life system. There are also various sub-strategies. For example, potential sub-strategies for selecting lower-impact materials as part of a sustainable design approach include opting for cleaner materials, renewable materials, recycled materials, and recyclable materials. Additionally, sub-strategies for reducing material usage as part of sustainable design encompass minimizing material weight and reducing volume in transportation. (Source for section: Varadarajan, 2017).

In the past, brands have raised doubts about the advantages of such a concentration, asserting that it would result in increased costs. However, researchers have proposed numerous potential benefits associated with adopting sustainable practices in product development. These advantages encompass the exploration of new markets, improvement of brand image and reputation, enhanced brand capabilities, augmented financial value (rather than decreased), and elevated performance of new products (Bstieler et al., 2018). In the modern world, it is clear sustainable approaches to product development can require a lot of effort, but when done right, business owners can reap substantial rewards, as well as a positive impact on the planet and people. Yet, we also know that consumer adoption of these sustainable innovation practices is critical for new product success, therefore, understanding consumer behavior and what influences sustainable consumption is required (Bstieler et al., 2018). Manager's Corner 2.2 presents a case of Unilever's sustainable living plan.

Manager's Corner 2.2
Unilever's Sustainable Living Plan (USLP):

Background Unilever Sustainable Living Plan—2010 to 2020—Unilever, a multinational consumer goods company, implemented a Sustainable Living Plan, demonstrating a commitment to addressing social and environmental challenges. The USLP was a first of its kind in terms of its scope and ambition. It set out three big goals to transform the business across the value chain, underpinned by over 70 time-bound targets. According to Unilever, key findings include: 1. Make sustainable living effortless 2. Business must drive systems change 3. New ways of measuring social impact 4. Be ready to move fast.

Changes Made Unilever has made significant strides in reducing its environmental footprint by improving the sustainability of its supply chain and investing in renewable energy sources. Additionally, the company has pledged to enhance the well-being of billions of people by promoting health and hygiene in its products.

Impact According to the Unilever report of our total of 1.3 billion people reached, PwC have independently assured the number of people reached on a cumulative basis by an intervention through our on-ground programs for handwashing, self-esteem, sanitation, oral health, skin healing, and safe drinking water; plus TV reach from oral health. This total was 729 million by the end of 2020.

Unilever's approach goes beyond simply creating sustainable products; it involves a holistic strategy to positively impact society. Through its Sustainable Living Plan, Unilever not only benefits the environment but also addresses global health and well-being challenges, showcasing a model for other companies to follow.

Reflecting on Unilever's Sustainable Living Plan—Examine the impact of Unilever's sustainability initiatives on consumer behavior. How have consumers responded to the company's commitment to responsible and sustainable products?

Please also consider the long-term viability of this approach—initially, it was set to take place 2010–2020—What is happening now? Have other companies adopted similar strategies? Sustainability initiatives often require long-term commitment and adaptability. Evaluate Unilever's approach to ensuring the long-term viability of the Sustainable Living Plan. How does the company demonstrate adaptability to evolving environmental and social challenges?

Information Sources: ► https://www.unilever.com/files/92ui5egz/production/16cb778e4d31b81509dc5937001559f1f5c863ab.pdf

2.12.2 Sustainability and Customer Accountability

Researchers have proposed that while a comprehensive model of co-design encompassing both the business producing the product and the behavior change of the consumer using the product may be absent, most approaches and tools suggest four fundamental principles to assist behavior change:

- Make it easier for people to adopt a desired behavior,
- Make it harder for people to perform an undesired/unsustainable/unethical behavior,
- Make people want the desired behavior,
- Make people not want the undesired behavior.

Habit changing through design and influencing behavior is a notion that is increasingly being used to encourage sustainable consumer behavior. To aid in this, a more nuanced aspect for marketers to deliberate is the existence of friction at all stages in the product purchase cycle that are intended to be seamless. Of these, choice is the most interesting—a bittersweet friction because it gives on the one hand and takes on the other. The plethora of options in contemporary life is frequently regarded as an unparalleled benefit: an average of 31,530 items adorn the shelves of a typical supermarket (▶ https://www.fmi.org/our-research/food-industry-facts, 2022). Edward, who writes for Marketing Week (see article friction of choice), who could find fault with such abundance? However, personal encounters often defiantly convey a contrary sentiment, occasionally dissuading consumers from making purchases altogether. Choice presents a challenge. Should marketers contemplate minimizing options by streamlining their product ranges? While this notion may prompt hesitation among many marketers, consumers exhibit the capability to assert control independently. A subtle yet expanding trend is leaning toward the adoption of a "uniform" wardrobe. Adherents wear the same thing every day. They buy multiple versions of the same kit and never vary their look. Even some of the most iconic leaders in the world have previously engaged in this activity, albeit for conserving precious brain energy rather than for anti-consumption purposes.

Steve Jobs famously wore the same black turtleneck (Issey Miyake), blue jeans (Levi) and New Balance sneakers every day; this quickly became his signature look as well as being part of the overall brand of Apple. Steve Jobs also understood that he had a finite capacity of brainpower to make well-thought-out decisions. A minute more a day using his brainpower to decide which t-shirt to wear is less brainpower than he would have to think about his company. There have been many instances where marketers assume that the more choices they offer, the more likely customers will be able to find just the right thing. For instance, offering 50 styles of jeans instead of two increases the chances that shoppers will find a pair they really like. Nevertheless, research now shows that there can be too much choice, and when there is too much choice, consumers are less likely to buy anything at all, and if they do buy, they are less satisfied with their selection. (See article—More is not always better—Harvard Business Review). Brands should consider streamlining the available options to simplify the customer experience. Ultimately, simplifying the process can lead to heightened consideration, increased engagement, and more sales. While proposing a reduction in a product portfolio might be a bold move for marketers, positioning a brand as a leader in aiding consumers to adopt sustainable habits and consequently purchase fewer products can be a prosperous approach.

Habit changing through choice has been at the heart of Patagonia's advertising style, driving forward a passion to change consumer behavior, with Patagonia urging consumers to buy their premium products less often. The start of these campaigns was strategically launched in the run-up to Christmas to encourage people to think about the environmental impact of consumerism and to buy only what they need. Previously, in the chapter, the term premiumization was discussed in a negative light. Traditionally, premiumization was a sales tactic used to encourage the consumer to consume more. Patagonia has reversed this concept to now encourage a trend that is characterized by the tendency to be selective, to invest in fewer, better products and experiences for the consumers.

Information sources for Patagonia's "Don't Buy This Jacket Campaign":

► https://eu.patagonia.com/gb/en/stories/dont-buy-this-jacket-black-friday-and-the-new-york-times/story-18615.html

► https://www.marketingweek.com/case-study-patagonias-dont-buy-this-jacket-campaign/

2.12.3 The Anti-Consumption of Products (Kotler, 2020)

We have come full circle to end this discussion. This chapter has considered how the contemporary growth-oriented world of consumerism emerged. Most assign this to the Industrial Revolution in the nineteenth century, which increased the production of products and resulted in the highest levels of consumption ever recorded. Equally, the significant impact consumerism had on culture and society has been documented. Consumption was adopted as a lifestyle and embedded into all corners of the planet, resulting in humans using brands as props to project a specific identity. This culture is changing. There is a new age of consumerism, which, according to Kotler (2020), can be described under the five principles of anti-consumption:

1. Life Simplifiers
2. Degrowth Activists
3. Climate Activists
4. The Food Choosers
5. Conservation Activists

Each of the consumer types impacts the production of products in circulation. According to Kotler (2020), life simplifiers follow a simplistic/decluttered lifestyle. The degrowth activists believe consumption will exceed the carrying capacity of the planet, and therefore, a drastic reduction in consumption levels is needed now. Climate activists care about their carbon footprint and maintain an eco-friendly lifestyle, whereas the food choosers believe that following a vegetarian, vegan or plant-based diet will have a long-lasting positive impact on both people and the planet. Finally, the conservation activist promotes the reusing and recycling of existing goods to benefit the planet and society. The circular economy has also been proposed as a response to assist with not only prompting reuse and recycling but also as an alternative business model to mass production. Targeting the eco-friendly consumer type (climate and conservation activists), marketers should devote attention to producing more reusable products. Adopting efficiency and productivity improvements reduces the impact of production and consumption on the planet. The principles surrounding the circular economy have evolved from some of the initial ideas around production in the car industry by Toyota in the 1940s. Within this process, resources used to make the product (a car in Toyota's case) are circulated constantly through closed loops, because of this, the life of materials/resources is prolonged through the means of recycling and then re-use. Has this been the answer to all the recycling problems? Unfortunately not, many critics of the circular economy claim that the process of recycling is still considered a continuum of the traditional growth-oriented system since it still promotes consumption-led growth based on material re-use of products and even though materials are recycled and re-used to curb demand for new materials, consumption of goods is not decreased, in addition, recycling processes require a considerable amount of energy with results in negative environmental consequences.

Circular economy is an economic model that aims to minimize waste and maximize resource efficiency by promoting the continual use, recycling, and repurposing of materials throughout their life cycles.

Eco-friendly materials refer to materials that are produced and used in ways that minimize their environmental impact and carbon footprint, often through sustainable sourcing, reduced energy consumption, and recyclability.

Waste Generation is the creation of redundant or discarded materials during the production, consumption or disposal of products.

It is hard to achieve a balance for the stabilization of the volume of products produced. However, the COVID-19 pandemic has prompted a shift toward better consumption habits worldwide. Regardless of which consumer type has emerged most popular, consumers are becoming more thoughtful about their consumption levels, (See Statista study 2023—Consumers more concerned about sustainability in 2023 than a year ago worldwide), and brands that promote ethical and environmental values will prosper. The products that will flourish in the future will be the ones presented not as commodities but as concepts—the brand as experience, as a (sustainable/healthy) lifestyle.

2.13 Product and Service Development for Sustainable Development Goals

In this chapter, we plunged into the persistent reality of mass production and consumption, where the relentless hum of machinery and the rhythmic dance of workers paint a vivid picture of an industrial landscape consumed by efficiency. The factory, a symbol of progress, purrs with a relentless determination, its gears and machinery orchestrating a symphony of production that echoes through vast and eerie spaces. The consequences of mass production and mass consumption were laid bare, creating a stark contrast between the efficiency of the factory floor to produce products and the environmental devastation unfolding beyond. Having set the scene, we then explored how traditional models for new product development have played a significant role in propelling us toward this precipice of relentless production. These models once celebrated for their efficiency and innovation, have inadvertently become enablers of a system that prioritizes quantity over sustainability. Considering these revelations, it becomes imperative to align our efforts with the United Nations Sustainable Development Goals (SDGs) to address the challenges posed by mass production and consumption. The SDGs, particularly those related to responsible consumption and production (SDG 12), climate action (SDG 13), and life on land and below water (SDGs 15 and 14), serve as beacons guiding us toward a more sustainable and equitable future.

Responsible consumption involves making mindful and ethical choices in purchasing and using products or services to minimize environmental and social impacts, promote sustainability, and contribute to a more equitable and responsible global economy.

In the first manager's corner, we explored Patagonia, a prominent outdoor clothing company which stands as a trailblazer in championing sustainability and ethical practices within the apparel industry. This section highlights Patagonia's commitment to various United Nations Sustainable Development Goals (SDGs), notably SDG 12 on responsible consumption and production, as evidenced by the company's pivot to recycled materials, reducing the demand for added resources and minimising environmental impact. Additionally, their commitment to fair labor practices throughout the supply chain aligns with SDG 8, promoting decent work and economic growth. By actively addressing challenges associated with clothing production on its website, Patagonia fosters transparency, contributing to consumer education and awareness. With 98% of styles in their Spring 2024 collection utilizing recycled materials, Patagonia exemplifies its belief that "Together, We Can Change How Clothes Are Made," highlighting a dedication to industry-wide transformation. In doing so, Patagonia not only creates value for sustainable development but also sets a compelling precedent for responsible and environmentally conscious practices in the apparel sector and the production of products, aligning with global goals for a more sustainable future.

Transparency in business involves honest communication about a company's practices, policies, and impacts, allowing stakeholders and consumers to make informed decisions.

We next explored an example of how Volvo use carbon-neutral energy sources, such as hydroelectricity and wind power, for their electrical needs to drive their closed-loop product lifecycle. The company addresses several United Nations SDGs through its environmentally conscious practices. Volvo's use of carbon-neutral energy sources such as hydroelectricity and wind power aligns with SDG 7, promoting access to affordable, dependable, sustainable, and modern energy. This approach demonstrates the company's dedication to reducing the carbon footprint associated with its electrical needs. The company's emphasis on modern, efficient equipment and industrial processes throughout the manufacturing process contributes to SDG 9 by promoting sustainable industrialization and fostering innovation. Volvo's focus on energy-efficient engines, designed to reduce pollution and CO_2 emissions, directly contributes to SDG 12 by promoting responsible consumption and production. The emphasis on efficient engines that require less fuel aligns with sustainable manufacturing practices. Volvo's commitment to energy efficiency and reduced emissions in its engines addresses SDG 13 Climate Action, as it actively contributes to the global effort to combat climate change by minimizing the environmental impact of its products. The value created for sustainable development in products and production by Volvo lies in its integrated approach. By incorporating carbon-neutral energy sources, efficient manufacturing processes, and recyclable materials, Volvo not only aligns with global sustainability goals but also sets an example for the automotive industry. As the industry shifts toward more sustainable practices, Volvo's approach highlights the importance of integrating environmental considerations into the entire product development process and the corporate strategy, contributing to a more sustainable and responsible future for automotive manufacturing.

Carbon footprint is the total amount of greenhouse gases (including carbon dioxide and methane) that are generated by our actions.

Climate change is a real and undeniable threat to our entire civilization. The effects are already visible and will be catastrophic unless we act now. Climate action refers to taking urgent action to combat climate change and its impacts.

Industrialization is the development and implementation of industrial methods (e.g., the assembly line) of production on a large scale.

We also showed in the service section of this chapter how responsible consumption and production are not confined to mere products. The Body Shop was used as a case study to demonstrate how they effectively use the 3Ps to bring awareness to responsible consumption. The Body Shop's utilization of the three additional Ps of marketing (physical environment, processes and people) to enhance its brand purpose and navigate the current economic climate addresses several United Nations Sustainable SDGs, displaying its commitment to sustainability and responsible business practices. The Body Shop's focus on sustainable fixtures made from repossessed wood and recycled plastics in its stores aligns with SDG 12 by promoting responsible consumption and production. The use of 100% recycled materials destined for landfills minimizes the brand's environmental footprint, contributing to more sustainable practices within the retail industry. What was also great to see is how they create value for SDG 4: Quality Education: The Body Shop's emphasis on education as a central aspect of up-skilling teams for local activism projects contributes to SDG 4 by promoting quality education. The brand's commitment to providing workshops with grassroots activist groups not only enhances employee knowledge but also fosters community engagement and awareness. The Body Shop's unique approach to recruitment, aiming to create a team of passionate activists, aligns with SDG 8 by promoting decent work and economic growth. By prioritizing employees who are committed to the brand's values and the goal of creating a fairer world, The Body Shop fosters a positive work environment and contributes to sustainable economic development. The value created for sustainable development of products and production by The Body Shop lies in its holistic integration of environmental and social considerations into its business model. By using recycled materials in-store fixtures, providing education and up-skilling opportunities for employees, and focusing on activism as a core brand purpose, The Body Shop not only meets consumer demands for sustainable products but also actively contributes to global goals for a more environmentally friendly, socially responsible, and economically inclusive world.

Toward the end of the chapter, we discussed sustainable innovation for new product development and ethical co-design with an emphasis on re-evaluating current approaches to product development and consumption. We introduced Lego Ideas and its exemplification of sustainable innovation for new product development and ethical co-design, the company addresses several United Nations SDGs, highlighting its commitment to sustainable practices and community engagement. Lego's utilization of crowd-sourcing ideas through LEGO IDEAS aligns with

SDG 9 by promoting industry innovation and the development of sustainable infrastructure. By involving users in the co-design process, Lego fosters a collaborative approach to product development that taps into diverse creative perspectives. Lego's commitment to involving users in the product ideation process contributes to SDG 12 by promoting responsible consumption and production. By actively seeking input from its customer base, Lego ensures that its final creations align with consumer preferences, reducing the likelihood of unnecessary waste and promoting sustainable consumption. The collaborative nature of Lego Ideas, where users submit designs and community members vote on submissions, reflects SDG 17 by fostering partnerships for the achievement of shared goals. Lego's engagement with its customer community strengthens ties and builds a sense of shared responsibility for sustainable product development. The value created by Lego for the sustainable development of products and production lies in its inclusive and participatory approach to innovation. By leveraging crowd-sourcing and user engagement through LEGO IDEAS, Lego not only taps into a diverse range of creative ideas but also builds a community that is actively involved in the design process.

Our final example displayed Unilever's ambitious Sustainable Living Plan, which ran from 2010–2020. Through Unilever's Sustainable Living Plan (USLP), groundbreaking impact on new product development and transforming consumer habits, the company addresses various United Nations SDGs, highlighting its commitment to sustainability and responsible business practices. Unilever's Sustainable Living Plan aligns strongly with SDG 12 by focusing on making sustainable living effortless. Through ambitious goals and over 70 time-bound targets, Unilever aims to transform its business across the value chain, encouraging responsible consumption and production practices. In addition to this, Unilever's commitment to improving the sustainability of its supply chain and investing in renewable energy sources contributes to SDG 7 by promoting affordable and clean energy. The company's efforts to reduce its environmental footprint demonstrate a commitment to environmentally conscious production. Unilever's ambition did not end there, good health and well-being emerged as important as Unilever's pledge to enhance the well-being of billions of people by promoting health and hygiene in its products aligns with SDG 3. By incorporating health and hygiene considerations into product development, Unilever contributes to the goal of ensuring good health and well-being for all. Unilever's Sustainable Living Plan, with its focus on driving systems change and introducing new ways of measuring social impact, aligns with SDG 9 by promoting industry innovation and the development of sustainable infrastructure. The value created by Unilever for sustainable development of products and production lies in its comprehensive and ambitious approach to addressing social and environmental challenges. Importantly, Unilever's commitment to promoting health and hygiene through its products reflects a comprehensive approach to sustainable development.

Renewable energy is energy derived from natural sources that are replenished at a higher rate than they are consumed.

Social impact refers to the impacts of an organization's activities, products, or services have on the well-being, development, and overall welfare of individuals and communities, comprising both positive and negative consequences.

As we navigate the complexities of the industrial landscape, the chapter highlights the importance of re-evaluating our approach to product development and consumption, emphasising the need for a paradigm shift that aligns with the principles of sustainability encapsulated in the SDGs. Only through a concerted effort to break free from the shackles of relentless production can we hope to build a future where efficiency coexists with environmental stewardship, which reflects a landscape rejuvenated by responsible choices and conscious consumption habits.

Summary

In this chapter, we have defined what is meant by products and discussed some of the classic models of product development. We also looked at how we have arrived at a growth-oriented system and some of the strategies the biggest brands in the world have used, which have contributed to the perpetual consumption of products. We explored how the economic systems and business models laid the foundation for the mass-produced product and the rapid development of the growth-oriented system. We reached services as a means of integrating the consumer into the co-creation process of products. Many businesses have struggled to adopt the sustainable production of products throughout the entire product lifecycle. According to the SDG update report in 2022 (UN Sustainable Development Goals Report, 2022), COVID-19 has wiped out more than 4 years of SDG progress. Businesses are in a perfect position to make an immediate impact on such issues. We discussed some of the potential ways forward in addressing such issues. Treated as a strategic priority for brands, sustainable innovation for products and services can be a solution for addressing both the production of ethical and responsible products and the ethical and responsible use of those products by consumers. Sustainability is no longer tolerated as a PR gimmick; it must be weaved through the entire brand strategy.

Review Questions

1. Define the concept of "triple bottom line" and explain its significance in sustainable product development.
2. Discuss the key principles of the Boston matrix and how it can be applied in product portfolio management.
3. Explore the stages of the product lifecycle and provide examples of products in each phase.
4. Outline the steps in the new product development process and explain the importance of each stage.
5. How did the era of mass production contribute to the evolution of product development processes?
6. Explain the role of capitalism in shaping product development strategies and priorities.

7. Elaborate on the challenges and benefits associated with incorporating sustainability into product development.
8. How does the concept of co-creation contribute to the development of products and services for sustainability?
9. Analyze the ethical considerations associated with product obsolescence and planned obsolescence.
10. Discuss the challenges and opportunities associated with circular economy principles in product development.
11. Explain the significance of stakeholder engagement in sustainable product and service development.
12. Discuss how capitalism has influenced the marketing of products, including tactics like upselling and cross-selling.
13. Provide examples of companies that have successfully implemented premiumization strategies in their product lines.
14. How can companies balance the pursuit of profit with responsible and sustainable product development practices?

Discussion Questions

1. Investigate the ethical dimensions of planned obsolescence in the context of product development. Assess how this practice aligns with or contradicts principles of sustainability and responsible consumerism. Use examples from the smartphone industry to support your answer, highlight instances of planned obsolescence and discuss the broader implications for environmental sustainability and consumer trust.
2. Critically evaluate the role of premiumization in contemporary product development and marketing strategies. Explore how the pursuit of premium status influences consumer behavior and brand loyalty, drawing on examples from luxury brands such as Rolex and Tesla. Discuss the potential challenges and ethical considerations associated with creating and marketing premium products and analyze how companies navigate these complexities.
3. Examine the challenges and opportunities of integrating sustainability into the product development process within a capitalist framework. Provide detailed examples, such as Patagonia and its commitment to sustainability, to illustrate effective practices. Analyze the impact on market positioning, consumer perception, and long-term profitability, considering how sustainability aligns with or challenges traditional capitalist motives.
4. Research the critical role of sustainability in the innovation of products and the ethical considerations involved in co-design processes. Utilizing examples from leading companies in sustainable product development and their commitment to circular design, discuss how these organizations integrate environmental, social, and economic considerations into their innovation strategies. Analyze the challenges and opportunities associated with ethical co-design, emphasising the collaborative efforts between businesses, designers, and consumers in creating products that align with sustainability goals. Consider the impact of such initiatives on consumer perceptions, market competitiveness, and the long-term viability of sustainable innovation.

5. Discuss the challenges and opportunities associated with leveraging service characteristics to address the broader societal and environmental goals outlined in the SDGs.

Practice Quiz

1. How did the era of mass production contribute to the evolution of product development processes?
   (a) It emphasised personalized and artisanal production.
   (b) It led to the standardization of products and assembly line techniques.
   (c) It had no impact on product development.
2. Explain the role of capitalism in shaping product development strategies and priorities.
   (a) Capitalism has no influence on product development.
   (b) Capitalism encourages sustainable and eco-friendly product development.
   (c) Capitalism influences strategies to maximize profit and market share.
3. Discuss the impact of consumer demand on the mass production era and the development of standardized products.
   (a) Consumer demand had no influence on mass production.
   (b) Consumer demand led to customized and unique products.
   (c) Consumer demand led to standardized products and assembly line production.
4. How does the concept of planned obsolescence relate to product development in a capitalist market?
   (a) It encourages long-lasting and durable products.
   (b) It promotes continuous innovation and frequent product upgrades.
   (c) It has no relevance to product development.
5. Explore the motivations behind companies engaging in premiumization and how it influences product development.
   (a) Premiumization has no impact on product development.
   (b) Companies engage in premiumization to offer affordable products.
   (c) Premiumization aims to create high-end, exclusive products for higher profit margins.
6. Analyze the ethical considerations associated with planned obsolescence in the context of consumer behavior.
   (a) Planned obsolescence is an ethical business practice.
   (b) It raises ethical concerns as it encourages wasteful consumer behavior.
   (c) Consumers prefer planned obsolescence for constant product upgrades.
7. Evaluate the role of innovation in driving product development strategies during the mass production era.
   (a) Innovation had no role in the mass production era.
   (b) Innovation was crucial for introducing new products and production techniques.
   (c) The mass production era discouraged innovation.

8. Discuss how capitalism has influenced the marketing of products, including tactics like upselling and cross-selling.
   (a) Capitalism has no impact on marketing strategies.
   (b) Capitalism encourages transparent marketing practices.
   (c) Capitalism promotes upselling and cross-selling to maximize revenue.
9. Explain the challenges and opportunities for sustainable product development within a capitalist framework.
   (a) Sustainable product development aligns perfectly with capitalism.
   (b) Capitalism poses challenges but also opportunities for sustainable product development.
   (c) Sustainable product development is not feasible in a capitalist system.
10. Provide examples of companies that have successfully implemented premiumization strategies in their product lines.
   (a) All companies avoid premiumization strategies.
   (b) Examples are not available as premiumization is an unsuccessful strategy.
   (c) Companies like Apple and Rolex have successfully implemented premiumization.
11. Discuss the role of advertising and branding in promoting premium products and influencing consumer perceptions.
   (a) Advertising and branding have no impact on consumer perceptions.
   (b) They play a crucial role in promoting premium products and shaping consumer perceptions.
   (c) Premium products do not rely on advertising and branding.
12. Explore the potential consequences of overconsumption in a capitalist-driven market.
   (a) Overconsumption has no consequences.
   (b) It leads to resource depletion, environmental degradation, and societal issues.
   (c) Overconsumption is encouraged for economic growth.
13. How can companies balance the pursuit of profit with responsible and sustainable product development practices?
   (a) Profit and sustainability are incompatible.
   (b) By prioritizing profit over sustainability.
   (c) By integrating responsible and sustainable practices into their business models.
14. Assess the impact of globalization on product development strategies in a capitalist economy.
   (a) Globalization has no impact on product development.
   (b) It encourages localized and region-specific product development.
   (c) It influences product development strategies to cater to a global market.

15. In which phase of the product lifecycle do sales typically stabilize, and the market becomes saturated?
    (a) Growth phase
    (b) Maturity phase
    (c) Decline phase
16. What does a "cash cow" represent in the Boston Matrix?
    (a) Products with high market share in a high-growth market
    (b) Products with low market share in a high-growth market
    (c) Products with high market share in a low-growth market

(You can find the correct answers to these questions by taking the quiz and then comparing your answers with the answers that are provided at the end of the book, under the "Practice Quiz Answers" part of the book.)

References

Bateson, J. E. G. (1992). *Managing services marketing: Text and readings* (2nd ed.). Dryden Press.

Belk, R. W. (1988). Possessions and the extended self. *Journal of Consumer Research, 15*(2), 139–168.

Borden, N. H. (1964). The concept of the marketing mix. *Journal of Advertising Research, 2*, 7–12.

Boston Consulting Group. (1968). *Perspectives on experience*. The Boston Consulting Group.

Bstieler, L., Gruen, T., Akdeniz, B., Brick, D., Du, S., Guo, L., Khanlari, M., McIllroy, J., O'Hern, M., & Yalcinkaya, G. (2018). Emerging research themes in innovation and new product development: Insights from the 2017 PDMA-UNH doctoral consortium. *Journal of Product Innovation Management, 35*(3), 300–307.

Dall'Olmo Riley, F., & De Chernatony, L. (2000). The service brand as relationships builder. *British Journal of Management, 11*(2), 137–150.

Day, G. S. (1990). *Market driven strategy: Processes for creating value*. Free Press.

De Chernatony, L., & Dall'Olmo Riley, F. (1998). Defining a "brand": Beyond the literature with experts' interpretations. *Journal of Marketing Management, 14*(5), 417–443.

Du, S., Bstieler, L., & Yalcinkaya, G. (2022). Sustainability-focused innovation in the business-to-business context: Antecedents and managerial implications. *Journal of Business Research, 138*, 117–129.

Elkington, J. (1994). Towards the sustainable corporation: Win-win-win business strategies for sustainable development. *California Management Review, 36*(2), 90–100.

Geurts, A., & Cepa, K. (2023). Transforming the music industry: How platformization drives business ecosystem envelopment. *Long Range Planning, 56*(4), 102327.

Harari, Y. N. (2014). *Sapiens: A brief history of humankind*. Random House.

Higgs, K. (2021). How the world embraced consumerism. BBC Future. Retrieved April 4, 2024, from https://www.bbc.com/future/article/20210120-how-the-world-became-consumerist

Hu, S. J. (2013). Evolving paradigms of manufacturing: From mass production to mass customization and personalization. *Procedia CIRP, 7*, 3–8.

Iyengar, S. S., & Lepper, M. R. (2000). When choice is demotivating: Can one desire too much of a good thing? *Journal of Personality and Social Psychology, 79*(6), 995.

Keller, K. L. (1993). Conceptualizing, measuring, and managing customer-based brand equity. *Journal of Marketing, 57*(1), 1–22.

Khan, L. M. (2019). The separation of platforms and commerce. *Columbia Law Review, 119*(4), 973–1098.

Klein, N. (2000). *No logo*. HarperCollins Publishers.

Kotler. (2016). Retrieved from https://www.marketingjournal.org/the-relationship-between-marketing-and-capitalism-phil-kotler/

Kotler, P. (1967). *Marketing management: Analysis, planning and control*. Prentice Hall.

Kotler, P., & Armstrong, G. (2000). *Principles of marketing*. Prentice Hall.
Kotler, P., Armstrong, G., Harris, L. C., & Piercy, N. (2013). *Principles of marketing PDF eBook*. Pearson Higher Ed.
Kotler's. (2020). Anti-consumption movement. Retrieved from https://www.marketingjournal.org/the-consumer-in-the-age-of-coronavirus-philip-kotler/
Lovelock, C., & Patterson, P. (2015). *Services marketing*. Pearson Australia.
Melewar, T. C., & Saunders, J. (2000). Global corporate visual identity systems: Using an extended marketing mix. *European Journal of Marketing, 34*(5/6), 538–550.
Prahalad, C. K., & Ramaswamy, V. (2004). Co-creation experiences: The next practice in value creation. *Journal of Interactive Marketing, 18*(3), 5–14.
Sheth, J. N., Newman, B. I., & Gross, B. L. (1991). Why we buy what we buy: A theory of consumption values. *Journal of Business Research, 22*(2), 159–170.
Singh, P., Brown, D. M., Chelekis, J., Apostolidis, C., & Dey, B. L. (2022). Sustainability in the beer and pub industry during the COVID-19 period: An emerging new normal. *Journal of Business Research, 141*, 656–672.
UN Sustainable Development Goals Report. (2022). Retrieved Accessed February 29, 2024, from https://unstats.un.org/sdgs/report/2022/The-Sustainable-Development-Goals-Report-2022.pdf
Varadarajan, R. (2017). Innovating for sustainability: A framework for sustainable innovations and a model of sustainable innovations orientation. *Journal of the Academy of Marketing Science, 45*, 14–36.
Vargo, S. L., & Lusch, R. F. (2004). Evolving to a new dominant logic for marketing. *Journal of Marketing, 68*(1), 1–17.
Vargo, S. L., & Lusch, R. F. (2008). Service-dominant logic: Continuing the evolution. *Journal of the Academy of Marketing Science, 36*, 1–10.
Wiesman, D. W. (2006). The effects of performance feedback and social reinforcement on up-selling at fast-food restaurants. *Journal of Organizational Behavior Management, 26*(4), 1–18.

Further Reading

Fatimah, Y. A., Govindan, K., Murniningsih, R., & Setiawan, A. (2020). Industry 4.0 based sustainable circular economy approach for smart waste management system to achieve sustainable development goals: A case study of Indonesia. *Journal of Cleaner Production, 269*, 122263.
Kelleci, A., & Yıldız, O. (2021). A guiding framework for levels of sustainability in marketing. *Sustainability, 13*(4), 1644.
Sun, J. J., Bellezza, S., & Paharia, N. (2021). Buy less, buy luxury: Understanding and overcoming product durability neglect for sustainable consumption. *Journal of Marketing, 85*(3), 28–43.
Takeuchi, H., & Nonaka, I. (1986). The new product development game. *Harvard Business Review, 64*(1), 137–146.
Van Doorn, J., Risselada, H., & Verhoef, P. C. (2021). Does sustainability sell? The impact of sustainability claims on the success of national brands' new product introductions. *Journal of Business Research, 137*, 182–193.
Zuboff, S. (2015). Big other: Surveillance capitalism and the prospects of an information civilization. *Journal of Information Technology, 30*(1), 75–89.

Pricing Strategy for People, Planet, and Profit

Contents

S. M. R. Shams et al., *Sustainable Marketing*, Classroom Companion: Business,
https://doi.org/10.1007/978-3-031-79130-7_3

Preliminary Vignette: Restaurant Pricing, Food Wastage and "Bottom of the Pyramid" Consumers

2.7 billion people globally survive on less than $2.50 per day—a socioeconomic group referred to as "Bottom of the Pyramid" (BOP). Malnutrition and food poverty are significant challenges to them, bringing ill health, premature mortality and unhappiness. Expanded academic and commercial understanding of this vital but underrepresented population segment, showing that, contrary to previous assumptions, BOP consumers are as brand—conscious as wealthier ones and also represent a profitable market, albeit one for which product and service offers should incorporate initiatives and features designed to address their basic needs ethically. In Sri Lanka, where a large proportion of the 22 million inhabitants suffer poor socioeconomic circumstances, the prevalence of food poverty is accompanied, rather ironically, by high levels of food wastage from stores and restaurants. In a recent research study by Apostolidis et al. (2021), participants indicated that they would welcome a mutually beneficial solution, recognizing not just the health and financial benefits to consumers and retailers, but also the broader environmental gains achievable by tackling the overproduction of food and the routine disposable of edible, high quality surplus.

The study explored the viability of establishing a smartphone app which would enable shops and restaurants with excess capacity (e.g., foods approaching the age when they will spoil) and economically deprived consumers to interact and transact with each other. The app would allow retailers to charge a small, heavily discounted price to needy consumers, usually at the end of trading hours when food is often thrown away. Most consumers were willing and able to pay a reduced price for what might be very high-quality food from restaurant dishes, although some were hampered by having to share smartphones with family and neighbors, thereby impeding their ability to respond to offers in a timely manner. Almost all retailers stated that they would rather recoup a small amount of money and help bring about a societal benefit than write off the stock completely. However, many of them also anticipated barriers to participating in such a scheme, such as the time investment, staff training, and technological prowess required to manage the system. Perhaps their major concern was the possibility of "cannibalization of sales"—a situation in which a seller launches a product or service which, instead of creating wholly incremental custom and revenue, switch-sells the firm's existing customers into another of its offerings, competing against itself and damaging its established custom. While this could potentially be an issue for some of the stores which sell relatively inexpensive food items, it seems highly unlikely that poorer consumers could be considered as potential restaurant diners, especially at more exclusive establishments. However, it is possible that restaurateurs were concerned with the reactions of their core clientele—wealthier diners who pay handsomely for fine products and may resent its cut-price distribution to others. Indeed, social class divisions within society may exacerbate such an impediment.

3.1 Introduction

Price is the only element of the marketing mix which brings money into an organization rather than spending it. The upper limit of a sustainable pricing strategy is dictated by what customers are prepared to pay or what they can afford, and the lower limit by how much the product or service costs to provide. However, these basic economic parameters constitute only the widest part of a decision-making funnel when selecting prices. The successful marketer must design pricing strategies to position the firm and its goods favorably against those of constantly evolving competitors, which may vary from one market to another. Marketers must serve the organization's broader marketing and business objectives while reinforcing the brand meanings conveyed within its marketing communications. They must also consider the needs and demands not only of customers and end users but of numerous, often contradictory, stakeholders. Finally, marketers must leverage pricing to increase sales volumes and profits while being starkly aware that a minor misjudgment could severely damage the firm's commercial sustainability and its relevant value propositions for societal, environmental and commercial stakeholders, including customers.

If pricing strategy is a balancing act, the crosswinds impacting upon it have been strengthened by the proliferation of digital media and the breathless expectations of globally interconnected, information-rich, time-poor consumers. In this chapter, we build upon "classic" strategic pricing approaches by exploring how they may be augmented by offering customer value propositions which contribute to the triple bottom line, i.e., people, planet, and profit (Elkington, 2013). To be people-oriented, pricing strategies should enable customer choice through clarity, contributing to a logical decision-making process in a manner which avoids coercion or the underhand manipulation of unconscious heuristics. To help protect the planet, pricing strategies should be employed which discourage the needless consumption of non-renewable or non-recyclable resources, instead supporting customers to buy, use and dispose of less harmful goods while deriving value from them. To ensure profitability, pricing strategies must be based upon constantly updated market intelligence, which captures real and potential changes within the micro- and macro-environments while also recognizing the unique strengths and limitations of the firm's internal environment. We will critique techniques which pitch prices according to customer perceptions of worth, dynamic competitor activities, and business needs that appeal to increasingly tech-savvy, ethically motivated audiences.

3.2 What Is Pricing?

Price is the amount demanded by a seller in exchange of a product or service. This amount is normally money, but on occasion, the price may instead be a reciprocated favor or the provision of a good or service of equivalent value. Price is one of the seven elements of the extended marketing mix and is one of the four original elements of Borden's (1964) marketing mix (i.e., product, price, place and promotion). Unlike the other elements, price brings money into an organization rather than using the company's cash for its activities. However, price does not exist in isolation from the other elements in the marketing mix but must work alongside them, and even overlap them, to provide a consistent and coherent brand message to customer and non-customer stakeholders. Price is not simply about generating profit. It also helps to define stakeholder perceptions of the brand (Steenkamp, 2017), the organization's position against the competition, and the customer value proposition. It is of paramount importance to price one's products and services strategically and to maintain a long-term orientation rather than switching to different approaches according to short-term considerations. To be reactive when pricing one's goods is to allow the marketplace, and particularly competitors, to dictate the direction and potential success of the firm. Perhaps even more damagingly, it erodes customer confidence in the organization. Imagine having bought a new car in a buoyant market, only to see the manufacturer apply deep discounts to the same model a few months later when the market becomes more depressed. Not only would this make most customers feel exploited, disappointed and angry, but it would also make them very nervous when deciding whether to trust that company's future price points (Lee et al., 2016).

While it is extremely important to be consistent and strategic in one's pricing approach, there are also reasons to make minor adjustments. Pricing is impacted by many variables, both internal and external to the organization. It is linked to the availability and costs of labor and to the raw materials and components which are used in the manufacturing of the product or production of the service. It may also be dependent upon the availability and cost of transportation and distribution, warehousing, appointing and managing other channel intermediaries such as resellers. Michael Porter's (2008) seminal five forces model shows that organizations come under threat from substitute products, new competitors entering the market, the increased bargaining power of customers, increased bargaining power of suppliers, and general competitive rivalry within the industry or product category. For these reasons, organizations may need to react to changes in the external market environment when deciding upon a marketing strategy, including the pricing strategy. Not only that, but these competitive forces should be constantly surveyed and analyzed and fed into a market information system which enables strategic leaders to make appropriate decisions for the commercial direction of the firm. Furthermore, pricing may be influenced by internal factors, such as the skills and knowledge within the workforce at any one point in time, the recruiting of staff, the financial strength of the organization, its capacity to manufacture in bulk or across a variety of product lines, its other marketing and business objectives, and other broad considerations. Pursuing an appropriate pricing strategy can often feel like walking a tightrope between adhering to a pre-agreed course and accommodating unforeseen forces.

The ubiquity of discounts and repriced goods and services indicates that many sellers find it difficult to achieve a steady continuity in their pricing. A brief glance at items in retail premises confirms this. Less apparent are the fluctuations in price, which occur within business sales. This is because not only does the organization have the ability to change tack, but many default an element of discretion to individual salespeople when undertaking their sales appointments and price negotiations. Consider the case of a salesperson who, on the final working day of the month, is one sale away from making her sales target and earning a healthy commission bonus. In the appointment, she assesses the sensitivity of the customer to changes in price and concludes that the buyer is much more likely to place an order immediately if the price is significantly lowered. She is perhaps nervous that, if she holds the line and maintains her current quotation, she will achieve the sale anyway, but perhaps at a later date. However, she wants the commission and is prepared to risk her customer's confidence in her firm's pricing strategy to achieve her commission bonus. Of course, this may actually be in the best interests of her firm, which may also desperately need that sale on the same day to achieve its own objectives. However, it is likely that, by chasing the deal at a lower negotiated price, the industrial buyer will perceive the salesperson as more flexible, and possibly weaker, on price, and it is almost certain that this will shape the buyer's future price expectations and undermine the profit potential of that customer-seller relationship. The salesperson would be well advised to remember that prices which are achievable by a selling organization are bound by the costs of manufacturing and selling the product and by the value which is perceived by customers in attaining that product. It can be tempting for salespeople and marketers to disregard the significance of, for example, a 5 or 10% discount from the listed price. However, on high-value goods, this often equates to a large sum of money. Furthermore, in certain industries, especially those which rely on low-profit margins and high volumes, such as supermarkets, reducing the transaction price by 5 or 10% may equate to a profit reduction of 30–100% or even trading at a loss.

3.3 Customers and Market Demand

Having highlighted some of the major considerations which impact upon pricing, we can begin to examine the relationships between pricing, customers, and marketplaces. Pricing decisions should be founded upon a customer value proposition (Kumar & Reinartz, 2016). If the customer believes that a sufficient level of value will be gained by purchasing the product or service, then they will pay the asking price, assuming that there are no other factors preventing or discouraging them from doing this, such as a superior competitor product or an inability to pay. To purchase goods and services, customers do not simply invest money. They also commit their time, energy, and mental capacity toward the decision-making process. Naturally, for low-cost, low-risk, routine purchases, the decision is simple, and the decision-making process is all but absent. When choosing a more significant purchase, however, such as a new car, a new home, or a wedding venue, the decision-making process is likely to be longer and more sophisticated, often including more members of the decision-making unit. Customers will confirm that your pricing is correct or incorrect through whether they purchase a product in sufficient numbers to meet the firm's volume and profit aspirations.

If the above explanation sounds familiar, that is because it stems from ideas of economic and political philosophy which have been with us for over two centuries. The Scottish philosopher Adam Smith put forward ideas of supply and demand, which are vital not only to pricing but to marketing and business more holistically (Norman, 2019). While his detractors have accused him of an overly customer-centric standpoint, which risks trampling the needs of sellers and employees underfoot, his ideas repeatedly play out in the marketplace. If the price of a product or service increases, this enables the seller to increase their profits. However, it also results in a decrease in demand as more customers begin to question its value for money. Conversely, if the price of a product or service decreases, the phenomenon of price elasticity is likely to create customer demand, as more buyers recognize the increased value for money represented. However, as the achievable profit per unit decreases, the number of suppliers who wish to make and sell the product decreases as their perceptions of profit opportunity dwindle. As a result of this, less product is made and sold, leading to an increased scarcity, which in turn enables the remaining suppliers to increase their prices without a significant loss of custom. In fact, Adam Smith saw markets as organisms which were constantly self-regulating and striving to achieve an equilibrium in which the quantity of goods made and sold is, eventually, approximately equal to the quantity of goods bought and consumed. Only in such a way did he believe that significant pricing fluctuations would disappear and stability be achieved.

There are four broad market scenarios in which suppliers find themselves:

1. Demand increases, but supply is static. Therefore, the price equilibrium is raised to a higher level as customers compete to buy available products;
2. Demand decreases, but supply is static. Therefore, the price equilibrium drops to a lower level as suppliers compete to sell surplus products;
3. Supply increases, but the demand is static. Therefore, the price equilibrium drops to a lower level as suppliers compete to sell surplus stock once more;
4. The supply decreases, yet demand is static. Therefore, the price equilibrium rises to a higher level as customers compete to buy scarcer products.

In certain situations, suppliers unite to nurture a market so that the price of their goods attains a level which is more desirable for them. This type of collusion is illegal in most circumstances, as it represents anti-competitive practice. Indeed, there have been numerous price-fixing scandals among major economies in recent years. The inter-governmental body, the Organization of the Petroleum Exporting Countries (OPEC), is an international cartel of governments who collaboratively coordinate the reserves of, and production of, oil and petrol to ensure maximum profitability over a medium to long term. In this way, they can safeguard their income even when oil production is problematic. This results in consumers often paying extremely inflated prices for a necessary commodity, especially when pump prices are increased further by local taxation.

Adam Smith's ideas on market forces were not intended to be absolutely applicable to real-life situations without consideration of mediating factors. Like many theories, they presented a simplistic and hypothetical landscape upon which the ideas could be played out. There are several limitations to Smith's theory of supply and demand. Perhaps the most famous is that it assumes the notion of "Economic Man." Most figures of the Enlightenment were wedded to the idea of human rationality but must have understood that people do not always act in the most rational manner (e.g., Albanese, 2015). For instance, it is very common for people to make choices which are not in their best interests or which are inferior to alternative choices which they could have made. People act upon egotism, loyalty, fear, and other emotions when choosing goods. Such considerations were outside the scope of major supply and demand theories, which also imply a level playing field of competition in which all suppliers sell products which can be compared precisely so that any differences in specification from one product to another can be accounted for financially within a product price. Naturally, this is not always the case. Indeed, certain products and services are presented to the market in such a way as to obfuscate the Points of Parity (PoPs) and Points of Difference (PoDs) between them and their competitor products. An example would be utility tariffs, which often impede direct consumer comparisons. Classical economic theory tends to suggest that people will buy in greater volume if the price is reduced, although this is clearly not the case when we consider examples such as engagement rings.

Volume discounts: discounts offered in return for bulk orders or buying high volumes over time.

Likewise, it assumes that all suppliers and sellers compete on a level playing field rather than in real-life circumstances, which invariably incorporate advantages and disadvantages for each individual player. Trade tariffs and import duties, in particular, bring about differing conditions between suppliers. In practice, humans do not simply buy the cheapest product unless the purchase is a pure commodity with no features to differentiate it from the competition. However, even with commodities such as petrol, the application of brand characteristics and the image and service levels of the retailer often result in customers preferring particular suppliers. Such preferences become significantly magnified as the products or services being considered are more differentiated and often when the decision is a higher value, higher risk, and less frequent. The choice of wedding venue, for example, is one of many where customers lose much of their price sensitivity due to emotion, the demands of others in the decision-making unit, and the expectations of society and other social units. Finally, classical theory does not reflect the fact that customers often lack information and time. A lack of information (or perhaps, conversely, a burdensome encumberment of information with which to make a decision) may lead to "analysis paralysis," avoidance of making a choice, or a flawed purchase decision. Likewise, a lack of time, especially in which to make a low-cost and low-risk purchase decision, may discourage the consumer from undertaking an economically literate decision-making process.

3.3.1 Elasticity of Demand

As we have discussed, products and services differ in the extent to which demand for them is sensitive to changes in price. Commodities like salt or petrol usually have low sensitivity because customers are simply unable to avoid purchasing them without a major inconvenience or change to their lives. In 2000, in the UK, many lorry drivers blockaded fuel stations and refused to replenish the stocks of bunkered gasoline, bringing widespread disruption to British roads and supply chains. The action was intended to protest the sharp rises in fuel prices experienced between 1993 and 2000, at which point drivers were paying around 80 pence per liter. Many private car drivers, although inconvenienced, sympathized with the sentiment. They noted that over 80% of the sales price was accounted for by taxation, and that fuel prices were among the highest in Europe. However, they were not in a position to stop buying fuel or to significantly reduce their fuel consumption, as they needed to power their cars for commuting, holidaymaking, and other social and domestic purposes. In the following two decades, fuel prices rose in the UK by approximately 50%, although this number fluctuates according to global factors (Alon et al., 2016), such as bull markets in oil-producing countries and increased demand from rapidly expanding economies. Despite the significant post-inflation price rises, there has been little sign of people surrendering their cars. We may, therefore, say that fuel is an inelastic commodity (Coglianese et al., 2017), at least until battery-powered vehicles become a more economically viable option for the majority of private motorists.

When a product or service is very sensitive to changes in price, meaning a small price increase will significantly reduce customer demand (Shiu et al., 2009), we call this elastic demand. Products and services tend to be elastic where certain factors exist:

1. If the product has a number of acceptable rival products at a similar price;
2. Where a product is of questionable quality, suffers relatively low brand esteem, or is ubiquitous;
3. When substitute products are readily available, and there is no particular switching cost to customers who change suppliers;
4. When customers can easily make direct product comparisons between competitors; and
5. When the purchase represents a significant investment to the customer and is likely to be granted a thorough decision-making process.

   It is, therefore, vital that firms consider the elasticities of their products or services when determining a pricing approach. Elasticity is usually calculated by dividing the percentage change in customer demand by the percentage change in price, expressed as a percentage. The smaller the resulting number, the less elastic the price, and the more the company is able to increase prices to boost profitability without sacrificing custom.

When pricing for people, firms are often acutely aware of their customers' tolerance to price increases. This can lead to the selection of customary pricing as a strategy when sellers strive to provide a product or service at an established pricing point, even if their own costs are increasing. To maintain profitability, rather than increasing the price, the seller instead reduces the amount or quantity of product to be sold. The rationale underpinning this approach is that their customers are more sensitive to price increases than they are to size decreases, or at least more aware. In 2018, food giant Mondelez had to revert to manufacturing its Toblerone chocolate bars in their original shape and proportions after suffering a customer backlash when they broadened the gaps between the chocolate triangles in the product. The redesign had been implemented to reduce the weight of a chocolate bar from 170 g to 150 g, thereby enabling Mondelez to maintain prices around the £1 mark in the UK. However, loyal customers noticed the shrunken nature of their much-loved treat, the change became a common talking point, and Toblerone suffered the negative effects of Word of Mouth (WOM) and Electronic Word of Mouth (EWOM) communications. When covert product amendments are detected in this way, customers may feel cheated by what they regard as a disingenuous and underhand tactic (Guardian, 2018). Of course, sometimes, a provider may wish to sell a product or service at a higher price to maximize profitability during a period of brief opportunity. In February 2022, Peterborough United, a football team playing in the third level of the professional league pyramid in England, found themselves in the fifth round of the FA Cup, the world's oldest surviving knockout football competition. Drawn against world heavyweight Manchester City in their limited capacity stadium, Peterborough's directors announced price rises to non-season ticket holders of £30–£42. However, over 2 weeks before the match was due to be played, they were forced into a climbdown, adjusting those prices down to a range of £23–£36. Peterborough's owner, Darragh MacAnthony, expressed regret at the initial decision and a desire to rebuild bridges with loyal fans who had perceived that they were being exploited (BBC Football Website, 2022). Naturally, it is not possible for a football club to reduce the length of a match or cut the number of players it fields for a game, but the case is illustrative of why firms like Mondelez may shy away from highly visible price increases.

3.4 Pricing Strategy Within the Organization's Broader Objectives

Just as the business objectives of an organization should align with the company's vision and mission, the choice of pricing strategy should align with the broader marketing objectives and business requirements of the firm. A company's objectives are dictated by numerous factors such as financial strength, positionality against the competition, age of product range, and market opportunity. During the 2007 to 2009 global economic crisis, a number of previously aggressive organizations decided to take a step back, reduce their surplus stock, cut overhead costs, look after their cash flows, and reduce investment in peripheral activities and enterprises. Their intention was simply to survive, and to await better times when they may be prepared to re-establish a more aggressive stance in the marketplace. An organization which finds itself in that position is likely to lower prices to encourage sales and liquidize as much stock as possible, assuming that there is sufficient price elasticity for its products. However, it also needs to protect its brand prestige and the brand expectations of its customers to avoid distressing the brand and thereby damaging its profit potential in the long run. If an organization has a new product or senses that a competitor is weak, it may decide to pursue a strategy of profit maximization, at least for a short period of time. If an organization considers that it has an attractive product range and is enjoying strong brand awareness, then it may decide upon a strategy of **market share leadership**, or at least to increase its market share while it is prominent in the minds of customers. However, an organization which builds a strong reputation based upon the strength of its offerings, and maybe the robustness of its products, is more likely to pursue a **product quality leadership** strategy, eschewing volume aspirations to reinforce its brand strengths (Olbrich et al., 2017). It is critical that the overall pricing strategy selected for an organization is consistent with its broader business objectives. Furthermore, decisions appertaining to price strategy must account for considerations encapsulating the design, distribution, and promotion of products, if they are to produce an efficacious marketing program.

3.5 The Three Major Approaches to Pricing

The three major approaches to pricing are normally considered to be cost-based pricing, customer-based pricing and competitor-based pricing. **Cost-based pricing** is preoccupied with manufacturing and distribution costs rather than prioritizing customer pricing needs. When implementing a cost-based pricing strategy, organizations normally start the process with the product, agree upon its production cost, and put into place a price which covers the costs and delivers profit while conveying value to customers. **Customer-based pricing** involves offering products or services to customers at the lowest viable price while maintaining a healthy profit. Firms using this strategy must remember to account for customer needs and wants. These may, of course, include a need for reassurance of quality through pricing rather than simply the lowest price possible. Firms which embrace this approach typically begin the process by considering their customers, ascertaining what value they must provide to those customers, agreeing upon the price which those customers are willing to pay for that value, and extrapolating backwards to ascertain the cost which they should target when producing the goods or service, and then proceeding to the product or service development stage. **Competitor-based pricing** is mindful of the actions of competitors and the effects which they have upon the organization. It incorporates a number of sub-strategies, most notably **meet-the-competition pricing**, in which competitor prices are closely reflected, **undercut-the-competition pricing**, in which the firm leverages price elasticity to gain customers, and strategies which involve pitching the firm's prices a certain percentage above the competition to reflect superior quality. **Price leaders**, the firms who have the lowest costs per unit due to economies of scale, are able to sell a product at the same price as competitors while maintaining a larger margin. This is akin to fighting a battle with a larger army and more machinery, as it enables the firm to attack more effectively when desirable, or to weather an onslaught from rivals. We shall now look at the three major pricing strategies in more detail.

◘ Price matching. Image source: ▶ pexels.com/royalty-free-images/

One of the UK's "Big Four" supermarkets, Sainsbury's, attempts to repel the growing popularity of a cut-price German competitor, Aldi, by pledging to match prices on any identical products sold there. Supermarket pricing is highly visible to competitors, and competitor pricing intelligence within the FMCG sector is extremely comprehensive and constantly updated. Therefore, when a supermarket makes this pledge, it has almost always proactively ensured pricing parity (or better) on all comparable products. Therefore, it serves to challenge customer perceptions that Aldi is cheaper, as Sainsbury's clearly believes that price elasticity could encourage customers to change supermarkets. Effectively, Sainsbury's is not challenging its customers to scour Aldi for cheaper products but saying, "There's a lot of noise about Aldi, but there's really nothing to see, and here is our proof – so stay with us!" It enables Sainsbury's shoppers to remain loyal, freed of the uncomfortable suspicion that they may be missing a better deal. It is worth noting that Aldi's low pricing is usually limited to own-brand products which are not stocked by its competitors and cannot, therefore, be subjected to direct price comparisons (◘ Fig. 3.1).

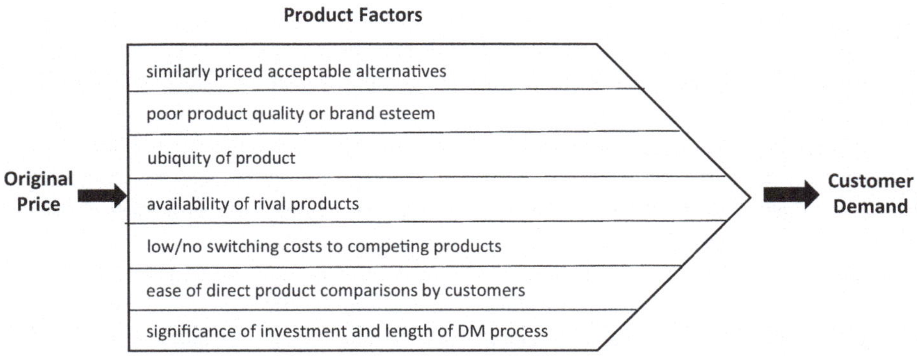

Fig. 3.1 Factors contributing toward price elasticity

Economies of scale: the extents of production and production costs, which increases savings from production costs by increasing the production capacity.

Manager's Corner 3.1: Determining the Price of Motorcycles to Enter a New Mar ket

Source: ► pexels.com/royalty-free-images/

Imagine that you are the marketing director for a US motorcycle manufacturer who wishes to enter the UK market after an absence of three decades. The firm is perceived as a heritage brand which harks back to a golden era of motoring in the 1950s and 1960s. As such, you do not intend for the brand to attack the mass market, but to trade upon its perceived prestige among older and less price-sensitive enthusiasts. You plan to sell these motorcycles through franchised channel intermediaries.

Initially, you will appoint 12 dealerships in major regional centers of the UK. It is worth noting that the motorcycles are already sold in the manufacturer's country of origin and in other key markets such as France, Germany, Italy, and Spain. You realize that setting an appropriate price for the first model to be launched in the UK is paramount to its chances of success. Too expensive, and motorcycle enthusiasts are likely to stay with more established brands which represent a lower risk. However, if the price is too low, customers will not perceive the brand as prestigious, brand associations with heritage will be undermined, and they will mistakenly perceive the product to be of inferior quality. It seems that you must adopt a customer-based pricing strategy, but how exactly? Perhaps a combination of sub-strategies which complement each other? Suggest one or two initiatives around which you would build your approach to this new market, providing your rationale for each and explaining the potential benefits and limitations of such an approach.

3.6 Cost-Based Pricing

When choosing a cost-based pricing strategy, the starting point is to establish a break-even volume and break-even price. The **break-even volume** is the number of items which the firm must sell to recover its costs and start generating a profit. It is calculated by dividing the fixed cost by the "price minus the variable cost." To do this, we need to understand the three basic types of cost:

(i) **Fixed costs** (e.g., factory rent, managers' wages, manufacturing machinery, marketing costs, heating, and lighting) are not affected by how many units the producer sells;

(ii) **Variable costs** (e.g., product components, raw materials used in manufacture, product packaging) increase with every additional unit sold;

(iii) **Total costs** are the sum of fixed and variable costs.

Cost-based pricing: setting prices relative to the costs of producing the product or service

An example of break-even volume is given here.

A landscape artist has fixed costs of £4000 per month (for things like studio rental and marketing).

She sells each painting for £240.

The variable cost of each painting (e.g., her oil paint, canvas, and frame) is £80.

Therefore, £4000 fixed cost/(£240 – £80) = £4000/£160 = 25 units.

In this example, the artist knows that she must sell at least 25 units per month just to cover her costs. After this point, she will start to earn profit. The profit will be the price minus the variable cost. Therefore, after the first 25 paintings are sold, she will be accruing £160 profit for each additional painting sold that month. If she wishes to begin making a profit earlier each month, she has two choices. The first is to raise the selling price of her paintings, which is dependent upon what customers are prepared to pay. The second is to lower her total costs, which may be achieved by lowering her fixed costs, variable costs, or a combination of the two. However, this is only possible if her suppliers are willing to grant a lower price or if she is able to secure a cheaper supply from one of their competitors.

- **Fixed costs:** costs which do not change according to the number of products sold, appertaining to overheads like factory rent, machinery and management wages.
- **Variable costs:** costs which increase as more products are manufactured and sold.

Now, let's assume that our artist is able to produce 100 paintings per month, and she has set that as a sales target for herself. She already knows that her monthly fixed costs are £4000 and that the variable costs attributable to each painting are £80. She wishes to ascertain the **break-even price** of each painting if she sells 100 each month. To do this, we divide the fixed costs by the unit sales and add to the variable cost.

Therefore, £80 variable cost + (£4000 fixed cost/100 unit sales) = £80 + £40 = £120.

In this example, the artist knows that if she sells 100 paintings per month, she must sell them at a minimum of £120 each to break even. With this knowledge, she is able to decide upon a sales price which will generate an appropriate level of profit. If she decides to sell each painting for £180, then she will realize £60 profit per painting, which represents a profit margin of one-third of the selling price. However, this calculation is contingent upon her selling exactly one hundred paintings per month.

- **Break-even price:** the price above which a sold item will make a profit, but below which it will produce a loss.
- **Break-even volume:** the volume of sales above which a sold item will make a profit at a predetermined price, but below which it will produce a loss.

Cost-based pricing is often used in organizations which are driven by accounting and finance decisions. In particular, such firms often adopt **cost-plus pricing**. This is simply covering one's costs before adding a predetermined amount of profit. The calculation of costs takes into account considerations such as product manufacturing costs, overhead costs, research and development costs, and the costs associated with transporting, storing and selling the product. By adding these together and then adding a fixed profit percentage, the firm arrives at a price. Firms taking a slightly more aggressive stance may choose **target profit pricing**. This approach requires the firm to calculate the number of units which it believes it can sell during one accounting period and then ascertain the difference between total costs and revenues received at this sales volume. Larger target profits are achievable by selling more units, reducing cost levels, or a combination thereof. In some industries, especially retail, **mark-up pricing** is very popular. A hat store buys a hat at the bought-in price, adds a predetermined percentage—the mark-up—and from this calculation derives a shelf price or selling price. The mark-up is always expressed as a percentage of their *bought-in price* (i.e., the price at which they bought the product), whereas firms using profit margin, instead of mark-up, express margin as a percentage of their *selling price*. Therefore, a mark-up of 100% is the same as a margin of 50%. Such an approach is popular with supermarkets and also retailers who sell a wide variety of product lines and lack sufficient time to price each line and product individually. Many product categories have a standard industry mark-up. Larger retailers rely on their suppliers to undertake market research and suggest viable selling prices.

- **Cost-plus pricing:** covering the cost of producing the service or product, then adding a predetermined profit.
- **Mark-up pricing:** adding a predetermined percentage onto the bought-in cost of a product before selling it for profit.
- **Margin:** the amount of profit on an item, expressed as a percentage of the selling price.

There are several limitations to pursuing a cost-based pricing strategy. First, it assumes that customers perceive that the product represents good value. If this is not the case, then even cost-based pricing, which is very mindful of price elasticity, would not be successful. Second, if customers believe that the product is exceptionally valuable to them, the firm risks creating such strong demand that they are unable to fulfil all the orders due to stock shortage. When this happens, it becomes easy for competitors to step in, benefit from the original firm's ideas and experience and take the surplus customers away. Third, many senior managers cannot resist the temptation to artificially overstate their costs to encourage or force their market-facing managers to set higher prices.

A further limitation of cost-based pricing strategies is presented by the **experience curve**, also sometimes called the **learning curve**. This model accounts for the decrease in costs which may be enjoyed by firms which have accrued significant experience in designing, making and selling a product. Imagine that our landscape artist has bought a factory so that she can make prints of her paintings. Her factory is designed to print 1000 paintings per week. If she produces fewer than this, then the cost per painting will be higher. If she produces approximately 1000 prints per week, her average costs will begin to decline because the fixed costs (factory rent, machinery, managers' salaries) are spread across a greater number of units. If she attempts to produce and sell more than 1000 prints per week, her average costs may increase because the factory may become inefficient as workers and machines become overstretched. Therefore, if she believes that she could sell 2000 prints per week, she should move to a larger factory and invest in more machinery so that she can run efficiently at this level of manufacturing. If she does so, producing 2000 prints per week would, therefore, cost less per print than if she were printing 1000 prints per week. If she acquired a plant which could achieve 3000 prints per week, that may bring even greater efficiencies. However, a plant which could produce 4000 prints per week may be less efficient because she would be paying for excess capacity. This limitation would be caused by these economies of scale, such as an accumulation of extra administrative work and the management of greater numbers of staff, both of which may slow down progress. Let us imagine that the painter moves to a plant which can produce 3000 prints a week. As her production workers become experienced in printing the paintings, they develop the capacity to do so more efficiently—for instance, by figuring out more streamlined processes and eliminating unnecessary processes. Given this higher volume, the factory becomes more efficient and enjoys economies of scale. Therefore, the average cost of each print decreases with accumulated production experience. Naturally, there must be enough customers willing to buy the prints to justify such a strategy.

To achieve the benefits discussed in the above case, manufacturers tend to price their products slightly lower than necessary, anticipating price elasticity, to increase sales volumes and produce lower costs through experience, which then facilitate subsequent price reductions. While the experience curve can be a powerful benefit to an organization which consolidates its position of acquired strength, it is not without its disadvantages. By selling the product slightly cheaper than would otherwise have been the case, the product image may be undermined and the brand distressed. Additionally, competitors are unlikely to sit back and let the firm take control. It is more probable that they will respond with their own price cuts, or perhaps by using lower-cost technology, which allows them to undercut the competition in return.

Distress a brand: to undermine customer confidence in a brand inadvertently, and therefore, the firm's ability to charge a high price through the use of poorly conceived marketing initiatives, usually focused on deep discounts.

3.7 Customer-Based Pricing

Customer-based pricing and value-based pricing are two very similar concepts which overlap significantly. By having a strong customer value-based proposition, organizations prevent the commoditization of their products, differentiate them from those of the competition (Davcik & Sharma, 2015) and are able to instill brand loyalty and build profit margins. As we have noted, customers rarely buy the cheapest product available on the market. There are very few people who buy the cheapest house in their town, go on the cheapest holiday possible and drive the least expensive car on sale within the local area. Instead, customers seek a combination of acceptable pricing and the delivery of value. Customer value is often understood in terms of the benefits which a product or service bestows (Payne et al., 2017). Therefore, customers are willing to pay a relatively higher price if they feel that the product or service offers superior value (benefits) in relation to their needs, wants and expectations. Although there are many types of benefits, we can categorize them into a small number of commonly recurring benefits. Products may save us time, save us money, save us space, save us effort, enable us to live our lives more efficiently, help us to feel good, help us to look good in front of others or bestow social status upon us. Examples of brands which convey these benefits effectively are shown in ◘ Table 3.1.

Assessing value can be difficult for sellers and buyers alike. Equating value to an equivalent cash value and accounting for it within a pricing structure can be even more problematic. However, this is a necessity if the firm is to price strategically and build a unique selling proposition within its brand. Often, firms need to sell products which are more expensive than those that the customer is already using. An example of this might be LED ceiling lights. Perhaps a salesperson may need to convince a local government authority to buy 10,000 LED ceiling bulbs to replace existing halogen bulbs in apartments and homes which it manages. Not only are all the proposed bulbs much more expensive than those provided by the incumbent

◘ Table 3.1 Examples of brands which focus on specific customer benefits

Customer benefit	Brand
Saves time	Dyson bagless vacuum cleaner
Saves money	Primark clothes
Saves space	Ikea furniture
Saves effort	TripAdvisor
Makes us efficient	Apple iPhone
Makes us feel good	Dove soap
Helps us look good to others	Louboutin shoes
Bestows social status	Mercedes car

supplier, but to be successful, the salesperson must persuade the decision maker to remove existing bulbs, which have residual value remaining in them while they are functional. This may sound like a difficult task, and so it may well be, but the salesperson can convey a number of powerful benefits to the government-employed buyer. The LED lights save a significant amount of electricity over a period of time. It is possible to calculate how long it would take for each bulb to pay for itself and begin generating savings. LED lights also emit less heat, last longer, and are safer as they pose a greatly reduced fire risk. The much longer life obviates the need to have them regularly replaced, thereby reducing maintenance costs. As we can see, buyers are often happy to replace a cheap product with a more expensive one if the benefits are readily perceivable and relevant to their needs. Of course, there will always be customers who do not appreciate value and instead choose to focus overwhelmingly on price. However, these customers may be so costly and problematic to keep happy that it is often preferable not to win their business in the first place.

Customer-based pricing (or Value-based pricing): shaping the firm's prices according to customer demands and needs.

Web Exercise 3.1

The Autotrader website is a useful resource for people who are planning to purchase a car or other vehicle. It enables them to search nationwide, and to apply filters to the search results to locate vehicles which match their purchasing criteria and fulfil their needs. For example, customers may filter the results by vehicle location, price, fuel efficiency, luggage space, fuel type, gearbox type, age, mileage, and many other important parameters. Imagine you are a private buyer who, 10 years into your career, wishes to upgrade to a more comfortable and modern car. Access Autotrader's website—► www.autotrader.com (Autotrader (USA), n.d.) in the United States, and ► www.autotrader.co.uk (Autotrader (UK), n.d.) in the UK—and search for an appropriate car within a reasonable distance of your location. Which filters did you apply, and which were the most important to you? Often, the most important filters are those which we apply first. What has this told you about your price expectations when buying a car? To what extent did you focus on the purchase price of the car, and how important did you consider insurance costs, fuel efficiency, servicing costs, and insurance premiums? Do you feel that you weighted these factors to reflect their relative significances? Explain your experience to a friend.

Let us now consider individual customer-based pricing approaches in more detail.

3.7.1 Good-value Pricing

Good value pricing entails combining good service quality and good product quality at a competitive, fair price. Sometimes, it is necessary for a brand to create a more affordable sub-brand, such as the Hampton by Hilton hotel chain, to capitalize upon a more price-conscious market segment. Brands which do this often find that they are sending very different messages to different customer segments, and for this reason, it is important to establish a difference between the brand and sub-brand, which customers can readily perceive. For example, under Hilton as the main corporate brand, they have two sub-brands: Hampton by Hilton and DoubleTree by Hilton. Marketers, as the architects of their brand (and sub-brands), need to be cautious here, as the cheaper sub-brand may sometimes cheapen brand perceptions of the original brand, undermining its brand prestige, ability to price higher, and profit potential.

Good-value pricing: charging a price which is noticeably attractive to customers, representing perceived good value, often through the use of budget sub-brands.

3.7.2 Everyday Low Pricing (EDLP)

Walmart and its UK subsidiary, Asda, are salient examples of organizations which use **everyday low pricing (EDLP)**. EDLP is a type of value pricing in which, rather than resorting to temporary discounts, the firm instead maintains constant low pricing. This encourages customers and other public to associate the brand with unconditional low-price promises and good value for money (Diallo et al., 2015). Customers know that they do not have to defer purchases or undertake private research to ensure they buy from the firm at the cheapest possible time. Instead, they know that the best deals are there whenever it is convenient for them to purchase, and this produces brand loyalty. To pursue an everyday low pricing strategy, a firm must enjoy everyday low costs.

Everyday Low Pricing (EDLP): using permanently low prices instead of sporadic discounts.

EDLP stands in stark contrast to **high-low pricing**, in which prices fluctuate significantly. Its principal limitation is that, by acculturating customers to low prices, it is very difficult for the firm to then increase its prices and, therefore, its profit margins. The more price-conscious a customer is, the more likely they are to look around for alternative products.

3.7.3 Value-Added Pricing

Firms use **value-added pricing** to increase the power which they derive from their pricing strategies. By emphasizing value and charging a premium for it, they gain control over the price-setting process rather than simply having to react to market forces. It also allows them to differentiate their products from those of the competition, thereby avoiding a direct price war which erodes profit margins (Sotgiu & Gielens, 2015). Often, the value proposition is a value-added service. For example, an extended warranty, industry-leading technical advice, or an enhanced recycling program for used goods would likely entice customers to buy at a higher price point rather than making direct comparisons on price with competitors.

3.7.4 Product-Mix Pricing

Product-mix pricing encompasses a number of pricing strategies used by organizations which market two or more products. The five main types of product mix pricing strategies are as follows: (i) product-line pricing, (ii) optional product pricing, (iii) product bundle pricing, (iv) captive product pricing, and (v) by-product pricing. We consider each next.

Product-mix pricing: a range of pricing approaches which consider the firm's products in the context of each other.

Product-line pricing is popular with organizations which sell a number of different products, such as motor manufacturers. Rather than making products which compete with each other and therefore cannibalize each other's sales, the organization employs a stepped system of pricing as it goes up the product range. Effective product line pricing should tempt customers to consider products and specifications which are higher up the range and, therefore, more profitable to the firm. For example, a motor manufacturer may make a small car, a medium-sized car and a large car. Ideally, the pricing for these three models should not overlap, or if they do overlap, they should only do so minimally. Any overlap would compromise the perceived uniqueness of each model. Therefore, careful pricing not only covers costs and generates profit but also shapes customers' perceptions of the brand and its individual products. Now, let us suppose that for each of these three car models, the manufacturer offers three different engine types and three levels of trim. The engine types are petrol, diesel, and electric, and the trim levels are L, LX, and GTX. To move from petrol to diesel, or diesel to electric, costs the customer $2000. To move from L to LX, or LX to GTX, costs the customer $3000. By operating such a system, each product has its own identity, and the customer can clearly see the benefits of upgrading to the next trim level or specification. It also helps to imply a fair rationale for the company's pricing and, therefore, builds customer acceptance of the strategy.

Product-line pricing: basing products' prices upon their relationships with other products in the seller's product range.

Optional-product pricing is the offering of non-compulsory options as specifications on larger products—for example, additional memory in laptops, or leather seats in cars. The selling firm enjoys an additional profit opportunity while safeguarding the competitive price point provided by the base product.

Optional-product pricing: pricing non-compulsory items, products and accessories to tempt customers who are making a larger purchase.

Product-bundle pricing is an approach used when firms offer a combination of products together for a deal at a cheaper price than a customer would pay for the items individually. Perhaps the most salient example of this is McDonald's Happy Meal, comprising a burger, fries and soda. Although, by offering such a deal, the seller reduces the profit margin for each of the individual items, she is likely to sell more items in total, thereby producing a higher overall profit. Additionally, firms are able to encourage customers to buy the most profitable items rather than just the ones which they had intended to purchase. For instance, although people go to McDonald's for burgers, they are tempted by fries and invariably end up buying soda. It is these oversized cartons of soda which carry the highest margin and use the least staff resources. Product bundles offer sellers a major benefit too, as they help to sell items which are purchased less frequently, thereby reducing wastage, especially in the case of perishable goods like food (Apostolidis et al., 2021). Naturally, the strategy would not exist were it not for a perceivable customer benefit, as in McDonald's Happy Meal, that it represents a more complete solution to their immediate needs at a competitive price. At its best, product bundle pricing should produce this kind of win-win situation, but to retain high school standards, sellers must beware not to sell too many unsolicited products which customers are unlikely to use and which, therefore, contribute to unnecessary manufacture and consumerism at the expense of the planet.

Product bundle pricing: offering several products at a lower price than would be charged for them individually.

Captive-product pricing is the approach in which the main item is sold relatively inexpensively because the seller knows that that item will need further purchases of more profitable goods to function as required. Perhaps the most obvious example of this is home printers, which are relatively inexpensive considering the technology they encompass but which invariably require ink cartridges which are proportionately much more expensive and need frequent replacement. It is common for manufacturers to design their machines to require components which are not compatible with the machines of competitors. Sometimes, this is for legitimate performance reasons, but on other occasions, it is perhaps a less ethical ploy to increase profit (Nagle & Müller, 2017).

Captive product pricing: pricing a product keenly in the knowledge that it must consume other, more profitable products to function as desired.

By-product pricing occurs when the manufacture of one product leads to a second product, which is superfluous and considered a by-product. The manufacturer is then able to price this by-product to provide additional profit, which helps to underwrite the cost of the main product. This can help the manufacturer to keep down the price of the main product, making it more competitive in the marketplace. Examples of by-products include sawdust from the making of wooden furniture, yeast extract from the brewing of beer, and feathers from the processing of poultry. The popular British product, Marmite, is a yeast-based spread made from the by-product of the beer brewing industry. Manufacturers of down jackets and sleeping bags may be able to fill their products with feathers, which are a by-product of poultry processing, therefore ensuring that no additional creatures have died to make the products.

By-product pricing: pricing a by-product in such a way that it supports the competitiveness of the primary product from whose manufacturer it was derived.

3.7.5 Customary Pricing

Customary pricing is an attempt by a firm to maintain a pricing level which is established in the minds of their core customers. As we have seen, Toblerone and other products have used customary pricing to varying degrees of success. To decide whether customary pricing is the most appropriate strategy, the firm must ascertain the elasticity of pricing and, therefore, how many sales will be lost for a specified increase in the selling price. Naturally, as costs of raw materials, distribution, and staffing increase, then it may become unavoidable to raise prices.

Customary pricing: striving to continue charging customers at an established price point.

Holding price points. Image source: ► pexels.com/royalty-free-images/

Sainsbury's understands that during a period of inflation, many of its customers are facing financial pressures and are therefore likely to be more price-sensitive, either shopping around multiple supermarkets for the best deals on specific items or choosing not to buy them at all if they are deemed to be unnecessary indulgences. In this photograph, Sainsbury's is telling customers that it has kept the price of this product at £1, protecting something of value from inflationary pressures to enable customers to continue buying it. In doing so, it is communicating that it understands the customer's situation, is on the side of its customers, and is making economic sacrifices of its own to act like a supportive partner to its customers. Although the signage may say "Price Lock," it conveys two much more personal messages to shoppers: "It's ok, you can still include this in your weekly shop, so go ahead!" and "We're here for you through rain and shine (so let's stay loyal to each other)." It is worth noting that Sainsbury's has chosen not to price the item at 99p, which is a common psychological pricing approach because in this instance, £1 is a more powerful message as it explicitly refers to a price which is likely to be a customer price point (above which the customer would not buy the product). In other words, "We know that £1 is your limit for this item, and we've managed to keep it at £1 for you."

3.7.6 Psychological Pricing

Psychological pricing leverages customers' responses to certain pricing characteristics and changes. It is almost always used in tandem with other pricing strategies. The most commonly experienced example of psychological pricing is when a supermarket product is priced at $0.99 or 99p instead of $1 or £1. It is thought that most customers unconsciously perceive a product priced in this way as much more than one penny cheaper. In fact, it increases sales by approximately 8% compared with what is achieved at a level dollar or pound. This is because customers perceive the giving back of change by the cashier in a similar manner to a discount or rebate (Boz et al., 2017).

Psychological pricing: a range of pricing tactics aimed at manipulating customer perceptions of the brand and the competitiveness of its pricing.

Other forms of psychological pricing include **odd-even pricing**, in which the seller mixes odd and even numbers to give the impression of precisely calculated prices which are appropriate to the value of the product, and **prestige pricing**, which prices a product more expensively than is necessary for the firm's profit aspirations, to build customer perceptions of prestige.

3.7.7 Second-Market Discounting

Second-market discounting occurs when a firm sells a product or product range at different prices according to which market it is being sold into. For example, an automotive manufacturer may sell cars at a higher price in Western Europe and at a lower price in the Middle East and North Africa. It may choose to do this because it realizes that customers' ability to pay Western European prices is not commonplace in those secondary markets. Therefore, the firm makes a pragmatic decision to accept smaller profits rather than negligible sales in those markets. Naturally, it is essential that the firm does not lose money there; therefore, it would probably sell the vehicles with a lower specification. For example, although all cars in North Africa would require climate control, the manufacturer may be able to sell cars there without blind-spot warning indicators, heated wing mirrors, rear window wipers and reversing sensors. A potential problem created through second-market discounting is the likelihood of goods being transported back from the second market to the primary one for resale. Such **gray imports** are likely to alienate existing customers in more affluent markets, as their more expensive purchases are subsequently found much more cheaply. It also creates customer uncertainty over price, undermining confidence and losing sales. For this reason, it is wise when using second-market discounting to adapt high-value products to the second market in such a way that they become unsuitable for sale in the primary market (e.g., through inappropriate specification levels or appearance). Sometimes, exporters must sell a product cheaper in one specific market as a reaction to localized competition.

Second market discounting: charging a lower price to a secondary market segment for the same product or service.

Another form of second-market discounting is to sell the same product or service to different customer segments in the same place. An example would be an art gallery giving student discounts for daytime visits, or perhaps a restaurant which gives discounts to pensioners and retirees who wish to eat there on weekday lunchtimes or other less popular hours. By employing this strategy, the restaurant is better placed to cover its fixed costs, which it would incur regardless of this extra custom. In doing so, a restaurant is able to generate incremental income and to spread its commercial risk across a broader demographic. By distributing the loading on the restaurant more thinly, staff and other resources are thereby used more effectively, rather than in a peak-and-trough pattern.

3.7.8 Price Skimming

Price skimming is a popular strategy among firms which sell technologically advanced products. We can consider its usage by placing it in the context of Rogers's (2010) Diffusion of Innovations model, which explains when, in the Product Life Cycle, different types of customers purchase products or engage with brands (Mitchell & Clark, 2019).

Price skimming: charging very high prices in the earlier stages of the Product Life Cycle, then gradually lowering them to sell to progressively more price-conscious customer segments.

It involves selling newly launched products at a high price to **innovators**, the first 2.5% or so of the customer market. These innovators are highly motivated by having the latest technology, are often extremely knowledgeable in their chosen product categories, and play an important role in testing new products and giving feedback to the firm. We saw that there were long queues of customers when the iPhone was launched for the first time in January 2007, as these customers wanted to buy the new technologically enhanced smartphone before anyone else could buy it. They are willing to accept the risk of the product being relatively impractical to be at the forefront of technological progress. To be a viable strategy, price skimming requires that the product has a significant technological advantage over its competitors, even though this advantage is likely to disappear in a short period of time as competitors catch up by embedding those innovations in their own products. For example, iPhones are sold at a relatively lower price today in comparison to when originally launched, as Samsung Galaxy and many other competitive smartphone brands started to offer the same or similar value that is offered by iPhone. Price skimming would not work with commoditized goods or ones which are relatively undifferentiated (Du & Chen, 2017).

Innovators: customers who buy products (or a category of product) very early in the Product Life Cycle, typically at launch, through pre-launch orders, or immediately afterwards.

After the firm has maximized its profits and recouped many of its research and development costs by charging a high price to innovators, it runs out of these customers and must then start selling to early adopters. **Early adopters** make up around 13.5% of the total market and are usually motivated to buy and own modern, stylish and fashionable items. Although they are willing to embrace an element of risk and pay a high price for innovative products, they are much more risk-averse than innovators and more price-conscious when buying technology (Jahanmir & Lages, 2016). They are typically people who like to have the latest model of car on their driveway, to wear branded clothes, and to enjoy eating out and foreign holidays. In short, they are quite materialistic and enjoy the respect and admiration of many of their peers. To trade with these customers, the firm must drop its price significantly from what it charged innovators. Early adopters are absolutely crucial when establishing a product in the marketplace, as they communicate the benefits to their peer groups. However, after the early adopters have bought the product, the firm then needs to start selling to the next customer segment.

Early adopters: customers who buy products (or a category of product) early in the Product Life Cycle, during its growth stage.

The next customer segment is the **early majority**, which makes up around 34% of the overall market. These customers are part of the mass market and may be considered broadly "typical." They are willing to pay slightly more, but not much more, than the average price of a product over the course of its entire life cycle. They are less image-conscious, more price-conscious, and more risk-averse than early adopters, and invariably wait for new products to become established in the marketplace before committing themselves to purchase. This is largely because they wish to hear peer feedback, and to see how the product fares when subjected to usage by other customers.

After this 34% segment is exhausted, the firm then has to drop its prices once more to sell to the **late majority** of customers, who also make up roughly 34% of the total market. The late majority are very similar to the early majority but slightly more cost-conscious, more risk-averse, and less motivated by owning modern goods.

Late majority: customers who buy products (or a category of product) in the late part of the maturity stage and in the earlier part of the decline stage of the product life cycle.

When this segment is also exhausted, the only remaining customer segment, which accounts for approximately 16% of the market, is the **laggards**. Laggards are extremely price-sensitive, are likely to have a very elongated decision-making process, often seek lots of product information and comparisons with competitor products, and are happy to wait several years before buying a major product to achieve a very low price and to ensure that the product is tried and tested by almost everybody else in society. Naturally, laggards represent a very small profit opportunity, and the firm may wish to discontinue selling the product rather than servicing their needs. This is especially the case as laggards are likely to suffer from post-purchase cognitive dissonance and buyer's remorse and to be very demanding of the firm's after-sales and service departments. Indeed, they may, through their custom, even make the brand less prestigious to its more lucrative customer segments.

Laggards: customers who buy products (or a category of product) very late in their Product Life Cycle.

It is worth noting that someone may be an innovator in one product category, such as mountaineering boots, crampons, and ice axes, but a relative laggard, or at least a late majority customer, in other product categories, such as cars or bicycles. However, consumers are driven by their deep psychological traits and often by their economic circumstances, so it is unusual for someone to vary drastically from one product category to another.

If we think about price skimming in the context of the Product Life Cycle (PLC), the **introduction** stage is part of the cycle in which innovators and some of the early adopters buy; the **growth** stage is when the early adopters are most likely to purchase; the **maturity** stage is one which appertains most closely to the mass market of early and late majority customers, and **decline** is the time when some late majority and all of the laggard customers tend to buy. For this reason, products are likely to require different marketing strategies throughout their life cycles, with highly technological and specification-driven communications at the beginning but reassurance of value and reliability later on in the cycle. It is possible to observe this phenomenon in action in the marketing of high-technology goods such as consumer electronics. In the third decade of this current century, as the sale of electric cars should progress from innovators and early adopters to the mass market, it will be interesting and informative to observe how marketing messages and forms of communication evolve.

3.7.9 Demand Pricing

Demand pricing is more market-oriented than any other method of pricing. It relies heavily upon up-to-date market intelligence, particularly on what customers are prepared to spend on a product or service. By ascertaining exactly what they perceive as value for money, marketers gauge where a price should be pitched to maximize profitability per unit while maintaining a healthy number of sales. It, therefore, demands detailed knowledge of the product's price elasticity or inelasticity within specific customer segments. The fall-off of customers—the rate at which customers stop buying the product—does not increase in a straight-line relationship as the price goes up. Rather, there are a number of "tipping points" beyond which many customers' tolerances to price changes are stretched to their limits. The information underpinning demand pricing is often incorporated into price-skimming strategies. Even where the product is not highly differentiated or technologically advanced, this type of intelligence is vital when launching a product to ensure that it is released to the market at the right price point.

Demand pricing (aka dynamic pricing or surge pricing): frequently adjusting prices to reflect the level of customer demand at any given point in time.

Marketing in Action 3.1: Pricing and Differentiation of Sub-Brands

◘ Source: ► pexels.com/royalty-free-images/

Brand positioning expert, Jack Trout (2005), analyzed the differences in prices between General Motors' brands in 1921, then upon the restructuring of prices by their CEO, Alfred P Sloan, in the 1950s and 1960s, and then the 2000 prices after more recent executives had forgotten the rationale behind Sloan's original strategy. In 1921, there was a significant overlap between the pricing of several of GM's brands, such as Chevrolet, Oldsmobile, and Buick. This meant that they cannibalized each other's sales and did not have a distinct difference from each other in the market. When Sloan reorganized the prices, Chevrolet had a small range of cheap cars, Pontiac was immediately above it, Buick had a larger range which was more expensive, Oldsmobile was above that, and the premium brand in the stable was Cadillac. As a result of these product and pricing changes, customers readily understood what each brand stood for and how the brands were clearly differentiated from each other. From this policy, General Motors maintained its strong sales volumes in the US car market but, unlike before, it became highly profitable and did not need a dominant enough market share to attract government intervention and the risk of an enforced divestment of brands. However, subsequent to Sloan's retirement, subsequent executives returned to the 1921 model, with a significant overlap between the pricings for Chevrolet, Pontiac, the now defunct Oldsmobile brand, and several others. Jack Trout referred to this as "schizophrenia." In effect, he said that General Motors had failed to let each of its constituent brands have their own distinct personalities, instead wanting them to be "all things to all people." This situation was brought about largely because of financial directors and accountants who saw an opportunity to take a car model by one brand and sell it re-badged by another brand. While this may have made good short-term financial sense—leveraging on synergies within research and development, design, components, and production to "kill two birds with one stone"—it was very bad news for the sustainability of the individual brands, and General Motors became loss-making as a result.

3.8 Competitor-Based Pricing

Competitor-based pricing is, as its name suggests, a pricing strategy whose chief concern is what rivals are charging for similar products and services. It assumes that, although a product may be highly differentiated from the competition, customers will still seek to make product comparisons and make judgments on the relative value for money of rival offerings. For consumers, much of this process may be unconscious, but this makes it no less significant in shaping their purchase decisions. When deciding which type of competitor-based pricing to use or whether an entirely different form of pricing strategy should be utilized, it is advisable for the organization to first ascertain the type of market in which it is competing (Fisher et al., 2017).

Competitor-based pricing: pitching prices against competitors' prices.

3.8.1 Pricing According to the Characteristics of the Competition

Pure competition markets contain multiple entities selling and buying the same types of products. Each buyer and seller has a small impact on the market price but cannot achieve dominance over other buyers and sellers (Cartelier, 2018). Marketing activities, such as research, pricing, marketing communications and product development, are relatively unimportant, as buyers have predetermined ideas regarding acceptable prices, and sellers would not be prepared to sell at a loss.

Pure competition: markets or product categories with many buyers and sellers, none of whom are dominant.

Monopolistic competition markets also have many customers and sellers, but commerce is undertaken at a broad range of price points, and the customer offers communicated by suppliers may be highly differentiated. By differentiating their products and encouraging customers to pay more money for stipulating that firm's products, each company is able to create a quasi-monopoly. Branding and marketing communications are used to differentiate products within the firm's chosen customer segments. As a result, firms are not too adversely affected by competitor pricing strategies.

Oligopolistic competition markets have few selling organizations, each of which is highly sensitized to each other's marketing and pricing. It may be the case that the products are highly differentiated, as is the case with televisions and automobiles, although they may be much more uniform, like timber or salt. The product category is likely to have very high barriers which prevent new entrants from competing against established suppliers. In a market such as this, a relatively small price cut by one firm is likely to produce a large number of conquest sales, as price-sensitive customers are attracted to switch from competitor products. To defend against such a move, competitors often retaliate by lowering their own prices or perhaps by increasing their service offerings to be equally competitive while avoiding a price war. Due to the lack of loyalty among customers, price cuts are only likely to produce temporary gains, as customers switch back to their original suppliers once non-discount pricing strategies are restored.

Oligopoly: a market in which a very limited number of suppliers compete with each other.

Pure monopoly markets have one seller only, which may be a private non-regulated company, such as Amazon, a private regulated company, such as a utility company, or a government-owned monopoly, such as a postal service. Governments often set artificially low prices when serving a large customer base through a monopoly, as they may need to subsidize less affluent customers or maybe stimulate economic activity. However, if they wish to generate profit for the state monopoly to reduce the rate of consumption, they may place prices artificially high. In any case, monopolies which are regulated, whether state-owned or private, aim to generate a fair return, setting prices accordingly. However, non-regulated monopolies are, within certain legal parameters and within the tolerances of the general public, able to charge what the market will bear. Nonetheless, they usually refrain from exorbitant pricing to avoid attracting new competitors, to mitigate the risk of government regulation and interference (Bös, 2015), and often to penetrate the market.

> **Pure monopoly:** markets or product categories in which customers can only buy from one supplier.

3.8.2 Penetration Pricing

Penetration pricing involves dropping a price by a small amount—perhaps 5%—to capitalize upon the elasticities of demand and to capture an increasing share of the market in a short space of time. It demands an intricate level of intelligence appertaining to competitors' pricing and relies upon the assumption that lower pricing will stimulate customers' perceptions of value. To undertake penetration pricing, profits must be partially sacrificed, necessitating that firms sell at a price which would not ordinarily be deemed acceptable. Time is of the essence, as the firm intends to make a significant number of conquest sales, poaching customers from its competitors, before its rivals have the chance to respond meaningfully. However, it is paramount that firms are not drawn into a sustained price war, as this may damage both the profitability and the brand image of all parties irreparably.

> **Penetration pricing:** dropping prices by a small percentage to gain market share by leveraging on customers' price sensitivity.

3.8.3 Predatory Pricing

Predatory pricing is often a gray area, ethically. In many countries, and within certain trading parameters, it may be illegal and risk heavy government sanctions. Predatory pricing occurs when an economically powerful organization triggers a price war in which it can tolerate the sacrifice of profit for longer than its weaker competitors. It is, in effect, a war of attrition in which the powerful force starves the smaller forces out of existence (Deshpande, 2018). The motivation for predatory pricing is to eliminate competition, thereby producing a monopolistic situation from which the powerful company can profit. Once the major competitors are eliminated, powerful firms invariably raise their prices above what they would otherwise have charged if there had been healthy competition, safe in the knowledge that customers cannot migrate to a rival supplier. This lack of competition may be extremely damaging for customers and is therefore often considered to be an anti-competitive practice, which is not good for an industry or a society, considering that the supply and demand are controlled by a single firm eventually.

Predatory pricing: adopting highly aggressive prices with the intention of damaging a competitor, and perhaps even eliminating them from the market.

Manager's Corner 3.2: Avoiding a Monopolistic, Anti-competitive Situation

Source: ► pexels.com/royalty-free-images/

As the Chief Executive Officer (CEO) of a motorcycle manufacturer, you are facing an unusual challenge. The motorcycle brands which your firm operates have been highly successful in recent years. The budget, mass market, and prestige brands within your stable each have a broad range of models. All of these models have proven popular, and combined you now enjoy a market share of roughly 60% in certain major markets such as the United States, the United Kingdom and Japan. This may sound like an ideal situation but presents a major problem. Your firm is in danger of being broken up by national governments, or being prevented from operating one of its successful brands, to prevent a monopolistic, anti-competitive situation from developing. You have several options open to you. First, you may sell one of the brands.

Second you may reduce customer demand for one or more of the brands, lowering sales volumes, perhaps by increasing prices. The third approach is to ensure that your budget, mass market, and prestige brands have clearly defined product ranges which do not compete with each other. At present, the mass market is contested by your mass market brand, the most upmarket of your budget brand's models, and the least expensive model in your prestigious brand's line-up. This has produced a stranglehold on the mass market, which you may be able to alleviate by differentiating the three brands more effectively. Which strategic choice would you take, and why? Explain the various benefits and limitations of each approach.

3.9 Recent Developments in Pricing Strategy

A number of new pricing techniques have emerged or become more popular during the twenty-first century. Many of these have gathered speed due to the proliferation of digital technology and interactive tools, which enable sellers and customers to communicate directly, often in real time. Another reason for their growing popularity is that customers increasingly demand an individualized, tailored experience.

3.9.1 Surge Pricing and Segmented Pricing

Segmented pricing is when a product or service is sold for different prices to different market segments. In one business-to-business example, many motor manufacturers sell cars to police forces and commercial vehicles to government bodies and hospitals at a greater discount than would be available to other customers even if they bought in the same volume. This is because the manufacturers acknowledge that these industries are providing a service for the common good and, therefore, do not wish to profit to the same extent from them. Of course, there are also less altruistic reasons. Such organizations are likely to buy in bulk, which helps bring efficiencies. Also, if a major fleet operator runs one manufacturer's cars or vans, this sends a powerful message to other potential buyers in the area about the product's viability and reliability. Another form of segmented pricing, this time in a business-to-consumer context, is when owners of tourist attractions, museums, art galleries, and historic monuments charge students and pensioners a smaller entry fee than other adults. The rationale for this seems to be that pensioners often do not have income, having retired from employment, and students often have no income or only a precarious income from part-time work and have significant outgoings relative to their income. As there is a seemingly equitable rationale, such segmented pricing approaches are normally legal. However, a disadvantage of such approaches is that the segments chosen do not capture the members' ability to pay. Many pensioners have considerable wealth, having inherited property and paid no mortgage for many years. Likewise, many students are from affluent backgrounds and may receive regular and generous payments from their families. Contrastingly, many people who are in full-time work are classed as suffering from in-work poverty, the phenomenon of working hard but not earning sufficient money to cover reasonable living expenses. Such mismatches between provision and need are unavoidably inherent within segmented pricing approaches, but such a strategy not only seeks to tackle financial inequality or account for it but also serves a commercial benefit for the seller, recognizing the differing elasticities of price among various customer segments.

Another example of segmented pricing which we often see from multinational firms, such as Unilever, is that the price they charge for one of their brands (e.g., Lux soap in North America or in Northern Europe) is different to the price for the same product in Africa or in South Asia. This is to take into account the lower purchasing power of consumers in those developing regions. To address this broader societal challenge at a global scale and to reduce inequality in consumption, companies try to source their raw materials from the local market so that they can lower the production price and, therefore, make the product affordable to customers in developing economic markets. We see a similar example in the printing industry, especially for textbooks. Publishers tend to publish books on lower-quality paper or in black and white ink to make the books affordable for students in emerging economic markets. Such geographic segment-specific pricing not only enables the multinationals to serve the different global markets profitably but also is instrumental in making the product available for a wider range of consumers in the global society who could not afford the higher quality product priced for Europe and North America.

Segmented pricing: applying different pricing approaches according to market segment or product type.

Surge pricing, also known as **demand pricing** or **dynamic pricing**, is a pricing approach in which the seller increases the price significantly at times of high demand. A high-profile recent example of this was the taxi service Uber increasing its prices on New Year's Eve. From time to time, we see negative news stories about sellers pushing up the price of bottled water during heat waves, thereby incurring the wrath of customers. Many organizations' prices surge at least a little when demand increases. When China was experiencing a strong economic boom during the early part of the twenty-first century—and therefore buying many raw commodities to build power plants, highways, and residential buildings—demand for steel and other commodities used in the construction industry increased significantly. As producers of these commodities had a ready market willing to pay a high price, this forced up prices for everybody else. You may have noticed how diesel and gas prices dropped during COVID-19 lockdown periods before rising sharply once more when lockdowns were lifted. This is because few people were buying gas frequently during the lockdown because they were unable to go anywhere, so the oil producers were prepared to sell at a lower price. So, is surge pricing or demands pricing not simply an extension of ordinary pricing, which takes into account Adam Smith's principles of supply and demand? To a degree, yes, but the distinguishing feature is that surge pricing appears as a deliberate attempt to capitalize upon increased demand rather than a defensive measure to mitigate the negative effects of lowered demand. Thirsty customers forced to pay $5 for a can of coke during a heatwave are likely to feel as though they have been exploited cold-heartedly, whereas a firm whose prices fluctuate by 5% or 10% for an item of undulating demand is unlikely to feel such a customer backlash.

3.9.2 English Auctions, Dutch Auctions, and First-Price Sealed-Bid Auctions

Auctions have been used for many centuries for the purpose of selling items. However, they have enjoyed a renaissance due to the increasing popularity of internet-based sales. An **English auction** starts with a low bid and often a reserve price under which the seller is not prepared to trade. Potential buyers then bid against each other until only one is remaining, the others having dropped out at price points above which they find the item uneconomical. **Dutch auctions**, on the other hand, begin with a high price, which gradually descends until someone is prepared to buy at that level. **First-price sealed-bid auctions**, otherwise called **blind auctions**, are popular in a number of contexts. For example, they are often used in the selling of private houses in Scotland. The technique demands that potential customers submit an offer in a sealed envelope, and at a predetermined time, the envelopes are opened to reveal the winning bidder. The advantage of this to the seller is that bidders have no knowledge of the competitors' price bids or, indeed, the number and identity of the rivals. Auctions are often used for the sale of antiques, used cars, fish, meat, and livestock. Sellers pay a commission to the auction house in return for their services in attracting customer footfall, processing the sales paperwork, and generating excitement during the bidding process to maximize the transaction price. Sometimes, auction houses will take bids from people who are not present in the building but are placing bids either themselves or through an agent via a phone call or on the internet. Elsewhere, auctions can take place wholly online, as is the case with most items listed on the online sales site eBay. On eBay, bidders do at least have the benefit of knowing if they are the current high bidder, if the reserve price has been met, how many bids there are, and how long is remaining until the auction ends. They are also able to see what the seller has previously sold and sometimes at what price. However, this visibility is two-way, and the bidder must accept that her credentials are also visible to others. Some customers have problematic auction behaviors, such as getting over-excited in the heat of the moment and bidding more than they anticipated. This is partially attributable to the "buzz" of the occasion but also to the indirect pressure applied upon them to abandon their pre-planned financial parameters in the hope that just a small breach will secure them a worthwhile benefit. Likewise, many people who use eBay and other online auction sites have problematic behaviors and, as people use online auctions at any time of day and are able to do so in any location where they have an internet signal, they may place bids while under the influence of alcohol, or in a situation where they do not have a cool head.

3.9.3 Freemium Pricing

Freemium is a portmanteau word combining free and premium. The idea of freemium pricing is simple and appears to present a strong benefit for society. An example of a freemium service is the language learning smartphone app, Duolingo. The app is free for all users, enabling people who could not ordinarily access language education to use the site and learn foreign languages. This in itself is a highly democratizing development which removes barriers to education and gives a level of equitability in a society. To fund this, Duolingo hosts adverts for sponsors. Such adverts can be rather intrusive—for example, when a 30-s video must be watched before the learner can progress to the next lesson or quiz. To remove the advert—and to enjoy other benefits of a premium service, such as extra lives and safeguards during the knowledge quizzes—customers are able to pay a nominal monthly fee. Not only do these premium users enjoy smoother, less cluttered online experiences, but they also help to finance opportunities for those less affluent than themselves.

3.9.4 Pay-What-You-Can, Pay-It-Forward, and Built-in Donations

A small but significant minority of organizations, often in the service industry or independent artisans, invite customers to pay what they can for their goods and services (Gross et al., 2021). Sometimes they may suggest a price but state that they would accept more or less. The actual amount tendered by the customer is often not known by the seller. On the Isle of Harris in Scotland's Western Isles, rangers from the North Harris Trust invite tourists to pay them whatever they feel appropriate at the end of a guided nature walk, suggesting that around £10 might be the norm. In this instance, the ranger takes the money directly from the customers in cash, so anyone wishing to pay a smaller amount, or perhaps using a small change, may feel an element of embarrassment, albeit not intended by the service provider. Sometimes, photographers or artists might have an honesty box for buyers of small items, inviting customers to pay what they think fit. Of course, they realize that a small minority of people are likely to take advantage of this situation by not paying anything. However, they also realize that other customers will be sufficiently impressed by their trust and goodwill to pay more than they otherwise would have done, which should more than cancel out the nonpayers. Even the popular internet app, Noom, invites new joiners to pay as much as they wish for the first 2 weeks' trial before going onto a direct debit contract. In part, this may be considered as an element of a puppy dog sales approach, in which the seller provides something of such value to the customer for a trial period that they do not wish to part with it when that period ends. However, it may also be considered an act of compromise or goodwill.

To "pay-it-forward" is to make a donation (e.g., equivalent to the price of your purchase) so that the selling establishment may provide economically underprivileged consumers with charitable assistance in the future. Edinburgh-based charity "Social Bite" operates five cafes in Scotland which train and employ homeless people, helping them back into a more comfortable situation. In addition to funding housing projects through charitable donations, it also runs a "pay-it-forward" initiative so that other homeless people can call in for a free meal, which has been funded by customers (Social Bite, 2022). Other organizations build donations into their purchase prices or offer customers the choice of opting into such an initiative.

3.9.5 Tariff Lock-Ins

At the time of writing, energy prices are climbing rapidly, and many households in the United Kingdom are facing gas and electricity charges, which are 40% higher than the previous year. Many customers choose an energy provider's service, tying themselves in for 2, 3, or 4 years. These commitments, despite preventing customers from shopping around in the near future, do guarantee that customers are rewarded for their loyalty by receiving rates which cannot rise in the agreed period (Money Saving Expert, 2022). There is clearly a benefit to the supplier, inasmuch as it can predict its income much better and reduce customer churn. However, if the prices it pays for gas and electricity rise much more sharply than anticipated, and they are unable to absorb the costs by hiking prices to customers who do not lock in their tariffs, then the suppliers may be placed in an untenable situation in which they cannot deliver the rates which they have contractually agreed with tariff lock-in customers.

3.9.6 Shared Ownership

Shared ownership is not a new idea. Since the earliest dawn of humanity, social groups have been pooling their resources and sharing the spoils of victory and of the hunt. Some industries, such as the maple syrup industry in Canada's Quebec province, pool their resources of production to enjoy the use of machinery and technologies which they individually would not be able to afford (McGinn, 2018). This brings efficiencies, which drive profitability onwards. At a more informal level, neighbourhood communities may group together to compile a tool library or machine library. Instead of each individual household having to buy an electric drill, an electric sander, cordless garden clippers, and other machinery which is used only infrequently, the tool library enables households to pool their resources and borrow each other's equipment. This brings a very real environmental benefit, as fewer products have to be manufactured to service householders' needs. Of course, it presents a thorny issue to manufacturers, who realize that their opportunities for selling in volume are compromised by such arrangements. However, they are free to provide a formal alternative, renting out tools directly or through intermediaries in the same way as plant hire companies. After all, publishers of books and music have for many years been generating royalty income from books, records, and compact discs lent to municipal library members.

3.9.7 "Usership, Not Ownership"—Leases, Hire Agreements and Loans

The idea of shared ownership is very closely related to the sentiment that it is often preferable simply to use a product than to own it. A case in point is motor vehicles. A car is nowadays an expensive item, conforming to an increasing array of environmental and safety criteria. Many people baulk at the cost of purchasing a vehicle, especially as they realize that it will depreciate sharply as soon as it leaves the showroom. The finance industry adage of "if it appreciates, own it, if it depreciates, lease it" appears very apt in this context. Private individuals who reside in metropolitan areas may need cars only sporadically, such as once or twice a year for long trips and holidays. Such infrequent usage does not merit the high cost of purchasing a vehicle, so they may be more likely to spot-rent a vehicle when necessary. However, many other people who wish to use a car on a very regular basis may also choose not to purchase it outright. This is often because they do not have enough money to buy the car and own it themselves. Instead, they are likely to have a contract hire agreement or other finance plan in which they enjoy the use of the vehicle for a predetermined period of time and for a specified mileage, for a pre-agreed deposit, monthly payment, and end-of-contract "balloon" payment. These agreements and similar long-term leases remove the residual risk from customers. That is to say, the risk is that the price of used cars will drop significantly during the period of the lease and that the customer will get insufficient trade-in money from the sale of the old vehicle to be able to afford a new one. This risk is mitigated by a lease. Furthermore, many customers are prepared to pay slightly more than they otherwise would have done to remove the risk of unwelcome surprises. Unfortunately, cars can be full of nasty surprises such as collisions, breakdowns, malfunctions and other unforeseen costs. By opting for a lease product which incorporates costs attributable to servicing, maintenance and repair, the customer can buy peace of mind. Like any other form of insurance, this is a financial product and, in its crudest terms, is a kind of bet. In the bet, the odds are weighted in favor of the insurance company when the individual bets are spread across a large enough population so that the insurance company makes a profit after all claims have been paid. However, individual customers also benefit, as it effectively socializes risk by pooling it across the same large population. In effect, no individual lease customer can get extremely lucky or extremely unlucky as a result of their car breaking down or not. Leases and other types of finance are extremely useful for organizations which need to sell "big ticket" items—which are bought infrequently for large amounts of money. If a motor dealership displays a car on its lot, displaying on the windshield that it costs $50,000, all that tells passing customers is that most of them cannot afford to buy the car. However, if it is instead displayed with a lease price, which is expressed as a weekly or monthly payment, prospects are then readily able to equate this number to a proportion of their monthly income and budget. Sellers can also use finance to make customers less price-sensitive. Social psychologists have shown in experiments that students attending university enrollment sessions are more likely to enroll on expensive initiatives if the logo of a credit card company is displayed prominently at the registration desk (Cialdini, 2008). Customers, it seems, are rather suggestible to the idea of finance, and its availability "loosens the purse strings." However, by leveraging on such a psychological fallibility, sellers could lead their customers into unsustainably expensive commitments.

Web Exercise 3.2

The website, ▶ www.moneysavingexpert.com, is popular among UK consumers. Founded in 2003, and attracting over 16 million visitors per month, it provides consumers with constantly updated, comprehensive advice on savings, insurance, mortgages, utility providers, credit cards, student debt, store deals and other issues around domestic finance. The main aim of the site is to flag up deals, combat consumer inertia and encourage people to shop around for the best rates. Most financial services organizations, acknowledging the popularity of the site within consumers' decision-making processes, are proactive in providing data to the site, while a few withhold data—often to the suspicion of consumers. Look through the website, or through others such as ▶ www.moneysupermarket.com (Money Supermarket, n.d.), to gauge the impact of the site's advice upon major companies which are active in the mass market. If you were the Marketing Director for such a company, how might you shape your price to ensure that, while maintaining strong profits, your firm attracts business from the site and avoids generating negative Electronic Word of Mouth (EWOM) for any of its price points or pricing practices? Discuss two or three ideas in depth with a prospective would-be colleague.

3.10 Setting Prices

Several models have been proposed for the price-setting process, but the following eight-stage process covers all the important elements combined in them:

1. **Developing pricing objectives**. This stage uses the overall business objectives of the firm to ascertain what pricing levels are likely to be necessary to deliver those objectives. For instance, if the firm wishes to push up-market, then it will need to raise its prices to reflect its increased prestige. However, if the firm wishes to increase its sales volumes to be a major player in the market, then it may have to pursue a policy of reducing prices, perhaps through penetration pricing, to increase market share. The firm's overall business vision and mission should be reflected in its pricing strategy so that the vision and mission statement can be conveyed to the firm's target audiences through its pricing strategy. For example, if the firm's vision is "reducing inequalities" (Sustainable Development Goal number 10), the pricing strategy could adopt freemium

pricing for some of its market offerings and its price ranges so that they can bring societal benefits by enabling the underprivileged segment to equally consume their market offerings.

2. **Gauging the target market's ability to purchase**. At this point, the firm undertakes market research to ascertain income levels and available expenditures for customers in their preferred market segment. Setting a price for essential food products (e.g., bread, milk, eggs) which reflects the disposable incomes of the target market could "reduce hunger" and support "good health and well-being" in society, which can contribute to the pursuit of Sustainable Development Goal (SDG) numbers 2 and 3. In general, a key context in the socioeconomic difference between the developed and developing economic countries is that in developed economies, the price of such essential food products is established in a way which can be more affordable by the mass market, in comparison to the way its price is offered in many developing economies. Simultaneously, setting a price that is affordable by the target market can increase the profitability of a market offering based on increased sales volume.
3. **Determination of demand.** During this stage, the firm will expand upon its previous analysis by measuring how demand is likely to fall or increase as prices rise or fall. By gaining more detailed insights into the elasticity of demand at various pricing points, it should have a clear idea of how many units it may win or lose for minor price changes at different points.
4. **Analysis of interrelationships between costs, profits and demand**. To remain financially sustainable, the firm must ascertain the cost of manufacturing and selling the product in comparison to the price for which it will be sold. To do this, the firm considers both fixed costs and variable unit costs based on different volume objectives.
5. **Evaluation of competitor pricing.** No organization's pricing can exist in a vacuum. Before arriving at a price, it is necessary to ascertain what pricing strategies and price points are employed by commercial rivals. By pitching prices slightly above rivals, it is possible to communicate prestige. By pitching prices slightly below, it is possible to communicate value. To ensure economic sustainability, firms must consider the potential of new entrants and disruptive competitors joining the battle and bringing with them new approaches to pricing (Porter, 2008).
6. **Pricing policy selection**. Having ascertained the possible threats arising from competitors' pricing policies (Hanna & Dodge, 2017) and considered the strategy best suited to mitigate those risks, the firm can then select a broad pricing policy. The initial decision is a relatively simple one regarding whether to adopt a cost-based, customer-based, or competitor-based pricing policy.
7. **Pricing method development.** Now that the broad pricing policy is in place, the firm is able to drill down into the details. Specific pricing mechanisms are required to convey and deliver the pricing policy and to determine the firm's stance in the market. Often, more than one mechanism is combined to ensure that the policy is not purely cost-focused but also considers the needs of specific customer segments. An example of this would be the use of psychological pricing in tandem with other approaches.

8. **Choosing a specific price point**. If the preceding seven stages have been undertaken thoroughly, the final stage should be much easier. It determines the price point at which the product should be marketed. Naturally, many products have different variables and specification levels; therefore, numerous price points will need to be decided upon for each product.

It is worth noting some of the features and limitations of price-setting processes within organizations. Perhaps the most significant is that prices are often set at the director level, and it is not unusual for marketers to be excluded from the final decision, undermining the extent to which the decision reflects current market demands and needs. In industries where prices and costs are constantly fluctuating, such as the oil industry, insurance, and airlines, firms may operate an entire pricing department, which in turn reports to senior executives. Other actors who typically influence prices and the price-setting process within organizations are production managers, finance managers, accountants, perhaps even sales managers, and, at a micro level, salespeople who have room for negotiation during their appointments.

Firms must consider which metrics they should use to ascertain the suitability of their pricing strategy. One of the most commonly used key performance indicators is Return on Capital Employed (ROCE). This can be calculated by multiplying the Return on Sales (ROS) with the activity or the stock turn. By increasing the stock term—that is, selling the stock inventory in one's warehouse as many times as possible during the course of a year, the company is likely to enjoy much better returns on capital employed. A second metric which is in widespread use for this purpose is Return On Sales (ROS), which is calculated by dividing net profit by the sales volume.

- **Return on Capital Employed (ROCE):** the amount of money which is realized from an investment.
- **Return on Sales (ROS):** a measure of the profitability of a company's sales.
- **Stock turn:** the number of times annually that a firm sells each item in its stock.

3.11 Adjusting Prices

Just as marketers should constantly scour the horizon, gauging the external environment, to check for opportunities and threats, they should also constantly gauge changes in competitor pricing and customer demands. Following this process enables the organization to adapt and survive rather than falling behind the curve and failing to notice changes in the elasticity of demand. Organizations which fail to do this invariably fall by the wayside. The process depends upon thorough market research and the use of both qualitative and quantitative data, which should be analyzed and presented to organizational decision-makers in a format which invites strategic change.

◘ Wine mark-downs and heuristics. Image source: ► pexels.com/royalty-free-images/

Most customers know a little about wines—their favorite grape varieties and whether they prefer sweetness or dryness (i.e., a more acidic flavor). However, few customers could accurately judge the merits of an individual vineyard or vintage (year of production). Therefore, we tend to rely on heuristics when buying wine. A heuristic is a "rule of thumb" or decision-making shortcut which enables us to make an appropriate purchase decision when we are short of time, information or expertise. A common approach to buying wine is not to buy anything for less than £5 a bottle, as it is more likely to be rough on the palate, but not to pay much more than £5, as £10 for a bottle of wine would be a little pricy for many people—a price worth paying when visiting friends but not when drinking privately at home. As there are many wines priced at £5–£7, another common customer tactic is to seek a wine which has been discounted heavily into the £5–£7 range. In the example shown, Sainsbury's has not reduced the price very heavily, but by a significant percentage nonetheless, and in two stages. Naturally, this approach could backfire if the customer interprets it as a sign that the supermarket is struggling to sell it. However, it is perhaps more likely that the customer would be attracted by the combination of the high original price (thinking, "It must be fairly good if they used to charge that price") and the current discounted price (thinking, "I've managed to stick just above £5, but get something which is less likely to taste cheap"). Of course, the original price, discounted price, and discount percentage are no indicators of a wine's quality, and supermarkets are smart enough to know how customers think, but they also realize that if they trick customers into less enjoyable products, they will lose their trust. As a result, fairness prevails.

3.11.1 Allowances and Discounts

Even in a calm market in which competitor actions and changes to customer demands are relatively slow and unpronounced, it is common for organizations to make minor price adjustments during the course of normal trading. These tweaks are aimed at manipulating demand, stimulating it during periods of opportunity, and ensuring that the business and marketing needs of the firm are addressed.

Discounts and allowances are one example of this. Many suppliers, especially in business-to-business sales, offer **quantity discounts** or **bulk discounts**. For example, a horticultural wholesaler who provides flower pots to a garden center is likely to sell 1000 pots at a cheaper price per unit than 100 pots. The benefit to the garden center is that, by buying in bulk, it can enjoy a larger profit margin. The benefit to the horticultural wholesaler is that it increases its stock turn and the amount of money which can be generated by its warehouse space. **Functional discounts**, also known as **trade discounts**, may be given by suppliers to channel intermediaries in exchange for certain marketing functions or administrative tasks being undertaken. Using the example of our horticultural wholesaler, it may provide a 5% discount on all goods sold in February to its best customer if that business advertises in the local press and runs an online promotion. Such trade discounts are commonly used to stimulate marketing activities, which will generate a benefit for both the channel intermediary and the supplying firm. In effect, these are very similar to the **promotional allowances** awarded by motor manufacturers for their dealerships' full participation in marketing, advertising, and sales support initiatives. **Seasonal discounts** are also popular with horticultural wholesalers, who need to carry on selling products throughout the winter despite the lack of consumer demand for hosepipes, plant pots and spades at that time of year. If the wholesaler were to suspend trading during the winter months, this would present difficulties in sustaining a permanent workforce, the cash flow would dry up, and warehouse space would become unproductive. In order to solve this issue, the wholesaler is likely to sell products in bulk for 25%, 30%, or even 40% discount during the off-season. In this way, the garden centers take some of the stock burden from the wholesaler.

Seasonal discounts: discounts offered to stimulate sales during periods of reduced customer demand.

In industrial sales, many suppliers offer **cash discounts** for early payment. For example, a wholesaler may discount 2.5% of the invoiced price in return for settlement of the invoice in full within 28 days. This has the benefit of encouraging clients to pay in a timely manner and helping the cash flow of the selling organization.

In product categories where the supplying organization sells something which has a residual value, such as the value remaining in an old car, it may offer **trade-in allowances**. This means that it is willing to buy a customer's old product, which is to be replaced and subtract the purchase price for this from the selling price of the replacement product. As an example, a car dealership which is selling a new car for

$50,000 may agree to pay the customer $10,000 for the old car which they are replacing. This $10,000 is then subtracted from the $50,000 selling price of the new car, meaning the customer only has to pay $40,000 and has alleviated the burden of having to sell the old car. Not only does this make the new purchase more affordable and easier, but it potentially provides the selling organization with good quality second-hand products to refurbish, prepare for the market and sell at a healthy profit. It is often the case that franchised motor dealerships achieve a greater profit per unit from their second-hand stock than from the sale of new cars, although one must also take into account the profit opportunity from servicing, maintenance, repair and warranty work when ascertaining the attractiveness of such business. The seller may choose to artificially inflate the trade-in allowance of the customer's old car, as private individuals tend to have a falsely inflated sense of their own car's worth. In fact, they are likely to be more sensitive to the trade-in allowance of their old car than they are to the selling price of the new car. Therefore, the dealership may increase the trade-in allowance but adjust the selling price of the new car accordingly to absorb the extra cost.

Trade-in allowances: money subtracted from the sales price of a new product in exchange for the customer's current product which is to be replaced.

3.11.2 Price Promotions

Selling organizations, and especially retailers, are likely to use temporary pricing initiatives to stimulate customer demand, increase market share, boost sales volumes and build the overall profitability of their sales operations. These constitute promotions, and often, these promotions have promotional pricing rather than simply leveraging other elements of the marketing mix. As we have already discussed, **loss leaders** can be extremely useful in generating footfall to retail premises, showing that profit-making items are bought on an unplanned basis by customers.

Loss leaders: loss-making items which are heavily promoted or positioned prominently to attract customer footfall, which ultimately brings sales of more profitable items.

Retailers may also use **special event pricing** to attract a greater number of customers to their stores. An example of this would be a furniture store or carpet retailer wishing to generate extra custom during January or other months which suffer from sluggish sales. Special event pricing does not necessarily consist of cash rebates or cash discounts but maybe some other form of financial benefits, such as extended warranties, free maintenance contracts, or the provision of 0% or low-interest finance. It is important that these are perceived by customers as having real benefits if they are to stimulate demand and result in incremental sales.

- **Special-event pricing:** applying pricing initiatives which are intended to support a special promotional event.
- **Cash-back deals/cash rebates:** refunding money to customers upon completion of a sale, often for "big ticket" items which will be paid for by consumers through monthly repayments.

3.11.3 Psychological Pricing

Although much of the **psychological pricing** which we see is based around the permanent use of simple techniques, such as selling a product for $4.99 instead of $5.00, Psychological pricing is also often used on an intermittent basis to stimulate demand and to guard against negative effects of competitor actions. A problematic element of this is the use of oversized packaging, for instance, in products such as Easter eggs, to create the impression among customers that the product is larger and, therefore, better value than is actually the case. Not only is this ethically dubious, as it arguably manipulates customer perceptions of value beyond a reasonable point, but it also generates unnecessary use of the earth's resources, many of which may not be recyclable. A variation of this technique is to over-package cheaper items and sell them next to more expensive or luxurious items, thereby using the "halo effect" from the latter to increase the price of the former. Psychological pricing can be used in tandem with promotional pricing. An example would be the introduction of a lavishly packaged item bearing logos and messages which reflect the company's sponsorship of a major sporting event.

Promotional pricing: short-term price adjustments to stimulate customer awareness, footfall and demand.

3.11.4 Segmented Pricing

Segmented pricing usually involves dividing the market into smaller sections and charging each segment different prices for the same products based on the attractiveness of the customers and perhaps upon their willingness or ability to pay. The practice of charging pensioners and students less money to enter a tourist attraction, discussed earlier in this chapter, is considered **customer-segmented pricing**. However, other segmented pricing approaches focus not on the customer segment but rather on the place in which the product or service is sold and the differences in characteristics between different forms of products and services. **Product-form pricing**, for example, enables the manufacturer to sell a product in different formats. An oil producer may be able to sell oil in large cans, intended for use in motor vehicles, but also sell the same product in smaller bottles with a spray cap, intended for household uses such as lubricating door hinges. **Location pricing** is in widespread usage by professional sports clubs. Tickets to watch a baseball game vary greatly in price according to whether the seat is in the bleachers, undercover and with a central position and good view of the field, or in a corporate hospitality box. Organizations may even use **time pricing**, which is when a product or service is priced according to whether the sale or usage occurs in peak or off-peak times. For example, holiday home rentals are often cheaper during the winter months, and golf membership is usually much less expensive for customers who are willing to play solely during working hours. This provides a welcome deal for customers who are willing and able to use a product or service during less popular times and also helps the seller to increase occupancy rates and usage during times when it is more difficult to sell the product or service.

Location and time pricing: setting prices according to where and when the product is sold.

3.11.5 Geographical Pricing

Geographical pricing is not to be confused with the much simpler location pricing. Instead, geographical pricing aims to mitigate the difficulties of trading internationally or across a broad geographical area and to ensure that the seller is competitive in a number of different markets. **Uniform delivered pricing** is perhaps the simplest form of geographical pricing, as it means that customers pay exactly the same price regardless of their location. For example, a customer in Anchorage, Alaska, would pay the same price to receive delivery of a truck as a customer in New York, Dallas, or Los Angeles. To achieve this, the organization must homogenize any freight charges. The risk inherent in uniform delivered pricing is that an unscrupulous, or at least opportunistic, competitor decides to cut the freight price for customers close to the center of manufacture. Another risk is that an organization which trades across a broad geographical area is unable to compete with smaller competitors who focus specifically on their local regions and, therefore, have lower average freight costs. To deal with this, some manufacturers use **FOB-origin pricing**, which is short for **Free On Board-origin pricing.** In this approach, the supplier charges all customers the same price for the product, regardless of location, and does not cover the cost of transport but avoids the cost of shipping by devolving this responsibility to customers. This may help to deal with the issue of absorbing freight costs to more distant customers and has the benefit of itemizing the prices so that the distant customer can see that a higher cost does not appertain to the product or benefit the supplier but is entirely attributable to freight charges. Some suppliers use **zone pricing**, which charges different amounts based upon where the product is to be transported but clusters geographical regions into different price bands. For instance, a firm in New York City may sell to its customers on the East Coast of America in price band A, to customers in the Midwest in price band B, to customers on the West Coast in band C, and to customers in Alaska and Hawaii in band D. Other organizations use **base point pricing**, or **basing point pricing**, which charges delivery costs from an intermediate center rather than the place in which the item was manufactured. Finally, some organizations use **freight-absorption pricing** to absorb some or all of the charges appertaining to transportation, and this is particularly common where the organization wishes to win a major new account which is distant from the place of manufacture.

- **Base point pricing/basing point pricing:** incorporates transportation costs from a regional distribution point rather than the place of manufacture.
- **FOB-origin pricing:** passes responsibility for transportation costs to the customer.
- **Freight-absorption pricing:** pricing which absorbs the cost of transportation into the amount to be charged, regardless of where the product is to be transported, so that the seller remains competitive in distant markets.
- **Geographical pricing:** charging different prices according to where the product is sold.
- **Uniform delivered pricing:** charging the same price to customers regardless of their distance from the seller and the shipping costs.
- **Zone pricing:** charging customers different prices according to the geographical zone in which they are located.

Marketing in Action 3.2: Kimberly-Clark's Pricing Strategy

■ Source: ▸ pexels.com/royalty-free-images/

Fast-moving consumer goods (FMCG) giant, Kimberly-Clark, will increase marketing spend to support price rises (Jefferson, 2022). Acknowledging that cost inflation—the increasing costs of materials, staffing, and transportation—will impact its business and those of its competitors, it plans to offset much of the problem by communicating the value of its products, such as Andrex, Huggies, and Kleenex. Despite a decline in volume largely associated with supply chain issues, and a commensurate decrease in marketing and other spends, the firm has enjoyed increased sales income through "significant" pricing action in support of its brands. Kimberly-Clark's CFO, Maria Henry, hopes that her price elasticity assumptions have been conservative, especially as their rival, Procter & Gamble, reported no discernible changes in customers' demand for its best-performing, premium-priced brands in the face of its own price rises. If you were the Chief Marketing Officer reporting to Maria Henry, what pricing decisions would you recommend appertaining to the sale of Andrex, Huggies, and Kleenex in UK supermarkets and other retail outlets? Assuming that product prices must rise by 10% year-on-year for two consecutive years—a hypothetic number for the purpose of this exercise—how would you implement and communicate them, with whom would you need to work, and how would Kimberly-Clark mitigate the potentially negative impact upon volumes?

(Source: Marketing Week article. Kimberly-Clark to support price rises with increased marketing spend. 26.1.22. Retrieved February15, 2022, from ► https://www.marketingweek.com/kimberly-clark-increase-marketing-price/)

3.12 Defending Against Competitors' Pricing

To defend against the effects of a competitor's price cut, a firm has several options. The simplest is to **reduce prices**, which may be a retaliatory approach or simply one brought about by nervousness and the desire to protect market shares. It may be effective, but presents a number of serious challenges. The firm may find it difficult to raise prices to their normal levels later on, as customers have become accustomed to paying lower prices. It may undermine the brand, distressing the products and inhibiting future profit opportunities. If the action is taken against a particularly strong competitor, it is likely that that rival will be able to sustain trading at a higher discount for a longer period of time, and by instigating or at least continuing a price war, the firm contributes to a spoiling of the product category and a reduction in its attractiveness to all suppliers (including itself). Of course, if margins were already tight, it may be impossible to reduce the price anyway without selling products at a loss. If this were the case and the firm still wishes to go ahead, it must subsidize sales of the discounted product by raising the profitability of other products in its portfolio, which may have the knock-on effect of making the firm less competitive in those areas. A more cautious approach that organizations sometimes take when they feel the need to offer lower prices is to **launch a low-price**

fighting brand. For instance, when the motor manufacturer Renault perceived that there was a strong market for inexpensive but low specced cars in Europe and that brands such as Suzuki were serving this market at the expense of Renault's overall market share, it wished to attack this market but not to undermine customer perceptions of its core brand. Therefore, it reintroduced the Dacia brand specifically as a fighting brand which would sell cars with relatively unsophisticated technology and low specification levels to very price-conscious customers.

For the reasons mentioned above, firms sometimes opt to do nothing or rather to **watch and wait**. This means that rather than risking a price war, trading at a loss, or distressing their brand, they simply allow the competitor to gain market share through heavy discounts, hoping that this is unsustainable, that the competitor will soon return to normal pricing, and that its customers will begin to consider alternative suppliers. An approach which is perhaps more commonplace is to attempt to raise the quality perceptions of customers for one's own products. Arguably, this is a central role of marketing activities at all times but is particularly important when the firm is subjected to attack by a competitor's pricing. Executed successfully, the defending firm may even decide to increase its own price at the same time as implementing improvements in the actual quality of its products and services. This **"increase quality and price"** approach helps to differentiate the brand from its attacker, moving it up-market into a territory populated by less price-sensitive customers who are likely to be more loyal and making the firm less dependent upon discounted prices. This may lead to an increase in profit per unit sold and move the seller away from the mass market, thereby producing a drop in sales volumes. It may also demand a heavy investment in research and development and after-sales services to trade with these more demanding customers at higher prices. Ultimately, when a firm decides how to defend against competitor pricing actions, it must consider the strength of the competition, the characteristics of the customers in the marketplace, and the ability and willingness of the firm to put into place certain defensive approaches and bear the consequences.

3.13 Pricing Strategy and Sustainable Development Goals

As we explored in the vignette which opened this chapter, firms may be able to alleviate certain social problems by taking flexible approaches to pricing, such as restaurants adopting a second-market pricing strategy to match excess capacity and food approaching its spoiling date with poorer, undernourished customers who would be willing and able to pay a small amount which the firm would not otherwise recoup. While there are logistical considerations for outlets that may wish to undertake such an initiative, the increased usage of smartphone apps and other forms of digital technology offers low-maintenance solutions which enable sellers to tailor pricing discretely to individuals according to specific supply and demand considerations. The implementation of a food wastage app helps to address Sustainability Development Goals (SDGs) 1, 2, and 3—"no poverty," "zero hunger," and "good health and well-being."

Later in this chapter, we explored how clothing manufacturers have resisted price-dominated customer offers by marketing products made from ethically sourced materials and often manufactured from recycled materials, thereby embracing and encouraging "responsible consumption and production" (SDG 12). Cleaner, less wasteful manufacturing, in turn, reduces the amount of plastic and chemicals which must be disposed of, thereby safeguarding "life below water" (SDG 14) and "life on land" (SDG 15) while taking—and encouraging others to take "climate action" (SDG 13). We also saw how the adoption of a "freemium" pricing strategy by Duolingo (and other online service providers) helps to remove barriers to education through an initiative which relies on other, wealthier customers to fund it willingly rather than on state intervention and taxation. This helps to address SDGs 4 and 10 by helping to facilitate access to "quality education" and producing "reduced inequalities" (or, at least, reducing the deleterious effects of wealth inequality).

If businesses globally are to realize the aspiration of inhabiting a fairer, cleaner and more sustainable planet and of preserving it for future generations, then it is important that they look beyond established pricing strategies which focus on reducing costs and prices as their core competitive vehicle, instead using emerging strategies—often facilitated by recent technological advances—which focus more heavily on other benefits such as product quality and longevity, ethical sourcing of materials, and aftersales support.

Summary

As we have seen in this chapter, price is not simply one element of the marketing mix but plays a holistic role in the success or otherwise of the firm. Many organizations struggle to achieve and maintain a price which satisfies both customers, stimulating demand to an optimal point, and provides sufficient profitability to meet the business needs of the organization. Setting prices is usually a long and deliberate process, and one which should be repeated on a continuous basis, using thorough and widespread market intelligence. Getting the price right is a little like walking a tightrope, not only because it generates income and profit but also because it heavily influences customer perceptions of the brand and dictates where the brand is positioned relative to its competitors. In setting prices, firms have a responsibility to the financial and economic sustainability of their own operations, as a wrong step can endanger the viability of the firm and jeopardize many jobs. It must also consider the needs of its external stakeholders. For instance, even though a manufacturer may wish to increase its sales volumes at the expense of profit per unit in order to achieve economies of scale, the retailers and other channel intermediaries who sell the product may instead need a higher profit per unit to satisfy their objectives. Finally, the organization must consider how to adopt a pricing strategy which is ethically appropriate and does not take an unnecessary toll upon the earth's resources. This may even entail, somewhat counterintuitively, discouraging customers from buying unneeded products but encouraging them to buy more expensive products which can be kept for longer and recycled. Such an approach has been used by manufacturers of mountain clothing such as Patagonia and Rab but is increasingly applicable to all consumer product categories. Treated thoughtfully by organizations, pricing strategies do not have to overpower other marketing decisions and customer considerations. Furthermore, firms which sacrifice some short-term profitability to trade in a way which is kind to customers, stakeholders and the environment often build high-value relationships which ultimately bring long-term financial sustainability.

3

Review Questions

1. What are the main types of competitor-based pricing, and what are the benefits and limitations of each when applied by an organization in a competitive market?
2. When using penetration pricing to increase market share, what are the risks faced by firms, and how might they mitigate those risks?
3. If General Motors were to cease manufacturing and selling vehicles and you were advising Ford, what pricing strategy would you advise them to take in the US market (i) immediately, (ii) for the next 2 years, and (iii) thereafter, and why?
4. If a restaurant notices that they waste lots of unsold stock each night, what pricing strategy would you recommend they take? Remember that it may be unfeasible for them to order or cook fewer of the affected items, as they must be able to service unforeseen "walk-in" demand.
5. You are advising the UK Government on how it might quickly accelerate customer trial and adoption of electric vehicles in readiness for 2030 when all new cars sold in the UK must be electric. How do you suggest stimulating this demand to bring about change?
6. What do you consider to be the main disadvantages or limitations of price skimming from the perspective of customers from each of the five main categories (e.g., innovators, early adopters, etc.)? Explain why.
7. Your consumer electronics firm manufactures a range of 12 different laptops, tablets and smartphones, but you find that they cannibalize an unacceptably high proportion of each other's sales. Suggest a broad strategy to address this. Remember that tough decisions may be necessary in the short term to realize long-term benefits.
8. As a hairdresser, your new salon is struggling to attract new customers despite enjoying a busy location and stylish décor, and you are also struggling to convert new customers into repeat clients.

Discussion Questions

1. A leading US university has started delivering online lectures by developing and using a "MOOC" (Massive Open Online Course) platform. It has three agendas: the first is to raise the public's awareness of the university; the second is to generate incremental tuition revenue from remote learners who would not otherwise study with the university; and the third is to address a Widening Participation agenda, which enables underrepresented groups to overcome barriers to education—particularly disabled students, those from economically

deprived backgrounds, students with caring responsibilities, parents, and mature students. You have been tasked with designing and implementing a pricing strategy which will fulfil these three objectives. Discuss your various options before suggesting one. Give a detailed account of your rationale and the intended outcomes.

2. In bad news for Mondelez, supply chain issues and climate events have driven up the price of cocoa beans, which means that it will need to increase the price once more of its Toblerone chocolate bars. Keen to avoid the customer backlash which followed its recent redesign of the product to make it smaller, it is uncertain how to tackle the issue. Suggest three possible strategies and, for each, explain the likely reaction from consumers and impact upon sales.
3. Global economies are facing wage stagnation, price inflation, hikes in production and employment costs, recession, limited government spending, and widespread precarity of employment. As a leading North American manufacturer of recreational vehicles (US: "RVs"; UK: "motorhomes"), you wish to be at the forefront of the industry in forging a new pricing strategy to deal with these issues in your home market and when selling to European and Asian markets. Create a Gantt Chart for the next 24 months, showing what activities you would undertake to achieve this, and prepare to discuss your plan.
4. Your company produces high-quality clothing and footwear for explorers, mountaineers, climbers, wilderness hikers, kayakers and survivalists. Recently, you have noticed an increasing volume of your product being purchased by young, urban consumers for "streetwear." You wish to preserve your brand ethos and core customer base of "outdoor enthusiasts" and are concerned that your products' adoption by skateboarders, BMX bikers and parkour enthusiasts may jeopardize this aspiration, as it once did for Timberland. What pricing strategy or strategies might you adopt to maintain your core customer base while discouraging the new, unintended customer segment?
5. If you were charged with taking a popular but pricey consumer electronics brand, such as Dyson, into a large and growing but developing economy such as Nigeria, what pricing approaches might you use to gain a foothold and ensure economic sustainability?

Practice Quiz

1. Which of these has a relatively "inelastic" demand?
   (a) Mortgages
   (b) Package holidays
   (c) Champagne
   (d) Salt
2. If everyone in a country must buy all of their water from the government's water authority, what sort of market is this?
   (a) Pure monopoly
   (b) Monopolistic competition
   (c) Pure competition
   (d) Oligopolistic competition

3. What is the calculation used to ascertain the break-even price of a product?
   (a) Fixed cost + (variable cost/unit sales)
   (b) Variable cost + (fixed cost/unit sales)
   (c) Total sales/(fixed cost + variable cost)
   (d) Fixed cost/(price – variable cost)
4. Which pricing strategy is most likely to maximize profit early in the life of a technologically advanced product?
   (a) Product-line pricing
   (b) Customary pricing
   (c) Price skimming
   (d) Second-market discounting
5. What are the three main categories of pricing policy?
   (a) Needs-, wants-, and demands-based
   (b) Market-, manufacture-, and means-based
   (c) Contraction-, consolidation-, and conquest-based
   (d) Cost-, customer-, and competitor-based
6. Which of these is an attractive offer which gains attention, generates customer footfall, and therefore enables sales of more profitable items?
   (a) Loss leader
   (b) False footer
   (c) Hidden header
   (d) Dollar drainer
7. Which pricing technique is used by Duolingo and other apps to encourage customer trial before increasing their profitability?
   (a) Honey trap pricing
   (b) Freemium pricing
   (c) Sprat-and-mackerel pricing
   (d) Ads-or-your-money pricing
8. If a store buys a suit for £200 and sells it for £400, what is the mark-up, and the margin, expressed as percentages?
   (a) 200% mark-up and 200% margin
   (b) 100% mark-up and 100% margin
   (c) 50% mark-up and 100% margin
   (d) 100% mark-up and 50% margin
9. Which pricing strategy is most closely associated with Walmart or its UK subsidiary, Asda?
   (a) Value-added pricing
   (b) Customary pricing
   (c) Everyday low pricing (EDLP)
   (d) Second-market discounting
10. Which websites collect content—including offers and prices—from other websites, enabling easier customer comparisons?
    (a) Aggregators
    (b) Consolidators
    (c) Reconcilers
    (d) Unifiers

11. What is the term for the drop in costs associated with accumulated production experience?
    (a) The Law of Many
    (b) The experience curve
    (c) The parallax error
    (d) The volume reward
12. What is usually the first stage of the price-setting process?
    (a) Development of profit targets
    (b) Market segmentation
    (c) Ideation
    (d) Development of pricing objectives
13. A McDonald's Happy Meal is an example of which of these?
    (a) Menu pricing
    (b) By-product pricing
    (c) Product-bundle pricing
    (d) Captive-product pricing
14. A pizza restaurant which offers a 30% discount to students at lunchtimes is using which of these pricing approaches?
    (a) Second-market discounting
    (b) Penetration pricing
    (c) Seasonal pricing
    (d) Odd-even pricing

(You can find the correct answers to these questions by taking the quiz and then comparing your answers with the answers that are provided at the end of the book, under the "Practice Quiz Answers" part of the book.)

References

Albanese, P. (2015). Inside economic man: Behavioral economics and consumer behavior. In *Handbook of contemporary behavioral economics* (pp. 25–45). Routledge.

Alon, I., Jaffe, E., Prange, C., & Vianelli, D. (2016). Setting global prices. In *Global marketing* (pp. 465–498). Routledge.

Apostolidis, C., Brown, D., Wijetunga, D., & Kathriarachchi, E. (2021). Sustainable value co-creation at the bottom of the pyramid: Using mobile applications to reduce food waste and improve food security. *Journal of Marketing Management, 37*(9–10), 856–886.

Autotrader (UK). (n.d.). Retrieved February 15, 2022, from www.autotrader.co.uk

Autotrader (USA). (n.d.). Retrieved February 15, 2022, from www.autotrader.com

BBC Football Website. (2022). Article on Peterborough United ticket pricing. Retrieved February 15, 2022, from https://www.bbc.co.uk/sport/football/60352564

Borden, N. H. (1964). The concept of the marketing mix. *Journal of Advertising Research, 2*, 7–12.

Bös, D. (2015). *Pricing and price regulation: An economic theory for public enterprises and public utilities* (Vol. 34). Elsevier.

Boz, H., Arslan, A., & Koc, E. (2017). Neuromarketing aspect of tourism pricing psychology. *Tourism Management Perspectives, 23*, 119–128.

Cartelier, J. (2018). *Money, markets and capital: The case for a monetary analysis*. Routledge.

Cialdini, R. (2008). *Influence: The psychology of persuasion* (5th ed.). Allyn & Bacon.

Coglianese, J., Davis, L. W., Kilian, L., & Stock, J. H. (2017). Anticipation, tax avoidance, and the price elasticity of gasoline demand. *Journal of Applied Econometrics, 32*(1), 1–15.
Davcik, N. S., & Sharma, P. (2015). Impact of product differentiation, marketing investments and brand equity on pricing strategies: A brand level investigation. *European Journal of Marketing, 49*(5/6), 760–781.
Deshpande, D. S. S. (2018). Various pricing strategies: A review. *Journal of Business and Management, 20*(2), 75–79.
Diallo, M. F., Coutelle-Brillet, P., Riviere, A., & Zielke, S. (2015). How do price perceptions of different brand types affect shopping value and store loyalty? *Psychology & Marketing, 32*(12), 1133–1147.
Du, P., & Chen, Q. (2017). Skimming or penetration: Optimal pricing of new fashion products in the presence of strategic consumers. *Annals of Operations Research, 257*(1–2), 275–295.
Elkington, J. (2013). Enter the triple bottom line. In *The triple bottom line* (pp. 23–38). Routledge.
Fisher, M., Gallino, S., & Li, J. (2017). Competition-based dynamic pricing in online retailing: A methodology validated with field experiments. *Management Science, 64*(6), 2496–2514.
Gross, H. P., Rottler, M., & Wallmeier, F. (2021). The influence of external reference price strategies in a nonprofit arts organization's "pay-what-you-want" setting. *Journal of Philanthropy and Marketing, 26*(1), e1681.
Guardian. (2018). Retrieved February 15, 2022, from https://www.theguardian.com/business/2018/jul/20/toblerone-to-revert-to-original-shape-but-with-bigger-size-and-price
Hanna, N., & Dodge, H. R. (2017). *Pricing: Policies and procedures*. Macmillan International Higher Education.
Jahanmir, S. F., & Lages, L. F. (2016). The late-adopter scale: A measure of late adopters of technological innovations. *Journal of Business Research, 69*(5), 1701–1706.
Jefferson, M. (2022). Marketing Week article. *Kimberly-Clark to support price rises with increased marketing spend.* 26.1.22. Retrieved February 15, 2022, from https://www.marketingweek.com/kimberly-clark-increase-marketing-price/
Kumar, V., & Reinartz, W. (2016). Creating enduring customer value. *Journal of Marketing, 80*(6), 36–68.
Lee, K. Y., Jin, Y., Rhee, C., & Yang, S. B. (2016). Online consumers' reactions to price decreases: Amazon's Kindle 2 case. *Internet Research, 26*(4), 1001–1026.
McGinn, B. (producer). (2018). The maple syrup heist. S1, E5, Dirty Money, Netflix.
Mitchell, S. L., & Clark, M. (2019). Reconceptualising product life-cycle theory as stakeholder engagement with non-profit organisations. *Journal of Marketing Management, 35*(1–2), 13–39.
Money Saving Expert. (2022). Retrieved February 15, 2022, from www.moneysavingexpert.com
Money Supermarket. (n.d.). Retrieved February 15, 2022, from www.moneysupermarket.com
Nagle, T. T., & Müller, G. (2017). *The strategy and tactics of pricing: A guide to growing more profitably*. Routledge.
Norman, J. (2019). *Adam Smith: What he thought and why it matters*. Penguin.
Olbrich, R., Jansen, H. C., & Hundt, M. (2017). Effects of pricing strategies and product quality on private label and national brand performance. *Journal of Retailing and Consumer Services, 34*, 294–301.
Payne, A., Frow, P., & Eggert, A. (2017). The customer value proposition: Evolution, development, and application in marketing. *Journal of the Academy of Marketing Science, 45*(4), 467–489.
Porter, M. E. (2008). The five competitive forces that shape strategy. *Harvard Business Review, 86*(1), 25–40.
Rogers, E. M. (2010). *Diffusion of innovations*. Simon and Schuster.
Shiu, E., Hassan, L. M., & Walsh, G. (2009). Demarketing tobacco through governmental policies—The 4Ps revisited. *Journal of Business Research, 62*(2), 269–278.
Social Bite. (2022). Retrieved February 15, 2022, from https://social-bite.co.uk/about-us/
Sotgiu, F., & Gielens, K. (2015). Suppliers caught in supermarket price wars: Victims or victors? Insights from a Dutch price war. *Journal of Marketing Research, 52*(6), 784–800.
Steenkamp, J. B. (2017). Global brand equity. In *Global brand strategy* (pp. 243–273). Palgrave Macmillan.
Trout, J. (2005). Schizophrenia at GM. *Harvard Business Review, 83*(9), 22.

Further Reading

Ansari, A., Siddarth, S., & Weinberg, C. B. (1996). Pricing a bundle of products or services: The case of nonprofits. *Journal of Marketing Research, 33*(1), 86–93.

Bertini, M., Suehler, S., Halbheer, D., & Lehmann, D. R. (2022). Carbon footprinting and pricing under climate concerns. *Journal of Marketing, 86*(2), 186–201.

Bolton, L. E., Keh, H. T., & Alba, J. W. (2010). How do price fairness perceptions differ across culture? *Journal of Marketing Research, 47*(3), 564–576.

Datta, H., Heerde, H. J., Dekimpe, M. G., & Steenkamp, J. E. M. (2022). Cross-national differences in market response: Line-length, price, and distribution elasticities in 14 Indo-Pacific rim economics. *Journal of Marketing Research, 59*(2), 251–270.

Huang, Y., Ellickson, P. B., & Lovett, M. J. (2022). Learning to set prices. *Journal of Marketing Research, 59*(2), 411–434.

Kübler, R., Pauwels, K., Yildirim, G., & Fandrich, T. (2018). App popularity: Where in the world are consumers most sensitive to price and user ratings. *Journal of Marketing, 82*(5), 20–44.

Lee, H., Lalwani, A. K., & Wang, J. J. (2020). Price no object!: The impact of power distance belief on consumers' price sensitivity. *Journal of Marketing, 84*(6), 113–129.

Li, X., Li, K. J., & Wang, X. S. (2020). Transparency of behaviour-based pricing. *Journal of Marketing Research, 57*(1), 78–99.

Wang, C. X., Beck, J. T., & Yuan, H. (2021). The control-effort trade-off in participative pricing: How easing pricing decisions enhances purchase outcomes. *Journal of Marketing, 85*(5), 145–160.

Zhang, K., Cai, F., & Shi, Z. (2021). Do promotions make consumers more generous? The impact of price promotions on consumers' donation behaviour. *Journal of Marketing, 85*(3), 240–255.

Brand Management, Marketing Communication, and Promotion Management for People, Planet, and Profit

Contents

S. M. R. Shams et al., *Sustainable Marketing*, Classroom Companion: Business,
https://doi.org/10.1007/978-3-031-79130-7_4

- **Key Terms**

Brands: A brand's value, or equity, comes from consumers' perceptions about its features, performance, name, and the company behind it. This distinguishes it from generic products by creating a unique image and connection with consumers.

Branding is the process of strategically shaping and managing the perceptions, emotions, and associations that consumers have with a company, product, or service. It involves creating a distinct identity and conveying its value proposition to target audiences to build recognition, loyalty, and trust.

Advertising is a form of marketing communication aimed at promoting products, services, or ideas to target audiences through various channels such as print, broadcast, digital, and social media. Objectives include persuading and influencing consumer behavior, driving sales, brand awareness, and customer engagement.

Promotion involves the use of various marketing tactics to communicate and persuade target audiences about the benefits of a product, service, or idea. It encompasses activities such as advertising, sales promotions, public relations, and personal selling, all aimed at increasing brand visibility, sales, and customer loyalty.

Differentiation is the process of distinguishing a product or brand from its competitors by highlighting unique features, benefits, or qualities that set it apart in the minds of consumers.

Distinctive assets are unique visual or auditory elements associated with a brand, such as logos, jingles, or packaging designs, that help to create strong brand recognition and recall among consumers.

A commodity is a basic good or raw material that is interchangeable with other goods of the same type, and commodities are uniform, they typically possess few distinguishing features or qualities that set them apart from competitors' offerings.

Brand management involves overseeing and strategizing the development, maintenance, and enhancement of a brand's identity, perception, and equity across all brand touch points to ensure consistency, relevance, and resonance with target audiences.

Brand storytelling is the art of using narratives, emotions, and experiences to convey a brand's values, mission, and personality in a compelling and memorable way, forging deeper connections with consumers. Storytelling establishes trust by humanizing the brand. It allows consumers to relate to the brand on a personal level, fostering a sense of authenticity.

Sustainability entails managing brand operations in a manner that preserves the well-being of the environment, society, and future generations while also ensuring the long-term viability and success of the brand.

Monopoly Power: A monopoly is a situation where certain brands have a significant amount of market share, leading to a brand controlling much of an industry.

Antitrust refers to laws and regulations aimed at promoting competition and preventing monopolistic practices and unfair business tactics, ensuring a level playing field for brands and protecting consumer welfare.

Competition refers to the rivalry between brands operating in the same industry or market, vying for customers, market share, and profitability.

Consumer habits, shaped by the workings of the brain and influenced by cognitive biases, represent the ingrained patterns of behavior and preferences that individuals rely on when making purchasing decisions, influenced by factors such as lifestyle, culture, convenience, and brand loyalty.

Big tech refers to large, influential technology companies that dominate various sectors (not a single industry) of the digital economy, wielding significant power in areas such as online search, e-commerce, social media, smartphone manufacturing, and cloud computing.

Consumer identities encompass the multifaceted ways individuals construct and express their sense of self, often leveraging brands as symbolic props to convey desired identities, aspirations, and social status to others.

Status anxiety is the discomfort individuals feel about their perceived social standing, often influenced by comparisons with others and exacerbated by consumer goods, which can serve as visible markers of status and contribute to feelings of inadequacy or insecurity.

Unethical concerns in branding arise when companies engage in deceptive practices, exploitation of labor or resources, or disregard for social and environmental responsibilities, leading to a loss of trust and reputation among consumers.

The triple bottom line refers to a business approach that considers not only financial performance but also social and environmental impacts, aiming for sustainable success across economic, social, and environmental dimensions.

Corporate Social Responsibility (CSR) is a business practice that involves integrating ethical, social, and environmental concerns into a company's operations and interactions with stakeholders, demonstrating a commitment to sustainable practices and positive societal impact.

Greenwashing refers to the deceptive practice of misleading consumers by falsely portraying a brand, product, or service as environmentally friendly or sustainable, often exploiting growing environmental concerns without making meaningful changes.

Consumer activism refers to the collective actions taken by individuals or groups of consumers to advocate for change in business practices, policies, or products, often aiming to promote ethical standards, environmental sustainability, or social justice.

Supply chain transparency involves accountability and the practice of openly disclosing information about the origins, processes, and impacts of products or services throughout their production and distribution.

Ethical branding entails aligning a brand's values, actions, and messaging with ethical principles and practices, prioritizing honesty, integrity, and responsibility in all aspects of business operations and interactions with stakeholders.

Stakeholder engagement involves actively involving and communicating with individuals or groups who have a vested interest or are impacted by a brand's decisions, actions, or outcomes, fostering collaboration, understanding, and mutual benefit.

Cultural sensitivity is the awareness, understanding, and respect for the cultural differences, norms, values, and practices of diverse individuals and communities, essential for effective communication, collaboration, and relationship-building across cultural boundaries.

Brand loyalty is the emotional attachment and repeat purchasing behavior exhibited by consumers toward a specific brand, often resulting from positive experiences, perceived value, and trust built over time.

Brand equity refers to the intangible value associated with a brand, including factors such as brand awareness, perceived quality, customer loyalty, and the overall strength of the brand's image and reputation.

Brand awareness is the level of familiarity and recognition that consumers have with a particular brand, reflecting its visibility, presence, and recall in the minds of target audiences, which is crucial for driving consumer consideration and purchase decisions.

Physical availability refers to the accessibility and presence of a product or service in physical retail spaces or distribution channels, ensuring convenient access for consumers and increasing the likelihood of purchase. Brands must be easily available when a consumer decides to make a purchase.

Mental availability refers to the likelihood of a brand coming to consumers' minds in relevant purchase situations, influenced by factors such as brand awareness, salience, and associations, crucial for steering purchase decisions.

Brand tone of voice refers to the consistent style, personality, and language used by a brand to communicate with its audience across various channels and touchpoints, reflecting its values, identity, and relationship with consumers.

Business models describe the framework through which a company creates, delivers, and captures value, outlining the core strategy and operational processes that drive its profitability and sustainability; essentially, it's how the brand makes money.

Representation in marketing involves authentically showcasing diverse individuals and experiences, serving as positive role models, and breaking down barriers, as when people see themselves reflected in various positions (i.e., in advertising and promotional materials), it inspires belief in the possibility of achieving similar success.

The Veblen Effect describes the phenomenon where the demand for a product increases as its price rises, often due to the product's perceived exclusivity or status symbol value rather than its inherent utility or quality.

Conspicuous consumption refers to the public display and consumption of luxury goods and services as a means of signaling social status, wealth, and prestige, influencing consumer behavior and brand preferences by associating brands with status symbols and exclusivity.

A logo is a visual symbol or graphic representation that serves as a distinctive identifier for a brand, product, or organization, often embodying its values, identity, and purpose in a simple yet memorable design.

A slogan is a brief and memorable phrase or tagline used in advertising and marketing campaigns to encapsulate a brand's message, values, or unique selling proposition, aiming to resonate with consumers and enhance brand recall.

Category entry point refers to the strategic positioning or introduction of a new product or brand into a specific market segment or niche, aiming to capitalize on unmet consumer needs, preferences, or emerging trends to gain competitive advantage.

Memory structures and brands are intertwined as brands aim to create strong associations and memorable connections in consumers' minds through consistent messaging, experiences, and emotional resonance, fostering brand recall and preference.

Neural pathways are interconnected networks of neurons that transmit signals and information throughout the brain, shaping cognitive processes, learning, and memory formation, including the associations and perceptions individuals develop toward brands and experiences.

Mental shortcuts, or heuristics, are cognitive strategies used by individuals to make quick and efficient decisions based on simplified rules of thumb, often influencing consumer behavior and brand perception through the unconscious processing of information.

Nudge's theory proposes that subtle changes in the presentation of choices can influence individuals' decisions and behaviors, often steering them toward more beneficial or desirable outcomes, shaping consumer choices and brand preferences through behavioral economics principles.

Prospect theory suggests that individuals' decisions are influenced by the potential gains and losses relative to a reference point rather than by absolute outcomes, providing insights into how consumer choices and preferences are shaped by perceived risks and rewards about certain brands in marketing contexts.

Bounded rationality posits that individuals make decisions based on limited information, cognitive limitations, and time constraints, leading to satisficing rather than optimizing behavior, influencing consumer choices and brand perceptions by shaping decision-making processes in marketing contexts.

Framing involves presenting information in a particular way to influence individuals' perceptions, judgments, and decisions, often shaping consumer attitudes and behaviors toward brands through the context and emphasis placed on different aspects of a message or experience.

Stereotyping involves categorizing individuals or groups based on oversimplified and often inaccurate assumptions about their characteristics or behaviors, and using it during segmentation can lead to biased targeting, exclusion, and misrepresentation.

Sweatshops are workplaces characterized by poor working conditions, low wages, long hours, and often exploitation of labor, with the term "sweat" reflecting the harsh and oppressive environment where workers typically toil under strenuous conditions for minimal pay.

Labor exploitation refers to the unethical practice of taking advantage of workers' vulnerabilities by subjecting them to poor working conditions, low wages, long hours, and other forms of mistreatment, violating basic human rights and ethical standards within the workplace.

The concept of the unpaid consumer in digital environments refers to individuals who provide valuable data, content, or engagement to online platforms without receiving monetary compensation, raising ethical questions about data privacy, fair exchange, and the commercialization of user-generated content.

Prosumption describes the phenomenon where consumers not only consume products or services but also actively participate in their creation, customization, or dissemination, blurring the lines between producers and consumers.

Persuasive technologies are digital tools and platforms designed to influence users' attitudes, behaviors, and decision-making processes through techniques such as persuasive design, nudges, and behavior change interventions, raising ethical considerations regarding user autonomy, manipulation, and privacy.

Dark patterns of design involve intentionally deceptive or manipulative user interface design techniques aimed at tricking users into taking actions they might not otherwise choose, often at the expense of their autonomy, comprehension, or well-being, highlighting ethical concerns about transparency, trust, and user empowerment in digital environments.

Artificial Intelligence (AI) refers to the simulation of human intelligence processes by machines, encompassing tasks such as learning, reasoning, problem-solving, and perception, enabling computers to perform tasks that typically require human intelligence.

Generative AI refers to a subset of artificial intelligence techniques that involve training models to generate new, original content, such as text, images, or music, based on patterns and examples from existing data.

Automation in marketing involves the use of software and technology to streamline repetitive tasks, such as email campaigns, social media posting, and data analysis, increasing efficiency and allowing brands to focus on strategy and creativity.

Algorithms are step-by-step procedures or formulas used by computers to solve problems or perform specific tasks, often relying on data inputs and rules to make automated decisions and predictions in various domains, such as machine learning, optimization, and data analysis.

Recommendation algorithms are computational techniques that analyze user data and behavior to suggest personalized content, products, or services, enhancing user experience and engagement on digital platforms such as e-commerce websites, streaming services, and social media platforms.

Algorithmic amplification refers to the process by which digital algorithms prioritize, promote, or magnify certain content, often based on user engagement metrics or predefined criteria, influencing the visibility and reach of information or media on online platforms, potentially shaping choices about brands.

Echo chambers are online environments or social networks where individuals are exposed predominantly to information, opinions, and perspectives that reinforce their existing beliefs with limited exposure to diverse viewpoints.

The Attention Economy refers to the economic system where the scarce resource is not goods or services but rather individuals' attention, driving competition among businesses and platforms to capture and monetize users' time and engagement through compelling content, advertising, and digital experiences.

4

Preliminary Vignette

In the heart of a bustling US city, the pulse of daily life reverberates through a marketplace where conscious consumer Maya embarks on a journey of mindful choices. Navigating aisles lined with promises of sustainability, she pauses before products with brand messages that beckon with a dual commitment to quality and a healthier planet. Each decision is a purposeful step toward a conscious consumer identity, a reflection of values woven into the fabric of her choices as Maya uses brands as props to achieve this desired identity. While Maya carefully inspects items adorned with eco-friendly insignias, the market becomes a haven for those seeking refuge from the grasp of big tech's monopoly power. A quiet rebellion takes shape in the rave between sustainable brands and the conscious consumer. It's a rebellion against the homogeneity propagated by tech giants and conglomerates, an assertion that personal (uninfluenced) choices matter, the planet is important, and individuals hold the power to dismantle monopolistic forces. Within this bustling marketplace, a collective hum of intentional living emerges. Brands that prioritize ethical, sustainable practices and help individuals with more sustainable habits become allies in the fight against status anxiety and the excesses of consumerism. Maya's choices, once confined to the realm of personal preference, become a ripple in a larger movement—a movement that extends beyond product labels and into a commitment to breaking free from the monopolistic hold of corporate giants. This chapter explores the nexus of conscious consumer identity and the role played by sustainable brands, unveiling hope for an optimistic narrative where individuals reclaim agency in the face of conglomerates. Through a building movement of intentional choices, consumers like Maya forge a path toward positive habits facilitated by brands, creating a sustainable future for themselves and the planet and challenging the monopoly power that has cast a shadow over the diverse mosaic of consumer voices.

4.1 Introduction

The purpose of the branding process is to identify one seller's products and services from another seller's products or services. Therefore, many assume that the process of branding and building a brand more generally sits within the "promotional" element of the marketing mix. However, in reality, branding is a strategy that influences and is influenced by all 4Ps of the marketing mix. As we've discussed in other chapters, the "4Ps" (product, price, place, promotion), also known as the "marketing mix," provide a conceptual framework employed by marketers to communicate the encoded meaning potentials of their brand to the most profitable consumer segments. Branding has emerged as an essential cornerstone of a business. According to the results of the Gartner 2020 CMO Spend Survey, 37% of CMOs cited brand strategy as vital to delivering a successful marketing strategy overall (see: ► https://www.gartner.com/en/marketing/insights/articles/marketers-guide-b2c-brand-strategy).

There are two main reasons for this, first, within the dynamic landscape of social media, the intense competition among brands inundates consumers with a diverse array of advertising messages, creating an overwhelming experience, therefore if businesses can cut through the noise with some clear messaging supported by a strong brand, they are far more likely to be considered in a buying situation. Second, branding is seen as the most important business function in a recession, why? Because consumers who have typically built a relationship with the brand have more trust in that brand during economic downturns. This could include situations whereby the brand has to increase the prices of products and services because the cost of materials is increasing. Consumers who are familiar with and regularly use the brand are far more likely to tolerate price increases and not switch to competing brands than consumers who have not built some familiarity with the brand (Campbell & Keller, 2003). Furthermore, consumers who are frequently reminded through branding activities that a brand exists are far more likely to remember a brand as an option in a buying situation. Developing an authentic and trustworthy brand is challenging. Building a consistent brand that is recognized, recalled, and at the top of a consumer's mind in a buying situation is difficult (Aaker, 2012). This takes time, and consumers may need to become familiar with a brand long before it regularly appears in the consumer's consideration set. As such, brands have adopted specific strategies to assist this process, and it is vitally important that brands position themselves around values and meanings that enable the consumer to use the brand's image and personality to construct a desired identity and relationship with the brand.

Source: ► https://unsplash.com/

One way in which brands have tried to build relationships with consumers in the past has been through the adoption of messaging that promotes responsible, ethical, and sustainable behaviors. Recent years have seen a significant increase in the adoption of ethical values as an identity for the conscious consumer. Because of this, sustainability, i.e., societally environmentally caring marketing strategies and messages, has become a common aspect of daily life for brands. The idea remains if a brand builds an identity around ethical values, then consumers will be more willing to align with that particular brand. Equally, for some time, there has been social pressure and government prompting of brands to send messages to the public to encourage people to think and act in a way that makes the world a better place. It is widely accepted that brands should engage with all stakeholders on corporate social responsibility (CSR) issues and regularly communicate messaging about their CSR programs, products, services, and the impacts with all concerned stakeholders. A sustainable approach to messaging and brand building aims to convince consumers to reward responsible businesses by selecting them; therefore, those brands that do not keep up will lose their competitive edge, largely due to increases in these regulatory requirements and social demands (Kelleci & Yıldız, 2021). However, although brands have engaged with some form of CSR since the 1990s, a considerable share of these actions has often taken the form of a symbolic gesture, primarily functioning as a public relations maneuver aimed at projecting a virtuous image, fostering favorable corporate perceptions, and building positive brand associations. In today's contemporary environment, the traditional growth-orientated system that has measured businesses' success via profit and financial terms is widely agreed as unsustainable. Brands are under pressure to strategically integrate ethics and sustainability into the business agenda, balancing traditional economic targets as well as social and environmental concerns. This chapter looks at the evolution of the use of CSR as a branding tool and sustainability strategies adopted by brands and aims to understand how brands in the contemporary era can incorporate ethical sustainability and position this centrally to their business strategy by focusing on the triple bottom line, i.e., People, Planet, and Profit. This includes not only generating profits but also protecting the environment and improving the lives of consumers so that a brand's societal and environmental contribution can underpin its commercial competitive advantage, as well. The adoption of sustainable lifestyles is on the rise, but consumers need more help with integrating this into their lives. Marketers of brands, pivotal in steering us into a situation of depleting the world's resources, have employed powerful marketing strategies to spur consumption, resulting in a level of overconsumption that surpasses our most apprehensive imaginings. Therefore, brands are not only ideally placed to initiate immediate change and help consumers create good habits; they have a moral responsibility to do this. This chapter will focus on how brands might do this and prosper.

4.2 What Is a Brand

We know that branding is vital to a firm's quest to differentiate itself from other competitors in the market or product category. Whether it is products or services, it is crucially important that a brand is recognized in a buying situation. We also know from the discussions of other chapters on marketing mix that a marketer creates real value for a customer via the 4Ps' marketing mix, but a company should not establish the marketing mix until it determines which customers it wants to reach, serve, and satisfy. Typically, a business can't win all customers unless it's a monopoly. A monopoly is a situation where an individual company has absolute control over a specific good or service in the marketplace. However, in the contemporary environment, absolute monopolies have been declining for some time. Most monopolies that exist today do not necessarily dominate an entire global industry, but they do control significant assets/holdings, perhaps in one country or region. We have to travel back to the 1800s for the introduction of monopoly regulation, which was an important juncture for brands and business growth strategies. In 1890, the Sherman Antitrust Act (Letwin, 1955) was introduced to break up monopolies in the USA, and this included the collapse of Rockefeller's Standard Oil and American Tobacco Company (by 1911). The reason that monopoly regulation was introduced and still exists is to prevent companies from engaging in predatory or exclusionary actions; this includes leveraging their market share by price discrimination and/or binding contracts on exclusive deals. There is an idea that the Sherman Act was designed as a consumer welfare prescription in response to a fear of concentrated power and antitrust. We will return to this point later in the chapter when we discuss brands (big tech) that have been accused of engaging in anti-competitive monopolistic behavior. However, because of monopoly regulation, a business must decisively select a particular category of customers and subsequently craft a robust marketing mix tailored to be both highly appealing to that customer segment and maximally profitable for the brand. According to Kotler (2016), a business can succeed best if it is aware of its competitors and it is aware of the community at large and its values/beliefs. The evaluation of a company by consumers centers on its ability to deliver superior products or services, increase customer satisfaction, and enhance overall value. These factors, among others, collectively contribute to creating a great brand.

So, how do we truly define a brand, and what is the difference between a brand, a business, or a product more generally? Our story commences in the distant past, highlighting that the concept of branding is far from novel. People have been branding "things" for an age, for example, Vikings in Norse times branded their swords with their initials to signal ownership. Furthermore, the signaling of ownership over time evolved in the agricultural era, with farmers branding their cattle. Again, the purpose of this was to signal ownership and to say, "This is my cow don't steal it." In the form we recognize today, the first brand-based products appeared around the same time as the factory. Markets were being flooded with uniform mass-produced products that were virtually indistinguishable from one another, and competitive branding became a necessity of the machine age during the Industrial Revolution (Source: Klein 'No Logo', 2000). The first task of branding was to grant proper names for generic goods such as flour, sugar, soap, and cereal, which had previously been scooped out of barrels by local shopkeepers. According to Klein, as early as the 1880s, branded corporate logos were introduced to mass-produced products like Campbell's soup, Heinz pickles, and Quaker oats cereal.

4

Source: ► https://unsplash.com/

Klein further emphasized that design historians uncovered a deliberate effort in crafting logos to instill a sense of familiarity. This strategic move led to the establishment of a nationwide lexicon of brand names, displacing the influence of small local shopkeepers. As consumers engaged with products through these distinctive names and logos, it laid the foundation for the consumer-centric branding paradigm we now acknowledge. Branding evolved into the consumer-based ownership we witness today. Manufacturers started to burn their identities onto products and anything that could carry the name. For example, the famous Guinness harp and the Bass red triangle are among the world's oldest recognized registered trademarks, being first registered in 1876 (Batey, 2008).

Source: ► https://unsplash.com/

An important reason why brand development began with logo design was because, at the time, in the 1800s, levels of illiteracy among populations across the globe were extremely high therefore, a graphic logo and trademarks were vital for people to distinguish a product and to become memorable, identifiable and ready to sell in a buying situation. But, of course, a brand is much more than a logo. A Harvard Business Review (HBR) article (see A Logo Is Not a Brand) wrote extensively about how a logo is not a brand. The article remarks that lots of people think that a brand begins and ends with the name and the logo, but the brand is much more than that, it is a strategy. It is the brand's products, service efforts, and product portfolio and the story the brand uses to sell those products/services. According to HBR, the brand is the business's call to action; it is the way the brand speaks the tone of voice (TOV), and messaging is central to this process, but messaging alone cannot fully constitute a brand; rather, a brand encompasses a comprehensive array of communication tools, particularly in the digital age, extending beyond mere messaging. Elements such as the user interface play a pivotal role. Moreover, a brand is intricately woven into the fabric of its workforce—the people who embody and represent the brand's values. The facilities that shape the brand's physical environment are about caring about the business at every level in a great amount of detail. It is this holistic fusion of marketing communication strategies, digital components, and the collective embodiment of values by the brand's personnel that truly defines and fortifies a brand's identity.

As with a lot of theoretical definitions in the academic literature alongside practitioner literature, many espouse the greatest definition of branding. Name and a trademark are a classic example of branding, however, in the contemporary environment, branding has evolved into something much more complex than its original purpose. Three key definitions of branding draw on similar features and are grounded in the seminal work on branding. The American Marketing Association's traditional definition of a brand includes "*a name term sign symbol or design or a combination of them intended to identify the goods of services of a seller or group of sellers and to differentiate them from those of competitors.*" In some detailed work on defining a brand, De Chernatony and Dall'Olmo (1998) found 12 broad definitions, these include a brand is a (1) legal instrument, (2) logo, (3) company, (4) shorthand, (5) risk reducer, (6) identity system, (7) image in consumers' minds, (8) value system, (9) personality, (10) relationship, (11) adding values, and (12) evolving entity. We also have the seminal work of Aaker (1996) and his brand equity model, which offers many unique paths to help clarify the brand strategy. Aaker (1996) suggests that his model provides a brand with a road map to achieve brand growth, he defines a brand as "*a brand vision should attempt to go beyond the functional benefits to consider organisational values a higher purpose brand personality and emotional, social, and self-expressive benefits.*" According to Aaker (1996), The customer-based brand equity model provides a framework for differentiating a brand by strategically managing and enhancing key elements such as brand awareness, loyalty, associations, assets, and quality. By optimizing these dimensions, a brand can create a unique position within its product category, fostering differentiation and competitive advantage.

4.2.1 Differentiation

Many of these definitions are rooted in a conceptual and theoretical basis; therefore, one of the most useful definitions of branding, particularly for practitioners and students trying to apply branding concepts to live situations, has come from Al Achenbaum (a distinguished marketing professor and seasoned marketing executive renowned for his impactful contributions to prominent brands). He suggests what distinguishes a brand from its unbranded commodity and branded counterpart and gives it its equity is the sum of consumers' perceptions and feelings about the product's attributes, how it performs, the brand name, and the company associated with producing it. This quote nicely captures the essence of what a brand should do by highlighting three areas vital to creating a brand. It talks about the equity of the brand; we know that Aaker, in his 2012 text "Building Strong Brands," describes brand equity as the value the consumer places on the brand. The quote also considers the consumer's perceptions about the product/service attributes and the meaning of the brand; we know that consumers ultimately position the brand. But most importantly, the definition sets up a nice continuum between a brand and a commodity.

■ Source: ► https://unsplash.com/

A commodity is the most basic of goods. Commodity features include a situation where supply exceeds demand, and there is no distinction between the commodity and similar others. Therefore, consumers perceive the product's functional benefit, i.e., it is consumed only to serve a core. Basic physical products are not meaningfully differentiated from one another; therefore, a brand is not the same as a good/product, and according to Kotler, "*If you are not a brand, you are a commodity. Therefore, price is everything, and the low-cost producer is the only winner.*" This sets up a continuum between a basic product and a branded product:

THE BRAND CONTINUUM

BRANDS ◄- ► **COMMODITY**

(A BRAND MEANS SOMETHING) **(ANYTHING CAN BE A COMMODITY)**

4.2.2 Differentiation Through Brand Storytelling

To demonstrate the brand continuum, let's take the basic commodity of water, which we used in ► Chap. 2. Water is an essential natural resource that all life on Earth depends on, and about 70% of the Earth's surface is covered with water (see Statista overview on Water & Wastewater). Furthermore, up to 60% of the human body is made up of water, with the amount changing slightly due to factors such as age, sex, and hydration levels (see Statista article on Average water consumption per person across India in 2021, by age group). Water is largely tasteless, odorless, and colorless. Recognizing that water is scarcer in some places, still in much of the world, water is delivered directly to people's homes at a relatively low cost and is easily available in great quantity. So how can some of the biggest brands in the world turn such a basic commodity into a recognizable brand that has meaning and is differentiated?

◘ Source: ► https://unsplash.com/

This happens through effective branding strategies that take a basic commodity from an undifferentiated product to a brand that has meaning. There are many differentiated approaches that premium water brands such as Voss and San Pellegrino have adopted; they include taste, flavoring, carbonation, color, the bottle design (packaging/labeling shape/color/functionality), ways to drink, size, price, bundling with other products, where it physically distributed and available. But most importantly, according to San Pellegrino, is the story that has emerged and evolved over the brand's historical time to include a focus on "*design, tradition and taste, that begins in the Alps to eventually arrive at your table*" (San Pellegrino, 2024). The brand has successfully leveraged the country of origin and the brand's heritage to tell a story that continues to resonate strongly with its consumers: "*This icon of Italian taste has been passed down to become our Tasteful Italian Story. This is a story in the making, and we will nurture it for another 120 years.*" This is a great example of how you can turn the most basic commodity form into a brand that has meaning, is differentiated, and can charge a premium for its products. Brand storytelling is an approach that involves using narrative elements to convey the values, history, and unique aspects of a brand; this goes beyond simply presenting facts and features, aiming to engage and connect with the audience on an emotional level to build relationships with consumers and differentiate themselves in the market. San Pellegrino has utilized brand storytelling to craft a narrative around its premium mineral water. Through reminiscent storytelling, they convey the rich history of the source, the picturesque town of San Pellegrino Terme in Italy. This storytelling strategy not only accentuates the brand's heritage but also creates a sense of authenticity, associating the product with a tradition of purity and excellence.

According to Kotler and Armstrong (2010), "*Differentiation is the act of designing a set of meaningful differences to distinguish the company's offer from competitors' offers.*" For decades, the literature in marketing has focused on the notion of "differentiate or die," which suggests consumers choose brands based on meaningful differences. In recent years, there has been debate over the use of the term differentiation, with some suggesting (Romaniuk & Sharp, 2016) that brands are, in fact, different from one another, and they are perceived as such by customers, but differentiation plays a small role in how brands compete. Instead, brands should focus on being distinctive as an alternative perspective (Sharp, 2010).

4.2.3 Distinctive Brand Assets

Distinctive brand assets (sometimes referred to as elements or brand codes) are used to create meaning. The brand minus the general equivalent equals the associations in your consumer's mind, this, in turn, amounts to the brand's positioning in the mind of the consumer relative to the competition. Brand elements/codes are a set of remembered qualities that help communicate information to the customer.

These qualities should differentiate one brand from the competition and, therefore, provide a reason to buy that brand over the other. According to Sharp and Romaniuk, differentiation has to be perceived and valued by the customer for a brand selection to take place, but distinctiveness is also required. Distinctiveness, or distinctive elements/codes, refers to any characteristics that serve the purpose of clearly and unambiguously conveying the brand's identity, aiding customers in easy recognition without confusion, particularly amid a competitive and noisy marketplace (Sharp, 2010). These include brand names, colors, logos, taglines, characters, or even advertising styles. An example of this in practice and how this has worked well is the Snicker's case study discussed by Mark Ritson.

◘ Source: ► https://unsplash.com/

Snickers is one brand that Mark Ritson showcases for having strong, distinctive brand codes/elements. According to Ritson, in 2009, Mars' Snickers brand was in decline—losing sales and share, the brand's campaigns had no clear messaging to differentiate them from rivals, and it was underutilizing the distinctive assets it had established over decades (see Marketing Week article—Ritson on how Snickers perfectly balanced differentiation and distinctiveness (n.d.-b)). Ritson suggests brands have three or four distinctive brand codes, which could include a logo, color, shape, pattern/design, or image. He states that Snickers has four clear codes: its logo, the parallelogram, the picture of the inside of the bar, and the brown Pantone. As soon as Snickers re-aligned efforts and started communicating these distinctive codes, sales increased. Skillful use of distinctive codes was an immediate success, with the sales decline first detained and then reversed. Sales increased by 15% globally between 2009 and 2011, and in 56 out of its 58 markets, sales were on the up again. Distinctive brand codes benefit both the marketer and the consumer. For the marketer, they can be used in any marketing activity to build, refresh, or reinforce the consumer's memory of the brand, for the consumer, these distinctive codes make it easier for the consumer to identify the brand in a buying situation (Sharp, 2010).

4

Manager's Corner 4.1: The Rainforest Alliance

Background The Rainforest Alliance is a globally recognized non-profit organization that works to conserve biodiversity and promote sustainable livelihoods by transforming land-use practices, business practices, and consumer behavior. Founded in 1987, the organization has since become a leading force in advocating for sustainable agriculture, forestry, and tourism.

Impact Since the brands started in Central America more than 30 years ago, they've built an alliance that includes farmers and forest communities, companies, and consumers—working together to conserve critically important forests and cultivate more sustainable livelihoods. Today, they have sustainable agriculture training and certification, as well as landscape projects, in more than 60 countries around the world.

Environmental Conservation The Rainforest Alliance has made significant strides in protecting forests, conserving biodiversity, and mitigating climate change through its rigorous certification programs. It works with farmers, foresters, and businesses to promote sustainable land management practices, reducing deforestation and habitat destruction. Social Impact—By promoting sustainable practices, the Rainforest Alliance has positively impacted the lives of millions of workers and communities around the world. Through its certification programs, it ensures fair wages, safe working conditions, and access to education and healthcare for workers in the agricultural and forestry sectors. Economic Development—The Rainforest Alliance's approach to sustainability not only benefits the environment and communities but also contributes to the economic viability of businesses. By helping companies adopt sustainable practices it opens up markets, enhances brand reputation, and fosters long-term resilience and profitability.

Distinctive Assets Analysis of Certification Program

How does the Rainforest Alliance's certification program contribute to its brand recognition and consumer trust?

What are the key criteria for certification, and how do they differentiate the brand from its competitors?

Analyze the impact of certification on consumer behavior and market demand for sustainably sourced products.

Information Sources:

- https://www.rainforest-alliance.org/
- https://www.rainforest-alliance.org/impact/

4.2.4 Differentiation vs. Distinctiveness

In essence, both differentiation and distinctiveness are important in defining and building a brand. Both of the main recent contributors in this area, Ritson and Sharp, have suggested that the marketing industry has placed too much emphasis on differentiation, positioning, and purpose rather than distinctiveness and building strong, recognizable brand elements. Both have suggested that marketing textbooks for decades have wildly overestimated the importance of differentiation and ignored the impact of distinctiveness. Sharp's work is often associated with empirical research and the Ehrenberg-Bass Institute's findings. He emphasizes that brand growth is primarily driven by reaching more buyers (i.e., increasing penetration) and not by differentiation. His research challenges the idea that differentiation is the primary driver of brand growth. Ritson relies on a broader range of marketing literature, including research that suggests that differentiation can be a valuable strategy in certain contexts. He highlights that while penetration is crucial, differentiation can help create preference and loyalty among existing customers. In reality—brands can use differentiation and distinctiveness to create a strong and memorable brand identity in the minds of consumers. Differentiation focuses on setting the brand apart from competitors, while distinctiveness involves creating a unique and easily recognizable brand image or element. In reality, a good marketer should focus on both when building a brand, as both assist the consumer's decision-making process.

Guidance in decision-making is imperative due to the inherent tendency of humans (who are consumers) toward making errors in their decision-making processes. When framing decisions regarding communication strategies with audiences, marketers frequently prioritize what Daniel Kahneman terms as system two thinking. This mode of thinking directs attention to demanding mental activities that emphasize deliberate, conscious decision-making, in contrast to the swift, intuitive decision-making characteristic of system one thinking, as described by Kahneman (2011). System one thinking is the fast, subconscious, automatic, effortlessly originating impressions, feelings, and thinking that drives approx. 95% of decision-making. System one thinking is mostly used by consumers to make decisions about brands. System one thinking also has important implications for understanding decision-making about brands and how consumers create responsible, ethical, and sustainable habits, particularly in the current environment where brands must formulate a strategy that strategically positions the facilitation of consumers in educating sustainable habits at its core. It, therefore, makes sense when building any brand to focus on distinctive codes because the consumer who typically uses system one thinking to decide on a purchase will find it easier to use distinctive codes in a buying situation. Accordingly, Mark Ritson suggests, "*Differentiation is important but distinctiveness and being there when those immediate system one choices are made is far more important.*"

4.3 Branding, Marketing, Advertising, and the Promotional Mix

Often, marketing, branding, advertising, and promotion are used interchangeably, and although related, they are distinct concepts. The goal of these concepts is to increase sales/profit and enhance a business's reputation, but essentially, branding and advertising are subsets of marketing. It is generally assumed that branding is what the marketing and communications team deals with, for example, the promotional tactics that a business produces, but branding is so much more, it is also not a substitute for marketing, both are needed. Marketing is the activity a company engages in to bring a product or service to market, branding is everything a company does to bring a business to life once they have decided what product or service to introduce to the market (Smith & Fischer, 2021). Brand, business product, or service are used frequently, sometimes to describe the same thing, but they are different, the business is the company or the organization that produces the products and services, and the brand is the identity, reputation, and image that is built in the consumer's mind about the business producing those products and services. The marketing promotional mix is often used to assist in marketing activity and building a brand. There are a variety of tools and media used to help engage an audience, and within this mix (sometimes referred to as the promotional mix or the communications mix) is where advertising is situated.

Advertising refers to any marketing communication that reinforces brand awareness and positive brand image and associations. This is often achieved via physical store atmosphere, packaging, endorsements, and product placement, as well as other traditional forms of mass media such as television, radio, magazines, newspapers, print, billboards, and, of course, in the contemporary environment, any online, digital, or social media communication. "Advertising" includes any message where the primary communications objective is brand awareness and consumer attitudes toward a brand, regardless of the media used (Rosenbaum-Elliott, 2020). The primary aim of brand advertising is to affect the buyer behavior of consumers (Sharp, 2010). There has been a lot of debate and discussion about how advertising works and its role in assisting branding. Understanding consumer behavior and how consumers respond to advertising communicated by brands has been central to this. The traditional approach to understanding how a consumer makes decisions about selecting a brand implies that consumers follow a sequence of stages. These stages recognize that the consumer has a need, they then gather some information to help make the decision, an evaluation of competing brand's offerings is considered, a purchase is then made, and then a post-purchase evaluation begins.

This "idealized" information processing model assumes a rational consumer who is motivated to make decisions with near-perfect memories. In reality, there is much divergence from this classic mode of thinking because advertising works by creating and refreshing memories about brands, and much thinking and decision-making is non-conscious and emotional, not completely rational (Sharp, 2010). Classic decision-making models often overlook the intricacies of leaky memories and the reality that consumers, particularly in certain product categories, may not always seek to make optimal choices; instead, a satisfactory "that will do" mindset can suffice. Understanding the nuances of decision-making, from leaky memories to the acceptance of satisfactory choices, becomes pivotal in probing into the theories behind brand decision-making. Theoretical influences on branding theory and practice are important if we are to truly understand how brands can be used to assist consumers in making better decisions about consumption activities that impact (ethics, sustainability, and responsibility) people and the planet. The next section explains some of the key theories that underpin branding, consumer behavior, and decision-making.

4.4 Theoretical Influences on Branding

4.4.1 Traditional Theoretical Influences on Brands: The Decision-Making Models

While consumption has emerged as a central feature of contemporary society, it is important to consider how the shift in paradigms has influenced the meaning of branding. Traditionally, an economic approach to branding was adopted, which assumed consumers make rational choices with perfect information. Seemingly, consumers were seeking utility maximization with limitless time. There was also an assumption that consumers were simply viewed as passive receivers of brand meanings with brand value managerially determined. This received view of brand meaning is profoundly embedded in the objectivist information processing and cognitivist theories of consumer behavior. Critiques often emerged relating to the information processing models, particularly regarding their often simplistic, linear nature, which oversimplifies the complex and dynamic reality of consumer decision-making processes. Within this frame, the brand holds ontological status explicitly as a cognitive understanding: the brand exists in the mind of the consumer in a linear and structured way. From this perspective, brand managers largely communicate to consumers with a focus on sending a consistent brand image/message using the marketing mix and the functional performance of a product is used as an advantage. This strategy was regarded as unsustainable over the long term, particularly with the shift in consumer perceptions in the current environment (Beverland, 2021). The emphasis was also on brands owning economic value, and this value should be protected at all costs. This idea in many countries was influenced by law and reflects a heritage of trademark laws, even the American Marketing Association once referred to trademark law in their definition of a brand.

At the time, this made sense because brands had economic value, and protection was sought to safeguard logos, taglines, colors, advertising styles, etc. As competition increased in markets, brand managers' attention was guided by safeguarding their brand. Copying became ubiquitous, if a brand had a profitable new product idea, competitors at some point tried to capture market share from the successful brand. Therefore, protecting the brand's most valuable assets became integral in the machine age to maintain a brand's distinctiveness (Smith & Fischer, 2021). A lot of traditional theoretical influences on branding have been broad and include wider business functions, such as brand management, the financial worth and assets of a brand, organization studies, and operations. The focus here was on the business balance sheet, return on investment, and share price. Traditional theories of branding include how consumers make choices based on linear cognitive information processing; these were limited in what they could explain about consumer behavior. In reality, information processing models do not include a greater awareness of how the human mind works. Theorists started to consider the psychological, cultural, and sociological impact on consumer decision-making, in essence, how do human minds work and how do they behave? And most important, for this chapter, how do brands exist in the human mind?

4.4.2 Contemporary Thought on Theoretical Influences: The Brain and Memory

It is worth remembering the discussion on defining a brand—a brand is the set of associations, perceptions, and images that exist in the consumer's mind. As such, every encounter (either a purchasing encounter, a consumption experience, or mere exposure to an advertisement) a consumer experiences with a brand serves to create, refresh, or maintain the consumer's overall perceptions of the brand (Sharp, 2010). For a long time, marketing focus has been on *how to persuade* a consumer to buy their brands through advertising, and because of this, the crucial role of human memory has been neglected. Persuasion is very important and continues to be very effective in buying situations. However, persuading a consumer to buy from a brand often involves influencing immediate purchasing decisions while using strategies to ensure the brand is stamped in the consumer's memory, creating a lasting and memorable impression for future recall, and brand loyalty has been overlooked by some seminal theorists.

Understanding how consumers process information to build stronger and more positive brand perceptions among a brand's target audience is crucial. Memory is a process of acquiring information and storing it over time so that it will be available when an individual needs to access that information (Bettman, 1986). Neuroscience researchers have found that memory serves as a generative, reconstructive, and dynamic entity, but memory is not a literal reproduction of the past; instead, it is an ongoing constructive process, and therefore, memories are modified and reconstructed repeatedly (Ofengenden, 2014). Retrieval of a past experience involves assembling scattered memory features from different neural locations. This assembling, according to neuroscientists, is not a simple process in which all the pertained features obey at the time of retrieval to form an exact representation of past occurrence (i.e., a memory). Rather, retrieval is a process in which features from the original memory are omitted or lost, and features from other associated memories are attached to the retrieved memory. The interaction between a cue and a memory representation can add or eliminate information from the encoded engram, import-related information from other engrams, change the order of encoded information, and so on. Thus, according to researchers, remembering something is not like storing a file on the computer, which remains the same from the moment we close the computer at night to the next morning when we open it again. Instead, every time we access a memory, we come up with different results. Schachter (1996, p. 71, cited in Batey, 2008) contests the traditional view that memory is simply an activated engram of past events, he disputes that a piece of information is stored away somewhere in our brain until the conscious experience of the memory results in this piece of information being activated. As such, memories are neither created from scratch nor do they constitute an activated picture of a past event. Rather, the brain stores information, events, and experiences by strengthening connections between different neurons that participate in the encoding of an experience. Encoding refers to converting information or a message into a meaningful code in such a way that the information enters the brain in a mode it will recognize, the brain mixes this new information with what is already in the memory and stores it until it is needed in the future. Therefore, brands rely on consumers to retain the information they collect about their products and services so that consumers can apply the retained information to new, future purchase decisions. This works within what is called a brand-associative network (Batey, 2008). In the network, new information and experience connect with existing networks, this new information about the brand passes down a neural pathway that modifies the brand information and adds to the associative network. As these connections between brand associations are reinforced over time, they effectively come to define the brand in the consumer's mind, and these collective associated meanings of a brand, therefore, resemble the structure of a neural network.

Section Sources: *Sharp* (2010), *Batey* (2008)*, Doidge* (2007).

For brand managers, this information is vital as memory significantly influences consumer choice and decision-making. An advertisement cannot build memory structures if it is not processed, and memory structures cannot generate sales if they are not associated with the brand that is being advertised (or even worse, evoke memory structures for competition) yet, according to Sharp (2010), most advertising exposure fails these two hurdles. The complex network of rational and, more frequently, emotional and unconscious needs that initiate consumer behavior is described as a "need state," and brands succeed or decline in their ability to meet these consumer needs (Batey, 2008). Emotional and identity-related needs are met by the symbolic benefits that brands offer consumers (Elliott & Wattanusuwan, 1998). Researchers have shown that the most vivid autobiographical memories tend to be of emotional events, which are likely to be recalled more often and with more clarity and detail than neutral events (Baddeley, 1982). Emotions emerge as reactions to events or experiences that hold significance for individuals' goals and motivations. Consumer engagement is intricately tied to motivations and emotional responses, serving as the inducing force driving a significant measure of consumption, but only if the brand can capture consumer attention. The way a marketer presents (or frames) the message influences the likelihood that a consumer will be able to recall it later. Early memory theorists suggest that memories simply fade over time, in the process of decay, the structural changes that learning produces in the brain simply disintegrate, and previous information is displaced as we learn additional information, which is why consumer's attention is so important for marketers (Doidge, 2007). According to Sharp (2010), advertising can indeed work without consumers paying much attention, at a subconscious level, we can notice things, but just because the brain is monitoring the environment does not mean it is letting much of this information into our long-term memories. The work of Doidge (2007) also supports this assertion, he advocates that brain space is finite, we have finite neurons, and we need to conserve both space and neurons for tough decisions, yet we often praise the ability to multitask—while we can learn when we divide our attention, divided attention does not lead to abiding change in our brain maps—change only occurs when paying attention. This poses the fundamental challenge: brand managers must grasp the workings of memory and attention to navigate the brain's screening process successfully, prompting a modest emotional response that refreshes the memory structures associated with the brand (Sharp, 2010). We know that emotion exerts a substantial influence on attention and memory, and a prime source of emotional involvement is the search for identity. But, if past events cannot be remembered, language, relationships, or personal identity can't develop. Yet, we must consider that identity isn't just constructed by the self, it is also constructed in a socio-cultural context.

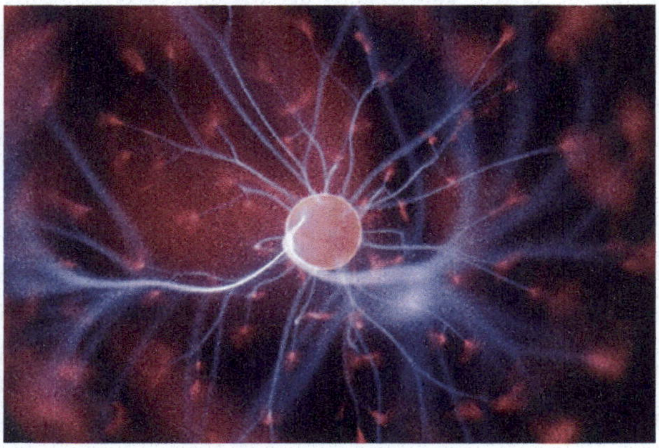

Source: ► https://unsplash.com/

4.4.3 The Socio-Cultural Context and Identity

The evolution of self-identity is intricately linked with the simultaneous development of collective social identity (Jenkins, 2014), and endeavors to create a person's self-identity often involve the consumption of products and services. The contemporary environment is rich in symbolic meaning. Through the socialization process, consumers learn to agree on shared meanings of certain symbols as well as develop individual symbolic interpretations of their own. Consumers use these symbolic meanings to construct, maintain, and express each of their multiple identities. Because of this, how humans make sense of themselves is greatly aided by symbolic resources, and as consumption plays a central role in supplying meanings and values for the creation and maintenance of their personal and social world, advertising is then recognized as one of the major sources of these symbolic meanings (Elliott & Wattanusuwan, 1998). Through advertising, cultural meanings are transferred to brands, and it is brands that are often used as symbolic resources for the construction and maintenance of identity (McCracken, 1986; Mick & Buhl, 1992). Brands also possess their own identity, and according to Holt (2004), this is how brands become cultural icons. Holt (2004) suggests iconic brands provide extraordinary value because they address the collective anxieties and desires of a nation, as such, people are constructing their identities in response to the same historical changes that influence an entire nation. Holt's work focuses on how big brands in the world become cultural symbols that hold deep meaning for consumers. He argues that successful brands go beyond traditional marketing and advertising strategies to establish themselves as cultural icons, here we use Nike as an example to demonstrate some of the key concepts Holt discusses and consumer culture more widely:

- Storytelling and Mythmaking: Nike has crafted a powerful brand narrative centered around the idea of "Just Do It." This slogan encourages consumers to overcome obstacles and pursue their dreams, effectively creating a myth around the brand that resonates with athletes and anyone striving for success.
- Cultural Orthodoxy: Nike aligns its brand with the prevailing cultural values of determination, excellence, and athleticism. It has a long history of endorsing high-profile athletes who embody these values, such as Michael Jordan, LeBron James, and Serena Williams, making Nike a symbol of sports achievement.
- Subcultural Authenticity: Nike has successfully tapped into various subcultures, including streetwear and urban fashion. Collaborations with designers like Virgil Abloh's Off-White and artists like Travis Scott have helped the brand stay relevant within these niche communities.
- Cultural Innovation: Nike continually innovates its products and marketing strategies. They were one of the pioneers in integrating technology into sportswear with products like Nike+ and the Nike FuelBand. This constant innovation keeps them at the forefront of sports and athletic culture.

- Rituals and Symbols: Nike's iconic "swoosh" logo and the Air Jordan line of shoes have become symbols of style, athleticism, and status. The act of lacing up a pair of Nike sneakers has become a ritual for athletes and casual wearers alike.
- Cultural Resonance: Nike advertisements often address social and cultural issues, such as gender equality, racial equality, and environmental sustainability. These campaigns not only resonate with consumers but also position Nike as a socially responsible brand.
- Brand Communities: Nike has built a massive global community of athletes and fitness enthusiasts who identify with the brand. Events like the Nike Women's Race and Nike Run Club provide opportunities for consumers to connect with each other and with the brand itself.

For this reason, Holt (2004) suggests brands become meaningful because they accomplish things in society for consumers' identities, and customers who make use of the brand's symbolic meanings for their identities forge tight emotional connections with the brand.

We have discussed the cognitive and information-processing models that impact consumer decision-making, as well as neuroscience and psychological considerations regarding how the brain works with memory and attention. The socio-cultural impacts on identity creation and how consumers use brands as props to project a desired identity (status) have been considered. There is one final (significant) theoretical influence that considers how consumers make decisions about brands. This also has roots in how the human brain works and includes some of the typical cognitive errors that disrupt human thinking and brand choice-making. This theory is underpinned by behavior sciences, and within this field, there are three main ideas, which include bounded rationality, prospect theory, and nudge theory. There is not enough room within the scope of this chapter to discuss this theory in depth as the focus here is on understanding how certain theories are more relevant to marketers of sustainable brands hoping to improve and shape consumers' habits. However, a brief overview and, in particular, some further detail on nudge theory is what we will discuss in this final section on the theoretical influences on branding.

4.4.4 The Influence of Behavior Sciences on Consumers and Brands

Bounded rationality views decision-making as a fully rational process but suggests the idea that humans take shortcuts that may lead to sub-optimal decision-making behavior, this is because humans have evolved to conserve energy. Behavior economists such as Herbert Simon suggest that humans do not assess all available options appropriately because they want to save on search and deliberation costs; therefore, because of this, decision-making is not always made in the sense of the greatest reward because limited information is available. Prospect theory includes some typical cognitive errors that the brain makes when making decisions. Key theorists in the area include Kahneman and Tversky (2013), who propose that the handling of gains differs from that of losses. Choices may be influenced by outcomes attained with certainty, as they tend to be given more weight compared to uncertain outcomes on the scale of the problem. They also suggested that prospect theory comprises two stages: the editing stage and the evaluation stage. The editing stage holds particular significance for brand marketers, as it involves navigating risky situations such as purchasing, often employing various heuristics for simplification. Kahneman's research established that common human errors can arise from heuristics and biases. From our brief discussion on this earlier on in the chapter, we know that system one (fast thinking) utilizers heuristics, we also know that decision-making around brands and products uses system one to undertake shortcuts, and sometimes it can provide an immediate solution, but often this mode of decision making is irrational and imperfect. A heuristic is a strategy that ignores part of the information (a cognitive shortcut), intending to make decisions more quickly (and conserving brain energy for other tasks). An example is the availability heuristic: people judge how likely something is by how easy it is to recall instances of that type. (See article on Availability Heuristic from The Decision Lab, n.d.-a). The availability heuristic leads to a cognitive bias where humans make a judgment about something based on how available examples are in their minds. This is important for brands because the idea implies if something can be recalled, it must be important, or at least more important than alternative solutions (e.g., other brands in the category) not as readily recalled, and therefore is inherently biased toward recently acquired information. This bias inherently favors recently acquired information, emphasizing the significance of brand visibility, memorable advertising strategies, and maintaining a consistent, positive presence in consumers' minds. Brands that successfully leverage the availability heuristic can potentially enjoy a competitive advantage in the marketplace.

Nudge Theory "A nudge, as we will use the term, is any aspect of the choice architecture that predictably alters people's behavior without forbidding any options or significantly changing their economic incentives" (Thaler & Sunstein, 2008). To be considered as a mere nudge, an intervention should be effortlessly implemented and cost-effective to sidestep (e.g., should not have financial implications for the user). Nudges are distinct from mandates in that they prioritize subtle influences over outright restrictions. Nudges over mandates are considered beneficial as they uphold individual freedom by employing subtle influences instead of imposing rigid restrictions. Nudges, through their non-intrusive nature, respect people's autonomy and allow for a more adaptable, effective, and socially accepted approach to influencing behavior. An illustration of a nudge would be strategically placing fruits at eye level to encourage healthier choices, as opposed to imposing an outright ban on junk food.

Source: ► https://unsplash.com/

Nudging is, therefore, increasingly used to "nudge" people to make better decisions (Thaler & Sunstein, 2008). Nudges address the reality of our limited brain energy, directing focus to essential matters amid the overwhelming array of potential options. While traditionally used to capture personal attention, there's a rising trend in employing nudges for the collective good. For instance, a climate-change nudge aims not to benefit the individual decision-maker but to contribute to broader goals like reducing greenhouse gas emissions. In Switzerland, for instance, people have been nudged to automatically enroll in clean-energy programs. If they don't want to, they can opt out, although the "dirtier" program may be more expensive. That nudge is designed to protect people from climate change generally, not necessarily to protect individual choice-makers.

Section Source: see Mckinsey article on much anew about nudging (n.d.).

Nudging contrasts with other ways to achieve compliance, such as enforcement. We will discuss in depth some of the negative associations levied at nudging later on in the chapter. To summarize this section for brands, we know consumers respond to nudges through framing effects, influenced by how information is presented. They also rely on heuristics for making satisfactory, if not flawless, decisions. Sharp (2010) contends that consumers adeptly utilize these mental shortcuts. For instance, the heuristic of associating higher prices with better quality generally holds true, contributing to effective decision-making.

4.4.5 Summary of Theoretical Influences

In essence, as well as the emotional impact on the brain and, in particular, memory, consumers are somewhat conscious of the symbolic benefits that certain brands bring to their lives. For example, status symbolism refers to a very deliberate act by the consumer to acquire high-priced and perhaps luxury goods to signal a certain position in society. Or, and possibly very relevant for this chapter's discussion, a consumer may find themselves aligned to a brand that embodies values the consumer would like to project—e.g., I'm a conscious consumer, I care about the environment. This section explored the theoretical influences and the multifaceted role of brands, examining their use as tools for projecting identity and the interplay between symbolic meaning and functionality. Ethics, sustainability, and responsibility become pivotal, particularly in the context of conscious consumerism. The evolving realm of behavioral sciences, encompassing various cognitive biases, was also highlighted, and it emphasizes that brand managers in the modern era can strategically leverage these biases to address sustainability challenges and gain a competitive edge.

4.5 Physical and Mental Availability

We have already discussed that creating a successful brand is a multifaceted endeavor, under the scientific laws of marketing, there are two critical components in achieving this goal: brands largely compete in terms of physical and mental availability. This holds significant importance in the context of our discussion in this chapter, especially when examining how brands can influence consumer habits. The primary objective of any marketing initiative for brands is to ensure the brand's accessibility for purchase, with all other considerations taking a secondary role (Sharp, 2010). We revisit the vital discussions we had about how the brain works here and how the brain holds and links pieces of information together. We know the brain works by creating associations between things; for the consumer to make a purchase, the brand must create links to the category entry point (a category could be coffee, for example), this helps to wrap the brand with the positive experience of the occasion. However, researchers in the area of physical and mental availability have argued that too much emphasis has been placed on persuading the customer:

> *A lot of advertising is geared to persuasion rather than building mental availability. Persuasion is about assuming you're in the room and you're arguing your point, whereas mental availability is about getting into the room. And so that's the big challenge: most organisations are failing to get into the room, but they're spending all their money on arguing as if they're already there.*—Prof. Jenni Romaniuk, International Director, Ehrenberg-Bass Institute (2022)

While the persuasive approach remains undeniably influential in advertising, an exclusive emphasis on persuasion overlooks the critical aspect of dedicating efforts to establish both physical and mental availability. Engaging consumers requires not only convincing them of a product's/service's merits but also ensuring its presence and prominence in their physical and mental spaces. This dual focus not only enhances the effectiveness of marketing efforts but also contributes to building lasting connections with the target audience by fostering a sense of accessibility and familiarity.

Mental and physical availability are often referred to as key drivers of brand success because they address different aspects of a brand's presence in the market and its ability to connect with consumers. Mental availability (brand awareness) and physical availability (accessibility and distribution) are both critical for brand success. Mental availability is a measure of the breadth and depth of perceptions of a brand. So, a buyer could be aware of a brand but know nothing else about it—which equates to low mental availability and, therefore, less chance of the buyer choosing that brand in a buying situation. High mental availability is when a consumer has high awareness of the brand and knows what it does and what it stands for. According to Ehrenberg-Bass, a higher level of mental availability makes people far more likely to buy. Mental availability helps create a strong brand image and ensures consumers think of the brand when making purchase decisions. Physical availability, on the other hand, ensures that consumers can easily find and purchase the brand's products, which is essential for converting awareness into sales and fostering brand loyalty. Physical availability refers to the brand's presence and accessibility in the real world, primarily through distribution and retail channels. It encompasses the brand's ability to be readily available when and where consumers want to make a purchase.

Section Source: Sharp (2010).

This all sounds very easy, if the goal is to change behavior, then simply make sure the brand is easy to think of and easy to buy in a purchasing situation, however, human behavior is complex, and the need for a transition toward a sustainable way of living is now more pressing than ever. For brands to be successful in their endeavors to produce change, this will not only require a solid understanding of how people behave in different buying situations but also an understanding of behavior in different contexts. We consistently underestimate context, and this is truly important if brands are serious about disrupting consumers' habitual behavior. According to Shotton (2018), brands need to adapt their communications to harness the power of consumer habits because nearly half of behavior is habitual, and habits are context-specific. Because habits are so hard to break, Shotton suggests a dramatic shift in one's environment is required to alter those habits; therefore, since habits are context-specific, if a consumer's environment changes, the habit becomes moveable, for example, when consumers undergo a life event their environment is changed enough to destabilize habitual behavior, this can include an event such as securing a new job, starting university, having a baby, or getting married.

Other research on choices shows that customer demand is influenced by the way choices are presented to them (Thaler & Sunstein, 2008). For example, restaurant menus can be (re-)designed in such a manner that they inform consumers intending to "nudge" more environmentally and societally beneficial food choices. We know that nudging builds on the understanding that not only the quantity or the accessibility of information provided to people matters but also how this information is presented, i.e., framed. Framing is the conscious phrasing of information (by brands) in a way that triggers certain values and attitudes of individual consumers, re-naming food items on a menu is a good example of this (Lehner et al., 2016). Furthermore, easy access to unhealthy food, for example, significantly increases the consumption of such food. Even the visibility and smell of unhealthy food impact consumption levels. Framing can also be used to nudge customers toward healthy options, as studies have found that labeling that adopts positive attribution is far more effective than negative. For example, 80% fat-free is more effective than 20% fat (Decision Lab (n.d.-b) article on 'The Framing Effect'). Focusing on the effect of size on food waste, Kallbekken and Sælen (2013) conducted a study among hotel guests in Norway. They reduced the plate diameter from 24 to 21 cm at restaurant buffets in seven hotels and found that, on average, food waste was reduced by almost 20%. Simplification means that information is made more straightforward and presented in a way that fits the information processing capabilities and decision-making processes of the individual. It is especially of value in complex products or services, e.g., financial or investment decisions—this is important information for brands and sustainability. Simplification is crucial for brand messages in the sustainability sector because it ensures that complex information about environmentally friendly products or services is made more accessible to a broader audience. The complexities of sustainable practices, eco-friendly technologies, and their impact on the environment can be challenging for consumers to grasp, particularly as the focus on a business impact on the environment is relatively new. By simplifying the messaging, brands can effectively communicate the benefits of their sustainable offerings, making it easier for consumers to understand and integrate eco-conscious choices into their decision-making processes. This not only enhances consumer awareness but also encourages widespread adoption of sustainable practices, contributing to the overall success of environmentally responsible brands.

Although many nudges aim to help individuals make better, ethically informed decisions, and because nudging interventions have become so easy to design, relying somewhat on behavioral insight, there are claims that sometimes nudges affect different people differently. Therefore, the use of strategies like framing and nudges to influence choices can be controversial due to concerns surrounding autonomy, transparency, and ethical implications. Critics argue that these techniques may manipulate individuals by subtly guiding their decisions without their full awareness or informed consent. Additionally, the lack of transparency in these persuasive tactics raises questions about the fairness of influencing choices without explicit acknowledgment. Some have suggested nudging, rather than educating or alerting, is a manipulative act of power that results in groups such as policymakers making decisions for other people rather than letting people take responsibility for their own lives, limiting autonomy (Zuboff, 2019). Striking a balance between influencing choices for positive outcomes and respecting individual autonomy remains a challenge, leading to ongoing debates about the ethical boundaries of behavioral interventions in decision-making processes.

4

Marketing in Action 4.1: Scenario Planning Case Study: Scaling the Tofoo Brand in the Plant-Based and Meat Substitute Market

Background Tofoo is a pioneering brand in the plant-based food industry, specifically focusing on tofu-based products. Established with a mission to provide delicious, nutritious, and sustainable alternatives to meat, Tofoo has gained significant traction among health-conscious consumers and those seeking environmentally friendly food options. Tofoo's branding is characterized by its commitment to quality, sustainability, and innovation. The brand's logo features vibrant colors, symbolizing freshness and vitality, while its packaging is designed to stand out on shelves (mental and physical availability), conveying a message of modernity and health consciousness. Moreover, Tofoo's branding emphasizes transparency, showcasing its commitment to using non-GMO, organic ingredients, and sustainable manufacturing practices. The website serves as a hub for engaging content, including recipes, nutritional information, and articles promoting the benefits of a plant-based diet. Tofoo aims to educate consumers about the versatility and health benefits (high in protein and essential amino acids) of tofu while fostering a sense of community among its target audience.

Objective The objective of this marketing task is to leverage Tofoo's branding strategies to scale the brand not only within the tofu market but also in the broader plant-based and meat substitute market. By effectively positioning Tofoo as a compelling alternative to traditional meat products, the goal is to capture a larger share of the market and compete with meat-based offerings on taste, convenience, and sustainability.

- **Scenario Planning Case Study**

Market Analysis: Conduct a current and up-to-date comprehensive analysis of the plant-based and meat substitute market, including current trends, consumer preferences, and competitor landscape.

Identify key opportunities and challenges for Tofoo within these markets, considering factors such as product differentiation, distribution channels, brand awareness and consumer perception.

Brand Positioning Evaluate Tofoo's current brand positioning and identify areas for improvement to effectively target both the plant-based and meat substitute market segments. How is the brand's Tone of Voice landing with target consumers?

Product Expansion Explore opportunities for expanding Tofoo's product line beyond traditional tofu-based products to appeal to a broader audience.

Marketing Campaign Design a multichannel marketing campaign aimed at increasing brand awareness and driving sales in both online and offline channels. Utilize Tofoo's informative and entertaining website as a central platform for content distribution, leveraging engaging recipes, blog posts, and user-generated content to attract and retain consumers.

Partnership Development Identify potential strategic partnerships with retailers, restaurants, or influencers to expand Tofoo's reach and enhance brand credibility.

Information Sources:
- ▶ https://tofoo.co.uk/
- ▶ https://www.warc.com/content/article/the-tofoo-company/119757

4.6 Managing Brands and Their Common, Ethical, Responsible, and Sustainable Concerns

With the growing awareness of society on problems regarding the planet, the relationship between business and sustainability has been thrown into the spotlight. A responsible brand is a brand that includes ethical, social, and environmental values as well as economic ones in its commitment to the stakeholders. As we have explored for most of this chapter, branding, when done well, offers a business a significant competitive advantage. We know that branding strategies exert considerable influence over consumer behavior, habits, and purchases, and because of this, ethical concerns invariably arise around strategies used by brands to influence behavior. According to Kotler, there once was a time when consumers were confused about ethical attitudes and brands:

» *Ethical consumption is perhaps the biggest movement in branding today. However, consumers are confused. The public is not sure what an ethical brand actually is. There are so many angles to being ethical.*

While there are so many angles to being ethical (responsible and sustainable), which we will address in the next section, there is certainly reduced confusion among consumers, as consumers today are more attuned to brands' business practices and, therefore, to become a fully sustainable brand, businesses need to consider the impact of everything they do and how the consumer perceives this. A 2021 report found that ethical consumer spending and finance in the UK in 2020 broke through the £100bn mark for the first time. It also revealed that all sectors had seen growth, but the most significant growth has been in eco-travel and transport and green home spending, such as green electricity tariffs and energy-efficient appliances. The report is produced in conjunction with the Co-op Group, and Steve Murrells, CEO of the Co-op Group, said:

» *Our Ethical Consumerism Report is a barometer on consumer behaviour—and shoppers are turning up the heat by boycotting businesses which fail to act on ethical or social concerns. The report is a warning to brands that they must do business a better way for workers, communities and the planet. But it also offers clear evidence to policymakers that they can positively influence change.* (Source: ▶ ethicalconsumer.org)

In the 2022 report, spending on ethical products grew by almost 35% in 2021, to £141.6 billion from £106bn the previous year. This is the largest jump in the value of the market since ethical consumers started tracking these figures in 1999. While this is very promising, it's important to note the 2023 report has shown us that the ethical market is not immune to the wider uncertain economic backdrop, with growth slowing significantly across many sectors for the first time since 1999. However, it's also important to note that the total ethical market value has remained resilient (Source for section see the Ethical Markets Report 2021, 2022, 2023 for the full picture).

These figures from the ethical market reports have implications for long-term brand strategy because while growth has understandably slowed during 2023, there continues to be a long-term trend toward ethical market growth in the UK. According to Beverland (2021), there are a range of ethical issues typically aimed at brands, and these can have a dramatic impact on a brand's long-term profit and market share. A shift to co-creation means that brands can no longer act without regard for ethics, as co-creation embodies a collaborative approach where consumers actively participate in shaping the products and services they engage with. In the context of the ethical consumer, co-creation empowers consumers to contribute their insights, preferences, and values, enabling a collective effort to develop sustainable and socially responsible products. This is particularly so for brands that take a strong aspirational or even moral stand or whose positioning is framed ethically. But brands also suffer from ethical challenges, ethical dilemmas typically emerge between various cultural, societal, institutional, organizational, temporal, and personal beliefs and rules—these require ethical reasoning to be engineered into marketing planning and brand strategy. Large corporations exert tremendous influence on society, and their practices can either threaten or help protect the planet. Thus, advertisers of brands must decide where to draw the line when making big promises about a product's features, capabilities, or value. In the next section and for the remainder of this chapter, we discuss some of the most important sustainability concerns that brands commonly face. This will then lead to (arguably) one of the most pressing issues of our time—the impact of the current underlying business models of brands on the environment in the digital realm.

4.7 The Triple Bottom Line, United Nations Sustainable Development Goals and Brands

As discussed in business sustainability literature, a substantial development within the CSR literature emerged in the 1990s when Elkington proposed a 3Ps model known as the triple bottom line (TBL), the TBL model emphasizes that profitability is the prime objective of any organization, but environmental issues, employees, customers, and communities are equally important for long-term sustainable development. The triad of People, Planet, and Profit, commonly referred to as the 3Ps, constitutes a foundational framework. The triple bottom line (TBL) concept, encompassing social, environmental, and economic dimensions, serves as a vital criterion for evaluating brands. Consequently, a strategic alignment between the TBL and the United Nations SDGs provides a useful position for a comprehensive approach to sustainable and responsible business practices. According to Beverland (2021), brands are accountable to shareholders and are judged on their ability to improve on last quarter's results, often through stimulating more and more consumption of products (we spoke about this in ► Chap. 2). To achieve this, brand managers have triggered status anxiety among consumers, resulting in ever-quickening fashion cycles in clothes, technology, food, toys, entertainment goods, etc., and this has implications for resource use, including emissions, climate change, and waste. Even goods that are branded as greener and environmentally friendly can have a larger carbon footprint than brands like to admit (we will discuss brand greenwashing later on in the section). The triple bottom line holds immense significance for brands as it forms the basis for supporting the United Nations Sustainable Development Goals (SDGs). These frameworks serve as valuable tools to navigate the critical challenges that brands must confront to achieve long-term sustainability, such as the unsustainable patterns of consumption and their impact.

4.8 Unsustainable Patterns of Consumption and Unethical Branding Practices

The association between unsustainable consumption patterns and SDG 12 (i.e., responsible consumption and production), aimed at establishing sustainable consumption and production, is evident. The exponential rise in resource utilization, driven by overconsumption, stands as a pivotal factor contributing to the triple planetary crisis. This crisis encompasses climate change, biodiversity loss, and pollution, with the world's unsustainable levels of consumption identified as the root cause. Addressing these patterns is paramount for brands in fostering a more sustainable and harmonious relationship between humanity and the planet.

Source: ► https://unsplash.com/

Fashion industry brands serve as a great illustration of the escalating issue of overconsumption. In the dynamic realm of fashion, these brands have adeptly adjusted to the rapid pace of change by embracing a fast fashion business model. This adaptation, while reflective of industry agility, has also contributed significantly to the intensification of overconsumption, necessitating a re-evaluation of practices to ensure long-term sustainability and ethical consumption within the fashion sector. Some fashion brand managers claim that this is the only option to compete in a capitalist-driven fashion market, but this has resulted in overproduction and overconsumption of fashion products. We know from our discussion in ▶ Chap. 2 how the basic supply and demand of products work, to compete fashion brands increase production, and in order to make a profit, they need consumers to buy fashion garments, often more than the consumer needs to. Due to these manufacturing processes, the fashion industry finds itself at the center of environmental damage, namely the carbon emissions the industry produces and the waste that is generated. This poses a direct link to landfill issues, thereby implicating SDG 13 (i.e., climate action) and 15 (i.e., life on land). SDG 13 calls for urgent action to combat climate change and its impacts, as the excessive production and disposal of fashion items contribute significantly to environmental degradation and carbon emissions. Furthermore, SDG 15, focused on combating land degradation, is implicated as the manufacturing processes, waste, and non-biodegradable materials from fashion brands contribute to landfills, exacerbating the global challenge of land degradation. The key problem here is when anything is produced in mass, then those products become more accessible because prices become cheaper (fast fashion, an industry based on high volumes and low prices), if this wasn't bad enough for the planet's resource use, brands that offer sustainable alternatives are largely criticized for being more expensive and therefore many question whether sustainable clothing is only for higher income earners.

4.8.1 Societal Inequality and Status Consumption Fueled by Brands

The fashion industry has also played a pivotal role in human identity construction, providing a way for the individual to express themselves and signal certain aspects of their personality through clothing. This has vastly aided what is known as status consumption. Status consumption is defined as:

> *The motivational process by which individuals strive to improve their social standing through the conspicuous consumption of consumer products that confer and symbolize status both for the individual and surrounding significant others* (Eastman et al., 1999, p. 42).

Linked to SDG 1 (End poverty in all its forms everywhere) and SDG 8 (Promote sustained, inclusive, and sustainable economic growth, full and productive employment, and decent work for all), status consumption emerged over concerns about one's position and identity relative to that of others in society and has not only motorized increases in consumption but has also amplified consumer debt. Using brands as props to project a desired identity is not a new phenomenon (see Veblen effect section, i.e., ► Sect. 4.8.2), brands have strategically played with consumers' fears over the status of being left behind for decades, recognizing that at the core of human needs is the desire to fit in. This desire to fit in even starts at home and refers to the comparison individuals exercise with one's neighbors as a benchmark for social class via the accumulation of material goods. A failure to keep up with one's neighbors was often perceived as socioeconomic inferiority, and this produced huge amounts of pressure on society as the demand for status goods, fueled by conspicuous consumption to satisfy consumer preoccupations with their relative social standing and prestige, forced people to live above their means and take on credit card debt (Mason, 2000). Research (Walasek & Brown, 2015) has suggested that people in more unequal societies spend more on status goods, work longer hours, and are more likely to get into debt. Status consumption has been reported to increase under greater income inequality societies (Walasek & Brown, 2016), and the impact of income inequality on increased consumption not only has implications for well-being and living conditions (with quality of life lower in more unequal societies) but also for sustainability.

■ Source: ► https://unsplash.com/

4.8.2 The Veblen Effect

In 1899, Veblen coined the term "conspicuous consumption" to explain spending on and the acquiring of luxury commodities (goods and services) explicitly as a public presentation of economic wealth. Veblen suggests that conspicuous forms of consumption enable displays of wealth because they signal to others the availability of disposable income. Interestingly, what is critical to this chapter is that Veblen proposes that the ability to signal status through consumption is closely tied to wastefulness. This is evident in the demonstration of wealth through excessive material purchases, which squanders financial resources, and in the indulgence of visible leisure activities instead of engaging in paid work, effectively "wasting" time. Resultantly, purchases of luxury goods have risen (considerably) over time, evident through the so-called Veblen effects, whereby demand for goods increases with their price, even when lower-priced substitutes are available. Luxury goods are not confined to the fashion industry; it would be unfair to focus purely on this market. Luxury industries include brands (not exhaustive) from the technology sector, automotive industry, tourism industry, and property industry. So, how does conspicuous consumption work in today's consumer-conscious environment? Using the Smartphone category as an example, at the height of a trend, a brand will release a product only to put out a newer, better one a few months later. This is commonly seen today with Apple products (see planned obsolescence discussion in ► Chap. 2). Just as a consumer finally gets their hands on an iPhone 14, for example, and feels satisfied that they have maintained their social standing, Apple introduces the new iPhone 14 Pro Max. The hunter-gatherer brain is naturally drawn to novelty, yet with socioeconomic inferiority, feelings of inadequacy may arise culturally, leading individuals to believe they no longer possess the latest gadget in the room. This causes "status anxiety" and forces consumers to buy (1) A product they do not need, and (2) A product they cannot afford. This also leads to staggering levels of waste, notably in the form of factory chemicals and wastewater indiscriminately dumped into the environment, exacerbating environmental degradation (SG15).

Manager's Corner 4.2: The Veblen Effect: Unveiling the Luxury of Identity with Apple

Background Apple Inc., founded by Steve Jobs, Steve Wozniak, and Ronald Wayne in 1976, has evolved from a computer manufacturer to a global technology powerhouse. Renowned for its innovative products, sleek design, and premium pricing strategy, Apple has transcended its status as a mere consumer electronics company to become a symbol of luxury, sophistication, and identity.

Impact as a Luxury Brand Exclusivity and Prestige: Apple's meticulous branding, minimalist design aesthetic, and premium pricing contribute to its perception as a luxury brand. Owning an Apple product signals status, exclusivity, and sophistication, fostering a sense of belonging to an elite group of consumers. Apple products serve as props through which consumers express their identity, aspirations, and social status. The sleek design, cutting-edge technology, and iconic logo reinforce a sense of individuality and sophistication, aligning with consumers' desired self-image. Apple's widespread popularity and cult-like following create a sense of social proof, where individuals perceive owning Apple products as a validation of their identity and belonging to a desirable social group.

Driving Status and Status Anxiety Apple products serve as symbols of wealth, taste, and social status, leading consumers to engage in conspicuous consumption to signal their affluence and sophistication. The pervasive influence of brand symbolism and conspicuous consumption can exacerbate status anxiety, where individuals feel pressure to maintain or elevate their social standing through material possessions, including Apple products.

How does Apple leverage its brand image and product design to facilitate identity signaling among consumers?

What strategies can brand managers employ to enhance the symbolic value of their products?

Discuss the role of conspicuous consumption in exacerbating status anxiety among consumers.

How can brand managers mitigate the negative effects of status anxiety while capitalizing on consumers' desire for social validation?

Information Sources:

Trigg, A. B. (2001). Veblen, Bourdieu, and Conspicuous consumption. *Journal of Economic Issues*, *35*(*1*), 99–115.

- https://hbr.org/2014/10/apple-luxury-brand-or-mass-marketer
- https://medium.com/@aditya-korngitnnaya/the-apple-iphone-a-veblen-good-5534ea46aa1f

4.8.3 Diversity, Equity, Representation and Stereotyping in Branding

Brands have long grappled with where to draw the line between stereotypes, manipulating vulnerable consumers, and the use of this for effective advertising. Referring to SDGs 5 and 10 (achieve gender equality, empower all women and girls, and reduce inequality within and among countries), the struggle results from brands having a specific budget for their advertising campaigns and, at the same time, needing to see a return on that advertising investment through an increase in sales (shareholder business model demands ROI). Typically, a brand will not engage in advertising activity if it's not profitable. Stereotypes emerge when brand managers begin campaign planning by identifying the most profitable segments they want to target. Quite often, segmentation overlaps with stereotyping, in fact, there's a very dangerous line between segmentation and stereotyping, particularly if the campaign is looking to connect with a specific demographic of consumers that has long held certain behaviors. For example, all too often, brands have fallen back on the stereotypical idea that women do all the cleaning in a household. The housewives stereotype has caused controversy for many years and has been used in many everyday situations, most commonly cooking, cleaning, and caring for children. One of the most recognizable advertising campaigns for causing such controversy includes the British brand Fairy Liquid.

Source: ► https://unsplash.com/

The stay-at-home young mum became a familiar symbol of Fairy's adverts since the brand's first television advertising campaigns back in the 1960s. The key selling point was not merely the cleaning power of the washing up liquid, but instead how gentle on the skin the soap is, this was linked to the famous tagline "*Fairy Liquid helps married hands keep that single girl look.*" It seems outrageous now that marketing teams produced such messaging, so why did brands from certain industries engage with such stereotyping and representation behavior? The brand, such as Fairy Liquid for a cleaning product, maintains a historical customer profile, documenting the purchasing behaviors and preferences of its previous clientele. Typically, in the planning stage of such campaigns, brand managers generate a consumer profile and target demographic based on historic appeal. When the brand knows the primary audience and decision maker for washing up liquid is a female between the ages of 25 and 50, it will cater to that audience. Therefore, the stereotype in advertising campaigns becomes enticing as it represents the most profitable customer base, even though a percentage of that customer base is also males in their early 30s or retired couples in their 50s/60s and beyond. Ultimately, the stereotype for the audience with the most buying power wins. In the UK, things have changed, using stereotypes to sell products/services is now against the law. Britain's Advertising Standards Authority (ASA) announced the ban in December 2018, with a 6-month buffer period before it went into effect. The ASA says adverts that imply a woman's role is in the kitchen "*can restrict the choices, aspirations and opportunities of children, young people and adults.*" Ads can't show men or women "*failing to achieve a task specifically because of their gender*" ("e.g., a man's incompetence to change nappies; a woman's failure to park a car"). While much has been done to eliminate these out-of-date ideals, they are still very much at play, although often now in a more subtle way.

The problem isn't simply confined to stereotyping, representation also matters. Representation does not merely include gender inequalities but also body image issues, physical disability, LGBTQIA+, sexual orientation, ethnicity, race and religion, and ageism among the core issues. Representation is crucially important for positive change because if a consumer sees someone like themselves included (in an advertisement campaign) and celebrated in a positive manner as part of society, it sends a huge signal about inclusion and belonging. This has a ripple effect on the most vulnerable in society, thus providing them with the confidence to aspire to such societal ideals. We know from our discussion on the theoretical influences on branding that from a socio-cultural perspective, brands and their advertising campaigns often shape cultural views and norms, and brands leverage the idea of the creative consumer searching for identity through consumption. The problem is that the default in media and advertising campaigns is non-disabled, white heterosexual, middle-class (often male) audiences. For a great proportion of advertising history, much of advertising by brands has not represented underserved audiences.

The housewife stereotype example from Fairy Liquid alienates a significant portion of today's audience by implying that women's roles are confined to domestic chores. Gender roles have evolved and are still evolving, yet many modern campaigns still misrepresent a substantial segment of the population. According to a World Advertising Research Centre (WARC) report, women's roles in society have changed and continue to evolve, they are the key decision makers in households and are responsible for between 70% and 80% of all purchases worldwide. And with more women in the workforce, they are creators of wealth themselves. At a time of great cultural and social change epitomized by movements such as #metoo and #unstereotype, old stereotypes are being challenged, and brands must adopt more relevant and appropriate ways of engaging with this vast and lucrative audience to supercharge this change.

Another commonly overlooked strategy in the realm of brand advertising pertains to the fixation on youth and the pursuit of engaging younger demographics. This is not a new phenomenon; it has been used for many decades, and in some cases, it has become the default advertising strategy. First, we've become accustomed to thinking that new trends and ideas that emerge are driven by young people, and hence, in order to be seen as relevant and modern, brands and marketing agencies have historically believed they must focus on keeping up with "the kids." It's easy to disregard that along with their dominant scale and purchasing power, the 50+ generation is a group for whom both trend-setting and activism have long been second nature. For example, the most radical actions of the Insulate Britain group have been carried out by retired or semi-retired citizens with the time, energy, experience, and confidence to stage high-profile stunts for a cause they believe in. According to the Office of National Statistics, data from the last census reported median wealth in the UK was highest for households whose head was aged 55 years to under State Pension age; the wealth of this group was 25 times higher than those aged 16–24 years.

4.8.4 Brands and Exploitative Labor Practices

Sweatshops It's no secret that products (notably clothing) from the biggest brands in the world are habitually made by workers with low wages and poor working conditions. Some may think that these issues are a thing of the past, but unfortunately, this is still an ongoing issue globally (see Statista "The Low Wages of Garment Workers").

Emerging in the 1800s as a result of the factory (see discussion on this in ► Chap. 2) and then globalization, the term sweatshop was used to describe a working environment that took place in a small (not well-ventilated) crowded room that was prone to health and safety hazards including fire risks and rodent infestation. According to Soyer (1999), people associate the term sweatshop commonly with the following factors:

- Poor working conditions
- Low wages
- Long hours
- Arbitrary power of the boss over the workers

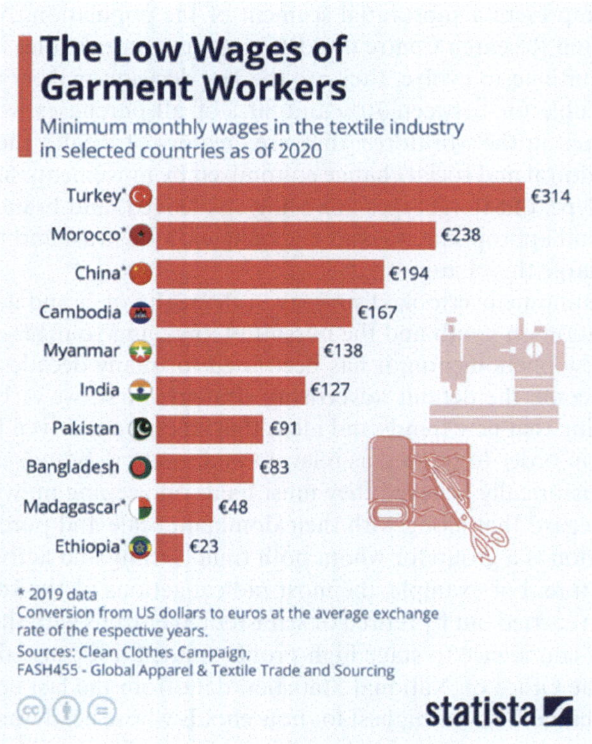

◘ Source: ► https://unsplash.com/

Soyer suggests the term sweatshop was in use from the 1890s and referred to the "sweaty" working conditions workers faced as a result of hard work, harsh discipline, and conditions that created a "sweating system." Sweatshops surfaced in big cities such as London and New York and typically attracted immigrants to those locations.

Source: ▶ https://unsplash.com/

Through a process of globalization, brands started to move their operations overseas to lower costs, increase profits, and exploit other countries by doing so:

> *In this buyer-driven system the owners/directors of the brand-named company, usually from the West, maximise their profits by forming a global commodity chain that undermines wages in developed countries and keeps wages low in developing countries* (Gereffi & Pan, 1994).

While labor laws were being introduced in developed countries, political and economic pursuits motivated multinational corporations to seek an overseas workforce. Government corruption and inadequate labor protection legislation in developing countries were widespread. It seems obvious then to position sweatshops on the opposite end of the continuum to fairtrade factories, and it won't come as a shock to learn that sweatshops have been the most effective way for fashion brands to make a profit from their fast fashion business models. Big brands like H&M, which continue to enjoy (see article on brands that use sweatshops) a monopoly over many consumers' shopping habits, rely on speedy supply chains and underpaid workers from factories overseas (see study published in 2019 by New York University). If this wasn't bad enough, such brands have started to hide the realities associated with manufacturing their products by putting "Made in America" or "Made in UK" on the clothing labels because countries like the United States and the UK are considered low-risk countries for poor labor conditions compared to countries with a poor economy and an immoral government where the cruelty of manufacturing is rife. This process forms part of a wider practice known as greenwashing. We will discuss greenwashing in the final section of the chapter.

When we refer to brands exploiting employees, this is not merely confined to the brand's immediate workforce but the entire supply chain. Cost advantages have led to complex global supply chains involving numerous suppliers/intermediaries, many of whom have been accused of unscrupulous labor practices, as well as health and safety violations. A report produced in 2020 by forced ▶ labour.org revealed that the supply chains of most major apparel brands and retailers are

4

tainted by forced labor. The report highlighted how the Chinese government has rounded up an estimated 1–1.8 million people based in an area within northwest China called Uyghurs. These people were kept in detention and forced labor camps, the largest interment of an ethnic and religious minority since World War II. Fast forward to 2024, and there are still claims being made about the region. Almost a tenth of the world's aluminum, a key material for car manufacturing, is produced in the Xinjiang Uyghur Autonomous Region (Xinjiang or XUAR), a region in northwestern China where the Chinese government is conducting a long-running campaign of repression against Uyghurs and other communities. Brands continue to source millions of resources from the Uyghur Region. Despite the potential threat of being involved in forced labor associated with the areas of aluminum, certain car manufacturers in China have yielded to governmental coercion to implement less stringent human rights and responsible sourcing criteria in their Chinese partnerships compared to their international endeavors. The majority of these brands have inadequately scrutinized their supply chains for aluminum components, failing to recognize and rectify any connections to Xinjiang Uyghur Autonomous Region (see article: Asleep at the Wheel Car Companies' Complicity in Forced Labor in China, 2024). Furthermore, major corporations claim that they do not tolerate forced labor by their suppliers but have offered no credible explanation as to how they can meet this standard while continuing to do business in a region where forced labor is rampant. What makes this worse is that forced labor worldwide also includes a staggering number of children.

4.8.5 Modern Slavery and Child Labor

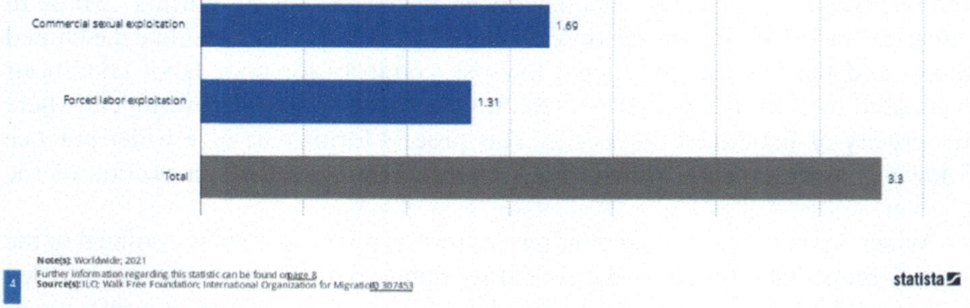

Source: Statista

Consumers in wealthy countries are knowingly or unknowingly buying goods worth billions that are produced by people trapped in modern slavery and forced labor. Humanitarian charities such as UNICEF work tirelessly to raise awareness of such shocking statistics: nearly one in ten children are subjected to child labor worldwide, with some forced into hazardous work through trafficking. And Fairtrade International suggests that of these children stuck in child labor practices, it is estimated that 112 million children and youth work in agriculture. There is a large amount of cross-over between the SDGs and Fairtrade's work, and of the 169 targets underlying the goals, there is barely a single one that isn't somehow related to food and farming. Furthermore, it is not just organizations such as Fairtrade and UNICEF who are trying to help tackle the SDGs, some brands in existence have placed issues such as slavery and child labor central to their business strategy. In seeking to tackle slavery and illegal child labor practices on cocoa plantations in West Africa, Tony's Chocolonely has created a brand for the sole purpose of calling out chocolate giants that benefit from hiding behind their supply chains and keeping the purchase price of cocoa inhumanly low:

» *The quirky story of Tony's Chocolonely started back in 2003. A Dutch journalist Teun van de Keuken learned of the widespread modern slavery and illegal child labor across cocoa farms in West Africa despite explicit commitments by world's leading chocolate brands to eradicate slavery and child labor from their supply chains. Teun decided to clean this up by being an investigative journalist on a popular Dutch TV program called "Keuringsdienst van Waarde" looking to unveil the sordid details of the chocolate supply chain. He followed this up in 2005 by launching 5000 limited edition chocolate bars into the Dutch markets under the brand name of Tony's Chocolonely. Since then, it grew from just €1m in 2011 to becoming a Fairtrade certified B Corp with worldwide revenues of €110m by 2021 (Section Source: WARC article, Marketing needs marketing: The curious case of Tony's Chocolonely).*

Source: ► https://unsplash.com/

A great example of a mission-led brand—Tony's did not start with the ambition to make the best chocolate in the market, the brand intended to help consumers make a conscious choice and raise awareness among retailers of the slavery problem that exists in the cocoa industry. According to a Marketing Week article from 2019 (see How Marketing and Supply Chain Combine to Create a "Mission Brand" (n.d.-c)), Tony's Chocolonely makes sure that they know exactly where their cocoa beans come from. They buy them directly from five partner cooperatives in Africa. Besides paying a Fairtrade premium to the farmers, they also give extra money called "Tony's Premium": $460 per kilo in Ghana and $520 per kilo in the Ivory Coast. They promise to work with these farmers for at least 5 years to help them improve their farms and make more cocoa. The brand wants to spread its mission in everything it does, like putting a 100% slave-free label on the packaging and making the chocolate bar look uneven to show the unfairness in the supply chain.

It is frequently claimed that many of the big brands don't know where their cocoa comes from, and this gets more complicated the further down the supply chain you go as traceability becomes complex and cocoa farmers are typically separated from cocoa exporters and by multiple intermediaries, this means it's really important for brands that want to improve to be transparent about their supply chain. This creation of awareness around supply chain transparency in the cocoa industry appears to be working. Nestlé, a brand that has been at the center of child labor controversies in the past, announced an initiative in January 2022 to tackle the issue. The biggest food producer aims to tackle child labor by implementing an "innovative" $1.4 billion plan, which involves directly paying African cocoa farmers to address the poverty that leads to this issue (see Financial Times article on Nestlé pays to eliminate child labor from chocolate supply chain). Furthermore, Nestlé produced a report at the end of 2021 to outline their plans going forward on how they aim to tackle some of the challenges the company has long been criticized for. These included a response to the COVID-19 pandemic and an aligned focus on the United Nations SDGs. These commitments emphasized minimizing Nestlé's impact on the world's resources, contributing to a future in which they are available for generations to come, with further objectives coming under climate, water, nature, biodiversity packaging, and circularity. With a specific commitment to improving cocoa sustainability, a Nestlé media release outlined their income accelerator program offers a novel approach to help support farmers and their families in their transition to more sustainable cocoa farming. It is hoped the incentives will encourage behaviors and agricultural practices that are designed to steadily build social and economic resilience over time. Nestlé's new initiative sharpens focus on the root causes of child labor.

Section Source: ► https://www.nestle.com/media/pressreleases/allpressreleases/tackle-child-labor-risks-farmer-income-cocoa-traceability

Also, see document—Nestlé's Creating Shared Value and Sustainability Report 2021.

Consumers' growing awareness of social issues, including child labor in cocoa fields, has led to heightened scrutiny of brands' social responsibilities. This awareness extends beyond isolated issues, as people are increasingly sensitive to human rights abuses worldwide. Whether it's forced labor in factories or on cocoa farms, consumers are urging brands to take action against any shady practices perceived to benefit their brand. Moreover, through online platforms, unpaid consumer reviewers play a significant role in amplifying concerns and holding companies accountable for their actions.

4.8.6 The Unpaid Consumer

While sweatshops, child labor, and poor working conditions have been a historical problem for brands (and continue to be for some), there's been a rise in new, less explicit developments that some suggest fall under exploitative labor practices. This has emerged as a result of the internet and is referred to as unpaid consumer work. According to Ardvisson, one of the most important trends in contemporary consumer society is the progressive inclusion of consumers in the process where value is produced around products and brands. A growing number of companies now involve consumers in the coproduction of brands, experiences, design, marketing strategies, and even product development (Zwick et al., 2008). The involvement of consumers is not merely confined to feedback on the design of new products; consumers' labor activities go beyond this in the contemporary digital environment. The emergence of Web 2.0 technologies (which refers to websites that emphasize user-generated content and allow users to interact and collaborate typically on social media sites), where internet sites became interactive, enabling collaboration between consumers and other consumers and marketers, has resulted in blurring the boundaries between production and consumption. Toffler (1980) described this prosumer phenomenon long before it became a consequence of the technological age. Toffler imagined prosumers emerging from technological changes to the nature and interaction of consumption and production and acknowledged the crucial role of brands in creating "presumption"—cutting costs and relying on consumers to undertake work and externalize labor expenses through what he termed "willing seduction," thereby blurring the line between production and consumption. In today's digital environment, much of this unpaid work takes place online, and customer review sites are a great example of this in action. According to Statista, online reviews comprise voluntary consumer-generated evaluations of businesses, products, or services by Internet users who purchased, used, or had experience with a particular product or service. Online product or brand reviews are a crucial part of today's digital information ecosystem that significantly impacts consumer welfare and brand performance:

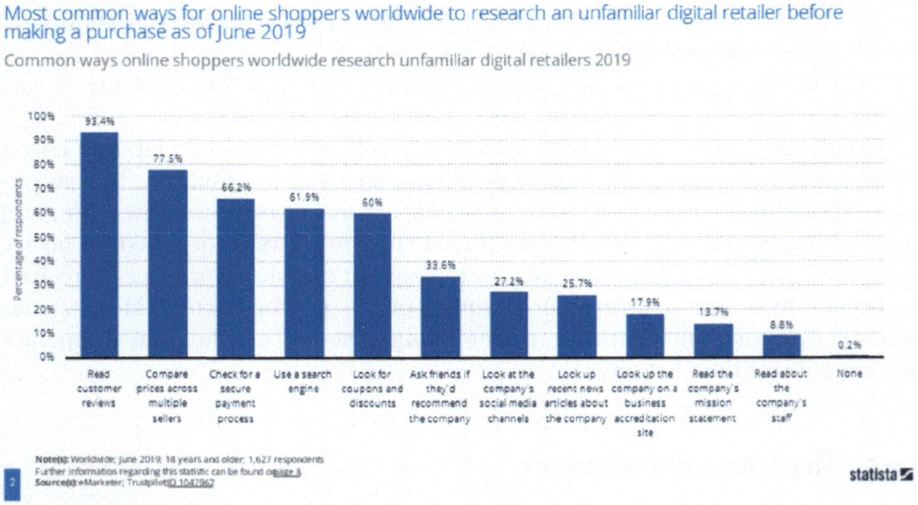

Source: ▶ Statista/

Statista reports that in the United States, product ratings and reviews have grown to become a major factor that influences internet users when deciding where to shop online. In a 2019 survey on the most common purchase influence among online shoppers in the United States, 62% of respondents stated that online customer reviews were very helpful. Web 2.0 brings about new forms of labor, what Terranova (2012) calls the "free labor" of consumers–producers, which is simultaneously voluntarily given, unwaged, enjoyed, and perhaps exploited. This type of user-generated content (or unpaid labor) doesn't just refer to branded review sites such as TripAdvisor, Yelp, Google reviews, or Amazon because non-branded blogs and social media sites influence consumers' behavior about a product or service they intend to buy. We must also consider other user-generated content, such as the type of information that is created on the popular social media platform Facebook and the world's most popular search engine, Google. Historically, the saying goes if something is free, "you" are the product (said by world-famous American sculptor and video artist Richard Serra in 1973) this appropriately describes the kind of activity and exploitation taking place on Facebook. Facebook carefully positions its brand as a networking platform to avoid accusations of being a media platform. However, to make money through the various advertising revenues that support Facebook's business model, Facebook needs content from users, once this content is published by the various users on their own profiles, Facebook captures this information data and turns it into insights to assist advertisers to better engage with audiences, this makes for more effective advertising return on investment yet reinforces the fact that "people are the product." Facebook produces no content of its own and completely relies on users to do this. On a larger scale, Google does the same. Google is not a social media platform but a search engine, and a search engine's job is to organize and locate content that has been created by others that is publicly accessible. We will discuss this in more detail in the next section when we specially discuss big tech brands.

Many have questioned whether the co-creation of user-generated content is designed to exploit producing consumers yet masked as entertaining them. Consumers are not generally paid for what they contribute to the co-creation of information on social media platforms, but this isn't the only unethical issue associated with big tech brands.

4.9 Big Tech Brands and Persuasive Technology

In recent years, there has been much written about the big technological companies of the world. It is useful at this stage to clarify what we mean by "big tech" because although the term is used frequently in the media and by society more generally, it's not the most accurate way to describe the world's most profitable businesses. According to a Conversation article (see "Big Tech" isn't one big monopoly), they are big because they have been listed at some point as the most valuable public companies, and because they largely do things with computers, there has been a temptation to bundle them all together, but this obscures the reality that they are different, specifically concerning the company, business models and practices. The general idea around this distinction is that properly understanding the context these businesses operate in allows us to understand the key revenue sources, which are often the root cause of the unethical practices these "big" companies have been accused of. For example, while Facebook makes money off of ad revenue from attention-grabbing content, Apple generates profit from selling electronic hardware. However, there are many similarities in the way these companies collect data, analyze behavior, and use algorithms. Many have suggested that these companies have made our lives easier through their innovation. But, some have a different perspective, in his work from 2018 on "The Four" (Facebook, Google, Amazon, and Apple), Scott Galloway presents an interesting narrative of big tech brands:

» *These tech giants have inspired more joy, connections, prosperity and discovery than any entity in history. They have created hundreds of thousands of jobs. They are responsible for an array of products and services and are entwined into the daily lives of billions. A supercomputer in your pocket, internet into developing countries, mapping the earths landmass and oceans. Many espouse they make the world a better place. However, consider a different view…A retailer that refuses to pay sales tax, treats its employees poorly, destroys hundreds of thousands of jobs but has been celebrated as an exemplar of business innovation. A tech company that withholds information about domestic acts of terrorism from the FBI yet is celebrated as a religion and cult. A social platform that analyses thousands of images of your children activates your phone as a listening device and sells this info to Fortune 500 companies. An Ad platform that commands, in some markets, a 90% share of the most lucrative sector in media—but avoids anti-competitive regulation and therefore prevents other such search engines from entering the market.* (pp. 1–2)

◘ Source: ► https://unsplash.com/

And if you aren't convinced by the impact these big four have had on most people's daily lives, try going one day without using one of these global brands. So how have these big tech giants created a situation where they are so central to humans' lives? Through carefully planned and well-executed brand-building strategies? Not quite. There's no denying that Facebook, Google, Amazon, and Apple have all built powerful iconic brands adorned with globally distinctive/recognizable logos and identity systems. They are often heralded for improving lives in many ways, notably convenience, comfort, and efficiency. However, there are some other less well-highlighted activities that all of these companies engage in.

A suitable place to begin is to consider the design of the persuasive technologies these big computer companies adopt. According to Fogg (1998), interactive technology changes a person's attitudes or behaviors. Fogg suggests that in considering technologies and persuasion, persuasion in this context should be defined as "*an attempt to shape, reinforce, or change behaviors, feelings, or thoughts about an issue, object, or action.*" Therefore, Fogg stated back in the 90s that technology (he used the term computers back in the 90s) has the potential to change attitudes and behaviors through different means. Fogg's discussion is, of course, dated and refers to computers and the media of the late 1990s; however, the term he coined, "persuasive technologies," could not be more appropriate in describing some of the design techniques adopted by engineers working for the big tech companies in the contemporary environment. The seminal work of Fogg is central to the discussion on the growth of the Big Four. Fogg is the founder of the Stanford Behavior Design Lab, which was initially known as the Persuasive Technology Lab. It's no secret that Stanford University is located approx. 13 miles away from Silicon Valley, where 3/4 of the Big Four have their headquarters. It's often reported that Stanford played a central role in the emergence of Silicon Valley, both through its academic programs and investment in the tech ecosystem (Adams, 2003). A powerful sense of regional solidarity accompanied the rise of Silicon Valley; thus, regionalism helped align Stanford's interests with those of the area's high-tech firms for the first 50 years of Silicon Valley's development. Therefore, it comes as no surprise that Stanford would create purposeful labs to encourage research that assists the development of new technologies, this began in the late 1990s/early 2000s era. Central to this was Fogg and his work on the ethical responsibilities of designers working with persuasive computers and the importance of educating about persuasion. Because Fogg understood the potential of behavior design and how this could be used to persuade certain actions of technology users, he was also very aware of the dangers that this could potentially bring about. Even with all good intentions, unfortunately, such information in the wrong hands isn't always used for good and has unintended consequences. Despite Fogg's efforts to inform students about the potential implications of these persuasive technologies, persuasive design took an unexpected turn. The Stanford Facebook class of 2007, taught by professors BJ Fogg and Dave McClure, gained attention for its innovative approach to teaching about social media platforms, particularly Facebook (sometime before Instagram, Snapchat, and TikTok emerged). The class focused on the practical aspects of building and launching Facebook applications, leveraging the platform's APIs (Application programming interface) and developer tools. Students collaborated on projects, gaining hands-on experience in application development and marketing strategies within the Facebook ecosystem. Despite its innovative approach to teaching emerging technology, the Stanford Facebook class of 2007 has faced criticisms, particularly regarding the unintended consequences of applying behavioral psychology principles to technology design. While BJ Fogg emphasizes promoting healthy habits, some argue that his emphasis on behavior hacks, such as making tasks easier rather than boosting motivation, has contributed to the proliferation of social media and smartphone addiction. Critics suggest that these tactics, while effective for engagement, have led to irresponsible design choices and potentially harmful consequences for users, raising concerns about the ethical implications of leveraging psychological techniques for technological manipulation. According to the New York Times (see article—The "Facebook Class" that Built Apps, and Fortunes (n.d.)), by teaching students to build no-frills apps, distribute them quickly, and worry about refining them later, the "Facebook Class" stumbled upon what has become standard operating procedure for a new generation of entrepreneurs and investors in Silicon Valley and beyond.

Many refer to this as the dark patterns of design; Brignall (see article—Dark patterns: inside the interfaces designed to trick you) suggests when consumers use a brand website/app, they don't read every word on every page—they skim read and make assumptions and brands are very aware of this. If a brand wants to trick a consumer into doing something, it can take advantage of this by making a web page look like it is saying one thing when it is saying another. Therefore, "dark patterns" are tricks used by brands in websites and apps that make consumers do things that they didn't mean to, like buying or signing up for something. Because of this, brands have engaged in deceptive behavioral data capture and numerous data breaches (Facebook's Cambridge Analytica scandal—see Guardian article—Revealed: 50 million Facebook profiles harvested for Cambridge Analytica in major data breach), this has largely been achieved through the development of sophisticated algorithms. An algorithm is a method of automated instruction designed to solve a specific problem or perform a particular task, it is sometimes referred to as a recipe a computer follows for a desired outcome and is employed in many ways, for example, Twitter trends or Instagram follow recommendations; Google personalized search results or Facebook newsfeeds (Willson, 2019). There are many types of algorithms; these include search, sorting, machine learning, recommendation algorithms, and pattern recognition algorithms (Gran et al., 2021). Given the widespread use of recommendation algorithms across various online platforms and services, consumers encounter these recommendation algorithms more often than most. An article from The Conversation (see article—Feedback loops and echo chambers: How Algorithms Amplify Viewpoints) nicely captures the relevancy of recommendation algorithms for our chapter and some of the ethical concerns they present. The article highlights that brands like Facebook, YouTube, Netflix, and Amazon developed recommendation algorithms to aid users in decision-making processes. These algorithms suggest various options to users, who then make choices that are incorporated into the algorithm's training data. However, this process often overlooks the fact that the user's choice itself originated from the algorithm's recommendations, creating a feedback loop where the algorithm's output influences its subsequent input. Consequently, users are frequently presented with recommendations resembling their previous choices, leading to a cyclic scenario. This raises the question: Did users click on content out of genuine interest, or were they influenced by the algorithm's recommendations? Because of this, there have been many questions raised about the ethical use of these algorithms. The use of algorithms in digital platforms has caused concerns about their potential to exacerbate societal divisions. Critics argue that algorithms may inadvertently reinforce echo chambers and polarize viewpoints. Moreover, one of the biggest ethical issues is the lack of transparency surrounding algorithmic decision-making processes, which undermine accountability and trust in digital platforms. This opacity hinders efforts to address ethical considerations and promote just, peaceful, and inclusive societies, which is linked to the United Nations Sustainable Development Goal 16 (i.e., peace, justice, and strong institutions). Urgent action is needed to enhance transparency and accountability in algorithmic systems to foster social cohesion and sustainable development.

Marketing in Action 4.2: Scenario Planning Case Study: The Ethical Digital Marketer—Crafting a Brand Strategy for Cognitive Responsibility

Background In today's digital age, social media has become an integral part of our daily lives. Platforms like Facebook, Instagram, TikTok, and Twitter have transformed the way we communicate, interact, and consume information. However, there is growing concern about how these platforms leverage our cognitive vulnerabilities to capture our attention and keep us engaged for extended periods. The Center for Humane Technology, founded by former tech insiders, has been at the forefront of raising awareness about the ethical implications of technology design, particularly regarding social media. They emphasize how these platforms exploit human cognitive biases, tapping into our innate desires and vulnerabilities to maximize user engagement and advertising revenue.

Objective The objective of this marketing-in-action exercise is to help marketers understand how social media exploits cognitive vulnerabilities and how marketing as a profession can address these ethical concerns.

Scenario Imagine you are a marketing consultant hired by a social media platform to increase user engagement and time spent on the platform. Your client is primarily concerned with maximizing advertising revenue and, therefore, growing their user base. However, you are also aware of the ethical implications of designing strategies that exploit cognitive vulnerabilities. Your task is to develop a digital marketing strategy for the brand that balances the need for user engagement and revenue generation with ethical considerations for user well-being, you must consider how this aligns with the entire brand strategy, such as the brand's values and tone of voice. The brand wants to do better and be more responsible in the digital realm; therefore, your strategy should aim to minimize the exploitation of cognitive vulnerabilities while still achieving your client's objectives.

4

Consider the following steps in developing your strategy:

Research and Analysis Start by researching the various ways in which social media platforms exploit cognitive vulnerabilities.

Identify Vulnerabilities Identify specific cognitive vulnerabilities that your client's platform may be exploiting, such as the fear of missing out (FOMO), social validation, or the novelty-seeking instinct brought about by Infinite Scroll.

Design Ethical Solutions Develop creative solutions that enhance user engagement without relying on manipulative tactics. For example, instead of using red notifications to trigger FOMO, consider implementing features that promote meaningful interactions and genuine connections among users, and think about different types of AI that might be used to identify common ground among users who otherwise disagree.

Implement Safeguards Propose mechanisms to safeguard users against excessive use and potential harm. This could include features like digital well-being tools, time limits, or prompts encouraging users to take breaks from the platform.

Measure Impact Define metrics to measure the success of your strategy beyond traditional engagement metrics. Consider indicators of user satisfaction, well-being, and long-term usage patterns.

Iterate and Improve Continuously evaluate the impact of your strategy and be prepared to iterate based on user feedback, emerging technology, and evolving ethical standards.

The persuasive technology and dark patterns of design discussion may seem like a responsible digital issue that belongs to the Digital Marketing chapter, but it's not solely confined to the digital sphere because these persuasive design techniques have infiltrated all domains of customers' daily lives with brands and this is why the discussion is paramount to this branding chapter. Brands must directly address pressing concerns, extending beyond big tech to encompass all entities leveraging persuasive design. This includes those fostering unsustainable, unethical, and irresponsible consumption levels (SDG 12). Enhancing transparency is integral to this endeavor.

We have discussed transparency previously when acknowledging brand authenticity. We know that customers are far more likely to build a relationship with a brand they trust, and we also know that transparency and authenticity are vital to trust. Brands are built on the messages conveyed through their values and brand tone of voice; in the contemporary environment, this has become increasingly central to a positive brand image. This is because there are emerging tools (The Fashion Transparency Index, Nielsen release their first Brand Sustainability Report in 2022, Sustainable Brand Index, to name a few) that consumers can now access and use to assess brands who use mission/purpose-led messaging in their communications but fail to put this into practice. But these are only just evolving and they are also currently only available in specific regions of the world. Similarly, just like regular brands that consumers come into contact with daily, social media platforms and big tech brands also have transparency challenges. This includes a whole range of complex considerations such as platform transparency, algorithmic transparency, and the freedom awarded to consumers to choose without undue influence of automation and algorithmic amplification. Many have rightly questioned the role of design in these processes. At the 2021 MIT Sloan Management School conference, the Social Media Summit, experts deliberated on maintaining transparency while safeguarding privacy and data security on social platforms. They discussed the challenge of navigating the "Transparency Paradox"—striving to simultaneously enhance openness and privacy/security. Key questions include identifying safe and secure models for transparency and understanding the trade-offs involved. A Harvard Business Review article (see article on "We Need Transparency in Algorithms, But Too Much Can Backfire") proposes technical transparency, such as disclosing algorithm source code, inputs, and outputs, as a means to foster trust. However, most algorithms today are registered assets guarded closely by for-profit companies, often remaining opaque.

The regulation of big tech, social media platforms, and the online environment has been a contentious topic of discussion for an extended period. Efforts to regulate have grappled with the rapid evolution of technology, the global reach of online platforms, and the complexity of addressing diverse cultural and legal landscapes globally. Issues such as transparency, misinformation, ad-driven business models, user privacy concerns, hate speech, the dynamic nature of content, algorithmic opacity, and algorithmic bias have created an urgency for effective regulation. However, attempts to enforce policies face huge challenges, including jurisdictional boundaries, resistance from tech companies, and the difficulty of shaping policies that maintain online safety without stifling innovation or infringing on individual rights.

4.10 Monopoly Power, Competition, Brands, and Sustainability

Related to issues highlighted around big tech's use of persuasive technology is also the notion that brands are using algorithms to show a preference for their own products/services. This has resulted in many claiming that competition in certain markets has been vastly affected by bad products, driving out good products while stifling innovation. If we look to explore this from an economic perspective, we can argue that the world's biggest tech brands have created their own monopolies. A monopoly is a situation where certain brands have a significant amount of market share, leading to a brand controlling much of an industry. This means that a single brand can have a huge amount of power over the price of products and services within that market/industry.

4

For example, let's look at the UK share of the online "search" market from September 2021:

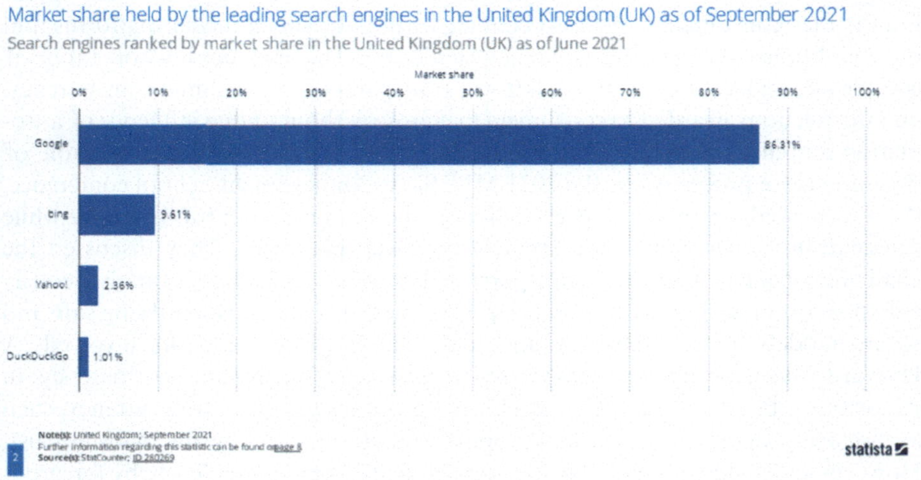

■ Source: ► Statista/

According to Statista, Google had the highest market share by far, with 86.31% of the desktop, tablet, and console search engine market in the UK. Following Google, we have Bing with a market share of 9.61%, followed by the Yahoo! search engine, etc. Google has held a steady dominance in the market since its emergence, and this has resulted in the most significant antitrust case for more than two decades (see Reuters article on Google antitrust case 2023 (n.d.)). For this chapter, our focus is on how monopolies hamper sustainable innovation for brands and, therefore, act with little regard toward people and the planet. How does this happen? From a market perspective, brands that operate as monopolies have very little competition and, therefore, have little incentive to innovate their products or services, this impacts consumer choice. In addition to this, a concentration of resources allows brands acting as monopolies to control capital, infrastructure, and technology, which limits the access of smaller to medium-sized brands, in turn stifling the growth of start-ups and new businesses. "Kill the start-up" is a term often used to describe how big tech goes about their business because, between them, the big tech brands have gobbled up hundreds of companies to prevent competing technology that might displace them. They have closed start-up platforms to ensure that users are fully dependent on "big tech" services. They have directly and unfairly competed with business users of their platforms by copying their products and self-preferencing their own products or services (e.g., Instagram stories were launched because Facebook was denied the opportunity to buy Snapchat).

Consumers and users are directly affected by these monopolistic/antitrust actions. There are many negative impacts on the users/consumers within markets operating as monopolies. Monopolies prevent new and diverse ideas from entering the market, and this results in far fewer choices for the consumer, moreover, this can lead to higher prices, with consumers picking up the increase. Society has encountered big brands before; we don't have to look far to understand the impacts of big pharma, the tobacco industry, and even the sugar industry. What makes this different is regulation; long-standing protection has been in place for industries that aim to protect consumers and the environment. We can also make some clear links between big tech, monopoly/antitrust activity, and the UN SDGs. We know that when brands participate in monopolistic practices, they limit market entry for smaller businesses and startups, impeding economic growth and intensifying inequalities (SDG 10: Reduced Inequalities). Furthermore, monopolies often exert disproportionate power, enabling them to manipulate prices and control essential goods and services, such as healthcare or education, undermining efforts toward SDG 3 (Good Health and Well-being) and SDG 4 (Quality Education). Additionally, anti-competitive behavior can restrict technological advancements and discourage sustainable practices by discouraging investment in research and development aimed at addressing environmental challenges (SDG 13: Climate Action). Therefore, tackling issues relating to monopolistic power and antitrust in the big tech industry is vital to helping achieve the UN SDGs, and a good place to start is turning our attention to the incentives that propel these brands to engage in such activity.

4.11 Big Tech Brands and Business Models

All of the tech giants, to a greater or lesser degree, operate a platform business model, meaning they connect sellers to consumers, whether it is connecting advertisers to a social network or search engine users, app developers to device owners, or retailers to online shoppers. By their nature, platform businesses benefit from network effects—the more consumers and users they acquire, the more appealing the platforms are to suppliers—and therefore leads to "winner takes all" outcomes. Facebook, for example, use an advertising platform business model, it provides content that attracts an audience and sells their attention to advertisers—just as television, radio, and print have done for decades. The key difference between Facebook and these legacy media businesses is where the content comes from. Rather than Facebook paying to create the material that attracts users, the users create it themselves for free, posting personal messages and sharing links. Another key difference is in the algorithmic amplification processes used to share divisive negative content. Algorithmic amplification has typically been adopted to spread information that is likely to go viral (often negative/divisive in nature) faster. This has resulted in what researchers and experts are now terming the "attention economy." This is how social media companies make money by exploiting user attention; users do not pay to use social media apps; users are the product being sold to advertising companies. Business models at the center of inquiry around responsible sustainability are not a new problem. For example, the shareholder value business model, which became prominent in the 1970s/1980s, focused on maximizing profits for shareholders through return on investment, regardless of demand, and with little consideration for the long-term social and environmental impacts of a company. However, how advertising operates on social media and other online platforms makes it vastly different from (1) Traditional media, and (2) Other business models. No other media has operationalized to engage human thoughts 24/7, and no other media has been able to personalize at scale with such precision, but the most pressing and rapidly evolving difference is the influence of AI technologies—no other media has used supercomputer functionalities to predict behavior and understand how to keep users constantly engaged.

Source: ► https://unsplash.com/

The launch of Generative AI tools/applications at the end of 2022 has brought about new and arguably more serious concerns about evolving technology and regulation. Transformer-based models such as OpenAI's GPT (Generative Pretrained Transformer) have demonstrated impressive capabilities in generating human-like text. We know that in marketing, certain aspects of artificial intelligence such as automation and algorithms have been used to assist brands with activities for a while, in activities such as content creation and targeting consumers. However, as these technologies begin to evolve, the temptation to rely on them more frequently to create content at scale or hyper-personalize, for example, brings a whole new set of problems for brands, people, and the planet. According to a World Advertising Research Centre (WARC) article (see How generative AI is redefining the game for marketers, and what can they do about it (n.d.)), the risks involved in using AI-generated content in advertising include inadvertent creation of harmful content, inconsistent brand tone of voice, erosion of creativity due to over-reliance on automation and environmental costs of producing more advertisements. What is clear is that as these technologies continue to evolve, the attraction of leveraging them for scalable content creation and hyper-personalization poses new challenges for brands, individuals, and the environment alike. While AI-driven marketing offers (and has done for some time) unprecedented opportunities, it also demands careful consideration of its ethical, social, and environmental implications. From mitigating algorithmic biases to minimizing carbon footprints, responsible AI implementation is pivotal in safeguarding diversity, sustainability, and inclusivity for brands. By aligning marketing practices with the principles of the UN Sustainable Development Goals, brands can navigate this evolving landscape with purpose, ensuring that innovation serves not only profit but also people and the planet.

4.12 Possible Ways Forward for Brands

Emerging Technologies and Brands In the context of the UN SDGs and the triple bottom line (i.e., People, Planet, and Profit), the integration of responsible AI practices in marketing promotional practices and brand management aligns with several key objectives:

People The importance of diversity and representation in marketing links directly to SDG 10 (Reduced Inequalities) and SDG 5 (Gender Equality). Brands that train marketing teams on diversity and representation, promote inclusivity, and ensure that AI algorithms do not perpetuate biases against certain demographics will foster a more equitable society and align with the SDGs' aim to leave no one behind.

Planet Many are not aware of the environmental impact associated with using AI models. Therefore, the focus must be on reducing the environmental impact of AI-powered content generation, which contributes to SDG 13 (Climate Action) and SDG 12 (Responsible Consumption and Production). Optimizing algorithms to minimize computational requirements helps to reduce energy consumption (Anjos et al., 2021) and carbon emissions, thus mitigating the environmental footprint of marketing activities. This aligns with the SDGs' call for sustainable practices to protect the planet for future generations.

Profit By utilizing the above strategies and focusing on people and the planet, brands can build a favorable image in the minds of consumers. By avoiding algorithmic biases, reducing fossil-fuel-based energy consumption, and investing more in renewable energy, brands can enhance their responsible reputation, attract a broader customer base, and improve operational efficiency, ultimately contributing to SDG 8 (Decent Work and Economic Growth), SDG 9 (Industry, Innovation, and Infrastructure) and SDG 13 (Climate Action).

Section Source: see WARC article on "How generative AI is redefining the game for marketers, and what can they do about it?."

Recognizing the multifaceted impact of emerging technologies on society, there is an increasing call for robust regulatory frameworks to govern their ethical and responsible use, these frameworks must also ensure alignment with societal values and sustainable development goals. Efforts to manage social media and address emerging tech and AI-related issues have seen significant legislative actions such as the General Data Protection Regulation (GDPR) and the Digital Markets Act (DMA). The GDPR, implemented by the European Union in 2018 and replaced by the Data Protection Act from 1998, aims to protect individuals' data privacy and reshape the way organizations approach data management. It introduces strict requirements for data handling, consent, and user rights, impacting how social media platforms collect and utilize personal information. Similarly, the DMA, proposed by the European Commission, targets large tech companies' market dominance, particularly focusing on platforms acting as gatekeepers (see Google antitrust lawsuit). It seeks to promote fair competition, transparency, and innovation in the digital market. These regulations address concerns regarding data privacy, market power, and accountability as social media platforms increasingly integrate AI technologies. However, ongoing challenges persist, including the need for continuous adaptation to rapidly evolving technologies and ensuring global cooperation to effectively manage the digital landscape's complexities. The problem lies in the inability to govern technology, which is evolving faster than societal understanding of the technology. Well-established legal systems have developed through a long process of trial and error, in general, they regulate relatively slowly-changing human societies; therefore, it is probably humanly impossible to explicitly formulate a highly complex set of detailed rules, have them apply a highly diverse set of circumstances and get it right on the first implementation (Bostrom, 2014).

According to a Forbes article (n.d.) (see "The Risk Of Disputes Could Be Huge If Businesses Don't Manage The Privacy Concerns Surrounding AI"), at a time when privacy regulations, security, and data ethics are growing concerns around the world, tech and tech brands more generally are in a unique position to address the issues we have highlighted in this chapter and other social concerns that affect communities—bringing positive and lasting change while targeting business goals. As discussed throughout this book, the triple bottom line framework can expand business success and systems of measurement to include the three bottom line categories: people, planet, and profit. This accounts for more than just the shareholders: this includes employers, customers, suppliers, and communities. By looking at the bigger picture, tech companies can ensure their longevity and become a force for good. TBL is an effective tool for integrating sustainability into the brand strategy, balancing traditional economic motives with social and environmental concerns. We know the foundation of all strategy is how the brand makes money; it is unrealistic to think that brands would exist without this. However, as we have highlighted in this chapter, this has often been achieved at the expense of the world's resources, and this can no longer happen. There are many potential ways forward; brands need to decide which way is best for them, the markets they serve, their consumers, and the most sustainable way for the brand in the long term. As discussed in ► Chap. 2, like other sustainable development models, circular economy addresses decoupling, resource efficiency, production efficiency, slower material flows rather than linear business/economic models, and lower resource extraction without reducing economic activity (McCarthy et al., 2018). Moreover, according to the World Economic Forum (see article on "Why businesses must embrace the circular economy to build sustainable success"), the benefits of the circular economy for brands include greater customer intimacy and retention, more attractive products, recurrent income sources, greater resilience, reduced production times, more extensive use of products, lower impact on the climate and environment with lower costs. Despite the circular economy being praised in its initial attempts to think differently about production and resource impacts on the planet, the recycling part of the system has been criticized as a continuum of the traditional growth-oriented since it still promotes a consumption-led growth based on material re-use and uses a lot of energy to do this. Moreover, the incentive to encourage consumers to possess a greater quantity of goods remains unchanged (Kelleci & Yıldız, 2021). In response, others have proposed a service economy, which is said to be the juncture where sustainable production and sustainable consumption meet, suppliers can make their business both more sustainable and more profitable by focusing on services that extend the efficiency and value of their products, thereby turning demand for reduced material use into a strategic opportunity (Rothenberg, 2007). Some argue that because services are more difficult to imitate than products, they are a greater source of competitive advantage for brands. Case studies introducing these business models and other sustainable development models are still in the early stages, making it premature to advocate for the adoption of any specific model as the prevailing business model going forward. Even second-hand markets/refurbished markets, which some of the big tech brands promote, such as Apple, may increase material consumption rather than decrease it. What is clear, though, is that society faces environmental limits to material consumption, and therefore, any enlightened business model brands adopt going forward must seek to minimize unnecessary consumption of material goods. The classic business models prioritizing profit maximization, which we witness in both the shareholder value model of the 1980s as well as the platform business models of present-day social media and big tech brands, do not serve society or the planet, in fact, we can go as far to say that we cannot let companies exist whose business models conflict with public health. Hence, business models centered around customer preferences, emphasizing lifestyle choices like downshifting or embracing simple living, foster a communal shift in mindset and tackle consumption levels, which are at the core of the current marketing sustainability challenges. This would also help change the marketer's mindset since consumers will have a new level of consciousness toward consumption (particularly after huge changes in the environment, such as the impact of COVID-19).

We have discussed issues regarding energy wastage and advancements in AI/evolving technologies, but there is also the existing problem of large, expensive, and wasteful advertising campaigns. Marketing can be wasteful, and marketers can also be accused of insincere actions where sustainability is concerned, for example, turning a blind eye to the amount of waste advertising campaigns generate while at the same time being tasked by the brand to send out sustainable messaging about the brand. According to a Marketing Week article (see Marketers must rethink campaign production to become more sustainable (n.d.-a)), the era of extravagant campaign shoots in distant locales or allocating substantial budgets to single-use brand materials may be drawing to a close. Marketers are increasingly recognizing their responsibility for the environmental consequences of their work, and brands risk being labeled as hypocritical if they don't adapt.

This critique is not solely confined to wasteful advertising campaigns, brands have long had an issue with transparency around their carbon footprint.

Greenwashing has been around for some time and threatens to undermine efforts being made. Greenwashing is a form of marketing communication that inflates a brand's environmental credentials. It is a form of misinformation used to entice the growing conscious consumer segment. Nielsen Media Research presents that 66% of global consumers are willing to pay more for environmentally friendly products, therefore, when these customers perceive brands as socially responsible, they may be more willing to buy the products from these firms at a higher price. The fashion industry is among the worst offenders of greenwashing. This includes brand giants such as H&M and Zara, the world's first fast fashion retailer (famously claiming they can turn a garment from the design stage to being sold in stores in only 15 days). At the center of most of these criticisms is the inability of these brands to control and be transparent about their supply chain. Avoiding greenwashing means honest and transparent communication about a brand's environmental impact. An interesting trend has emerged relating to the complexities around communicating a brand's sustainable efforts. According to a WARC article (see "Washed Green" or "Transparent Grey"? Understanding why honesty pays off in environmental communication), social psychologists have long investigated how consumers react to negative information and have concluded that consumers generally process it more carefully and trust it more than equivalent positive information (a similar point was made about social media using negative information to exploit our cognitive vulnerabilities) What this means is that if brands are transparent about the environmental problems that their pro-environmental actions seek to address (e.g., the reduction of polluting emissions) and avoid presenting such actions as "beneficial" to the environment, they might have a better chance of being trusted and eliciting favorable intentions and behaviors. Therefore, marketers should not be afraid of revealing deficiencies in terms of environmental sustainability, and research shows that "negatively-framed" eco communications can build trust (Pino et al., 2022). It is clear that consumers want to do more but need better education and signposting. Marketers are ideally placed to help with this, and marketing messaging needs to be more about informing rather than selling. If we don't address the biggest brand's foundational issues associated with social media platforms and AI more generally, then as a society, we cannot come together to solve the bigger issues impacting sustainability and climate change. We risk being divided, isolated, and distracted, and we cannot possibly tackle the world's biggest challenges in this state. Moreover, the prevalence of greenwashing, the urgency of working toward the UN Sustainable Development Goals, and the interconnected principles of people, planet, and profit highlight the seriousness of addressing unethical practices by brands. Without addressing these concerns, including managing brands and common ethical, responsible, and sustainable concerns, we jeopardize our ability to unite and effectively tackle the world's most significant challenges.

4.13 Brand Management, Marketing Communication, and Promotion Management for Sustainable Development Goals

The chapter opened with an exploration of how consumers employ brands as tools for identity projection, emphasizing the development of the conscious consumer identity. Individuals increasingly associate themselves with brands that mirror their values and beliefs; this is particularly true in the realm of sustainability. Within this framework, effective brand management should not be merely about product promotion but encompass a broader strategy that integrates sustainable development goals into branding initiatives. Marketing communication and promotion management, therefore, become crucial components in articulating a brand's commitment to sustainability and engaging with consumers authentically. By fostering transparency and meaningful dialogue, brands can forge deeper connections with their audience while advancing the collective pursuit of sustainable development objectives.

Building upon the foundation of brand identity and its association with sustainability, the chapter then explored the interplay between brands and societal well-being, drawing clear connections to the United Nations Sustainable Development Goals (UN SDGs). The chapter highlights unsustainable patterns of consumption perpetuated by certain brands, which contribute to environmental degradation, resource depletion, and social injustices. From exploitative labor practices to the exploitation of natural resources, these unsustainable behaviors emphasize the urgent need for brands to realign their practices with the principles of sustainable development.

Moreover, the chapter sheds light on past unethical branding practices that exacerbate societal inequalities and perpetuate harmful stereotypes. By examining instances of unequal representation and the reinforcement of stereotypes in marketing communications, the chapter highlights the detrimental impact of such practices on social cohesion and inclusivity. In doing so, it emphasizes the relevance of UN SDGs related to gender equality (SDG 5), reducing inequalities (SDG 10), and promoting inclusive societies (SDG 16). The discussion progressed to include the prevalence of exploitative labor practices within global supply chains, where brands often prioritize cost-cutting measures at the expense of worker rights and dignity. This exploitation not only violates fundamental human rights (SDG 8) but also perpetuates cycles of poverty and inequality, hindering progress toward broader sustainable development objectives.

The concept of the "unpaid consumer" is examined, highlighting the unseen costs borne by individuals and communities (often online) due to the externalities of production and consumption. These hidden costs highlight the need for a more holistic understanding of value creation and consumption patterns. This section stresses the relevance of SDGs related to responsible consumption and production (SDG 12), as well as the importance of partnerships (SDG 17) to address systemic challenges and foster sustainable business practices. In essence, by critically examining the intersection of brands with societal issues such as inequality, exploitation, and unsustainable consumption, the chapter reveals the imperative for brands to embrace their role as agents of positive change and actively contribute to the realization of the UN SDGs. Through ethical branding practices, inclusive marketing strategies, and responsible supply chain management, brands can not only enhance their reputation and profitability but also advance the collective agenda of sustainable development and societal well-being.

The discussion on big tech brands, persuasive technology, monopoly power, competition, and antitrust measures is intricately linked to several SDGs. First, addressing monopoly power and promoting fair competition aligns with SDG 9 (Industry, Innovation, and Infrastructure), which calls for the development of resilient infrastructure, the promotion of inclusive and sustainable industrialization, and fostering innovation. By ensuring a level playing field in the digital marketplace, antitrust measures contribute to fostering innovation and promoting economic growth that benefits all stakeholders. The conversation intersects with SDG 10 (Reduced Inequalities), which aims to empower and promote the social, economic, and political inclusion of all. The concentration of power among big tech brands can exacerbate economic inequalities and limit opportunities for smaller businesses and entrepreneurs. Effective antitrust enforcement can help mitigate these disparities and promote a more equitable distribution of resources and opportunities. Furthermore, discussions surrounding data privacy and consumer rights in the digital age resonate with SDG 16 (Peace, Justice, and Strong Institutions), which calls for ensuring public access to information (reducing algorithmic opacity) and protecting fundamental freedoms in accordance with national legislation and international agreements. Strengthening regulatory frameworks to safeguard data privacy and uphold consumer rights is essential for building trust in the digital ecosystem and fostering a fair and transparent digital society. One of the most pressing issues which was highlighted in this chapter is the insidious nature of some persuasive technology. Advocating for responsible consumption and production, as outlined in SDG 12, intersects with discussions on persuasive technology and its impact on consumer behavior by emphasizing the importance of ensuring that people everywhere have the relevant information and awareness for sustainable development and lifestyles in harmony with nature. By promoting transparency and empowering consumers to make informed choices (without algorithmic manipulation), regulatory measures can support sustainable consumption patterns and contribute to achieving SDG 12 targets.

In the "Manager's Corners" and "Marketing in Action" tasks, the discussion on Apple products as symbols of wealth, taste, and social status highlights the phenomenon of conspicuous consumption, which can contribute to status anxiety and societal pressure for material possessions. This discussion intersects with several SDGs. Primarily, the emphasis on conspicuous consumption and the pursuit of material possessions align with SDG 12, which calls for substantially reducing waste generation through prevention, reduction, recycling, and reuse. By promoting a culture of mindful consumption and reducing the emphasis on materialistic symbols of status, efforts can be made to minimize the environmental and social impacts associated with excessive consumption patterns. This isn't so simple given that Apple has strategically built its brand around status anxiety and symbolic purchases, particularly evident in the cycle of consumers feeling compelled to upgrade to the latest iPhone models. This emphasis on conspicuous consumption presents a significant obstacle in efforts to promote sustainable consumption patterns, as it perpetuates a culture of consumerism and materialism that runs counter to the goals of SDG 12 (Responsible Consumption and Production).

In contrast, the discussion on Tofoo, a pioneering brand in the plant-based food industry, aligns with several SDGs, primarily SDG 2 (Zero Hunger) and SDG 3 (Good Health and Well-being). Tofoo's mission to provide delicious, nutritious, and sustainable alternatives to meat contributes to promoting food security, improving nutrition, and supporting sustainable food systems (SDG 2). Additionally, by offering environmentally friendly food options, Tofoo contributes to mitigating the environmental impact of food production and consumption, supporting SDG 13 (Climate Action) and SDG 15 (Life on Land). Moreover, Tofoo's branding, characterized by its commitment to quality, sustainability, and innovation, resonates with SDG 12 (Responsible Consumption and Production). By promoting sustainable production practices and offering consumers environmentally friendly choices, Tofoo exemplifies the principles of responsible consumption and production outlined in SDG 12.

As we navigate the complexities of unethical brand practices and marketing communication, the chapter emphasizes the importance of reassessing our approach to branding and consumption. It emphasizes the necessity for a paradigm shift that aligns with the principles of sustainability embodied in the SDGs. Only through a concerted effort to move away from branding strategies that focus purely on profit or use sustainability as a gesture PR tactic can we envision a future where efficiency aligns with environmental stewardship. This future would be characterized by a landscape revitalized through responsible brand choices and conscientious consumption behaviors.

Summary

In this chapter, we have introduced and defined the concept of a (sustainable) brand. Many organizations struggle with the branding process, particularly building a long-lasting, easily recognizable brand that stays true to its values. Traditionally, we know that marketing scholars encouraged brands to craft unique products and services in order to differentiate their offerings from competitors to attract and retain consumers. However, in recent years, research has demonstrated that consumers rarely perceive brands as being meaningfully different, and therefore, the ultimate aim for a brand is to own some distinctive elements such as a logo, tagline, or colors that help the consumer to notice and recognize the brand in a buying situation. We have considered some key theoretical influences on branding, notably the impact of consumer psychology and behavior science stimuli on brand decision-making. From a sociological perspective, we know that consumers use brands as props to project a desired identity. The conscious consumer identity has evolved, and consumers are very aware of brands they can purchase from to enhance this identity. Creating a strong brand is important, but managing a brand in the short and long term is vital for brand growth. This chapter has discussed some of the ethical challenges brands need to manage in the contemporary environment. The triple bottom line (People, Planet, and Profit) and the Sustainable Development Goals are important for brand management in the current environment, and there is now an expectation that brands will move forward and focus on de-prioritizing purely making decisions on what will maximize profit and instead also consider reporting societal and environmental metrics for growth.

Toward the end of the chapter, we focused on what some have claimed is the biggest threat to our society, the big tech brands and evolving AI technologies. Plagued with accusations relating to monopolistic behavior, stifling innovation and competition, exploitative practices, free speech, and transparency issues, the big tech brands' business models are not fit for purpose.

In moving forward, businesses must evolve their foundation revenue models to align with the changing incentives of our times. This involves a departure from viewing sustainability merely as a public relations tactic aimed at enhancing brand image. Instead, brands must champion a paradigm shift toward fostering better consumption habits rather than promoting endless consumption. By embracing this ethos, brands can carve out a meaningful role in shaping a sustainable future for both our planet and society. This requires a commitment to innovation, transparency, and genuine dedication to the principles of sustainability. Only through such concerted efforts can brands truly contribute to a future that prioritizes the well-being of people, planet, and profit alike.

? Review Questions

1. Define what a "Brand" is and explain its significance in the contemporary marketplace.
2. What is the difference between "Branding" and "Advertising"?
3. How does "Promotion" contribute to building brand awareness and consumer engagement?
4. Explain the concept of "Differentiation" in the context of branding strategies.
5. What are "Distinctive assets," and why are they important for brand recognition? How do they help the memory retrieve brand information?
6. Define what a "Commodity" is and discuss how brands strive to avoid being perceived as such.
7. Describe the role and responsibilities associated with "Brand Management."
8. How does "Brand Storytelling" contribute to a brand's identity and consumer connection?
9. Discuss the importance of "Sustainability" in contemporary brand management.
10. Explain the concept of "Monopoly Power" and its potential impact on competition.
11. How do "Consumer Habits" influence brand success and market dynamics?
12. Analyze the role of "Big Tech" in shaping modern branding practices.

13. Explore the impact of "AI and Branding" on consumer experiences and brand strategies.
14. Discuss the significance of "Brand Loyalty" in sustaining a brand's success.

Discussion Questions

1. Explain the role of sustainability in contemporary brand management. Provide examples of brands that have successfully integrated sustainability into their brand strategy rather than using it as a public relations gesture. Discuss how this has influenced consumer perceptions and loyalty.
2. Analyze the impact of monopoly power on competition in the market. Extensively research the extent of Amazon's monopoly power, discuss the consequences for consumers and competitors, and explore potential strategies for promoting fair competition.
3. Investigate the evolving relationship between big tech, artificial intelligence (AI), and branding. Research Microsoft and analyze how AI is being employed in its branding strategies. Discuss the implications for consumer experiences, ethical considerations, and the potential for monopoly power.
4. Explore the concept of brand loyalty in the context of changing consumer habits. Consider the idea of whether true brand loyalty exists in the current competitive landscape. Choose a brand that you believe has successfully built and maintained brand loyalty, and discuss the strategies employed. Evaluate the impact of shifting consumer habits on brand loyalty and propose recommendations for brands to adapt.
5. Assess the role of stereotyping and representation in brand management. Choose a brand that has effectively demonstrated cultural understanding in its marketing and branding efforts. Discuss the importance of understanding diverse cultures, potential pitfalls to avoid, and the impact of cultural sensitivity on brand perception and success.

Practice Quiz

1. What term refers to the unique elements that make a brand easily recognizable?
   (a) Sustainability
   (b) Differentiation
   (c) Distinctive assets
   (d) Monopoly power
2. How does "Brand Storytelling" contribute to a brand's identity?
   (a) By avoiding consumer engagement
   (b) By establishing a brand as a commodity
   (c) By sharing a narrative that resonates with consumers
   (d) By encouraging unethical practices
3. What is the significance of "Sustainability" in brand management?
   (a) It has no impact on consumer preferences.
   (b) It is essential for building brand equity and loyalty.

   (c) It leads to greenwashing.
   (d) It promotes monopoly power.
4. What does "Monopoly Power" refer to in the context of brands?
   (a) Healthy competition
   (b) Unique brand storytelling
   (c) Dominance and control in the market
   (d) Ethical branding practices
5. How do "Consumer Habits" influence brand success?
   (a) They have no impact on brands.
   (b) They shape purchasing decisions and brand preferences.
   (c) They discourage brand loyalty.
   (d) They lead to supply chain transparency.
6. What is the role of "Big Tech" in shaping modern branding practices?
   (a) It has no influence on branding.
   (b) To date, unethical branding practices have been witnessed.
   (c) It contributes to sustainable practices.
   (d) It significantly shapes and influences brand strategies.
7. How does "AI and Branding" impact consumer experiences?
   (a) It has no impact on consumer experiences.
   (b) It enhances personalization and engagement.
   (c) It discourages brand loyalty.
   (d) It promotes greenwashing.
8. What term refers to the emotional connection and commitment that consumers have toward a brand?
   (a) Monopoly power
   (b) Brand awareness
   (c) Brand loyalty
   (d) Triple bottom line
9. Which of the following is associated with the responsibility of managing and enhancing a brand's reputation?
   (a) Supply chain transparency
   (b) Brand management
   (c) Greenwashing
   (d) Unethical concerns
10. What does "Triple Bottom Line" consider in evaluating business success?
   (a) Profitability only
   (b) Social and environmental impact, in addition to profitability
   (c) Competition and monopoly power
   (d) Cultural sensitivity
11. How does "Greenwashing" relate to branding?
   (a) It promotes environmental sustainability.
   (b) It discourages sustainable practices.
   (c) It involves false or exaggerated sustainability claims.
   (d) It encourages ethical branding.

12. What term refers to the dominance of a brand or a few brands in a specific market?
    (a) Brand equity
    (b) Status anxiety
    (c) Monopoly power
    (d) Consumer activism
13. What concept involves involving various interested parties in decision-making processes and brand initiatives?
    (a) Ethical branding
    (b) Cultural sensitivity
    (c) Stakeholder engagement
    (d) Brand equity
14. What term refers to the unique elements that make a brand easily recognizable?
    (a) Logo
    (b) Status anxiety
    (c) Mental availability
    (d) Business models
15. What is the significance of "Brand Equity" in evaluating a brand's strength?
    (a) It measures the social and environmental impact.
    (b) It assesses the brand's profitability only.
    (c) It evaluates the brand's perceived value and strength.
    (d) It considers monopoly power only.

(You can find the correct answers to these questions by taking the quiz and then comparing your answers with the answers that are provided at the end of the book, under the "Practice Quiz Answers" part of the book.)

References

Aaker, D. A. (1996). Measuring brand equity across products and markets. *California Management Review, 38*(3), 102.

Adams, S. B. (2003). Regionalism in Stanford's contribution to the rise of Silicon Valley. *Enterprise and Society, 4*(3), 521–543.

Anjos, J. C., Gross, J. L., Matteussi, K. J., González, G. V., Leithardt, V. R., & Geyer, C. F. (2021). An algorithm to minimize energy consumption and elapsed time for IoT workloads in a hybrid architecture. *Sensors, 21*(9), 2914.

Asleep at the wheel. (2024). Retrieved April 10, 2024, from https://www.hrw.org/report/2024/02/01/asleep-wheel/car-companies-complicity-forced-labor-china

Baddeley, A. (1982). Reading and working memory. *Bulletin of the British Psychological Society, 35*, 414–417.

Bettman, J. R. (1986). Consumer psychology. *Annual Review of Psychology, 37*(1), 257–289.

Bostrom, N. (2014). *Superintelligence: Paths, dangers, strategies*. Oxford University Press.

Campbell, M. C., & Keller, K. L. (2003). Brand familiarity and advertising repetition effects. *Journal of Consumer Research, 30*(2), 292–304.

Doidge, N. (2007). *The brain that changes itself: Stories of personal triumph from the frontiers of brain science*. Penguin.

Eastman, J. K., Goldsmith, R. E., & Flynn, L. R. (1999). Status consumption in consumer behavior: Scale development and validation. *Journal of Marketing Theory and Practice, 7*(3), 41–52.

Fogg, B. J. (1998, January). Persuasive computers: Perspectives and research directions. In *Proceedings of the SIGCHI Conference on Human Factors in Computing Systems* (pp. 225–232).

Forbes. (n.d.). Retrieved April 10, 2024, from https://www.forbes.com/sites/garydrenik/2023/12/14/the-risk-of-disputes-could-be-huge-if-businesses-dont-manage-the-privacy-concerns-surrounding-ai/

Gereffi, G., & Pan, M. L. (1994). Garment industry. In *Global production: The apparel industry in the Pacific Rim* (p. 126). Appelbaum.

Gran, A. B., Booth, P., & Bucher, T. (2021). To be or not to be algorithm aware: A question of a new digital divide? *Information, Communication & Society, 24*(12), 1779–1796.

Jenkins, R. (2014). *Social identity*. Routledge.

Kahneman, D., & Tversky, A. (2013). Prospect theory: An analysis of decision under risk. In *Handbook of the fundamentals of financial decision making: Part I* (pp. 99–127).

Kallbekken, S., & Sælen, H. (2013). 'Nudging' hotel guests to reduce food waste as a win–win environmental measure. *Economics Letters, 119*(3), 325–327.

Kelleci, A., & Yıldız, O. (2021). A guiding framework for levels of sustainability in marketing. *Sustainability, 13*(4), 1644.

Klein, N. (2000). *No logo*. HarperCollins Publishers.

Kotler. (2016). Retrieved April 10, 2024, from https://www.marketingjournal.org/the-relationship-between-marketing-and-capitalism-phil-kotler/

Kotler, P., & Armstrong, G. (2010). *Principles of marketing*. Pearson Education.

Lehner, M., Mont, O., & Heiskanen, E. (2016). Nudging—A promising tool for sustainable consumption behaviour? *Journal of Cleaner Production, 134*, 166–177.

Letwin, W. L. (1955). Congress and the Sherman Antitrust Law: 1887–1890. *University of Chicago Law Review, 23*, 221.

Marketing Week. (n.d.-a). Retrieved April 10, 2024, from https://www.marketingweek.com/marketers-rethink-campaign-production-become-more-sustainable/

Marketing Week. (n.d.-b). Retrieved April 10, 2024, from https://www.marketingweek.com/ritson-snickers-balanced-differentiation-distinctiveness/

Marketing Week. (n.d.-c). Retrieved April 10, 2024, from https://www.marketingweek.com/marketing-supply-chain-mission-brand/

Mason, R. (2000). Conspicuous consumption and the positional economy: Policy and prescription since 1970. *Managerial and Decision Economics, 21*(3–4), 123–132.

McCarthy, A., Dellink, R., & Bibas, R. (2018). *The macroeconomics of the circular economy transition: A critical review of modelling approaches*. OECD.

McCracken, G. (1986). Culture and consumption: A theoretical account of the structure and movement of the cultural meaning of consumer goods. *Journal of Consumer Research, 13*(1), 71–84.

Mckinsey article. (n.d.). Nudging. Retrieved April 10, 2024, from https://www.mckinsey.com/capabilities/strategy-and-corporate-finance/our-insights/much-anew-about-nudging

Mick, D. G., & Buhl, C. (1992). A meaning-based model of advertising experiences. *Journal of Consumer Research, 19*(3), 317–338.

Ofengenden, T. (2014). Memory formation and belief. *Dialogues in Philosophy, Mental and Neuro Sciences, 7*(2), 34–44.

Pino, G., Viglia, G., Nataraajan, R., Peluso, A. M., & Pichierri, M. (2022). How to shape consumer reaction to corporate environmental communications: Accentuating the negative to build trust can elicit favorable intentions and behaviors. *Journal of Advertising Research, 62*(1), 62–86.

Reuters. (n.d.). Google antitrust case. Retrieved April 10, 2024, from https://www.reuters.com/legal/us-wraps-up-antitrust-case-against-google-historic-trial-2023-11-16/

Rosenbaum-Elliott, R. (2020). *Strategic advertising management*. Oxford University Press.

Rothenberg, S. (2007). Sustainability through servicizing. *MIT Sloan Management Review*.

San Pellegrino. (2024). Retrieved April 10, 2024, from https://www.sanpellegrino.com/identity-history/our-history

Schachter, D. L. (1996). *Searching for memory: The brain, the mind, and the past*. Basic Books.

Shotton, R. (2018). *The choice factory: 25 behavioural biases that influence what we buy*. Harriman House Limited.
Smith, A. N., & Fischer, E. (2021). Pay attention, please! Person brand building in organized online attention economies. *Journal of the Academy of Marketing Science, 49*, 258–279.
Soyer, D. (1999, April). Garment sweatshops, then and now. In *New labor forum* (pp. 35–46). Labor Resource Center, Queens College, City University of New York.
Terranova, T. (2012). Free labor. In *Digital labor* (pp. 33–57). Routledge.
Thaler, R. H., & Sunstein, C. R. (2008). *Nudge: Improving decisions about health, wealth, and happiness*. Penguin.
The Decision Lab. (n.d.-a). Availability heuristics. Retrieved April 10, 2024, from https://thedecisionlab.com/biases/availability-heuristic
The Decision Lab. (n.d.-b). The framing effect. Retrieved April 10, 2024, from https://thedecisionlab.com/biases/framing-effect
The New York Times. (n.d.). The 'Facebook Class' built apps, and fortunes. Retrieved April 10, 2024, from https://www.nytimes.com/2011/05/08/technology/08class.html
Toffler, A. (1980). *The third wave* (p. 544). Morrow.
Veblen, T. The theory of the leisure class. In *The collected works of Thorstein Veblen* (Vol. 1. 1899 Reprint, 1994, pp. 1–404). Routledge.
Walasek, L., & Brown, G. D. (2015). Income inequality and status seeking: Searching for positional goods in unequal US states. *Psychological Science, 26*(4), 527–533.
Walasek, L., & Brown, G. D. (2016). Income inequality, income, and internet searches for status goods: A cross-national study of the association between inequality and well-being. *Social Indicators Research, 129*(3), 1001–1014.
Willson, M. (2019). Algorithms (and the) everyday. In *The social power of algorithms* (pp. 137–150). Routledge.
World Advertising Research Centre. (n.d.). How generative AI is redefining the game for marketers. Retrieved April 10, 2024, from https://www.warc.com/content/paywall/article/bestprac/how-generative-ai-is-redefining-the-game-for-marketers-and-what-can-they-do-about-it/en-GB/153504?
Zwick, D., Bonsu, S. K., & Darmody, A. (2008). Putting consumers to work: Co-creation and new marketing govern-mentality. *Journal of Consumer Culture, 8*(2), 163–196.

Further Reading

Aaker, D. A. (2009). *Managing brand equity: Capitalizing on the value of a brand name*. Simon and Schuster.
Aaker, D. A. (2012). *Building strong brands*. Simon and Schuster.
Batey, M. (2008). *Brand meaning*. CRC Press.
Beverland, M. (2021). Brand management: Co-creating meaningful brands. *Brand Management*, 1–424.
De Chernatony, L., & Dall'Olmo Riley, F. (1998). Defining a "brand": Beyond the literature with experts' interpretations. *Journal of Marketing Management, 14*(5), 417–443.
Elliott, R., & Wattanasuwan, K. (1998). Brands as symbolic resources for the construction of identity. *International Journal of Advertising, 17*(2), 131–144.
Galloway, S. (2018). *The four: The hidden DNA of Amazon, Apple, Facebook, and Google*. Penguin.
Holt, D. B. (2004). *How brands become icons: The principles of cultural branding*. Harvard Business Press.
Iglesias, O., & Ind, N. (2020). Towards a theory of conscientious corporate brand co-creation: The next key challenge in brand management. *Journal of Brand Management, 27*(6), 710–720.
Jung, J., Kim, S. J., & Kim, K. H. (2020). Sustainable marketing activities of traditional fashion market and brand loyalty. *Journal of Business Research, 120*, 294–301.
Kahneman, D. (2011). *Thinking, fast and slow*. Macmillan.
Klein, N. (2000). *No logo*. HarperCollins Publishers.
Romaniuk, J., & Sharp, B. (2016). *How brands grow. Part 2: Including emerging markets, services and durables, new brands and luxury brands*. Oxford University Press.

Schultz, D. E., & Block, M. P. (2015). Beyond brand loyalty: Brand sustainability. *Journal of Marketing Communications, 21*(5), 340–355.
Sharp, B. (2010). *How brands grow*. Oxford University Press.
Tierney, K. D., Karpen, I. O., & Westberg, K. (2016). Brand meaning co-creation: Toward a conceptualization and research implications. *Journal of Service Theory and Practice, 26*(6), 911–932.
Zuboff, S. (2019). *The age of surveillance capitalism*. Profile Books.

Distribution Management for People, Planet, and Profit

Contents

S. M. R. Shams et al., *Sustainable Marketing*, Classroom Companion: Business,
https://doi.org/10.1007/978-3-031-79130-7_5

Preliminary Vignette: Avon Cosmetics

Avon Cosmetics was launched in the late nineteenth century by a door-to-door book salesperson who noticed that many of his female clients were more motivated by the free samples of perfume that he gave away than by his core product. He also noted that many of them showed potential as salespeople, so he introduced a channel strategy of employing independent agents who were not employed directly by the company, but registered as agents for a small fee, receiving brochures and samples for their prospecting activities. As door-to-door selling was commonplace in the USA well into the twentieth century, Avon persisted with this approach. Although there was no barrier to men becoming agents, Avon became known for its "Avon Ladies" and for the phrase "Avon Calling." Both of these personified the friendly, feminine, and local face of Avon—a brand that came to be seen as approachable, reliable, and caring. Selling to an overwhelmingly female market, and predominantly 25- to 50-year-old, there was very little psychic distance between customers and agents as they were, in effect, the same demographic. A visit from the "Avon Lady" became not just the vehicle for a purchase transaction, but a regular social occurrence when neighbors could exchange news. By building a discussion of cosmetics, skincare, and haircare products into that occasion, it became an enjoyable, indulgent "me time," and an escape from work and domestic routine.

By the 1980s, Avon's rival cosmetics brands were selling successfully through retailing partners' department stores, pharmacies, and supermarkets. Avon's doorstep approach retained some advantages over its in-store competitors: Avon representatives could take their time in customer appointments without fear of a queue building at the till, therefore nurturing and individualizing customer relationships; by visiting customers at home, their clients were likely to be more relaxed; the lack of competitor products enabled agents to take their customers out of the market by satisfying their needs without being subjected to detailed in-store price and product comparisons; Avon only paid commission on goods sold, thereby minimizing financial wastage; and it did not run the risk of its agents switch-selling customers into competitors' products. However, in-store competitors also had several key advantages over Avon: They could provide an in-store experience with point-of-sales materials, a broad variety of stock, and store atmospherics; they could rely on customer footfall which had been generated by the stores and the other brands that they represented; they could meet and sell to a greater number of prospects; they were more able to process transactions and fulfil orders immediately; and they may also benefit from some incoming switch sales.

For the reasons discussed, Avon decided that it needed a presence in stores, but it faced a significant dilemma: If it started selling its products in stores, that would be likely to alienate its door-to-door agents, possibly making them redundant completely over time.

Disintermediation—manufacturers selling some products directly to customers instead of through their downstream intermediaries, which is one of the chief causes of **channel conflict**, which is destructive, ill-feeling between channel partners.

Intermediaries often feel betrayed by this and suspect that the manufacturer is "cherry-picking" the easiest sales to cut them out of the chain and reduce costs. This may be true, but it may also be the case that the manufacturer needs to make some sales directly for logistical reasons, or that it perceives that it has no choice (as was the case with Avon) if it is to respond to an evolving marketplace, survive and thrive. Avon's solution was to devise a premium range which would be sold solely in stores, thereby avoiding cannibalization of its agents' sales. However, strategies such as this invariably suffer from two major issues: Both agents and stores would effectively only be able to sell part of Avon's range, causing confusion and frustration among customers; and to produce premium products would raise the prices of the products sold in store, just at the same time as the manufacturer would incur costs from developing an additional route to market. Perhaps for these reasons, it was unsurprising that Avon became an early adopter of online retailing, beginning a serious internet presence as early as 1997, stepping up its online marketing and use of digital transactions in 1999, and maintaining a strong internet presence thereafter. Avon nowadays uses an omnichannel strategy, recognizing that different customer segments welcome different ways of experiencing the brand and buying from it. By providing personalization and interaction, online retailing has enabled Avon to stay true to its roots.

5.1 Introduction

Within the marketing mix, "Place" refers not only to the location where a product or service is sold to the customer but to all the activities related to locations selection, movements, and processes involved in getting goods from the producer to the buyer and end user. Rather than focusing solely on logistical factors, it is primarily concerned with the relationships and interactions which occur between various channel intermediaries, such as wholesalers, retailers, dealers, brokers and agents. As we shall see in this chapter, distribution channel partners do not simply manage flows of items and money, but interconnect and collaborate, contributing to the marketing process and adding value to the customer proposition. The responsibilities of channel intermediaries often overlap, and their business priorities to some extent misalign, and this may result in conflict which can undermine the brand, sales volumes, and customer satisfaction. Furthermore, increasingly well-informed customers are scrutinizing how organizations move their products and the ethical credentials of partner firms. Therefore, the selection, appointment, and management of appropriate channel members are crucial to the development of a sustainable marketing strategy.

To be people-oriented, distribution strategies must meet the increasingly stringent needs of modern consumers, often by adopting multichannel or omnichannel approaches. To help preserve the planet, organizations must pursue distribution strategies that seek to reduce their carbon footprint by identifying and tackling unnecessary transportation, and by helping consumers to adopt streamlined shopping behaviors. And to maximize profit, firms must minimize wastages suffered due to the perishability of stock, the expense of transportation, and the lack of liquidity associated with inefficient stock control. We will explore the roles and behaviors expected of channel partners, identify social, ecological, and economic factors impacting their ability to collaborate ethically, and analyze strategic approaches to maximize value within distribution channel management.

- **Multichannel retailing**: selling through more than one channel (e.g., retail stores, own website, and comparison/aggregator website)
- **Omnichannel retailing**: selling in a coordinated manner through different channels according to the immediate needs of the customer/prospect, often combining several channels within the sales process

5.2 Sustainability in Supply Chains

Retailers and manufacturers increasingly need to demonstrate ethical conduct as part of their sales propositions. Today's consumer is information-rich, having access to a wealth of product information via the internet. During the 2007–2009 Great Financial Crisis, many consumers had to save harder to buy what they wished. Following the COVID-19 pandemic starting from early 2020, and the subsequent geopolitical tensions around international trade, consumers' affordability (buying power) has once again been impacted in a way that is not desirable. Therefore, consumers demand that their suppliers redouble their efforts to add additional value. One of the ways in which customers sought, and continue to seek, value is in terms of brands' ethical stances. For example, during the last 15 years in the first two decades of this current century, consumers have placed much more emphasis on manufacturers of insulated jackets sourcing their geese down from farms that keep their geese ethically, on supermarkets to discontinue stocking eggs from battery hens, and on coffee brands to ensure that plantation workers are paid a fair wage. Broadly speaking, most consumers would like their brands and suppliers to treat workers and stakeholders with respect, to avoid any unnecessary harm to animals, to refrain from polluting the environment or depleting non-renewable resources unnecessarily, and to pursue alliances and relationships only with partners who behave ethically.

5

◘ Beverage companies must select suppliers who farm coffee sustainably and treat their workers ethically. Image source: ► pexels.com/royalty-free-images/

Fast fashion and clothing are problematic industries. The manufacture of garments places a large toll on the earth's resources and contributes heavily to pollution. Therefore, it is far preferable to buy a few high-quality clothes that are long-lasting than many lower-quality clothes that will wear out quickly. Naturally, many fashion firms rely on trends going out of style to drive customers into making more frequent purchases. In fact, they may build in this obsolescence through their own design and marketing activities, and by launching new products which make those of previous seasons look old-fashioned. By seeking to reduce their overheads, many mass manufacturers of clothing ceased making goods in higher wage economies in the late twentieth century, instead having clothes made for them by subcontractors in lower-wage economies such as Bangladesh. However, many factories in less economically developed countries pay their workers pitifully low wages for unreasonable long hours in very poor working conditions. When Dhaka's Rana Plaza garment factory collapsed in 2013, more than a thousand workers died and, despite the introduction of better safety standards, many factories are still unsafe. In 2021, at least 52 people died in an inferno in a juice-making factory southeast of Dhaka. It was found that volatile chemicals were stored dangerously and fire escapes were blocked or sealed shut. Sadly, waiting for businesses to raise their ethical standards voluntarily is often not viable, as the owners are discouraged by the prospect of increased overheads making them uncompetitively priced against rivals. Governments may be similarly slow to introduce legislation that could make their manufacturers more expensive than those in other countries, thereby risking the relocation of firms and a subsequent rise in unemployment. Therefore, the quickest and most effective way to bring about positive societal change is often through consumer power, by consumers applying economic pressure to firms through their purchase choices. While making one's own brand more ethical can be a challenge, it may be harder still to ensure that one's allies and partners always act with utmost integrity. This is especially the case where a firm is reliant upon a large number of suppliers spread across many countries, as this can make the suppliers' practices less visible. Köksal et al. (2018) suggested that the key to "cleansing" a firm's channels is to appoint agents who are specifically tasked with finding, appointing, and managing ethical partners—just as a firm might employ Public Relations and marketing agencies to manage elements of its corporate image.

5.3 What Is Distribution Management?

Most organizations that make products rely upon other organizations to make those products available to their customers. These partner businesses inhabit a space which is referred to as the **channel**, and they are known as **channel intermediaries**. They perform numerous important functions to make products available to customers and undertake tasks that would be commercially and logistically unviable for the manufacturer to do. It would clearly be impractical for a furniture manufacturer to deliver its products directly to individual consumers in huge trucks, to allow those consumers to visit the factory to see and test products before purchase, and for the manufacturer to deal directly with customer queries and aftersales issues. Instead, the furniture manufacturer operates warehouses and retail outlets which may be manufacturer-owned, third party-owned, or a mixture thereof. As we go through this chapter, we will analyze the functions of these intermediaries in more detail. Of course, some producers sell products that require a direct interaction between themselves and end users or make products which are simple to send directly to their consumers. Other organizations may be service providers, such as lawyers, dentists, or ski instructors, in which case they may have no need for intermediaries at all. However, firms usually have downstream channel partners, and these share the burden of marketing, transporting goods, and the risks associated with holding stock. Indeed, many organizations also use upstream channel partners. While **downstream channel partners** are positioned between the manufacturer and its end users, **upstream channel partners** supply raw materials, machinery, finance, human resources, and other goods and services to the manufacturer itself. Both downstream and upstream channel partners add value to the supply chain and help to ensure a satisfactory customer experience.

- **Downstream channel partner/channel intermediary**: a channel partner located between a firm and its customers
- **Upstream channel partner**: a channel partner that is further away from the firm's customers than the firm is—it usually supplies materials, machinery, finance, or other resources

When considering the different types of channel structures that a manufacturer may use, it is important to remember that many firms use a combination of channel structures. For example, a motor manufacturer may sell to retail customers through its dealerships, but it may also sell to them through leasing companies, which in turn supply the vehicles through dealerships. Therefore, there may be one or two channel intermediaries between the motor manufacturer and the retail customer. In the business-to-business context, the motor manufacturer may sell vehicles directly to organizations that operate large fleets. However, it may also sell to businesses' leasing companies, dealerships, or even both. Therefore, the motor manufacturer may have no intermediaries, one, or even two, when serving those customers. The term **channel length** describes how many intermediaries exist within the channel (i.e., including all upstream and downstream channel members), and **channel width** refers to the number of partners at each stage of the channel (e.g., how many wholesalers, retailers, or dealerships). For example, a motor manufacturer that sells through 300 different franchised dealerships, without the assistance of agents, wholesalers or brokers, would have short channel length but a large channel width.

Manager's Corner 5.1: Marketing Channels for New Homes

Source: ► pexels.com/royalty-free-images/

Imagine that you are the Sales and Marketing Director for a large construction company that builds and sells around 1000 new homes per year in your home country. These homes are a mixture of apartments, detached houses, and semi-detached houses. All have two, three, or four bedrooms, an ensuite to the master bedroom, fully fitted kitchens and bathrooms, gas-fired central heating, and solar panels.

They are clustered in ten developments of approximately 100 homes, each with a communal playground, lawned gardens, and hedging. The developments are usually situated at the edge of small or medium-sized towns, and are priced to be suitable for young families, retirees, and other potential buyers who earn a broadly average income. Each development has a show home which your salespeople use when promoting the houses to prospects while the developments are being built. Although your sales staff can negotiate slightly if sales are slow—for example, by offering upgraded kitchen appliances or an automated garage door—your company charges fixed prices rather than seeking offers. Unless sales are very swift, you also use other channels. You display houses on your company's website, advertise in selected regional newspapers and magazines, allow your real estate to be listed with competitors' properties on aggregator/comparison websites and, most importantly, your homes are marketed by local realtors (UK: estate agents). Using your knowledge of realtors (or looking at their websites to observe them), make a list of attributes and practices which you would like to see, and those which you would not wish to see, in your preferred realtors, and be ready to discuss why. How would you elicit loyalty and a strong performance from them, so that they prioritize selling your homes and representing your brand positively? Consider several approaches, highlighting the potential benefits and limitations of each.

5.4 Push-and-Pull Strategies

Producers of goods may use push strategies, in which the downstream channel intermediaries are incentivized or persuaded to hold stock. By paying for this stock, at least partially, the intermediary suffers stock pressure which arises from the need to liquidate stock by selling it to turn it back into cash. **Push strategies** are so called because the manufacturer pushes the product down through its downstream channel to the market.

Many producers of goods employ **pull strategies**, in which they use advertising, other elements of the marketing and marketing communications mix, and branding to directly appeal to prospective customers. By raising awareness of their brand, and making it more desirable to prospects, potential buyers are persuaded to visit the organization's retail premises—which are often operated by channel intermediaries—to view, order, and try products. For example, by persuading the end users by offering value-added benefits, the producer of a particular brand based on a pull strategy tries to confirm that the retail intermediary must stock their product on the retail shelf more, considering that the end users of this brand and its competitive brands look for this particular brand in retail store more than the competitive ones, because of the brand's added value. Having benefited from this increase in customer footfall, the retailers or dealers must order stock from the manufacturer to fulfil customer orders. It is called a pull strategy because the manufacturer pulls demand from the market up through its downstream channel. Here, marketers have the opportunity to add sustainability features to their products, e.g., creating a socially caring brand image as a value-added feature to pull the end users to the brand, as end users are more concerned now for socioecological sustainability than ever before.

In practice, the overwhelming majority of organizations employ a combination of push-and-pull strategies. For instance, a motor manufacturer uses digital marketing to reach its preferred customer segment, advertises using billboards, television slots, and radio stations, but also employs a regional manager to ensure that the dealerships are sufficiently stocked and playing a robust role in creating sales opportunities.

Footfall: the number of potential customers entering sales premises

5.5 Upstream Channel Partners

Upstream channel partners may be involved in the supply of raw materials (Srivastava Dabas et al., 2012) such as sheet metal, hardboard, and plastic pellets. These materials may be ingredients of the product to be manufactured but may also be used in the manufacturing process but not appear in the core product. An example of this would be a supplier of oxygen or helium to be used by automotive, welding, or healthcare organizations but not seen by the customers. Some upstream partners supply machinery, especially if they are dealing with clients operating within heavy industry, such as petrochemicals, mining, or manufacturing organizations. Some upstream channel partners also supply components that are incorporated into the end product. For example, a manufacturer of ventilation systems or air conditioning units may provide equipment that is used by construction companies in the houses and factories which they build for their customers. Some other upstream partners supply services rather than goods. Examples of these are commonplace in finance, insurance, and information technology (Palmatier et al., 2016). For example, a manufacturer of plastic goods may require specialized information on the properties of those plastics from an upstream provider of technical expertise. It may also require legal expertise to ensure that its products are compliant with governmental standards. These partners are considered to be upstream because they are located further from the end users than the manufacturer is.

5.6 Downstream Channel Partners

Downstream channel partners intermediate between originator firms, which are normally manufacturers, and the end users in a marketing channel (Claro & Claro, 2010). The most common intermediaries are wholesalers and retailers.

■ This food and drink seller is an independent retailer. Image source: ▶ pexels.com/royalty-free-images/

5.7 Retailers

Retailers are invaluable to customers, as they provide expertise, product knowledge, and advice on the sales process (Dukes & Geylani, 2016). They help to attract footfall through their localized marketing activities and by having a highly visible presence in areas which shoppers frequent. They may hold eye-catching promotions to help support product launches, or to compensate for slower trading periods. They order stock from the manufacturer or from wholesalers to ensure that production mirrors market demand, and receive small deliveries, which they handle and break apart to display a wide variety of individual products, and to prepare them for customers. They also hold backup stock on their premises, or perhaps in a regional storage facility, from which they can replenish stock in the store. They may coordinate the inward delivery of stock to their premises and may also arrange for the onward delivery of bulky items to end users. For example, a department store might arrange for a fridge freezer or oven to be delivered to a retail customer.

Retailers: firms or individuals which sell to end customers, typically on premises designed to capture passing trade and to facilitate the storage, display and selling of a variety of goods and services

Customers often require detailed information to assist them in their decision-making process. However, it would not be satisfying to the prospect or in the best interests of manufacturers if these customers were simply provided with lists of technical specifications and left to negotiate the choice themselves. Customers require guidance from people who are experts in the type of product that they intend to buy. They need to roll out alternative choices that are not appropriate to their needs, and to undertake direct comparisons of products based upon their suitability in fulfilling their requirements. Retailers are able to facilitate this level of interaction between the customer and a salesperson charged with representing the product or brand. In serving customers and answering queries, retailers may provide demonstrations and even talk through related areas of importance such as the provision of finance.

Not only do retailers offer variety and create enticing displays for prospects, but they may also be responsible for the successful rolling out of promotional activities at a local level. Let us consider franchised motor dealerships as an example, as they are a type of retailer. While the motor manufacturer may advertise on television and in other forms of mass communications media, to build brand awareness and desire among customers, it cannot then deal satisfactorily with prospects who react to those marketing actions, because it has insufficient staff to deal with such a large number of potential customers, and those staff are already undertaking functions other than dealing with individual members of the public. The local dealership, therefore, will not only ensure that national brand promotions are pushed at a local level but will also accommodate the additional customer footfall which results from those activities. It will build, cleanse, and maintain a database of customers and prospects within its local area, which will then be used as the basis for future sales and marketing activities. Not only this, but it will also process the sales paperwork, warranties, and perhaps any lease agreements. Additionally, it will provide aftersales support, and will perform maintenance and servicing tasks in accordance with the manufacturer's agreed standards.

5.8 Wholesalers

Wholesalers are often located in industrial parks near major highways, and usually have a very large storage building with a smaller office block attached. A core function of wholesalers is to accommodate deliveries of stock from individual manufacturers. These deliveries are typically made by very large trucks, so wholesalers need large amounts of external space in which they can maneuver and unload, and also large warehouses to store the goods. The accommodating of large incoming deliveries is logistically challenging, as it involves not only planning warehouse capacity but also the availability of staff and machinery and the coordination of incoming stock with outgoing custom orders. Wholesalers are often called upon to accumulate products over a longer period of time, even though this often costs them money in stocking charges. This is especially the case in product categories that experience high degrees of seasonality. For example, a manufacturer of hosepipes enjoys a high demand for their products during the summer when many people are watering their lawns and maintaining their gardens but has almost no demand during the winter. This would cause a significant problem for the manufacturer, as it produces very large peaks and troughs in its cash flow, making it very difficult to employ staff members throughout the entire year. Wholesalers help to mitigate this problem by taking delivery of stock during autumn, storing the stock over winter ready to sell in the spring, and taking some of the financial risks away from the manufacturer. Typically, a horticultural wholesaler would enjoy a manufacturer discount of 30% or more if they were required to keep items that they were unable to sell for 6 months.

Wholesalers: channel partners located downstream of the originator (manufacturer) and upstream of the retailers. They deal with manufacturer deliveries, manage stock, break bulk, offer variety, make retailer deliveries, undertake administrative tasks, and share risk with suppliers

Wholesalers undertake a task called **allocation**, which is the breaking of bulk into smaller quantities. To remain competitive, manufacturers must deliver their products on pallets and in large containers. The wholesaler then breaks down these outers into smaller amounts of stock which it can then sell to retailers. Manufacturers would not be able to make such large deliveries to retailers, who are often situated in crowded city areas, and they would not be able to maneuver their trucks there. Therefore, the wholesaler's role in this area is vital. By offering variety and enabling retailers the flexibility of ordering at short notice and in small quantities, they enable retailers to respond nimbly to the evolving market demands that they experience locally (Rosenbloom & Andras, 2008). By operating a fleet of smaller delivery vehicles, wholesalers are able to react quickly to suit retailer demands, making fast, last-mile deliveries.

Wholesalers manage storage, stock inventory, and distribution. Image source: ► pexels.com/royalty-free-images/

5

Wholesalers negotiate terms with their suppliers, not just to share the stock burden of tying up their money in products for several months, but also to decrease the effects of seasonality which we have considered. They are often responsible for driving the promotional activities of downstream intermediaries such as retailers. For example, although a manufacturer may employ a salesperson or sales-zone (territorial) manager to ensure the compliance of retailers in marketing initiatives, manufacturers also routinely build such tasks into the agreement with wholesalers, so that the wholesalers play an important role in motivating and managing retail outlets. Wholesalers also manage the stock inventory, ensuring that the manufacturer is producing and supplying the right quantities of the right product to meet local market demand. It may also credit check retailers to ensure their long-term viability as partners and provide information on competitors' pricing and activities. However, where a wholesaler manages the products of two or more rivals, it is ethically bound not to share sensitive information between them—although it may gather and disseminate information independently, which it may then provide to the manufacturer or to retailers to help them in making strategically important commercial decisions (Boyaci & Gallego, 2004).

Stock inventory: the products held by a channel member, and often a list of these

5.9 Franchisees

Franchisees are independent organizations that represent a manufacturer's brand within a defined territory. As we have discussed, motor manufacturers often employ franchised dealerships, which are a type of franchised retailer. The fact that they are franchised means that they have committed to represent the manufacturer in a number of ways. They may be required to have certain furniture, fixtures, and fittings, including corporate signage, which is consistent with the manufacturer's brand elements. They may also be required to carry a certain amount of stock for sale, and additional display and demonstration stock to enable customers to see and try the product before making a purchase decision. Franchisees are often required to employ a certain number of salespeople, and to attend specific training sessions, to ensure that those salespeople maintain a high level of product-specific expertise. In return, the manufacturer is likely to contribute to commercial initiatives which are mutually beneficial, such as local marketing activities, open-day events, and special promotional discounts. The franchisee also receives certain resources that would not be available to non-franchised retailers. These include sales leads and databases provided by the manufacturer, interest-free stocking charges and discounted demo and display stock, the benefit of carrying the manufacturer's corporate branding, the channeling of highly profitable aftersales and warranty work by the manufacturer to the franchisee, the status of being the only franchised retailer within the local market area, and the peace of mind and prestige which comes from being associated with a famous manufacturer (Croonen & Broekhuizen, 2019).

◻ Most motor manufacturers appoint franchised dealerships to sell vehicles to retail customers. Image source: ▶ pexels.com/royalty-free-images/

Franchisees may take bulk deliveries, share the risks and burdens of marketing and storing stock, provide the manufacturer with local market knowledge, generate customer footfall, provide customer service, supply salespeople who will negotiate the best terms during sales appointments, order appropriate amounts of stock, increase sales volumes, and manage the relationships between the manufacturer and its customers.

Franchisees: third-party organizations which agree to represent a brand by working on its behalf, usually in a sales and marketing capacity. The originator brand in such an agreement is a **franchisor**, and the contract tying the parties together is a **franchise agreement**

5.10 Brokers and Agents

Brokers and agents are often considered 'middlemen'—although the gender-specific nature of the term alone makes the idea obsolete—who simply take a slice out of the manufacturer's profit margins. However, their influence is not restricted to such a negative contribution, as they also help to locate and attract customers in both business-to-business and business-to-consumer markets. A well-known example of an agency is TKTS (► https://officiallondontheatre.com/tkts/), which sells late discounted tickets for musicals in the West End of London. Due to its distinctive branding and high level of visibility within tourist hotspots, it benefits from high levels of footfall to sell tickets to tourists and Londoners who would not otherwise have bought them. As a result, West End theatres do not have to stray from their areas of expertise to tackle the issue, and they benefit from incremental sales and increased occupancy rates. A less visible example of agents occurs frequently when organizations wish to enter a new, foreign market. In doing so, the manufacturer may have very little market intelligence or contacts in positions of influence, and relatively scant knowledge of local customs and culture. Indeed, the manufacturer may be attempting to enter a market in which purchasers expect to be approached through intermediaries. For this reason, a broker or agent may be indispensable in opening doors and bringing new clients on board.

Agents and brokers: intermediaries between a product/service provider and the end user. Agents usually help customers by "shopping around" for a good deal from a number of potential suppliers, while brokers usually find customers for the originator firm and sell on their behalf.

5.11 Finance, Leasing, and Insurance Companies

Some manufacturers make high-value, "big-ticket" items that customers purchase infrequently, and which cost a high price. Examples of big-ticket items are lounge furniture, cars, and conservatories or sunrooms. Such purchases represent a considerable risk to consumers, who must often suffer the consequences for a long time if they make an inappropriate purchase decision. Most customers cannot afford to pay cash for expensive items such as cars. Indeed, if a car dealership advertises a car at $30,000, that simply tells prospects that they cannot afford what is being sold and disengages them. A useful way around this problem is for sellers to offer expensive goods through a lease or other finance product. Over half of all retail customers buy cars in this way. Typically, they would pay a deposit of 20% or 25%, followed by a monthly direct debit for 36 or 48 months, followed by a payment balloon of 20% or 25% at the end of the contract. By paying this balloon payment, the customer acquires ownership of the vehicle which they have been leasing, and can therefore keep it, sell it, or use its value as a down payment on their next vehicle. While the total amount paid over the contract period exceeds the outright purchase price, the customer derives several key benefits from the arrangement, including affordability. Types of financial products that fall into this category are lease purchases, operating leases, and Personal Contract Plans (PCPs).

The most common form of PCP in the motor industry is contract hire. Not only do these financial products enable customers to enjoy the usage of high-value products which they could not otherwise afford, but they also help to give the customer peace of mind, as they usually include servicing, maintenance, and repair charges, servicing costs for a predetermined period of time, and sometimes even insurance and the replacement of certain components. Therefore, although the customer pays more over a long period of time, they benefit from a lack of "nasty surprises." From the perspective of the seller, the use of finance products is very helpful in customer retention. If a car dealership has sold a car to a customer using a three-year contract hire agreement, and the database tells them that that customer is 2.5 years through that contract, then the dealership can approach that customer to arrange their next vehicle before the customer shops the competition. In effect, the customer is captive, and this is a significant boon to the incumbent supplier (Chen & Popovich, 2003).

Home furniture retailers also use finance to offer buy-now-pay-later incentives to their prospects, and these may include the provision of interest-free credit. It is worth noting that manufacturers and retailers usually earn commission from independent finance companies for introducing their customers. This incremental revenue stream can sometimes be more profitable than selling the core product. For example, motor dealerships often sell GAP Insurance, which insures against the customer having more money outstanding on the lease than the remaining value of the car if the car is written off after an accident. GAP insurance is particularly profitable, and many dealers make more commission from selling it than they do from selling the car on which it is offered. Manufacturers and retailers may benefit from finance and leasing partners, and insurance companies, as those organizations may sell customers into the product of specific retailers and manufacturers with whom it enjoys good relationships. Therefore retailers and manufacturers can enjoy incremental sales that have been generated by a partner's salesforce rather than their own.

5.12 Marketing Agencies

Marketing agencies are a type of service provider rather than a supply chain intermediary. However, they are situated between the organization and its customers (Mullin, 2018), and they play an important role in increasing trade. A commonly used type of agency is a field marketing agency. These provide temporary salesforces for their clients, particularly for manufacturers who are launching new products or who wish to increase their footprint within a certain market segment. For example, when Camelot won the contract to operate the National Lottery in the UK in the late 1990s, it employed a field marketing agency, whose staff helped to raise awareness of the lottery and its potential benefits to newsagents and other retailers. When Fiat wished to penetrate the UK fleet market in the late 1990s, it turned to the same field marketing agency, whose salesforce visited major fleet operators, raising brand awareness, offering demonstration vehicles, and providing information on technical specifications and product characteristics. Also, energy service providers and mobile phone and internet network service providers often employ such marketing agencies as a third party in some countries, such as Australia. The marketing agency through their field salesforce promotes the services of the original service providers of such energy or mobile phone network services to help increase the original service providers' sales, while earning commission from each sale. This form of outsourcing allows the manufacturer great flexibility. In times of need, it can add appropriately trained and experienced members of staff very quickly to fulfil its commercial requirements. However, in periods of low demand, it does not have the responsibility of paying these staff, and contributing to healthcare plans, pensions, and other employment costs. Other marketing agencies provide important services such as the creative design of artwork, the coordination of promotional activities, and the securing of advertising space on billboards, radio stations, websites, and television.

5.13 Logistics Service Providers and Other Specialist Channel Intermediaries

There are a number of other downstream channel partners who handle logistical issues such as deliveries and transportation, often working in tandem with wholesalers (Aykol & Leonidou, 2018). Logistics service providers are a good example of this, providing data and expertise appertaining to the movement of products and the capacities of various channel intermediaries. Although these service providers are channel partners, title (ownership of products) rarely passes to them, and it is unlikely that they have to shoulder any of the burden associated with stocking charges and seasonality. Nonetheless, they are considered as service providers, as is also the case with importers, exporters, shipping companies, and haulage companies. Certain channel intermediaries work in sectors modifying original products before they are resold in a different form to end users. This is particularly common within the motor industry, as conversion companies transform base vehicles into ambulances, recreational vehicles, mobile libraries, and other forms of mobile applications. In doing so, they benefit from the expertise and market presence of the originator brand, and they also provide it with incremental sales opportunities and an additional route to market.

Web Exercise 5.1

According to Statista (2022), the ten biggest online retailers in the UK in 2020 were as follows:

Retailer	Business type	e-commerce revenue ($m)
Amazon (▶ https://www.amazon.co.uk/)	e-commerce specialist	14,700
Tesco (▶ https://www.tesco.com/)	Supermarkets	7600
Argos (▶ https://www.argos.co.uk/)	Catalogue retailer (owned by Sainsbury's)	5400
Sainsbury's (▶ https://www.sainsburys.co.uk/)	Supermarkets	4600
Curry's (▶ https://www.currys.co.uk/)	Electrical retailer	4100
John Lewis (▶ https://www.johnlewis.com/)	High-end department stores	3500
Ocado (▶ https://www.ocado.com/)	Online only grocery	2700
Marks and Spencer (▶ https://www.marksandspencer.com/)	Retailer of clothing, homeware, and foods	2400
Asda (▶ https://www.asda.com/)	Supermarkets	2400
Morrisons (▶ https://groceries.morrisons.com/webshop/startWebshop.do)	Supermarkets	2300

Amazon has, in recent years, become a ubiquitous presence, selling all categories of new and used goods online for home delivery from its numerous "fulfilment centers." Of the UK-based supermarkets, Tesco has embraced e-retailing most enthusiastically, using it as a way to maximize the benefits derived from running its highly successful Clubcard scheme, and to diversify into additional product and service categories such as movie rental, banking, and insurance. Argos built its business model around catalogue shops—customers would look through its printed catalogue of products, either in store or at home, submit a written order at an in-store counter, and wait for the order to be picked from the warehouse. In recent years, it has moved toward a paperless, "click-and-collect" service. Curry's is a popular store selling consumer electronics and electrical goods; Marks and Spencer is known for high-quality but conservative foods, garments, and housewares; and John Lewis is the UK's dominant chain of prestigious department stores. Ocado's presence as a retailer of grocery items is entirely online, whereas the remaining supermarkets all had well-established networks of physical stores before gaining an online presence.

Choose one of the two e-commerce specialists (Amazon or Ocado), one of the four traditional supermarkets (Tesco, Sainsbury's, Morrisons, or Asda), and one of the four specialist retailers (Argos, Curry's, John Lewis, or Marks and Spencer). Visit the websites of the three companies which you have selected. What are the main similarities and differences between them? Create a table to make quick comparisons. In particular, compare each company's messages, tone, website design, products and services offered, and order process. What examples of best practices could each of your three companies adopt from each other? How do they enhance the customer journey and add value? Make some rough notes to ascertain areas for potential development.

5.14 Value Between Channel Intermediaries

When marketing academics conceptualize channel partners, they usually do so as a supply chain or a demand chain. The concept focuses on how organizations make and sell goods with the assistance of partner organizations which can push products through the supply chain to customers. Invariably, the idea of "stock pressure" (e.g., an intermediary being under pressure to sell due to having much of its money tied up in stock), and of the manufacturer motivating and managing the intermediary, is dominant within the conceptualization of a supply chain. However, academics who understand the phenomenon in terms of a demand chain place an emphasis on organizations being responsive, proactive in their research, listening, and sensible before making decisions regarding value and collaboration (Emmett & Crocker, 2016). Porter (1985) famously envisioned a value chain in which he demonstrated how organizations ought to add value at every stage of the manufacturing and supply process. In it, the firm's activities that support the generation of value are shown in vertical columns, its primary activities are represented by horizontal rows and by scrutinizing efficiencies and identifying ways in which operations can be streamlined and leveraged upon synergies, the firm may gain a competitive advantage in the marketplace by operating more efficiently than its rivals. By doing so it may be able to charge a higher price for its products, eliminate waste, improve product quality, differentiate itself positively from the competition, and increase sales volumes.

Marketing in Action 5.1: Trade Fairs as a Tool of Distribution Channel

Source: ▶ pexels.com/royalty-free-images/

Have you ever been to a trade fair or show in which products were being displayed to retail and corporate customers, prospects, and enthusiasts? Maybe a motor show at a large exhibition center, or a powerboat show at a marina? If so, you probably noticed that there is a very large variety of people at work there—salespeople and managers, hospitality staff from agencies, technical experts, cleaners, stand managers, logistics staff who assemble, disassemble, and transport equipment, and many more. Furthermore, although some of the sales and management staff work for the manufacturer, others may be employed by wholesalers, dealers, retailers, and finance companies who work with the manufacturer as channel intermediaries. Now, imagine that you work for the superyacht builder, Sunseeker (▶ www.sunseeker.com), and you have been tasked with organizing their presence at next year's Monaco Yacht Show (▶ www.monacoyachtshow.com). Draw a table to show which colleagues, and which staff from channel partners and strategic alliances, you will need to be involved to ensure that the event is a success for you. Make a note of how many of each staff type you will need, what functions they will be expected to undertake, and any other resources that you or they will need.

5.15 Intensity, Exclusivity, and Selectivity within Channels

It is of paramount importance that organizations select channel partners with scrupulous care. To do this, they must consider their own commercial requirements, the sustainability and ethical concerns in their distribution channel management, the needs of customers, and the characteristics of the marketplace. One of the first major decisions, especially when trading in consumer markets, is whether to establish an intensive, selective, or exclusive network of channel partners (Frazier & Lassar, 1996). In an **intensive network**, the manufacturer decides to achieve blanket coverage of a large sales region, such as a country or continent. It does this by allowing a very large number of retailers to stock and sell its products to end users. It seeks very little control over these retailers, perhaps providing some signage, promotional materials, or point-of-sales materials to facilitate the store's sales and to encourage feelings of reciprocity. However, there is unlikely to be a binding supply agreement tying the retailer to a specified sales volume, and the retailer is able to sell rival manufacturers' products and, indeed, promote them more actively. Perhaps the best example of an intensive network is those used to sell chocolate and confectionery across a wide geographical area. A manufacturer such as Hershey's or Cadbury's is likely to use a variety of retailers for this purpose, from large out-of-town supermarkets to medium-sized outlets in cities, and much smaller retail operators in towns and villages. As the products are low in value and are routine purchases—often an impulsive purchase by consumers—no product, sales, or service expertise is demanded of the channel intermediaries. Instead, the manufacturer knows that, by having its product available in as many places as possible, it should secure an appropriate market share.

Intensive network: a very large network of retailers providing blanket coverage of a market territory

The opposite of an intensive network would be an **exclusive network**. In this model, a manufacturer sells its products in a small number of carefully selected locations. These are likely to be major cities such as New York, Chicago, and Los Angeles. Its local representation in those areas may be through a third-party retailer, or perhaps retail outlets owned by the manufacturer itself. Where third parties are used, they are selected very carefully, and the agreement between the two channel partners is likely to be extremely stringent, with much collaboration between manufacturer and seller. The manufacturer may provide the retailer with signage, branded promotional goods, local advertising financial support, point-of-sales equipment, display stock, extensive product training, and other benefits. In return, the retailer is likely to employ highly qualified and experienced staff and will probably sign an exclusivity contract that bars it from representing any other manufacturers at that site. As a result, the manufacturer gets sole representation, which means that its brand is not diluted by that of one of its rivals. Such an agreement may be put into place by a highly prestigious motor manufacturer such as Rolls Royce, or a very high-value fashion brand.

Exclusive network: a very small number of channel partners chosen to represent a brand only in high priority locations

A **selective network** is neither exclusive nor intensive, but broadly speaking is halfway between those two approaches. The manufacturer carries out a careful analysis of market data to ascertain where the most important sales opportunities occur. It is likely to conclude that it requires representation in major cities and perhaps tier 2 or tier 3 cities, to capture enough market share to fulfil its commercial objectives. A mainstream motor manufacturer such as Chrysler is likely to operate a selective network. It knows that people are willing to travel a larger distance to view and purchase one of their cars than the consumer would undertake to purchase an everyday consumable item such as soap powder or a chocolate bar but would not travel a significant distance than would be the case for a Rolls Royce or Ferrari customer. It is likely to have a stringent contract in place with its franchised dealers, but often it will not have an exclusivity agreement in place, so the dealer may represent more than one manufacturer, perhaps even at the same site. While the manufacturer enjoys good local representation in return for financial and technical support, it is aware that its intermediary may favor other brands and their products if they represent an opportunity to achieve better sales volumes or margins. Therefore, it must monitor, motivate and manage its selective network constantly and carefully.

Selective network: a network of downstream intermediaries, usually retailers, which provide widespread but not blanket coverage, and are located in most significant locations according to local market potential

5.16 Levels Within Channels

A **channel level** constitutes one layer of intermediaries between the manufacturer and its end users. In a small minority of cases, the manufacturer may choose not to use any intermediaries because it is able to sell directly to end users. In this respect, it may be considered a direct marketing channel. As we have discussed, some organizations have different numbers of channel levels according to the type of customer they are targeting. When selling commercial vehicles to very large fleet operators, a motor manufacturer may sell directly and therefore have no channel intermediaries at all. However, when selling to individual retail customers, it may sell through dealerships, through leasing companies, or through a combination of both. As a result, it has either one or two levels of intermediary. Similarly, the number of channel levels appointed by a manufacturer may vary from one market to another. In its dominant, home market, it may appoint wholesalers and retailers, whereas in secondary markets, it might allocate only agents.

Intermediary: an organization that deals with customers, end users, or other channel partners on behalf of its supplier/s

5.17 "Flows" Between Channel Members

5

To ensure that marketing channels work effectively in adding value, there are a number of important flows between the channel members. These all the flow of products, payments, ownership, promotional activities, and information. **Products** flow downstream from the manufacturer via trucks to wholesalers and their large depots, then downstream once more to retailers which operate smaller premises in customer-facing locations. **Payments** float upstream from retailers, who usually pay wholesalers for products received, and the wholesalers in turn pay the manufacturers for products that have been delivered to them. **Ownership** flows from the manufacturer to wholesalers at discounted rates which take into account the costs incurred through storage, transportation, marketing, interest, and cash flow. Wholesalers then apply a profit margin before selling the products onwards and passing ownership to retailers. Most **promotional activities** flow downstream from the manufacturer to wholesalers and retailers. Manufacturers employ the marketing communications mix to present the product offerings as attractively as possible before compelling their channel partners to promote the product to consumers or to subsequent downstream intermediaries through more localized initiatives such as local trade fairs (Mehta et al., 2002). **Information** typically flows in both directions, from manufacturer to downstream intermediaries as technical, product, and general market expertise, and from retailers and wholesalers upstream as customer feedback, intelligence, and data on the local market.

5.18 Types of Channel Structure

When choosing a channel structure, the manufacturer should not simply decide whether to operate an intensive, exclusive or selective network. It should also consider the requirement of its targeted customer segments and of its channel intermediaries. Furthermore, it must place its own business and commercial objectives front and center of any strategic decision. Each customer segment is likely to value different competencies and characteristics of channel partners. For example, Generation Z and millennial consumers are likely to expect a greater degree of interaction with retailers than consumers from Generation X or the baby boomer generation are. This is because millennials and Generation Z consumers are "digital natives" rather than "digital immigrants," having been born and raised into a digitally enabled world in which information and interaction are readily available. If this is the case, the manufacturer may need to deliver goods to consumers within specified timeframes, and to provide real-time communication, influencing its strategic channel choices (Goering, 2012). Manufacturers may also need to increase profitability by controlling their costs wherever possible, but not to the detriment of customer satisfaction. In addition to intensity and exclusivity, they must consider the following when selecting and designing a channel structure:

The **characteristics of the product being sold** and the **conditions prevalent in the marketplace** impact upon channel decisions. If the market is buoyant and the manufacturer aspires to grow, it is likely to require more intermediaries to provide an increased level of contact with its prospects and customers. Launch products may be more difficult to move around, and to store. Expensive products are usually bought less frequently by customers and tend to require longer-term storage and a greater level of support during the decision-making process. Perishable goods, predominantly within the fast moving consumer goods (FMCG) category of products, often need to be sold quickly before they reach an expiry date. They may need to be stored at certain temperatures or refrigerated, which has implications for transport and warehousing. If the product is volatile or dangerous, it will require specific expertise when being transported, stored, and sold. Organizations that make technologically advanced products, or those with stringent legal requirements, must factor in the extra level of customer guidance which will be placed upon downstream channel partners. Prestigious, high-priced products demand representation by channel intermediaries whose brands are as aspirational as the manufacturer's. If the manufacturer is differentiated from the competition by its high levels of customer satisfaction and service, then it must select partners who are willing and able to convey and transmit this within the channel. In a monopolistic market, which has no significant competition, the manufacturer may enjoy more freedom in its choices of how to trade with its customers. However, in a pure competition market, or one in which competitors continually seek to attack the manufacturer on price, service, or product specification, the manufacturer must appoint channel intermediaries that assist in the mitigation of those threats by building their own competitive advantage, perhaps through the levels of interaction with customers or an engaging service environment.

Customer location can be a critical factor in channel design. If the manufacturer is suffering from open points (significant geographical areas without representation), it will lose market share in those places and rely on other locations to counterbalance this disappointing effect. Manufacturers should also consider **how much territory an intermediary is expected to cover**. If the sales territory is heavily populated, the retailer or dealer is likely to need large premises and a large workforce to penetrate the market sufficiently and to offer good service levels to customers. The **costs associated with distribution, storage**, tariffs, and other costs often influence the speeds at which products may be transported between the manufacturer and the customers. This impacts upon the number of intermediaries which is viable. Furthermore, the firm and its partners must ascertain whether goods are **to be transported by air, rail, or road**, or a combination thereof, and consider how this impacts upon preparing these products for market and fulfilling customer needs. The final consideration is choosing the **number of channels required**. By selling directly to major customers, organizations are able to keep more of the profit margin for themselves, but in doing so take on a significant amount of burden in dealing with those customers, especially in terms of communications, aftersales, logistics, and finance.

Selling whiskies and cigars together can help retailers to serve their target audience. Image source: ► pexels.com/royalty-free-images/

5.19 Strategic Agreements Between Channel Partners

It is a common error for organizations to appoint intermediaries too hastily when filling open points, and this often results in a mismatch between manufacturer and retailer. While it is understandable that the originating firm wishes to address the shortfall in sales volumes and market share arising from an open point—a geographical area without representation—it must adopt a long-term orientation when selecting and appointing an intermediary to represent it in the marketplace. The appointment of unsuitable partners can impede the development and performance of each party for many years. Therefore, both prospective partners should undertake due diligence on each other, and ensure that their commercial aspirations, business objectives, capabilities, and motives are sufficiently aligned to sustain a viable collaboration (Carson & Ghosh, 2019).

The first of two major areas in which both parties should agree is **price expectations**. As we have discussed in ► Chap. 4. Pricing strategies not only bring money into an organization, but also position the organization, its brand, and its products and services against those of the competition. Therefore, pricing is a highly emotive subject for organizations. Some businesses wish to prioritize higher profit margins and higher prices over sales volumes, and this is particularly true of organizations that intend to build brand prestige and differentiate themselves from rivals through product and service quality. However, other organizations may prefer to sell at a low price to achieve large sales volumes and economies of scale, being less concerned with adding value or building prestige. If a high-value manufacturer appoints low-price intermediary, or vice versa, it is almost certain to be an unhappy alliance in which the two parties work actively against each other's interests.

The second major area in which channel partners should be in agreement is on the **sales and service territories** which will be granted by the manufacturer to the wholesaler or retailer. Once more, this is an extremely emotive issue. Manufacturers invariably need to maintain a certain sales volume within specific local territories if they are to fulfil their broader commercial objectives regarding market share, sales volumes, and profit. Many intermediaries seek the largest possible sales territory because it provides them with a greater number of sales prospects and leads and prevents (or at least discourages) rivals from competing against them to sell the same manufacturer's product within that area. In other words, the larger the sales territory, the lower the risk of having one's sales cannibalized by another of the manufacturer's retailers, and the easier it is to sell at a viable price. However, if territories are too large or heavily populated, the retailer's salespeople and service representatives are unlikely to cover the territory to a satisfactory level. This might result in customer dissatisfaction, lost sales opportunities, and other negative effects of the under-resourcing of a territory. In turn, this is likely to hinder the broader commercial objectives of the manufacturer. A related issue is that, within many countries, including those in the European Union, it is illegal for a manufacturer to stipulate to a channel intermediary that it may only sell goods in a specific area, as this represents anti-competitive practice (Iacobucci & Winter, 2017). In the EU, block exemption rules outlaw such practices.

Manager's Corner 5.2: Sales Channel Management in the Soccer Industry

Source: ▸ pexels.com/royalty-free-images/

Soccer or Association Football (known outside North America simply as "football") is big business. The most high-profile clubs such as Real Madrid, Manchester United, Liverpool, and Bayern Munich have global followers who watch games via satellite television and internet streaming services. These clubs are able to attract the world's finest players by paying huge transfer fees and wages.

Nowadays, the most renowned clubs earn only 20% or so of their income from gate receipts (i.e., money charged to spectators for entry to the stadiums for matches), with the rest of their income generated by image rights and intellectual property, the sale of branded merchandise, corporate sponsorship, stadium naming rights, television and broadcast commissions, and prize money from league and cup competitions. Imagine you are the chairperson of a Major League Soccer club in the USA, a Premier League club in England, or another globally prominent club. Make a list of all the companies, and types of companies, whom you would consider to be your downstream intermediaries and strategic alliances. What function do each of them undertake, what value does your club give them, and what value do they give to your club? Try to think beyond purely monetary transactions to other ways in which value may be transferred or co-created through their association with your club.

5.20 Channel Management

The objectives of two or more channel partners are never exactly aligned. There are always areas in which they have different aspirations, whether it be price, profit margin, sales volumes, market segmentation, or other areas of marketing strategy. The role of channel management is to recruit, monitor, motivate, and manage organizations that represent one's brand and sell one's goods. It is a particularly difficult and strategic role of the organization to nurture fruitful channel relationships. This has brought about a large amount of academic research, and many of the most prominent roles in an organization appertain to this function. The firm needs to ensure that its partners are committed to representing it properly, are incentivized to develop, and are motivated to push the business forward. However, if an organization pushes its partners too aggressively, it may alienate them, demotivate them, and discourage them from investing in joint operations. Moreover, it presents a real risk that the partner will instead collaborate with a rival.

The first core function of channel management is the **selection and recruitment of channel partners**. To do this, the organization will scrutinize market data to ensure that there is a sufficient commercial opportunity to represent a brand in a specific market territory. Assuming that this is the case, it will research the available partners within that location, and assess which are likely to be most compatible in a long-term strategic alliance. At this point, one firm may make a formal approach to the other, and conversations will commence to test the water. If sufficient commonality of strategic aspirations is found, then the two parties will seek to formalize the relationship contractually. Before officially appointing a local channel partner, the manufacturer is likely to approve a site, stipulate conditions appertaining to its standards of representation, and provide support to its new partner in meeting and maintaining those stipulated standards. For example, it may provide signage, branded point-of-sales materials, training for salespeople and service personnel, and financial support for learning initiatives (especially for launch events). In return, the local partner will commit to stock, display, and sell a certain volume and mix of the manufacturer's products, and to undertake certain activities such as training, marketing, and the provision of local market analysis.

Having appointed a local channel partner, the manufacturer or dominant party must then motivate its partner by issuing sales targets that are achievable, but which encourage the partner to fulfil their potential. These can be linked to commission or bonus payments, which may be payable to the partner organization, its members of staff, or a combination thereof. Targets are likely to be issued appertaining to monthly, quarterly, and annualized objectives, and these usually focus on sales volumes, market share, profit margins per sale, customer satisfaction scores, adherence to the manufacturer's standards, product mix, and the percentage of sales in which the customer also buys finance and insurance products. Some organizations, depending upon their immediate business needs, may wish to place targets upon other criteria, such as the retention of customers, the conquest of customers from rival manufacturers, the number of off-site appointments or demonstrations carried out by sales staff, or other parameters. Managers and salespeople may be motivated by money, the recognition of their peers, competitions and the gamification of work, awards, career progression, and job satisfaction.

Training programs are designed to maximize the benefit of having a channel intermediary and to protect the firm's reputation by ensuring that anyone who represents it is customer-oriented and has extensive product knowledge. Without this, there is a risk that an intermediary's representative may provide a customer with inaccurate advice, exposing the manufacturer to a risk of litigation. Moreover, salespeople who are well trained are more likely to recognize the competitive advantages of the products that they are selling, and therefore be confident and enthusiastic when presenting them to prospects. Furthermore, it enables them to understand how their products are positively differentiated from those of rivals, helping them to explain to customers how they add value. In doing so, they should be able to steer conversations away from price and toward the provision of benefits, making customers less price-sensitive and more profitable.

It is important that each channel partner monitors the other constantly, analyzing their performance and the contribution they make to its commercial objectives. Having targeted retailers or dealerships with key performance indicators that help to deliver the objectives, the manufacturer must continue to monitor its intermediaries' performance levels across multiple metrics. If a partner begins to underachieve, the manufacturer must drill down into the data to analyze the reasons for this shortfall, and to make timely interventions to help the partner get back on course.

5.21 Channel Conflict

Channel conflict is a negative phenomenon that can occur in marketing channels due to the misalignment of aspirations and objectives between channel partners (Eshghi & Ray, 2019). For example, if a car dealership wishes to sell fewer cars at a higher profit per unit, but the manufacturer and its representative wish to penetrate the market and increase market share at the expense of unit profitability, then the two parties are likely to have irreconcilable differences in their business aspirations and commercial strategies. Additionally, channel conflict sometimes arises due to interpersonal clashes, even within partnerships whose objectives are closely aligned. Therefore, channel managers must work hard to maintain interpersonal and inter-organizational relationships by being friendly, approachable, well trained, personable, and skilled in the avoidance and resolution of conflict. Conflict can be the result of **domain dissensus**, which occurs when each individual or organization is uncertain of where their responsibilities stop and those of their partners start. In such situations, a manufacturer and its retailer may leave marketing activities to each other, or expect the other to pay, leading to an under-representation of the brand in the territory. For this reason, it is important to draw up formal contracts which clearly stipulate the responsibilities of each partner.

5

Channel conflict: disagreements and underlying hostility and negativity between channel partners

5.22 Integration in Channels

Vertical integration, where a firm acquires every intermediary positioned between itself and the market, is one way in which channel conflict can be avoided or controlled. However, this can be rather a cumbersome approach, which diverts the manufacturer's resources away from its core capabilities. For example, a manufacturer of wheelbarrows may need to purchase all the horticultural wholesalers that represent it to achieve vertical integration. In contrast with this, **horizontal integration** occurs when one organization buys rival organizations that operate at the same level within marketing channels. For example, a horticultural wholesaler made a wish to buy other horticultural wholesalers that compete against it in a specific territory in the sale of garden equipment. A third strategy would be to enter a **co-marketing agreement** in which two organizations combine their brand equities, marketing expertise, and resources, representing each other's brand as well as their own (Takata, 2016). This approach is not limited to channel partnerships, but often works well within that context.

- **Vertical integration**: the process of a firm acquiring its downstream channel intermediaries by buying them
- **Horizontal integration**: the process of a firm acquiring rival firms which undertake the same function within a channel

5.23 Power in Channels

French and Raven's (1959) seminal theory categorizing types of power is particularly relevant to channel management. To detect and lessen the effects of channel conflict, it is important for organizations to consider which power types are at play within the channel, and what effects they are likely to have.

Referent power is the result of channel partners attempting to emulate each other, most commonly where one partner perceives the other as the leader. An example of this would be a retailer which has recently been appointed as a local representative of a very famous, well-established, and prestigious manufacturer. The retailer's members of staff, and indeed management, may be somewhat in awe of the reputation of the manufacturer and may therefore defer on important decisions to its representative. Of course, this can be perfectly justified where there is a gulf in experience, but it may also bring about a relationship that is deleteriously deferential, and harmful to one or both of the parties.

Referent power: power derived from one partner's reputation and standing, and other partners' willingness to emulate them or allow them to lead due to their reputation or standing

Legitimate power is the power that one partner enjoys over the other as a result of contractual, legal, or hierarchical imbalances. While a collaborative and cordial relationship may be maintainable between two such parties without resorting to legalistic pleas or leveraging contractual obligations, the "power of power," or the implicit knowledge of each party's respective strength, often influences partners' actions. In this respect, the holding of power is not overtly coercive, nor is it wielded ostensibly, but it is tacit, unspoken, and no less powerful.

Legitimate power: power derived from the law or a contractual agreement

Coercive power uses direct force to achieve one partner's wishes at the expense of the other partner. While it may be wielded covertly, for instance by a manufacturer deliberately making late deliveries or incorrect reimbursements to a retailer, it is more usually overt as the stronger party wishes the weaker party to recognize the wielding of power so that it has more influence in subsequent interactions. Coercion is not collaborative, democratic, or collegiate, but the explicit exploitation of a power imbalance.

- **Coercive power:** power which derived from the use of force
- **Reward power:** power derived from incentivizing and rewarding others

5

Reward power is the use of incentives to positively reinforce actions and behaviors in another party. It is very commonplace in business, and perhaps more so within channel management. The payment of commissions and bonuses, the provision of corporate hospitality, the use of competitions and recognition among peers, and other initiatives, encourage channel partner organizations and their members of staff to represent the manufacturer robustly to earn something of value to them. In managing this, the manufacturer's representative must be careful. They need to demonstrate consistency across their network of channel partners, and they also need to ensure that they do not cut across the managers and owners of their channel partners when incentivizing their staff members. For example, if a motor manufacturer wishes to incentivize dealership staff to sell more cars, this may upset the owners of the dealership if they have already incentivized their salespeople on profit per unit rather than units sold, as they may believe that the manufacturer's incentive introduces ambiguity and weakens their own incentive.

Expert power is very common within supply chains and is achievable by organizations, or more commonly individuals within those organizations, who earn respect due to their extensive knowledge, professionalism, or expertise in a particular area (Reimann & Ketchen Jr, 2017). An example of this would be a manufacturer's salesperson who has extensive technical knowledge of the products which she sells and can advise retailers on ways in which those products can be demonstrated to best show off their advantages against the competition. By adding value to the relationship between manufacturer and retailer—in this case, by helping the retailers to differentiate the product against competitor offerings and therefore make more profit—the expert member of staff is held in high esteem and is therefore trusted more than others in a similar position.

Expert power: power derived from the trust and respect which others have in a person or firm's expertise and knowledge

A seller of insurance products would normally have expert power. Image source: ► pexels.com/royalty-free-images/

5.24 Exclusivity in Channels

As we have discussed, a manufacturer must decide upon the level of representation that it requires from its intermediaries. To do this, it may wish to pursue an intensive, selective, or exclusive network coverage strategy, designed to deliver upon the needs of the product and of the marketplace. Exclusive distribution strategies involve appointing very few partners who are able to sell manufacturers' products against relatively little competition. This makes them more likely to invest resources in the brand and dedicate more time to its representation. **Idiosyncratic investments,** which are specific to one partnership rather than relating to all of the manufacturer's intermediaries or all of the retailer's suppliers, are more powerful than more generalized investments, as they are more targeted to the partner's needs and likely to result in feelings of reciprocity. Decisions appertaining to exclusivity in channels are normally influenced by the number of customers, the frequency with which the products are purchased, the complexity of the customer's decision-making processes, and the extent to which the products are complicated and require detailed explanation. However, there are two additional mechanisms through which the manufacturer may grant exclusivity. The first mechanism is called **exclusive dealing**. In an exclusive dealing agreement, the manufacturer prohibits its channel partner from representing competitor brands (Nocke & Rey, 2018). Therefore, if a motor dealership enters into an exclusive dealing contract to supply Volkswagen vehicles in Chicago, it would not be able to represent Ford, General Motors, Chrysler, or any other manufacturers within that area. A key benefit of this approach to manufacturers is that they do not suffer the consequences of retailers switch-selling customers into more profitable, attractive, or easy-to-sell competitor products. Of course, the reverse is true, and they are less able to benefit from retailers switch-selling customers into their products. As the retailer becomes more dependent upon the success of its chosen manufacturer, it is likely to commit more to actions that are mutually beneficial. However, in return, it may expect a greater level of support and compensatory benefits from the manufacturer, such as local marketing support, interest-free stocking, and heavily discounted corporate branded materials.

- **Exclusive dealing agreement**: A contractual agreement in which the downstream channel partner agrees not to represent its supplier's rival brands
- **Exclusive distribution**: A strategy in which the originator brand or firm appoints a very small number of carefully selected downstream channel partners, concentrated in key locations
- **Tying agreement**: A contract in which a downstream channel intermediary, usually a retailer, agrees to stock and sell its supplier's entire product range, rather than simply the most attractive or profitable items.

The second additional mechanism for granting exclusivity is called a **tying agreement**. These are often used by manufacturers who also impose exclusive dealing agreements upon their channel intermediaries. Tying agreements stipulate that a channel partner must stock, market, and sell all of its supplier's product range, rather than selectively cherry-picking the most profitable and easiest to sell products. In practice, this means that the manufacturer is likely to enjoy sales across its entire range, including those products that are likely to be more profitable, than if it had simply incentivized partners through commission payments to sell a random mix of products and models. Retailers and wholesalers may baulk against such an arrangement as it gives them less commercial freedom, but in return, manufacturers are usually willing to reflect this commitment by granting additional discounts and other support.

Web Exercise 5.2

Choose a prominent manufacturer (e.g., Procter & Gamble or Unilever for consumer goods, or Ford as car manufacturer or a manufacturer from any other industry) of a product that is usually sold through a retailer. Visit the website of the manufacturer and the website of the retailer. (If you are struggling for inspiration, try choosing a motor manufacturer and the dealership through which it sells its products in your local area.) What are the similarities and differences between the presentation of a specific product on the manufacturer's website, and on the retailer's site? Are the messages, tone, style, and branding represented consistently between them, or do they seem to have different priorities or objectives? For example, are both organizations trying to sell the product by emphasizing the same benefits (product quality, reliability, price, service back-up, brand, etc.)? To what extent do their marketing communications overlap? Do they complement each other, reinforce each other, or perhaps contradict each other? How would you change things if you were the marketing director for the manufacturer with the power to influence retailers' marketing approaches?

5.25 Logistical Considerations

Channel intermediaries, and manufacturers' representatives have many logistical issues to consider on a daily basis. Their roles are not only confined to managing the flows of products, but also to payments, information, and other valuable resources. The logistical issues fall into three main categories, which are inventory management, warehousing, and the management of logistical information.

Inventory management ensures that an intermediary always carries enough stock to show and demonstrate to prospective customers, and to fulfil customer orders. Retailers incur charges to do this, and it can be tempting for them to take shortcuts. For example, by placing cars, commercial vehicles, or recreational vehicles on the perimeter of their premises for customers to see and try, a motor dealership commits to tie up a large amount of cash that could otherwise be utilized for activities which generate a quick return on investment. However, without this visibility, customers are less likely to become aware of the dealership and its products, less likely to visit to seek further information, and as a result, sales are likely to suffer. Intermediaries must also place product orders with the manufacturer in response to customer orders, deal with the sometimes-complicated administrative processes, input the data, and follow-up to arrange delivery and keep the customer aware of timescales. Intermediaries must also ensure that they pay their upstream suppliers within agreed time parameters and process customer payments swiftly to ensure a healthy cash flow (Gallino et al., 2016).

Warehousing maximizes the efficient use of storage capacity within retail and wholesale premises. It helps to ensure that there are sufficient staff members, available loading bays, and forklift trucks to unload deliveries from trucks, break bulk deliveries into smaller-sized bundles of items (usually categorized by product type), and to move them and store them appropriately within the available space. Warehousing and logistics intermediaries must also manage the financial costs inherent in storing goods, and this involves making payments to their suppliers and taking payments from their customers. Warehouse operators—normally wholesalers—must also service the needs of the market, and in particular, the customers who order stock from them. This means making deliveries in vehicles that are small enough to negotiate the streets around retail premises and to enter their delivery yards. Additionally, wholesalers help to absorb seasonality within the supply and demand of the products which they handle. We have spoken about how a horticultural wholesaler may mitigate some of the cashflow difficulties that a manufacturer of hosepipes or wheelbarrows may face each autumn and winter. However, this is not confined to product categories such as this, but also appertains to any goods which are more likely to be bought by customers at specific times of the year—Christmas decorations, woolen hats, or tennis rackets being examples.

Logistical information management is an ever-expanding area of responsibility among channel intermediaries. The proliferation of digital technology and the use of big data means that business partners can drill down ever deeper into the data that drives their businesses. Channel intermediaries must decide which market intelligence is likely to be most useful to its suppliers, removing superfluous data and presenting information in an intuitive and user-friendly manner, so that suppliers can use the data for strategic marketing decisions. Wholesalers and retailers also analyze detailed information on stock rotation and appertaining to critical metrics such as return on sales, stock turn (how many times in a given period a channel member sells and replaces their entire stock), and return on capital employed (ROCE—how much income an investment delivers). It is commonplace for intermediaries to analyze data from previous sales periods to predict future sales and order levels. However, they must be careful to take into consideration emerging factors and evolving trends. For example, a downturn in the economy and below-inflation wage rises may undermine consumers' ability and willingness to make certain purchase decisions comparative to the previous year. Less predictable factors, such as the weather, or even the impact of overseas wars on oil prices, may have a much more unforeseen influence on customer demand from 1 year to the next. To ensure that logistical difficulties are addressed, wholesalers in particular liaise with other logistics companies to ensure the availability of transport and vehicles. They also seek regular updates on the status of ordered stock to ensure that it is being manufactured in an acceptable timeframe to fulfil customer orders and replenish warehouse stock.

5.26 Retailing Decisions

To make effective retailing strategies, manufacturers must take into account numerous considerations. They must decide what sort of retail outlets they require. For example, do they need the outlets simply to sell goods, or to provide a customer experience and demonstrations of sophisticated products? Manufacturers must know the potential market in specific locations, ensuring that the retail premises which they operate there are of a suitable size and standard. By understanding regional variations in market demand and customer requirements, they can appoint channel partners who will fulfil needs and deliver upon customer expectations. Next, we outline some of the major considerations when selecting, recruiting, and appointing a retailer.

Manufacturers must **position** their retail networks in relation to those of their rivals, not simply in terms of corporate image, brand, and prestige, but also to be differentiated across other criteria. In doing so, a manufacturer's retail network can add customer value (Raff & Schmitt, 2016) and support the brand. **Choosing a suitable location** is paramount to the success of a retailing strategy. It is commonplace for prestigious brands which sell higher-value products to operate exclusively in more affluent areas. For example, the British supermarket and grocery chain, Waitrose, only operates retail premises in more affluent, middle-class neighbourhoods. Other brands, such as FMCG brands aiming at a broad demographic, may demand blanket coverage of a country or region. Manufacturers of certain products may require representation in locations that are specific to the products they make and the audiences who buy them. For example, a ski wear manufacturer might choose to have outlets by Lake Tahoe, Aspen, Chamonix, Lillehammer, and Banff, rather than in New York City, Toronto, Paris, and Oslo. The **pricing policies** within retail strategies may be driven partially by the economic circumstances inherent in local markets, the strength of rivals, and the extent to which those two factors result in the elasticity of demand. Pricing is also influenced by the overall business and marketing objectives of the organization. For instance, if the firm wishes to increase market share as its main priority, it is likely to price less expensively, whereas if it wishes to reposition its brand upmarket, it will almost certainly pursue a strategy of pricing more expensively.

The **image, atmospherics, and overall characteristics** of stores help to convey the firm's brand (Blazquez et al., 2019). If a location is considered downmarket, perhaps due to the presence of off-brand, budget competitors, the firm may wish not to have representation there, even if to do so is to miss easy custom. Instead, the manufacturer should take a long-term perspective, and consider how capitalizing upon local profit opportunities may present broader challenges to the sustainability and prestige of the brand, therefore undermining future profit opportunities. Manufacturers must decide what products or models are to be sold by a retailer, and in what volumes. This is likely to influence the size of the store and its required stock capacity. It is common practice for supermarket chains to offer different **product mixes** depending upon the area in which the store is located. For example, in highly cosmopolitan areas with a diverse ethnic mix, they are more likely to offer extensive ranges of foods from different cultures. In wealthier neighbourhoods, they are likely to offer a broader range of more expensive and luxurious foods, such as sun-dried tomatoes, sushi, and kimchi. In areas where many customers suffer from economic disadvantage, supermarkets are more likely to offer a range that focuses on staple foods at more affordable prices. Finally, manufacturers and suppliers must agree on **the service levels and expectations** for retail outlets. While it is important for luxurious brands to deliver a sophisticated customer experience, doing so comes at a cost, as plush surroundings and additional services can prove expensive. When operating a budget brand or serving more price-conscious customers, it may be necessary to target reduced service levels in order to stay within a reasonable budget.

Store atmospherics can attract, relax, and delight retail customers. Image source: ► pexels.com/royalty-free-images/

5.27 Types of Retailers

While it is tempting to think of retailers as stores and shops, there are additional ways in which manufacturers sell products to consumers. These are chosen based upon local market potential, characteristics of the products being sold, and the needs of customers within specific targeted segments. Many organizations choose a multichannel or omnichannel retailing strategy, combining a mixture of retailer types. Such a decision is driven by the belief that customers are not homogeneous enough to demand one single type of customer experience and therefore one single type of retail outlet. Instead, they are thought to use different types of retailers and channels based upon the needs and expectations they have in relation to that specific purchase at that point in time. Some organizations use non-store retailing such as trading door to door to reduce overheads and build closer relationships with their customers, and reduce the psychic distance between the brand and its audience. Still more organizations choose to conduct their business remotely and digitally through online retailing. After the initial start-up costs, this can be a low-cost and highly responsive way of dealing with individual customers. Even organizations that maintain a physical presence through stores and shops are nowadays almost certain to maintain an online presence too, enabling customers to buy online and receive home deliveries, to buy online and then collect in store ("click and collect"), or simply to browse products online in preparation for a store visit and purchase.

5.28 Distribution Management and Sustainable Development Goals

In this chapter, we have considered the role which distribution management plays in strategic marketing, and the contributions which it can make to sustainable business if managed ethically. Distribution management strategies influence how and from whom a firm sources its components or raw materials, such as diamonds for grinding equipment or goose down for insulated jackets. Manufacturers can help to safeguard the natural environment and vulnerable workers by appointing suppliers who extrude and process materials in an ecologically sound and socially humane manner. For example, the use of diamonds from unlicensed mines can perpetuate modern slavery and precarious employment, dangerous working environments, child labor, deforestation of sensitive habitats, the decline of endangered species of flora and fauna, and the assault upon indigenous peoples and cultures. Therefore, ethical manufacturers should audit their supply chains to ensure that each upstream intermediary, employer, and partner adheres to strict legal and ethical standards. In doing so, they can directly contribute to the UN's SDG numbers 8 (decent work and economic growth), 12 (responsible consumption and production), and 15 (life on land). Depending upon the individual situation, they may also contribute indirectly to several other SDGs.

Ethical marketers must also look down the marketing channel at their downstream intermediaries to ensure that they are representing the brand to customers and end users in a manner that is fair and free from exploitation. For instance, a manufacturer of alcoholic beverages should ensure wherever possible that the bars that stock its products are licensed and do not encourage irresponsible or underage drinking. Likewise, when selling to supermarkets, they should satisfy themselves that the retailers are not manipulating consumers into problematic drinking behaviors through heavily discounted sales promotions. This directly feeds into SDG 12 (responsible consumption and production).

In essence, every organization should do due diligence on each of its partners to ensure that they are working to ethical standards that meet its own criteria and the demands of its customers, stakeholders, and society more generally.

Summary

In this chapter, we have considered how channels to market are composed of business partners and the activities that they undertake to move products from the originator, which is usually the manufacturer, to the consumer or customer. They are also responsible for a number of services that add value within the supply chain and facilitate the easy transaction of business between the firm and its audience. By collaborating with intermediaries, the customer proposition is enhanced, and channel partner organizations are able to focus on activities that demand their specific expertise. The relationship between channel partners should be mutually beneficial, and rather than simply moving and storing products, partners should also share the risks involved in the marketing, movement, storage, and selling of products. Marketing channels do not only comprise wholesalers, dealers, agents, brokers and retailers. In addition, channel intermediaries include service providers such as transport and logistics companies, and marketing agencies. It is paramount that organizations choose their channel partners wisely based upon long-term objectives rather than reactive and short-term considerations. Channel conflict can be reduced, and the channel collaborations made more fruitful through the intelligent training and incentivization of partners, and through the allocation of resources aimed at nurturing relationships at both an interpersonal and interorganizational level.

Review Questions

1. As the owner of a household furniture manufacturer, you have open points in 12 of the largest 20 cities in the United States. You can afford to supply five new retail outlets. What types of data would you like when choosing which five of the 12 largest cities to target, and what is the significance of those data?
2. Having chosen the five US cities in which you wish to supply furniture retail outlets, you must now choose one retail partner in each city to supply. What are your main considerations when choosing a suitable channel partner? Select around seven or eight types of data, explaining why each is significant to your decision.
3. A motor manufacturer has signed an agreement with a dealership group to represent it in Manchester, UK. While the retail site enjoys plenty of passing traffic and has ample space for display stock and customer parking, it is also used to represent two rival car franchises. How can the manufacturer's regional Network Manager reduce the risk of the new dealer switch-selling prospects into the other two brands?
4. An American university maintains a presence in Qatar, Bahrain, and Saudi Arabia by offering a selection of its business and management programs through a local university. However, as the US institution wishes to charge more for its programs while the local partners wish to recruit more students with lower course fees, channel conflict has developed. How should the US university and its three overseas partners address this and re-establish their previously strong relationship?

5. Having appointed a dealership to sell its products in Wyoming and Montana, a manufacturer of agricultural machinery must arrange training for the dealer's sales and servicing staff who will be representing it. What sorts of training should they provide to each type of staff member? Construct a Gantt chart showing the timings of a comprehensive and ongoing training program.
6. A US manufacturer of high-quality backpacks and mountain clothing wishes to gain a presence in the UK market. Its first decision is whether to pursue an intensive, exclusive, or selective network strategy. Compare the relative merits and pitfalls of each approach, then choose one and briefly explain how the brand should undertake it.
7. As the regional manager of a motor manufacturer, you have some budget available to develop your 25 franchised dealerships. You wish to spend that money wisely so that you see a healthy return on the investment. Compare four or five different ways in which you could spend the money and, for each, explain the intended benefits before choosing one specific approach.
8. Your firm, a manufacturer of lawnmowers, wishes to run a marketing campaign to raise awareness of the brand and drive customer footfall into your franchised stockists. What types of marketing communications would you use, to whom would you target them, and how would you ensure that your stockists are ready to capitalize upon these incremental sales prospects?

? Discussion Questions

1. Channel management is closely related to sales management, relationship marketing, and services marketing. Sales management because channel intermediaries are appointed to sell products on behalf of manufacturers, and their performances in doing so must be managed; relationship management because a healthy, long-term relationship between channel partners and their staff members must be built and nurtured to maximize the effectiveness of the alliance; and services marketing because channel partners provide an array of services to each other. Which of these three areas of marketing do you consider to be most closely related to channel management, and why? Prepare to state your case in a debate, using real-life examples where possible.
2. For several decades, theorists have explored what factors support successful channel partnerships, and marketing relationships more broadly. The main themes analyzed have been commitment, trust, relationalism, satisfaction, reciprocity, cooperation, interdependence, power, empathy, justice, value, and communication. Which three of these do you think would be the most important to a pharmaceutical manufacturer when appointing downstream channel intermediaries to represent them in a new foreign market? Explain why each of your choices is likely to be critical.
3. Anderson and Weitz (1992) wrote about "idiosyncratic investments"—investments made by channel partners in initiatives that build value for their specific partnership rather than each member's broader interests. These idiosyncratic investments were found to be powerful in the nurturing of productive long-

term partnerships. Imagine that you have been hired to advise the directors of a household furniture manufacturer on how to increase sales volumes and profits through its network of retailers. What idiosyncratic investments might you recommend that they make, and why? Consider three or four examples and explain why they might offer a strong return on investment.

4. Imagine that you are the Network Development Manager for a manufacturer of GPS systems and personal tracking devices that are used by sailors, canoeists, and wilderness walkers. You are in advanced talks with a chain of 100 stores that sells adventure clothing and equipment across the USA and Canada. It represents a large number of high-profile brands, including your direct competitors, Garmin and Suunto. How will you ensure that your products are represented to your satisfaction? What terms and conditions do you believe are realistic and desirable when drawing up the supply contract? Explain your rationale.
5. As the regional manager of a prestigious motorcycle manufacturer, you are concerned that you are losing sales from your franchised dealer in Miami to retail customers. The dealer operates four motorcycle brands from the site, which is located at a busy intersection popular with bikers and commuters alike. Sales of your motorcycles have declined from around 3000 to 2000 in the last 12 months, although your parts, servicing, and aftersales custom has not yet experienced a decline there. Your monthly visit to see the dealer's managing director is imminent. What questions will you ask her, and what other data will you seek, to ascertain possible reasons for the decline in sales, and why?

Practice Quiz

1. Ill-feeling and behavior between channel partners are known as what?
   (a) Channel dissent
   (b) Channel friction
   (c) Channel conflict
   (d) Channel force
2. Which of these distribution strategies is likely to have the largest number of retailers?
   (a) Exclusive
   (b) Intensive
   (c) Selective
   (d) Proactive
3. Which channel partner deals most capably with storage, breaking bulk, and offering variety?
   (a) Retailer
   (b) Dealer
   (c) Broker
   (d) Wholesaler
4. What is the term for investments that are made in one specific channel partnership?
   (a) Idiosyncratic investments
   (b) Bespoke investments

   (c) Unique investments
   (d) Reciprocal investments
5. Which of these is usually charged with finding supply for the customer, rather than finding custom for the supplier?
   (a) Retailer
   (b) Dealer
   (c) Agent
   (d) Broker
6. Which of these strategies may entail making products available in store, on the firm's website, and on third-party websites?
   (a) Optichannel
   (b) Omnichannel
   (c) Interchannel
   (d) Megachannel
7. What is footfall?
   (a) Prospects and customers who enter sales premises
   (b) A manufacturer's retailers within a given district
   (c) The average distance a customer is prepared to travel to a store
   (d) The speed at which a store serves its customers
8. What is the name of the phenomenon in which two channel members disagree about whether a task is their responsibility or that of their channel partner?
   (a) Cognitive dissonance
   (b) Buyer's remorse
   (c) Domain dissensus
   (d) Caveat emptor
9. What sort of integration occurs when a manufacturer acquires its downstream channel partners?
   (a) Diagonal
   (b) Horizontal
   (c) Vertical
   (d) Tangential
10. Which kind of power comes from legal and contractual requirements?
   (a) Coercive
   (b) Referent
   (c) Expert
   (d) Reward
11. What do we call the goods held by a channel member?
   (a) Stock initiative
   (b) Stock innovation
   (c) Stock inventory
   (d) Stock iteration
12. Which key performance indicator expresses how many times in a given period a channel member sells and replaces their entire stock?
   (a) Stock turn
   (b) Stock churn
   (c) Stock burn
   (d) Stock learn

13. Which of the following is a contractual agreement in which the downstream channel partner agrees not to represent its supplier's rival brands?
    (a) Monogamous retailing agreement
    (b) Loyal selling agreement
    (c) Universal supply agreement
    (d) Exclusive dealing agreement
14. Which of these entails selling to customers without using one's downstream channel partners?
    (a) Deselection
    (b) Disintermediation
    (c) Antidisestablishmentarianism
    (d) Indiscretion selling

You can find the correct answers to these questions by taking the quiz and then comparing your answers with the answers that are provided at the end of the book, under the "Practice Quiz Answers" part of the book.)

References

Anderson, E., & Weitz, B. (1992). The use of pledges to build and sustain commitment in distribution channels. *Journal of Marketing Research, 29*(1), 18–34.

Aykol, B., & Leonidou, L. C. (2018). Exporter-importer business relationships: Past empirical research and future directions. *International Business Review, 27*(5), 1007–1021.

Blazquez, M., Boardman, R., & Xu, L. (2019). International flagship stores: An exploration of store atmospherics and their influence on purchase behaviour. *International Journal of Business and Globalisation, 22*(1), 110–126.

Boyaci, T., & Gallego, G. (2004). Supply chain coordination in a market with customer service competition. *Production and Operations Management, 13*(1), 3–22.

Carson, S. J., & Ghosh, M. (2019). An integrated power and efficiency model of Contractual Channel governance: Theory and empirical evidence. *Journal of Marketing, 83*(4), 101–120.

Chen, I. J., & Popovich, K. (2003). Understanding customer relationship management (CRM) people, process and technology. *Business Process Management Journal, 9*(5), 672–688.

Claro, D. P., & Claro, P. B. O. (2010). Collaborative buyer–supplier relationships and downstream information in marketing channels. *Industrial Marketing Management, 39*(2), 221–228.

Croonen, E. P., & Broekhuizen, T. L. (2019). How do franchisees assess franchisor trustworthiness? *Journal of Small Business Management, 57*(3), 845–871.

Dukes, A., & Geylani, T. (2016). *Dominant retailers and their impact on marketing channels* (pp. 137–156). Edward Elgar.

Emmett, S., & Crocker, B. (2016). *The relationship-driven supply chain: Creating a culture of collaboration throughout the chain*. Routledge.

Eshghi, K., & Ray, S. (2019). Managing channel conflict: Insights from the current literature. In *Handbook of research on distribution channels*. Edward Elgar Publishing.

Frazier, G. L., & Lassar, W. M. (1996). Determinants of distribution intensity. *Journal of Marketing, 60*(4), 39–51.

French, J. and Raven, B. (1959). The Bases of Social Power. In Studies in Social Power, D. Cartwright, Ed., pp. 150–167. Ann Arbor, MI: Institute for Social Research.

Gallino, S., Moreno, A., & Stamatopoulos, I. (2016). Channel integration, sales dispersion, and inventory management. *Management Science, 63*(9), 2813–2831.

Goering, G. E. (2012). Corporate social responsibility and marketing channel coordination. *Research in Economics, 66*(2), 142–148.

Iacobucci, E., & Winter, R. A. (2017). European law on selective distribution and internet sales: An economic perspective. *Antitrust Law J, 81*, 47.
Köksal, D., Strähle, J., & Müller, M. (2018). Social sustainability in apparel supply chains—The role of the sourcing intermediary in a developing country. *Sustainability, 10*(4), 1039.
Mehta, R., Dubinsky, A. J., & Anderson, R. E. (2002). Marketing channel management and the sales manager. *Industrial Marketing Management, 31*(5), 429–439.
Mullin, R. (2018). *Promotional marketing*. Routledge.
Nocke, V., & Rey, P. (2018). Exclusive dealing and vertical integration in interlocking relationships. *Journal of Economic Theory, 177*, 183–221.
Palmatier, R., Stern, L., & El-Ansary, A. (2016). *Marketing channel strategy: Instructor's review copy*. Routledge.
Porter, M. E. (1985). *Competitive advantage: Creating and sustaining superior performance*. Simon and Schuster.
Raff, H., & Schmitt, N. (2016). Manufacturers and retailers in the global economy. *Canadian Journal of Economics/Revue Canadienne d'économique, 49*(2), 685–706.
Reimann, F., & Ketchen, D. J., Jr. (2017). Power in supply chain management. *Journal of Supply Chain Management, 53*(2), 3–9.
Rosenbloom, B., & Andras, T. L. (2008). Wholesalers as global marketers. *Journal of Marketing Channels, 15*(4), 235–252.
Srivastava Dabas, C., Sternquist, B., & Mahi, H. (2012). Organized retailing in India: Upstream channel structure and management. *Journal of Business & Industrial Marketing, 27*(3), 176–195.
Statista. 2022. https://www.statista.com/forecasts/870307/united-kingdom-top-online-stores-united-kingdom-ecommercedb.
Takata, H. (2016). Channel integration: An explanation according to David Teece's theory of the boundaries of the firm. In *Looking forward, looking back: Drawing on the past to shape the future of marketing* (pp. 176–176). Springer.

Further Reading

Chan, T. Y., Narasimhan, C., & Yoon, Y. (2017). Advertising and price competition in a manufacturer-retailer channel. *International Journal of Research in Marketing, 34*(3), 694–716.
Cui, T. H., & Mallucci, P. (2016). Fairness ideals in distribution channels. *Journal of Marketing Research, 53*(6), 969–987.
Draganska, M., & Vitorino, M. A. (2023). Decomposing the effect of advertising: What happens in the retail channel? *International Journal of Research in Marketing, 40*(1), 226–247. https://doi.org/10.1016/j.ijresmar.2022.07.005
Goić, M., Jerath, K., & Kalyanam, K. (2022). The roles of multiple channels in predicting website visits and purchases: Engagers versus closures. *International Journal of Research in Marketing, 39*, 656–677. https://doi.org/10.1016/j.ijresmar.2021.12.004
Gurnani, H., Singh, S., Tang, S., & Wang, H. (2022). Service provision in distribution channels. *Journal of Marketing Ressearch, 59*(5), 926–940.
Kolay, S. (2015). Manufacturer-provided services vs. retailer-provided services: Effect on product quality, channel profits and consumer welfare. *International Journal of Research in Marketing, 32*(2), 124–154.
Kuksov, D., & Liao, C. (2018). When showrooming increases retailer profit. *Journal of Marketing Research, 55*(4), 459–472.
Sands, S., & Posch, R. J., Jr. (1982). A checklist of questions for firms considering a vertical territorial distribution plan. *Journal of Marketing, 46*(3), 38–43.
Shankar, V., & Kushwaha, T. (2021). Omnichannel marketing: Are cross-channel effects symmetric? *International Journal of Research in Marketing, 38*(2), 290–310.
Sharma, A., Kumar, V., & Cosguner, K. (2019). Modeling emerging-market firms' competitive retail distribution strategies. *Journal of Marketing Research, 56*(3), 439–456.

Digital Marketing for People, Planet, and Profit

Contents

S. M. R. Shams et al., *Sustainable Marketing*, Classroom Companion: Business,
https://doi.org/10.1007/978-3-031-79130-7_6

Preliminary Vignette: Green Web Hosting for a Sustainable Future

Brand website, social media, online customer forum, mobile apps, and other digital platforms are now integral parts of businesses in private, public, or in the third sector. Running these digital platforms requires a lot of energy, which is produced from fossil fuel. For example, "running the internet accounts for approximately 1.7% of total global energy demand. 40% of the energy used by data centres is for cooling... Carbon emissions associated with operating and accessing the internet are similar to the emissions released by the global airline industry" (Future Energy, 2024, np). Thus, sustainable businesses are moving toward green web hosting to reduce carbon emission. For example, the Alphabet's (Google's) greenhouse gas emission in 2018 was 13.64 million metric tons, which was reduced to 10.18 million metric tons in 2022 (Statista, 2023a).

To achieve the global net-zero goal, Google declares that "our goal is to achieve net-zero emissions across all of our operations and value chain by 2030" (Sustainability, 2022, np). Net zero is "a target of completely negating the amount of greenhouse gases produced by human activity to be achieved by reducing emissions and implementing methods of absorbing carbon dioxide from the atmosphere" (Net Zero, n.d., np). To achieve this goal in order to address the challenges associated with global warming, United Nations (n.d.) declare that global carbon emissions must be reduced to 45% by 2030, and must be reduced 100% (the net-zero goal) by 2050. Other regional governing bodies and national governments are also working to pursue this global net-zero goal 2050. For example, the European Union "aims to be climate-neutral by 2050—an economy with net-zero greenhouse gas emissions. This objective is at the heart of the European Green deal, and is a legally binding target" (European Commission, n.d., np).

Source: ► pexels.com/royalty-free-images/

Green web hosting can help contribute to our global pursuit of the net-zero 2050 goal. It is because that "green web hosting is a greener, more environmentally friendly way to host websites. Most green web hosting companies aim to provide high-quality services while doing everything they can to limit their carbon emission" (Bustillos, 2024, np). Green web hosting is also a necessity today for a brand to survive commercially. Eighty-eight percent of customers do not want to be loyal to a brand, if the brand cannot improve its customers' societal and environmental footprint

(Townsend, 2018). Thus, "eco-conscious (customer) behaviour is now mainstream" (np), for example, 74% of people today bring their own bags for shopping, which is a +3% increase since 2019, "including almost half of people using fabric bags" (Kantar, 2023, np). Therefore, sustainable digital marketing is a necessity for brands today to not only help our global population to address the societal and environmental challenges but also to survive and prosper commercially.

The number of eco-friendly customers is increasing. Green web hosting is one of the sustainable digital marketing strategies that can help a brand to attract more eco-conscious customers, while enabling your brand (Digital Sustainability, 2024):

- To protect natural resources, because a lot of electricity which is produced by fossil fuel is required to run the World Wide Web.
- To protect people, wildlife habitats, and the planet.
- To improve the ranking and optimize your website performance.
- To contribute to the life of future generations.

"In 2020 there were 4.1 billion people using the internet, over half the global population, and this number is expected to increase! As the demand for the internet increases, the energy demand also increases" (Future Energy, 2024, np). As a climate action, green web hosting can help sustainable marketers to address the challenges associated with carbon emissions and climate change, while improving their websites' performance and attracting more eco-friendly customers. "Eco-conscious consumers are increasingly seeking out brands that walk the talk on sustainability. So, if your brand or organization prioritizes environmental responsibility, reinforce your message and values in every aspect of your digital presence" (Kovacevic, 2024, np).

In fact, sustainable digital marketing, e.g., "crafting strategies that resonate with eco-conscious values, (for example) choosing green web hosting, and creating content that speaks to the soul of sustainability" (Kovacevic, 2024, np), is therefore essential today for brands to contribute to people and planet, in order to make more profit by attracting more eco-conscious customers, who are increasing in number around the world.

6.1 Introduction: Background, Definitions, and Scope

Following the inception of the World Wide Web, the internet, the contemporary and embryonic aspects of these digital technologies, tools, and platforms, and the availability of many of these technological inventions to mass consumers in recent decades introduced the era of digital age. Such digital and mobile technologies, tools, and platforms include but are not limited to social media, artificial intelligence (AI), robotics, augmented reality, virtual reality, big data, database marketing and advanced analytics, cloud computing, blockchain, internet-of-things, search engine optimization, and other contemporary and embryonic technologies and tools, which already have a significant impact on marketing dynamics. This technology-enhanced marketing environment introduced the term digital marketing. See ► Sect. 1.3.1 (Historical and embryonic nature of marketing dynamics) and ▶ Table 1.2 for the discussion on the evolvement of marketing dynamics in different eras, the market influences, and marketing's response to the changes over time. But, we as marketers cannot ignore the traditional and extended marketing mix (see ▶ Fig. 1.7 for the 22Ps' extended marketing mix), but apply this mix to the contemporary mixed-zone marketplace. The mixed-zone marketplace consists of the traditional marketplace (i.e., the physical shops), a synchronized market (i.e., the mix of customers from different generations, some of them are very tech-savvy, some are less and some are not tech-savvy at all), and a metaphorical space. The metaphorical space is symbolic of marketers' and customers' interactions from distance, based on different contemporary technologies, such as social media, mobile platforms, emails, digital ads, etc., and also based on the embryonic aspects of fuzzy technologies, such as AI, robotics, data analytics, meta (among others) that aim to offer a more symbolic experience to customers, e.g., client meeting can take place online, but both clients and marketers will have a feeling of in-person interactions. The AI technology, investments in cryptocurrency, and the ongoing evolvement of the virtual space are signifying the development of this metaphorical space as a newer form of marketspace (i.e., the internet-enhanced digital marketplace) that is making the already closer global market further close.

Digital age—"The current era in which all kinds of information–numbers, text, sound, and video–can be put into digital form such that it can be stored, processed, and forwarded by any computer" (Czinkota et al., 2021, p. 103).

In this embryonic mixed-zone market, we apply the marketing principles in a balanced way to our target market/s across the traditional marketplace, synchronized market, and metaphorical marketspace. For example, while designing a product or service, marketers keep the options for digital version (e.g., the electronic version of this textbook), and its hard copy as the traditional version. Similarly, marketers include delivery costs in the pricing strategy for online sales. For customers, in many cases, they have options to purchase a product both from the online and traditional stores. Marketers also promote their market offerings today through traditional channels (e.g., newspapers, radio, television, billboard, telephone marketing), as well as through contemporary digital channels (e.g., social media, mobile technologies, web commercials, digital billboards, email marketing, online customer forums, websites, SMS marketing, search engine marketing, online live help, etc.). In these metaphorical marketing dynamics, marketers rely on different contemporary and embryonic technologies, such as database marketing, AI, robotics, or social media analytics to understand the needs of their different customer segments, forecast market demand, and design, deliver, and evaluate their products and services through the traditional marketplace, as well as the digital marketspace.

- **Marketplace** is the traditional place, e.g., a physical shop where two parties meet to exchange value.

- **Marketspace** is the internet-enhanced digital businesses, e.g., digital superstore where buyers, sellers, and other third parties (such as suppliers) supply, sell, and purchase goods and services.
- **Synchronized space** is the mix of customers from different generations, some of them are very tech-savvy, some are less and some are not tech-savvy at all.
- **Metaphorical marketspace** is symbolic of marketers' and customers' interactions from distance, based on different contemporary technologies, such as social media, mobile platforms, emails, digital ad, etc., and also based on the embryonic aspects of fuzzy technologies, such as AI, robotics, data analytics, meta (among others) that aim to offer a more symbolic experience to customers, e.g., client meeting can take place online from distance, but both clients and marketers will have a feeling of in-person interactions.
- **Mixed-zone market** consists of the traditional marketplace (i.e., the physical shops), a synchronized market (i.e., the mix of customers from different generations), and a metaphorical space (i.e., the technology-enhanced virtual space).

The technology-enhanced, more specifically, the information technology (IT)-enhanced virtual marketspace brings enormous opportunities for human civilization to communicate, establish and maintain relationships, or promote a social or ecological cause while promoting a product or service in less possible time and cost, and reaching to the most possible number of people around the world. Viral marketing through social media is a popular digital marketing strategy that helps marketers to reach out to enormous number of people based on an appealing message in less possible time and promotional cost. The appeal here can be triggered by celebrity endorsement, or promoting a societal or environmental cause, or both. For example, The ALS ICE Bucket Challenge campaign went viral, as the campaign invited participants to pour a bucket of ice water over their heads and challenge others among family and friends to do the same, or to donate to amyotrophic lateral sclerosis (ALS) research. The campaign quickly received momentum and became a wonder across the world to raise millions of dollars for the ALS association. The experts clarified the underlying reasons of the success of this ALS viral marketing campaign through social media, which include (PC Social, 2023):

- Strong emotional appeal related to a societal cause: Creating awareness and research fund for amyotrophic lateral sclerosis (ALS), a destructive neurodegenerative disease.
- Easily sharable content that is engaging.
- Social media amplification.
- Celebrity endorsement.
- Peer pressure and a healthy competitive environment, as participants challenged their family and friends, either to participate in the game or to donate to the ALS research.
- Feel good factor, based on donating to a social cause, i.e., ALS disease research.
- Community involvement.

- Timing and momentum.
- Organizational support, and
- Fund-raising impact.

Web Exercise 6.1

- Google "viral marketing based on a societal or environmental cause."
- Identify a few successful viral marketing campaigns which are directly related to a societal or environmental cause. While you Google "viral marketing based on a societal or environmental cause," many results of viral marketing campaigns would appear, which are not directly related to a specific societal or environmental cause. So, carefully select the viral marketing campaigns, only, which have a societal or environmental appeal.
- Analyze and share the reasons with your classmates that you think brought success for these viral marketing campaigns.

Share your views with your classmates, on how you think these societal- or environmental cause-related viral marketing campaigns are contributing to our global pursuit of one or more Sustainable Development Goals (▶ https://sdgs.un.org/goals).

There are two types of viral marketing: Organic viral marketing and amplified viral marketing. "Organic viral marketing is a type of marketing that spreads quickly and naturally through user-generated ▶ content and word-of-mouth, with little or no involvement or control from the brand" (PC Social, 2023, np). For example, celebrity endorsement to promote a brand message can attract word-of-mouth from the celebrity's huge fan base on social media, which could be an intrinsic motivation of people to participate, rather than being no or little control by the brand, for example, the case of ALS viral marketing campaign. On the contrary, the "amplified viral marketing is a controlled campaign where marketers have a more active role in shaping the campaign ▶ goals and methods of transmitting the viral message" (PC Social, 2023, np).

- **Viral marketing** through social media is a popular digital marketing strategy that helps marketers to reach out to enormous number of people based on an appealing message in less possible time and promotional cost.
- **Organic viral marketing** is a type of digital marketing that spreads quickly and naturally through user-generated ▶ content and word-of-mouth, with little or no involvement or control from the brand.
- **Amplified viral marketing** is a controlled digital marketing campaign where marketers have a more active role in shaping the campaign ▶ goals and methods of transmitting the viral message.

See more discussion in ► Sect. 6.3.2 to explore how successfully viral marketing campaigns could be designed, implemented, and evaluated to achieve sustainable digital marketing goals.

Because of this cost-effectiveness, global target audience, and the time-saving feature of digital marketing, we (the global population) today are engrained into numerous digital platforms and contents designed by marketers. For example, every minute in the year 2023, the internet generated the following engagements of brands with their target audiences (Austin, 2023):

- 2.4 million Google searches.
- 694,000 video hours viewed.
- 347,222 tweets.
- 3.47 million snaps created.
- 10.4 million viewing minutes on Instagram.
- 11,035 fake accounts removed from Facebook.
- 6.3 million total Zoom meeting minutes.
- 34,247 Slack messages.
- 11,834 chats on MS Teams.
- 6.94 million emoji sent.
- 3.02 million photos created with smartphones.
- 271,309 IOS and Android app downloads.
- 22,831 visits to ChatGPT.
- 241.2 million emails sent.

These digital dynamics enables marketers to influence their target audiences through the digital and mobile platforms. In this context, digital marketing can be defined as applying the traditional, contemporary, and embryonic marketing principles to influence a brand's target audience through different digital platform. The digital marketing dynamics have transformed the traditional media of reaching the target markets into a much more cost-effective and interactive business practice. For example, many of these digital platforms offer live chat with customers and enable marketers to promote their products and services more conveniently, in comparison to the traditional postal mails, newspapers, or television and radio commercials. In support of this view, generally, it can be argued that even at the end of the twentieth century, there was no real advertising medium for small businesses to promote their offerings to their wider target markets, based on their small budget. Today, even a next-door shop or a small bread and breakfast accommodation service provider can easily reach out to their broader target markets through various free web applications, such as social media and other free or low-cost websites that provide information. Another virtue of internet is that marketers are now able to interactively communicate with their customers via online chat, which is now a common practice for the telephone, gas, and electricity marketers to instantly answer their customers' questions, in order to help them to choose a competitive deal. In fact, IT-enhanced digital platforms enable innovative marketers to co-create value to ensure win-win outcomes, based on minimal resources, in comparison to their return on investment than ever before. For example, today, the world's (Kennedy, 2015; Shams, 2019):

- Largest accommodation provider owns no real estate, which is Airbnb.
- Largest taxi company owns no taxis, which is Uber.
- Most valuable retailer has no inventory, for example, Amazon and Alibaba.
- Largest phone companies own no telecom infrastructure, which are Messenger and WhatsApp.
- Fastest growing banks have no actual money, such as SocietyOne.
- Most popular media owner creates no content, which is Facebook.
- Largest software vendors don't write the apps, such as Apple and Google.
- Largest movie house owns no cinemas, which is Netflix.

These are well-established examples that demonstrate how digital marketing brings opportunities for entrepreneurial marketers, where irrespective of the marketers' initial available resources, the sky can be the only limit for them, in terms of their return on investment, and societal and environmental contribution. These brands are also good examples of sustaining their commercial competitive advantage. These brands applied the digital marketing concept involving online technologies to promote products and services. The way customers interact with brands has also changed since the beginning of the twenty-first century, because of the launch of different digital media and platforms to interact. Also, the COVID-19 pandemic has permanently increased further the significance of digital marketing for both customers and marketers.

◘ Source: ► pexels.com/royalty-free-images/

Similar to traditional marketing, digital marketing needs to consider the societal and environmental concerns in marketing strategy, as part of sustainable marketing initiatives. For example, websites become an integral part today to run a business. However, "one internet search produces 0.2 g of CO_2. The average website produces as much as 1.76 g of CO_2 per page view. An average website with 10,000 page views per month could generate 211 kg CO_2 per year, which is the same as would be absorbed by 105 trees in a year" (Jardine & Thorn, 2023, np). To address these environmental sustainability concerns, green web hosting is becoming popular to reduce CO_2 emissions, in order to achieve the net-zero goal. The Opening Vignette of this chapter discusses more about green web hosting as a sustainable digital marketing initiative. "**Sustainable digital marketing** balances achieving marketing goals through making a positive impact on our planet and society" (Kovacevic, 2024, np). It means the goal of sustainable digital marketing is "ensuring that every digital action, from clicks to campaigns, contributes to a more sustainable future" (Kovacevic, 2024, np).

- **Digital marketing** can be defined as applying the traditional, contemporary, and embryonic marketing principles to influence a brand's target audience through different digital platforms.
- **Sustainable digital marketing** balances achieving marketing goals through making a positive impact on our planet and society.

"Net zero is the international agreed upon goal for mitigating global warming in the second half of the (twenty first) century. The (Intergovernmental Panel on Climate Change) IPCC concluded the need for net zero CO_2 by 2050" (Net Zero Climate, n.d., np). Considering that the digital platforms produce CO_2 emissions, sustainable marketers need to be cautious and follow the responsible ethos to host green web platforms to reduce CO_2 emissions, in order to achieve the net zero goal, and to protect our environment.

6.2 Digital Environment and the Digital Marketing Dynamics[1]

6.2.1 The Technological Environment

The impact of contemporary technology is a major factor in the marketing dynamics. The growth of the use of mobile technologies, cloud computing, big data analytics, and blockchain technology demonstrates the evolution of unrestricted communication flows. Data communications are exploding, propelled by the internet

1 ► Section 6.2 (Digital environment and the digital marketing dynamics) is sourced from Czinkota et al. (2021), and reprinted based on the permission of copyright owner, © Springer).

and the rapid convergence of data and video with traditional voice traffic. Data have already overtaken voice on telephone networks. Concurrently, the availability of information to be communicated is increasing dramatically. Because all this information includes details about lifestyles, opportunities, and aspirations, the implications of this technology revolution for marketers are major. For example, a growing global awareness of products is likely to ease international product introductions. At the same time, such greater ease will also encourage the emergence of greater global competition for market share both at home and abroad. On the other hand, a growing detrimental image of a product in international markets would also impact its usage, production, and marketing in local markets. For example, "the excessive production and consumption of plastic has serious consequences on the environment and human health" (Heidbreder et al., 2019, p. 1077). As a result, the growing global awareness on the impact of plastic and plastic products is urging the reduction of plastic bag production, and many countries have already started to invent and use biodegradable packaging items to replace plastic-made shopping bag in their local markets. For instance, "a Thailand supermarket came up with a genius way (in early 2019) to reduce plastic packaging: wrap its produce in banana leaves instead" (Nace, 2019, np). Another example is in Bangladesh, where a newly invented jute-made 100% biodegradable shopping bag's production begins a new era of plastic bag usage and marketing to reduce the production and marketing of the non-biodegradable plastic bags in local market, and gradually in international markets (The Independent, 2018). In this way, in the contemporary local and international markets, the enormous flow of data and information across borders, and growing awareness and interests in a particular product or service in both local and international markets influence marketers' decision-making.

6.2.2 The Impact of Information Technology on Marketers

The use of **electronic data interchange** (EDI), merchandise scanning, and increased data sharing are restructuring the ways firms do business. In fact, technological capabilities influence a firm's operational and innovation capabilities (Peerallya et al., 2019). Consequently, technology affects the entire marketing strategy dynamics and covers the relationships with suppliers and manufacturers, the internal organization of the firm, and the firm's interaction with its customers. Even though the technology has been available for some time, we are entering a new era in terms of its application. As Bill Gates, co-founder of Microsoft Corporation, stated:

> The tools and connectivity of the **digital age** now give us a way to easily obtain, share, and act on information in new and remarkable ways. For the first time, all kinds of information—numbers, text, sound, video—can be put into a digital form that any computer can store, process, and forward. Standard hardware combined with a standard software platform has created economies of scale that make powerful computing solutions available inexpensively to companies of all sizes. And the "personal" in personal computer means that individual knowledge workers have a powerful tool for analyzing and using the information delivered by these solutions. (Gates, 2009, as cited in Czinkota et al., 2021, p. 103)

Because much of marketing is a dialogue, and information technology both enhances and restructures the existing forms of communication, marketers will be affected by information innovations more than ever before.

Electronic data interchange An electronic data entry system that allows manufacturers, logisticians, and vendors to communicate to maximize efficiency in production, delivery, and sales.

6.2.3 Upstream Relationships

The key to technology application rests with the building of information exchange relationships that convey vital knowledge accurately and rapidly and enable the recipients of the knowledge to respond by adapting their activities to the market realities. Most supermarkets and drugstores in the United States use electronic data entry and share data with vendors or develop special forms of cooperation in order to become more efficient in their market operations. For example, Hewlett-Packard has signed on with RFID (radio frequency identification) technology which is used to place radio tags on shipments in order to track and record their movement (RFID Technology, 2004). Suppliers and vendors can jointly use this technology to reduce costs and availability. For international shipments, the technology can also ensure that the shipment has not been opened and manipulated or altered by unauthorized parties. In an era of major international security concerns, such joint tracking can relieve much uncertainty from the shipping process. Some firms have gone even further in their upstream relationships by fully integrating some marketing aspects with their vendors. For example, Ahold Zaandam in the Netherlands has made one of its vendors, Heineken, fully responsible for maintaining supplies at its warehouses and for determining when and how to make its deliveries. As a result, the firm holds as little as 4 h demand worth of inventory at the warehouse (Fallon, 1994).

The use of information technology has also had distinct implications for the power relationships between manufacturers and their channel members. Increasingly, the point of purchase represents a crucial node of information. This is the place where critical information is accumulated regarding consumer behavior and promotional response. The available information lets retailers know what they are selling in each of their stores, how much money they are making on each sale, and who their customers are. Retailers no longer need to be saddled with stock that does not sell or run out of items that customers want to buy (Economist, 1995). Consequently, retailers have amassed so much power over manufacturers and other supply chain members that they can dictate the structure of the relationships and are able to force their vendors to either implement or adapt to information systems or lose the account. In light of the global sourcing policies of corporations, the participation of vendors in information systems will become increasingly important for manufacturers around the world. Firms in countries that are not able to provide an adequate information infrastructure will be severely handicapped in their ability to compete internationally.

6.2.4 Internal Effects of Information Technology

The information revolution also has major effects on the internal operations of firms. Leading-edge users of technology are able to develop cost advantages that permit them to assume leadership positions in their industry. For example, Tesco, one of the largest food retailers in the United Kingdom, has used information technology to control its ordering and stocking levels, with the following results (Czinkota et al., 2021):

- Major reduction of total stocks at its distribution centers
- Lead times of 1–2 days on fresh foods
- Out-of-stock conditions reduced by 30%
- Reduction of storage space from 40 to 20% of total store floor space
- Productivity benefits of more than $31.7 million

Such improvements are at the core of developing a position of strategic advantage and are therefore crucial to the marketer. These types of improvements enabled Walmart to surge ahead of its competitors Kmart and Sears to become the leading retailer in the United States.

Information also has an effect on the structure of firms and their hierarchies. For example, with better information flow, smaller organizations may be able to achieve economies of scale more quickly. In larger organizations, information availability can transform corporate structure by encouraging horizontal communications—for example, via electronic mail—and reducing the need for the traditional hierarchical organizational structure. New ways of handling information can also be decisive in determining the location of the firm's activities. For example, the ability to conduct banking via automatic teller machines has enabled financial institutions to greatly expand their reach of customers. With the increasing use of banking by telephone and internet, this expansion has already started to diminish the importance and value of old-fashioned bank buildings.

Economies of scale refers to a firm's better use of its resources and capability to reduce its overall production and operational costs, and increase its production to sell more products, based on less production cost of per unit product.

In terms of internal branding, for example to attract and retain employees, updated data and information on a firm's successes and advantages are also helpful to proactively plan, implement, and monitor internal branding campaigns. For instance, the experience of the UK-based mobile phone operator, Vodafone's internal branding campaigns on providing up-to-date information to employees, in order to ensure effective internal knowledge transfer within the company is relevant to further demonstrate the internal effect of information and communication. Vodafone regularly undertakes internal branding campaigns, in order to ensure "the confidence of employees is built through giving more knowledge of how the company operates, and how the company can be differentiated from competitors. As these employees are the direct communication line between the company and the customer, their confidence and the value of the brand in their mind is important" (Peterson, 2016, np).

6.2.5 Information Technology and the Customer

Technology also permits marketers to work with customers in more efficient ways. By linking purchases with specific customers, marketers can develop promotional programs based on specific purchase behavior. At the checkout point, for example, coupons can be automatically issued based on the scanned purchases. A customer who purchases pampers may get a coupon for baby food. Similarly, by using frequent shopper cards, customers can be identified and promotions can be targeted. For example, a family with a child can receive information on coupons for diapers, baby formula, or breakfast cereal depending on the age of the child (Clemons & Row, 1992). Alternatively, given the proper pieces of information, mail-order companies can compose individual catalogs based on specific customer profiles (Mazur, 1994). The information technology-enhanced social media networks and analytics also enable marketers to develop novel insights on customers' preferences, as well as to promote brands to a wider target audience in a cost-effective way. The "social media platform has created an exemplary scope for any brand to advertise its product through exposure, attention and perception; to develop opinions; and to create values" (Shareef et al., 2019, p. 58). Information will also offer major new opportunities for customers. For example, television shopping decreases the need to visit stores, and online product information makes price comparisons easier and reduces the ability of marketers to differentiate prices based on location. Simply put, information provides greater transparency in the market and therefore enhances the customers' ability to make better decisions.

6.2.6 Other Technology Factors

Information is not the only technology driver, of course. The development of superconductive materials and composite materials has made progress possible in fields such as transportation and electric power, pushing the frontiers of human activity into yet unexplored areas such as outer space and the depths of the oceans. Concerns about the environment are leading to new perspectives on products, production processes, and consumption and are likely to lead to new technologies that will provide substitute products and enhanced environmental "friendliness." The development of biotechnology is already leading to revolutionary progress not only in agriculture, medicine, and chemistry but also in manufacturing systems (Fukukawa, 1989). For example, California-based Calgene (part of the Monsanto Company) has planted fields of cotton that are already colored red, blue, black, and tan by the use of color-producing bacteria rather than toxic dyes. IBG of Baton Rouge, Louisiana, has cultivated a fat-eating strain of bacteria to produce McFree, which is sold to the McDonald's Corporation to clean french fry fat from restaurant drains. Proteins are beginning to be used to "stone wash" blue jeans and chinos (Day, 1995).

All these changes in technology are likely to introduce major discontinuities into the marketer's life. Although no one can foresee all changes, the marketer must at least be open-minded enough to be ready for them. In addition, it is important to avoid four common errors that can lead to the downfall of the firm (Foster, 1988):

- The belief that an evolutionary approach to technology will be enough.
- The assumption that there will be ample early warning about a coming discontinuity.
- The expectation that customer contact and input is sufficient preparation for change (customers may well be wrong in predicting what they want, particularly when new technology is concerned).
- The conviction that one's firm can always react fast enough to change.

6.3 Sustainable Digital Marketing

Before we discuss further digital marketing or sustainable digital marketing, let us clarify a few other relevant terms, such as digitization, digitalization, and digital transformation. **Digitization** refers to a system that transforms analogue information into a digital format as a prerequisite for the subsequent stages of consumption (Verhoef et al., 2021). For example, hardcopy books are transformed into digital PDF version or transforming old songs from cassettes into an online system for streaming and listening to music. **Digitalization** refers to converting a traditional (e.g., paper-based) business system into a technology-enhanced system (Verhoef et al., 2021). For example, brands today use social media, online customer forum, digital billboards, and different digital applications, instead of or in addition to the traditional means of communication with their target audiences. **Digital transformation** refers to a situation when digitalization takes place to a large extent and leads to a major company-wide or a specific business-wide transformation, ultimately based on a fully digitalized new business model to communicate with all internal and external stakeholders (Verhoef et al., 2021). For example, Volvo Cars move to car-sharing business, and brand it as Volvo M (or Volvo Mobility). Volvo On Demand, established in 2019, is a car-sharing company that offers rent-a-car and car-sharing services (Aberg, 2022). As a technology-based (App-based) service provider,

> Volvo Car Mobility is a mobility technology company, providing the smart mobility service Volvo On Demand. The service gives users flexible access to Volvo and Polestar cars and is highly appreciated by customers with a 4.8 out of 5 rating on App Store. Using the power of its propriety AI-supported mobility platform, Volvo Car Mobility is part of shaping sustainable cities by reducing the number of cars and enabling more people to move freely and sustainably. (Volvo On Demand, n.d., np)

In all businesses today across the world, we see digitization, digitalization, and digital transformation to different extents; some are fully digitally transformed (e.g., no paper-based communication) and some are keeping a balance in their business system between the extents of digitalization and the traditional method of communication. However, irrespective of the extent of digitalization in a business process, sustainable marketers should consider the key considerations to design, implement and evaluate a sustainable digital marketing strategy, which is discussed in ► Sect. 6.3.1.

- **Digitization** refers to a system that transforms analogue information into a digital format as a prerequisite for the subsequent stages of consumption.
- **Digitalization** refers to converting a traditional (e.g., paper-based) business system into a technology-enhanced system.
- **Digital transformation** refers to a situation when digitalization takes place to a large extent and leads to a major company-wide or a specific business-wide transformation, ultimately based on a fully digitalized new business model to communicate with all internal and external stakeholders.

6.3.1 Key Considerations for Sustainable Digital Marketing

To design a sustainable digital marketing strategy, sustainable marketers should consider:

- Digital marketing goals—Alignment with the triple Ps' stakeholder interests.
- Digital channels—The mix of digital and traditional media to determine the extent of digitalization.
- Digital audiences and value proposition centered on People, Planet, and Profit.
- Digital data and digital touchpoints.

Source: ► pexels.com/royalty-free-images/

Digital Marketing Goals—Alignment with the Triple Ps' Stakeholder Interests The marketing goal (e.g., creating awareness about the value of a product) should be aligned with its societal and/or environmental commitment, reflecting on its triple Ps' (People, Plant and Profit, or societal, environmental and commercial) stakeholders' interests. For this alignment, marketers should carefully select a societal or environmental cause that would be a good fit with the product or service. A good example of such brand- and cause-related alignments for a sustainable marketing strategy is Jeep and their promotion based on supporting the wounded members (and the families of the soldiers) in the United States military forces who were injured or died in overseas to protect their national interests (Jeep, 2014; Gastelu, 2022). Please, see more discussion on brand fit with a particular societal and/or environmental cause in ▶ Sect. 1.1.

◘ Source: ▶ pexels.com/royalty-free-images/

Digital Channels: The Mix of Digital and Traditional Media to Determine the Extent of Digitalization Who the audiences are? What their demographics (age group, gender, ethnic background, educational background, profession, income level, etc.) are? In general, marketers need to answer these questions before designing a traditional or digital marketing campaign. In addition, for a digital marketing campaign, marketers also should consider whether they need to include any traditional media in support of their digital marketing campaign. For example, TV commercial or newspaper ad to create initial awareness for a forthcoming digital campaign. ◘ Figure 6.1 lists some traditional and contemporary digital media that usually are considered by marketers to determine the extent of digitalization of a particular marketing project.

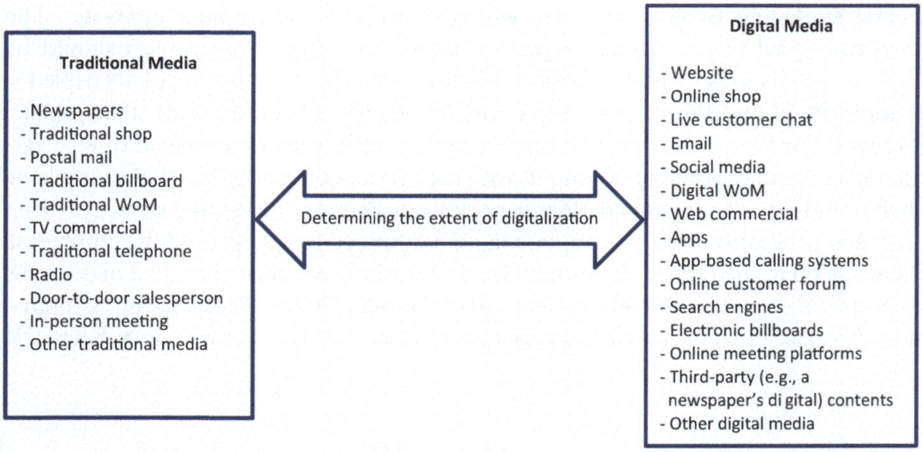

Fig. 6.1 The mix of traditional and digital media to determine the extent of digitalization

Source: ▸ pexels.com/royalty-free-images/

Digital Audiences and Value Proposition Centered on People, Planet, and Profit A sustainable digital marketing project should clearly communicate the value propositions that the project aims to contribute to its commercial, societal, and/or environmental stakeholders. Balancing in between the multifarious interests of a brand's commercial, societal, and/or environmental stakeholders is key to plan for a feasible value proposition for the triple Ps' stakeholders.

Source: ► pexels.com/royalty-free-images/

Digital Data and Digital Touchpoints Marketers need to determine the type of stakeholder data that would be collected and evaluated through a digital marketing campaign. What are the verification sources of veracity of this data? What would be the value of this data that would be generated from a digital marketing campaign (Shams & Solima, 2019)? How data would be analyzed to generate new insights, in order to influence the target audience in the near future? What will be the ethical considerations (e.g., how the privacy and interests of the target audience will be protected)? In addition to answer these questions, marketers also need to consider the customers' digital touchpoints, i.e., the digital platforms (e.g., a social media channel or an app, or a brand's website) based on what marketers share information digitally, and customers receive that information (e.g., product quality, price, after-sales service, etc.) about the brand. See more discussions on customers' digital touchpoints in ► Sect. 6.3.3.

6.3.2 Digital Channels and Sustainable Digital marketing Strategies

Digital Channels

Figure 6.1 lists some digital channels, which are generally used by marketers for digital marketing. These digital channels include but are not limited to a brand's:
- Website
- Online shop
- Live customer chat
- Email marketing
- Social media marketing

- Digital word-of-mouth (WoM)
- Web commercials
- Digital apps
- App-based calling systems (e.g., WhatsApp)
- Online customer forums
- Search engines
- Online meeting platforms (e.g., Zoom or MS Teams)
- Third-party contents (e.g., a newspaper's digital contents about a brand)

A brand's website is an important digital resource for the brand to communicate with its target audiences. In general, a website should include information, in relation to the traditional (and where relevant the extended) marketing mix. For example, the product and/or service that is offered by the brand. The product and/or service ranges, different features, and the customer segment(s) that are relevant to each product or service type should be specified. Price ranges and payment methods, e.g., online payment option, credit option, installment option, etc., should be described. Information about the product and/or service delivery options, such as online shopping and home delivery, postcode-based search options for nearby physical store(s), etc., should be provided on the website. The website also could include the link(s) of any web commercial, the brand's social media links, and contact information (e.g., email address and postal address). Optionally, the website could keep a live chat option for customers for instant question and answer. But, for keeping the live chat option, it is better to consider whether a brand's (especially if it is a small organization) available resources permit it to keep the live chat option. For example, sometimes we see that some websites have the live chat option, but customers do not receive instant response or need to wait long to receive information through the live chat option. In this case, it is better to not keep the live chat option if the brand's resources do not permit; as keeping this live option but not responding promptly to customer inquiry can devalue the brand's reputation.

The website of a sustainable brand should keep the information about the brand's corporate social responsibility, related to its societal and environmental commitment and contribution. ▶ Manager's Corner 6.1 describes McDonald's societal and environmental commitment and contribution, based on the information that is shared on their website. In addition, the brand's website should include a brief history, its people, its vision and mission. In addition to the basic information for a brand website that is discussed here, if the brand operates a dedicated online superstore, the brand should provide all relevant information to its customers that may include but not limited to warranty, return policy, suppliers of products or services offered by the brand, among others. See ▶ Web Exercise 6.2 to explore the information that should be provided by an online superstore to its customers.

Manager's Corner 6.1: McDonald's Societal and Environmental Commitment and Contribution

Source: ► pexels.com/royalty-free-images/

"It's a responsibility we take seriously—**because we know when our communities thrive, so does our business**. By thinking globally and acting locally, we are better equipped to effectively support our customers, our employees and the communities we are all part of" (np). Based on the sentence, "because we know when our communities thrive, so does our business," we see here that McDonald's have clearly stated the benefit of contributing back to people (society) and planet (environment), in order to underpin their commercial competitive advantage to grow their business across the world. Thinking globally (e.g., from a global viewpoint) and acting locally (e.g., contributing back to local communities and environment) are also stated in McDonald's website in their societal and environmental commitments section. For example:

- They donated $500,000 to earthquake relief efforts in Turkey in 2023 through the Save the Children, World Central Kitchen, and local nonprofits. They also provided 100,000 free meals to those impacted in Turkey, and in association with local partners, they chartered buses to transport people to their restaurants, in order to address the post-earthquake challenges.
- McDonald's have also ensured a $five million commitment to the Obama Presidential Center in their corporate head office in Chicago to support global programming and initiatives between 2022 and 2023.
- They initiated a Chicago-based investment strategy supporting local communities that offer a $3.5 million investment with 40 local nonprofits.
- McDonald's contributed more than $five million in 2022 as part of their employee assistance and international relief support led by the Red Cross in Ukraine and across Europe.

6

- Partnered with the American Red Cross Annual Disaster Giving Program, McDonald's also supported with $one million to pragmatically contribute to the USA and global disaster relief efforts. This contribution ensures that McDonald's is supporting the needs of emerging disasters, even before they occur.
- They offered $850,000 in community grants and sponsorships in the USA. Also, one penny from every Happy Meal at participating McDonald's restaurants ($5.9 million in total) was donated to Ronald McDonald House Charities® (RMHC®).
- McDonald's signed a partnership agreement with the Smithsonian Institution to become a founding $one million donor in support of the building of the National Museum of the American Latino.

In addition, on their Impact Report 2022–2023 (which is publicly available online), McDonald's have clarified in detail about their commitments and contribution to people and the planet to underpin their commercial competitive advantage. For example, commitments and contributions related to:

- Our Planet
  - McDonald's climate action
  - Responsible packaging and waste management
  - Nature, forecasts, and water
- Food quality and sourcing
  - Food safety
  - Nutrition and responsible marketing practices
  - Responsible sourcing
  - Animal health and welfare, and responsible antibiotic use
- Jobs, inclusion, and empowerment
  - Diversity, equity and inclusion
  - Talent and benefits
  - Human rights
  - Safe and respectful workplaces
- Community connection
  - Community impact and philanthropy
  - Ronald McDonald House Charities

Browse McDonald's and different other brands' websites across different industries to explore, how these brands are promoting their societal and environmental commitments and contributions through their digital channels, in order to make more profit.

Information sources: Community Impact & Philanthropy. (2022). With more than 40,000 restaurants, we have a unique connection with communities the world over. We believe that comes with a responsibility to help support the people within them. *McDonald's*. Retrieved May 6, 2024, from ► https://corporate.mcdonalds.com/corpmcd/our-purpose-and-impact/community-connection/community-impact.html; Impact Report. (2023). 2022–2023, Our purpose and impact report: McDonald's Corporation. *McDonald's*. Retrieved May 6, 2024, from ► https://corporate.mcdonalds.com/content/dam/sites/corp/nfl/pdf/McDonalds_PurposeImpact_ProgressReport_2022_2023.pdf.

Email marketing is another important channel for digital marketing. For email marketing, the email subject line and its contents are important to attract and retain the interest of recipient. For example, when we check our postal mailbox, upon looking at each of the mails, within few seconds we decide which one is a junk mail and which is an important one. Similarly, upon reading an email subject line, its recipients decide almost instantly whether to read the email further or to delete it. Therefore, if the subject line of an email fails to attract initial interest of its recipient, it will be really difficult to pursue success from an email marketing campaign. Also, after attracting readers' interest through the email subject line, the contents of emails could be structured in a way that could keep the readers' interest intact until the end of the email.

Since mid-late of the first decade of this current century till today, how many digital social media have been enticing the global population, and engaging people around the world to interact with each other from distance? To answer this question, we see there are Facebook, X (the former Twitter), Instagram, Snapchat, Tinder, TikTok, ClubHouse, Reddit, YouTube, Telegram, Messenger, Weibo, Kuaishou, Viber, LinkedIn, WeChat, Tumbir, Twitch, Quora, Athlinks, Pinterest, QQ, Discord, Douyin, beBee, and the list can go on. In terms of purposes of these social media, they can be classified into different categories. For example, some are for career purposes, some are better for sharing photos and videos, some are for sharing small text, and some are for chatting with family, friends, and colleagues. But, one thing is common across all these digital social media platforms, i.e., all these social platforms attract people from all over the world. As these digital platforms attract huge crowds, the social media brands promote them, mainly based on the contents generated by their users. The users of these digital social platforms include individuals as well as businesses. Businesses create contents about their market offerings to engage with their social media followers, as well as their followers' reposts and comments as further user-generated contents as digital word-of-mouth attracts further engagements on the posts of businesses. ◘ Figure 6.2 shows that the number of social media users in 2017 was 2.73 billion, which is 5.17 billion in 2024, and is expected to grow to 5.85 billion in 2027. It demonstrates how digital social media are becoming an integral part of the contemporary and future global marketing dynamics.

Web Exercise 6.2

- Browse ► amazon.com/, alibaba.com/ (or their local websites in your country), and the websites of other online retailers, and also browse the websites of your local supermarkets' websites, such as ► asda.com/, walmart.com/, target.com/, tesco.com/, aldi.com/ etc.
- Make a list of information from their websites that is shared by these online retailers and supermarkets (that have both online and in-person shopping option).

- Compare the information provided, and the information you may want to know if you were their customer, and suggest any additional information that you think these online retailers should provide to their customers.
- Justify your answer: Why do you think sharing any additional information that you have suggested would be instrumental to satisfy customers' needs more affluently?
- In addition to information for products and services, have you found information about these online retail brands' societal and/or environmental commitments and contributions?
- Do you think the information that is shared about these brands' societal and/or environmental commitments and contributions help our global community to pursue one or more Sustainable Development Goals?

Web commercial is another popular digital channel for marketers today to communicate with their target audiences. Especially, considering the expensiveness of traditional TV commercial, web commercial has received marketers' interest to reach out to their target audiences with relatively longer video contents in comparison to TV commercials. The production cost of video contents could be same for both TV and web commercials, but airtime for TV commercial is very expen-

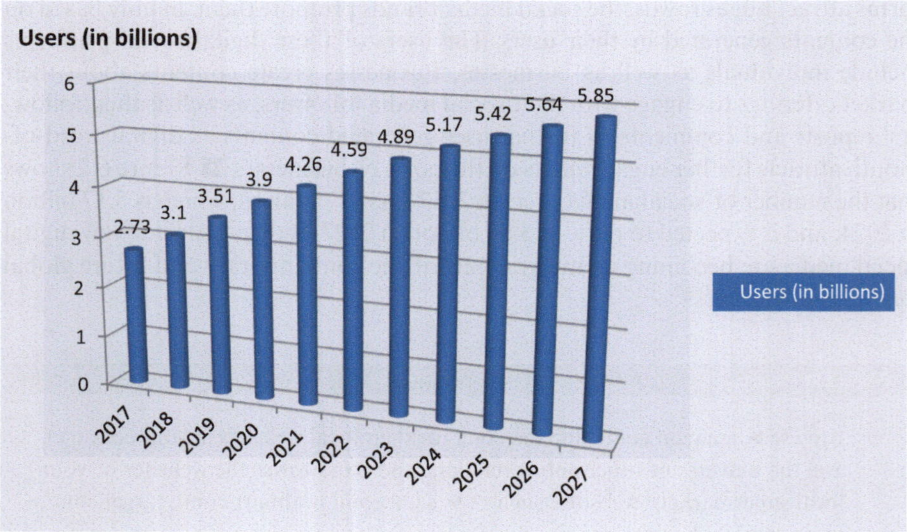

■ **Fig. 6.2** Number of social media users across the world (in billions) from 2017 to 2027 (Source: Statista, 2023b)

sive, whereas, marketers can communicate with their target audiences based on web commercials almost free of telecasting cost (unlike TV commercial) through their own website and social media channels. Digital apps, app-based calling and messaging services, online meeting platforms, and search engines are also other digital channels that are widely used in contemporary marketing practices to communicate with internal and external stakeholders of a brand. Online customer forums, such as Airbnb's online reviews, where customers can share their (satisfactory or dissatisfactory) remarks after they use the service of an Airbnb host. Such online customer forums help customers to select the best possible service providers within a particular destination. Also, third-party contents, for example, news about a brand published as a digital content in a mainstream newspaper or its social media can provide information to customers about the brand. ► This current Section discusses about the application of these digital channels for marketing.

Applications of Digital Channels for Marketing

■ Source: ► pexels.com/royalty-free-images/

In general, any marketing campaign, including digital marketing campaigns would have three purposes, and the digital channels discussed in this chapter can be used for these three purposes:

- Creating awareness.
- Clarifying our value proposition, i.e., why doing business with us is better for you instead of doing business with our competitive brands.
- And influencing and forming a public perception/opinion about our brand.

Creating awareness about a brand's presence in a market as much as possible, or as much as the brand expects to create, is a preliminary purpose of a digital marketing campaign. It is to clarify for the brand that we are here with an alternative and distinct value proposition to satisfy your needs. Creating awareness in relation to the marketing mix is useful. For example, what the product or service features are, what the price is or the price ranges are, from where customers can purchase it, and how can customers contact the brand should be communicated to create awareness. Also, communicating one or more authentic reason(s) clarifying the distinctive features of the brand's value proposition, related to its distinct product features, price or delivery options in comparison to the competitive brands will be valuable to influence customers, while creating awareness.

6

The goals for creating awareness and clarifying the distinct value proposition include creating *interests* in customers' mind to *consider* the brand's product, in order motivate customers to take *action*, i.e., to purchase it (Shankar et al., 2022). The third purpose of a digital marketing campaign (or any marketing campaign, e.g., through the traditional media) is influencing and forming a public perception/ opinion about our brand. For example, what comes to your mind *first* when you compare the core value offered by ALDI, the German multinational supermarket brand with its competitive supermarkets across different global regions, such as Woolworths or Coles in Australia; Tesco, Sainsbury's, or Asda in the UK; Kroger or Target in the USA? In response to this question, a number of postgraduate marketing students across different cohorts answered between mid-2019 and mid-2024 that ALDI is more affordable or slightly cheaper than its competitors, which is exactly what ALDI's marketers wanted their customers to think about them. Indeed, influencing or forming the perception of our target audience about the value of our brand is the art of marketing. It is not easy to influence or form that perception of a brand's target audience. It needs a consistent integrated marketing communication and branding strategy to convey the brand message to its target audience about the brand's value and promises to its customer. In such efforts of influencing and forming the target audience's perception, the most important consideration is ensuring consistency in what a brand promises to its customers and non-customer stakeholders, and what it delivers over a long period of time.

The digital channels that are discussed in this Section, all or most of them can be used to design, implement, and evaluate the following marketing applications:

- Advertising medium: For example, advertising through social media or web commercial.
- Direct-response medium: Online live chat, comments from social media, or online customer reviews are useful to directly respond to customers.
- Platform for sales transaction: For example, the online retail stores.
- Lead generation medium: The online meeting platforms can be used for client meeting from distance that can end up for lead generation. Also, positive digital WoM from social media or online customer forums can create new leads.
- Distribution channel: For example, online retail stores offer home delivery.
- Customer service mechanism: Responding to customers' enquires through email, online messaging apps, live chat option, or social media platforms can help brands to quickly answer and serve their customers further.
- Relationship-building medium: Many brands have a huge loyal fan base (followers) on social media that helps the brands nurture their ongoing customer relationships.

The ultimate goal of a sustainable digital marketing campaign using the digital channels is "ensuring that every digital action, from clicks to campaigns, contributes to a more sustainable future" (Kovacevic, 2024, np). For example, green web hosting, creating awareness about societal and environmental concerns, creating public perception about a brand as a socially and/or environmentally caring brand, based on communicating the brand's societal and/or environmental commitments and contribution over a long period of time.

The 7Ds of Effective Digital Marketing

◘ Table 6.1 discusses 7Ds of digital marketing. Considering these 7Ds while planning, implementing, and evaluating a digital marketing campaign would be instrumental for the success of the campaign.

Sustainable Digital Marketing Strategies

Different digital marketing strategies are popular in the digital marketing dynamics to influence the target audience of a brand, such as:

- Search engine marketing (SEM) and search engine optimizer (SEO)
- Mobile marketing
- Social media marketing
- Content marketing
- Celebrity endorsement
- Influencer marketing
- Viral marketing
- Programmatic advertising
- Omnichannel marketing

Table 6.1 The 7Ds to plan, implement, and evaluate effective sustainable digital marketing campaigns (Adapted from Chaffey & Smith, 2023)

The 7Ds	The significance and application of the 7Ds for effective digital marketing				
Digital goals	Brand goals, e.g., creating a socially and/or environmentally caring brand image	Objectives, e.g., creating awareness, clarifying the value proposition, or forming a public perception	An integrated approach across goals, audiences, devices, platforms, and media to reach the target audience	Digital transformation, i.e., determining the extent of digitalization (see Fig. 6.1)	Business models for all triple Ps' stakeholders: People (Societal), Planet (Environmental), Profit (Commercial)
Digital audiences	Consumer-to-consumer audience, e.g., WoM through online customer forums or social media	Target market segment(s), e.g., a geographical or demographic segment		Business-to-business audience	Business-to-customer audience
Digital devices	Tablet	Smartphone	Desktop	Smart speaker, e.g., Siri	Digital TV or digital radio network
Digital platforms	A social media, e.g., Facebook	An online retailer, e.g., Amazon or Alibaba	A search engine, e.g., Google	An operating system, e.g., Microsoft or Apple	Other sector-specific platforms, e.g., a specialized online platform developed only for the shipping industry
Digital media	Owned media, e.g., when a brand owns its web server to host its website	Paid media, e.g., paying to a search engine optimizer agent as an outsourced service	Earned media, e.g., earned through customers' positive digital WoM	Apps	Digital content, e.g., user-generated contents

The 7Ds	The significance and application of the 7Ds for effective digital marketing				
Digital data	Customers' behavioral data, e.g., the information that can influence a customer	Customer demographic data, e.g., age, income range, gender, ethnic background, education level, etc.	Value of an individual customer, e.g., how frequently an individual customer purchases from a brand and how much the customer spends each time	An individual customer's communication preferences, e.g., the type of information they prefer, the media, e.g., email, phone, SMS they prefer for communication	Big data analytics of a number of customers over a period of time to determine the trend in customer behavior
Digital technology	Martech, e.g., the tools that optimize the marketing efforts. For example, a content hub can be used to create contents	Marketing automation system, e.g., an AI-enhanced online live chat system	Augmented and virtual reality	Different other software and hardware are used to run the entire digital marketing dynamics for a particular marketing project	

■ Source: ► pexels.com/royalty-free-images/

Search Engine Marketing (SEM)) and Search Engine Optimizer (SEO) Let's assume that a consumer is googling for information for their next holiday, e.g., to find the best accommodation, flight, options for entertainment and sightseeing, etc., and pay for the best possible option as per the budget and preferences. A Google search can bring a wealth of information about alternative competitive holiday service providers. The question is whether our brand will appear on the top or at least on the first page of a Google search, so that the consumer can easily find our offering and compare it with other competitive offerings. In order to bring a brand's website and other digital information at the forefront of an online search, marketers apply SEM and SEO strategies to make the website more visible and to bring more web traffic (Almukhtar et al., 2021). Different online search engine includes but may not be limited to Google, Bing, AOL, Qwant, Brave Search, Gigablast, Lycos, Sogou, DuckDuckGo, Yahoo, Yandex, StartPage, Swisscows, Dogpile, Mojeek, Wikipedia, Baidu, Ecosia, Ask, Naver, WolframAlpha, Gibru, Boardreader, etc. As part of SEM and SEO strategy, using these search engines, marketers try to leverage various aspects (e.g., design, light web contents in terms of size of the contents, etc.) of their website to optimize the sites more appealing to search engines and relevant to users' keyword-based search items (Berman & Katona, 2013). Such SEM and SEO can be paid or generic, non-paid. For example, when you pay to a search engine to bring your web contents to the forefront of search results, it is a paid SEO and SEM. But, if your website automatically appears on the first page of search results because of the synchronization of your web contents, e.g., its lightness and compatibility with a search engine's requirements, it is then a generic or non-paid SEO and SEM. If you are a marketer of a search engine, green web hosting is key to make your search engine brand environmentally responsible and attractive to the environmentally conscious business customers who want to use the paid service of an environmentally friendly search engine to optimize the performance of their websites. If you want the website of your brand will appear at the forefront of generic web searches, making eco-friendly web contents, lighter web contents, and making your web contents compatible with the requirements of major search engines is useful.

Search engine optimization and search engine marketing refer to making a brand's website and other web contents appealing and compatible with the requirements of online search engines, in an attempt to bring the website and web contents at the forefront of an online search result.

Mobile Marketing and Social Media Marketing Mobile marketing can be defined as using different wireless and portable devices, e.g., smartphones, tablets, or smartwatches as a medium to convey marketing message, in order to influence a brand's target audience. In the same vein, social media marketing refers to using a social media platform to convey marketing message, in order to influence a brand's target audience. For both, mobile marketing and social media marketing, analyzing each individual customer's or user's feelings, expressions, and behavior using web analytics software and methods is crucial for effective mobile and social media marketing (Lamberton & Stephen, 2016). In this context, both mobile and social media marketing are instruments for digital marketing research as sources of marketing intelligence (Lamberton & Stephen, 2016).

- **Mobile marketing** is using different wireless and portable devices, e.g., smartphones, tablets or smartwatches as a medium to convey marketing message, in order to influence a brand's target audience.
- **Social media marketing** refers to using a social media platform to convey marketing message, in order to influence a brand's target audience.

Content Marketing Creating web contents, e.g., text, audio, video, and images to communicate a brand's target audiences mostly using digital marketing channels, is content marketing. Such contents could be generated by a brand, and also be generated by the brand's followers on digital channels, e.g., social media. For example, brands create and post contents on social media, which attracts users' (e.g., the brand's social media followers') engagement, e.g., users' comments, like, and share on the post, and such engagements also bring more web traffic for the brand to reach to more digital audiences. In these web dynamics, the contents, e.g., the original post that is shared on social media by the brand may have been created by the brand's marketing team, and the subsequent contents (e.g., the like, share, or comments) that are generated by the brand's followers on social media are known as user-generated contents as word-of-mouth (WoM) for the brand. Thus, content marketing is a marketing communication strategy to create and publicize contents to a brand's target audiences, with an aim to generate audiences' interest and engagement, in order to influence their behavior (Stephen et al., 2016). In this context, creating contents relevant to the needs of target audiences, i.e., competitive value proposition, societal and environmental commitments and impact of a brand, and differentiation with other competitive brands, in order to satisfy the target audiences' needs most lucratively is important to generate target audiences' interest and engagement with the content to influence their behavior, e.g., brand loyalty.

- **Content Marketing**—Creating web contents, e.g., text, audio, video, and images to communicate with a brand's target audiences mostly using the digital marketing channels is content marketing.
- **User-Generated Contents**—In web dynamics, the contents, e.g., the original post that is shared on social media by a brand may have been created by the brand's marketing team, and the subsequent contents (e.g., the like, share, or comments) that are generated by the brand's followers on social media are known as user-generated contents as word-of-mouth for the brand.

Programmatic Advertising "Programmatic advertising is a digital advertising technique that uses automated technology and algorithms to buy and sell ad space in real time. It involves using software and data-driven processes to streamline the ad buying and placement process" (Shiqun et al., 2021; as cited in Rosário et al., 2023, p. 13). The key stakeholders of programmatic advertising include the buyer (e.g., the advertiser), ad exchange, publisher, and the demand-side platform (DSP) to analyze the demand of ads based on an algorithm—based automated process to buy and sell digital ad space (Rosário et al., 2023). To facilitate the process (Rakic & Rakic, 2013; Rosário et al., 2023):

- A buyer or advertiser first sets specific criteria, such as their preferred digital target audience, promotional goal, budget, and other relevant requirements.
- Then, based on these criteria, the algorithm enables advertisers to reach their target audience at the right time.
- In this process, the ad exchange enables the publisher to offer their web platform to the advertiser.
- The DSP enables the ad exchange to facilitate bid for advertising based on real-time demand from the active users (digital target audience), and the top bidder (advertiser) wins the digital ad space to promote their product or service.

In 2017, the global market of programmatic advertising was US$187.83, which reached US$616.85 in 2024, and is expected to grow to US$724.84 in 2026 (Mironov, 2024). It indicates the popularity of programmatic advertising because of its effectual utility for marketers to reach out to their digital target audiences,

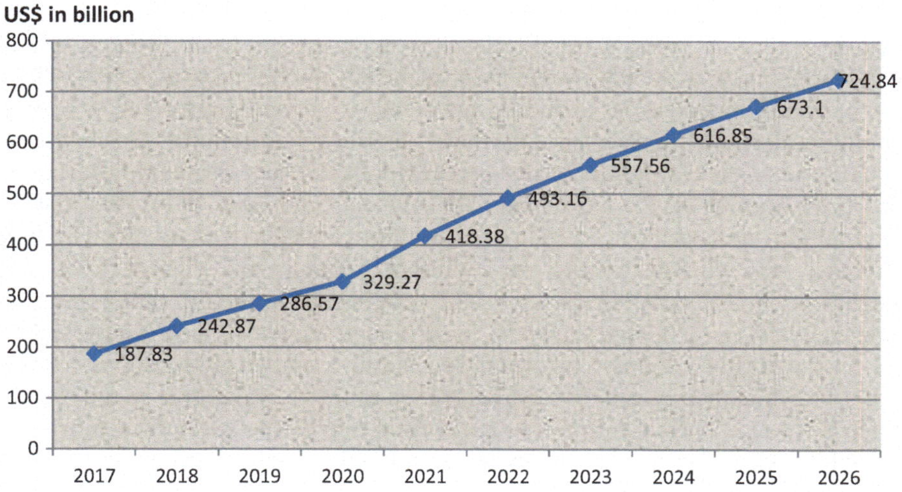

Fig. 6.3 The actual and projected growth of the global programmatic advertising industry (Source: Mironov, 2024)

based on real-time data of active digital audiences. Figure 6.3 demonstrates the growth of the global programmatic advertising industry. Following the discussion thus far on programmatic advertising, programmatic advertising can be defined as an automated algorithmic process to understand the real-time demand of digital ad based on recognizing the active digital target audiences at a given time, in order to enable advertisers to reach out to their digital target audiences, relying on real-time data of active digital audiences.

> **Programmatic advertising** is an automated algorithmic process to understand the real-time demand of digital ad based on recognizing the active digital target audiences at a given time, in order to enable advertisers to reach out to their digital target audiences, relying on real-time data of active digital audiences.

Celebrity Endorsement Celebrity endorsement to promote commercial products or services, and societal and ecological cause is a traditional marketing strategy. For example, Wedgwood, the producers of chinaware (i.e., the traditional Chinese ceramic artifacts) employed royal endorsement in the 1700s as a marketing strategy to promote their product. Lillie Langtry, the London socialite and West End stage actress became the poster girl in 1882 while promoting Pears soap, which marked her as the first-ever celebrity to promote a commercial product or service (Morin, 2002; Richards, 2022). Celebrities have a huge fan base, and because of the technological progress, e.g., launch of television, radio, and then more recently the internet-enhanced digital platforms, today, many celebrities have a global fan base. Fans consciously or subconsciously maintain a relationship with their celebrities, even without meeting with their celebrities ever in their lifetime.

Fans follow the lifestyle, fashion, and values of their celebrities in their life, because in general, fans trust their celebrities, imitate them from distance, love to spend more time to read news about their celebrities, and love to watch more television programs or web series including ads in television, newspaper, web or in digital billboard which are performed by their celebrities. For example, the childhood hero of one of the co-authors of this core textbook is Diego Armando Maradona who passed away on 25 November 2020. During the 1986 FIFA World Cup Football that took place in Mexico, the co-author was just 8 years old. The time difference between Mexico and the co-author's country of origin is 12 h. The matches of the Mexico World Cup used to commence in the afternoon and evening, e.g., between 2.00 and 8.00 p.m. local time in Mexico, which is at least 2.00 a.m. in the author's country of origin. No family member of the co-author is a sports fan, but the co-author used to tell his mum to wake him up before the matches started. The co-author still remembers one of the sweetest memories of his childhood, which is when his parents and sibling were sleeping, and he was watching Maradona's live football sitting on the carpet, and turning off the sound of their 24-inch black and white TV! Because of this influence of Maradona on the co-author and his global fan base, it is highly likely that Maradona's fans would be watching a TV commercial in full which is performed by Maradona, even if the product or service that is promoted by Maradona in that commercial is not liked by a fan at all! As a result, celebrity endorsement is a very popular strategy to promote a brand. Besides celebrity endorsement, incorporating a slice of target audiences' sweet memories from their life into a commercial is a popular advertising strategy too. This is why, often we see promotional commercials, for example, a commercial of a tea brand shows that people are enjoying an afternoon tea with their family or friends (a slice of life) while drinking and praising the value of that tea brand to their life. The purpose of correlating a sweet memory, a slice of time from the life of a brand's target audience is attempting to create an emotional bond with the target audience, so that once the target audience watches a brand's commercial, they remember their life memory. Also, the target audience would be able to recall a commercial spontaneously when they have a time in real life (e.g., hanging out with friends) if the commercial shows a relevant slice of time related to the real-life environment of the target audience.

6

Because of the influence of celebrities to effectively promote a brand or a message, public, private, and third-sector organizations have been endorsing celebrities to promote a societal or environmental cause, while promoting their brands. For example, Adidas endorses Lionel Messi to promote a societal and environmental cause of the need of recycling and reusing materials, proclaiming that "Messi's boots today, recycled into yours tomorrow" (Sustainability, 2020, np), while promoting Adidas boots. ▶ Marketing in Action 6.1 describes a sustainable marketing campaign based on celebrity endorsement. During the COVID-19 pandemic, the British Government (2021) endorsed British celebrities who have a positive image, in order to promote the benefit of having the COVID-19 vaccine. To promote peace (a societal cause) and address climate change (an environmental cause), the United Nations (2014) appointed Leonardo DiCaprio as its Messenger of Peace.

Influencer Marketing Influencer marketing is an extended version of celebrity endorsement in our contemporary digital marketing dynamics. The rise of social media since the first decade of the twenty-first century enables mass people to convey their message to the wider local, national, international and global audience more conveniently than ever before. As a result, in the recent decades, we have seen the rise of many local (e.g., within one or few cities), national, international (e.g., within a country, subcontinent or continent), and global (across the world) influencers who are followed by a considerable number of people on social media platforms, e.g., Facebook, Instagram, Tiktok, LinkedIn, among others. Similar to the celebrities, these digital influencers have an impact on the lives of their followers. Therefore, brands also promote their market offerings (e.g., products and/or services) through such digital influencers, as well. Besides large brands, such influencer marketing campaigns are also helpful for small- and medium-sized enterprises (SMEs) to effectively promote their offerings within their limited budget. For example, Daniel Thomas in Nottingham in the UK launched a business selling hot tubs, the "Danz Spas Hot Tub" when he was just 18 (BBC, 2014). Daniel Thomas as an SME entrepreneur promoted his business through Sophie Ellis-Bextor, who as a singer is an influencer on social media:

> The singer (Sophie Ellis-Bextor) was said to be looking to buy a hot tub, and was wondering if she and Danz Spas could come to a business relationship—she would be happy to endorse and promote the company, if it could offer her a good deal. Mr. Thomas, 26, was quick to realize that he had just been offered a fantastic opportunity (to promote his business, cost effectively through celebrity/influencer endorsement). (BBC, 2014, np)

◘ Source: ► pexels.com/royalty-free-images/

Based on the extent of the number of followers on digital platforms, Campbell and Farrell (2020) categorize different types of digital influencers, which are described in ◘ Table 6.2. Because of the cost-effectiveness and effectual impact of influencer marketing, the global industry of influencer marketing has expanded to US$24 billion in 2024 from US$ 1.7 billion in 2017 (see ◘ Fig. 6.4).

Table 6.2 Different types of influencers for influencer marketing (Source: Campbell & Farrell, 2020)

Types of influencers	Number of followers
Nano influencer	0–10,000
Micro influencer	10,001–100,000
Macro influencer	Up to 1000,000
Mega influencer	More than 1000,000
Celebrity influencer (rich and famous)	More than 1000,000

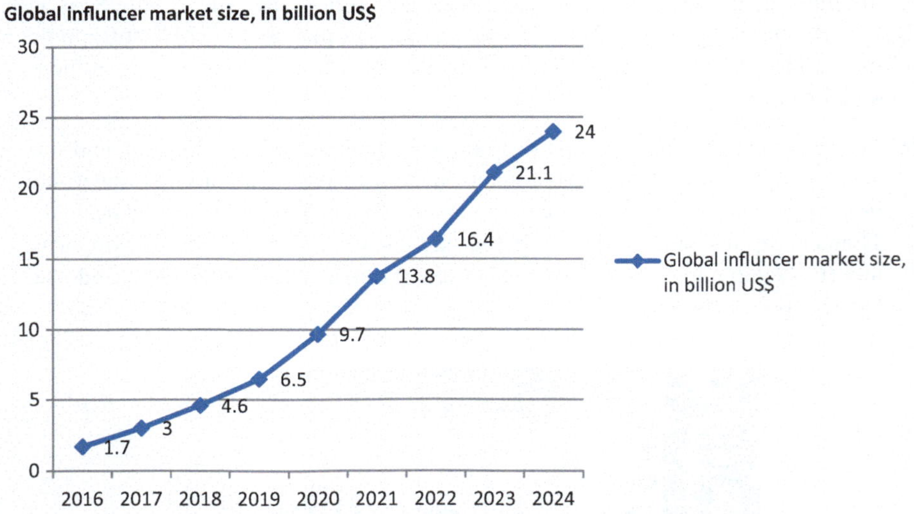

Fig. 6.4 The growth of the global influencer marketing industry (Source: Dencheva, 2024)

- **Celebrity Endorsement**: Promoting a brand relying on a celebrity's popularity, based on a formal business relationship with the celebrity.
- **Celebrity Endorsement for Sustainable Marketing**: Promoting a brand's societal, environmental, and commercial commitments and contributions, relying on a celebrity's popularity, based on a formal business relationship with the celebrity.
- **Influencer Marketing**: Promoting a brand relying on an influencer's influence on their followers, based on a formal business relationship with the influencer.
- **Sustainable Influencer Marketing**: Promoting a brand's societal, environmental, and commercial commitments and contributions, relying on an influencer's influence on their followers, based on a formal business relationship with the influencer.

Marketing in Action 6.1. Lionel Messi's New Sneakers Come from the Recycling Bin—Celebrity Endorsement of Adidas for Sustainable Marketing

Source: ► pexels.com/royalty-free-images/

Recycling and reusing materials, as part of the circular economy concept, are important to reuse limited existing materials, in order to reduce environmental impact of materials and reduce production cost. ► Section 1.3.1 and ► Fig. 1.5 discusses more about circular economy. "Clothing brands now produce almost twice as much clothing as they did 20 years ago. What's more, demand is expected to keep on growing. But clothes that are no longer worn are being dumped en masse into the environment without being recycled" (np). To reduce environmental impact of raw materials and reduce production cost, Adidas and H&M recycle and reuse their production materials, and promote their environmental contributions through digital channels. Both brands "are taking part in a trial (of a sustainable cotton project, funded by the European Union) to create a circular recycling model for the clothing industry" (np). For this sustainable cotton industry project, the European Union awarded €seven million to make the cotton industry more environmentally friendly. As a partner of this sustainability project, Adidas "is striving to create clothing and possibly shoes that are partly made of recycled fibers from the Infinited Fiber Company" (np).

As part of this sustainability project, Adidas aims to produce their apparels and sportswear in a way that would be more eco-friendly. For example, the sportswear that they produce for top professional athletes and amateur athletes who are their customers across the world, are now recycled more than before. To make an awareness about this environmental commitment and contribution, Adidas endorses Lionel Messi, and promotes it through digital channels, so that such promotions would not only be valuable for Adidas to create an eco-friendly brand image but also be instrumental to propel the competitors of Adidas to produce their sportswear that would be more eco-friendly:

> I am proud that Adidas is working to make sure that all of their boots, including mine are being made in a way that protects the environment. For me, this is the future of football—Leo Messi.

In 2015, Adidas announced a plan for recyclable and zero-waste sporting products and promoted this plan, based on the tagline, "Messi's boots today, recycled into yours tomorrow" (np). According to Gerd Manz, Vice President, Technology Innovation at Adidas:

> This is a game-changing development for football fans. Over the next three years, Sport Infinity aims to end the days of throwing away football boots. Instead, every pair of boots is not just recycled but reimagined to the consumer's most personal specifications.

As part of this sustainability project, Adidas has also partnered with Parley to contribute to the sustainability of life below water (i.e., Sustainable Development Gaol number 14).

> Following the announcement of SpeedFactory and our partnership with Parley for the oceans, Sport Infinity is the next step in our commitment to innovation and sustainability. This project will close the sustainability loop, creating a high-performance product that can always be recycled—Glenn Bennett, Executive Board Member, Global Operations at the Adidas Group.

Customers around the world today are more socially and environmentally conscious than ever before. To satisfy these socially and environmentally conscious customers, Adidas introduced their sustainability project, based on the circular economy principles. Eventually, the project not only helps Adidas to reduce their production cost by recycling and reusing production materials and satisfy their global customers, but also helps them to contribute to the Sustainable Development Goals 17 (partnership for the goals, i.e., the partnership of Adidas and Parley), 13 (climate action, i.e., by ensuring zero waste in their value chain), 12 (responsible production), and 9 (industry innovation, i.e., based on the technological innovation at Adidas).

Information sources: Adidas Group. (2015). Messi's boots today, recycled into yours tomorrow: Adidas reveals plans for custom, recyclable, zero-waste sporting goods. Retrieved May 12, 2024, from ► https://www.adidas-group.com/en/media/press-releases/messis-boots-today-recycled-into-yours-tomorrow; Kieckens, E. (2020). Lionel Messi's new sneakers come from the cycling bin. *Innovation Origins.* Retrieved May 12, 2024, from ► https://innovationorigins.com/en/lionel-messis-new-sneakers-come-from-the-recyling-bin/.

Viral Marketing Viral marketing through social media is a popular digital marketing strategy that helps marketers to reach out to enormous number of people based on an appealing message in less possible time and promotional cost. As sustainable digital marketing balances achieving marketing goals through making a positive impact on our planet and society; the purpose of sustainable viral marketing is to create awareness about a brand, its market offering, and societal and/or environmental commitment and contribution, based on a well-aligned societal and/or environmental brand message to cost-effectively attract the interest and engagement of an enormous number of population, mainly through digital platforms. ◘ Figure 6.5 discusses how marketers could design, implement, and evaluate effective sustainable viral marketing campaign.

Firstly, why you need to launch a sustainable viral marketing campaign needs to be determined. What is your marketing goal? As part of your marketing goal, would you like to create awareness about your brand and its value proposition, based on promoting a societal and/or environmental cause, or would you like to create a socially and/or environmentally caring brand image, or both? Upon determining your marketing goal, the societal and/or environmental goal of a viral marketing campaign needs to be determined too. For example, would you like to create public awareness about a societal and/or environmental cause, while promoting your brand? Besides creating public awareness about a societal and/or environmental cause, would you like to call the public for action to address the societal and/or environmental challenge, e.g., recycling plastics? The goal setting will be instrumental to envisage a roadmap to design a sustainable viral marketing campaign. For example, the message that you may need to spread out, the action you may want to incorporate into your viral marketing campaign, stakeholders and target audience need to be engaged, among others, can be planned, based on the goal of your viral marketing campaign.

> **Sustainable viral marketing** is creating awareness about a brand, its market offering, and societal and/or environmental commitment and contribution, based on a well-aligned societal and/or environmental brand message to cost-effectively attract the interest and engagement of an enormous number of populations, mainly through digital platforms.

Secondly, the brand message that you want to be viral needs to be determined. For example, would you like to communicate your brand's societal and/or environmental commitment and contribution, in order to create public awareness inviting them to address the societal and/or environmental concern? How the value delivery process of your brand contributes to the societal and/or environmental challenge (e.g., ► Marketing in Action 6.1 describes that Adidas aims to ensure zero waste in their process), can be incorporated into the overall message of a sustainable viral marketing campaign.

Thirdly, upon determining the goal and brand message, designing (planning) a sustainable viral marketing campaign should be guided based on its goal and brand message. During the design stage, sustainable marketers need to consider:

- Brand fit: Your brand and the societal or environmental cause that you focus on should be well-aligned. For example, promoting cancer research by a tobacco brand could bring criticisms for the brand, considering that critics would say,

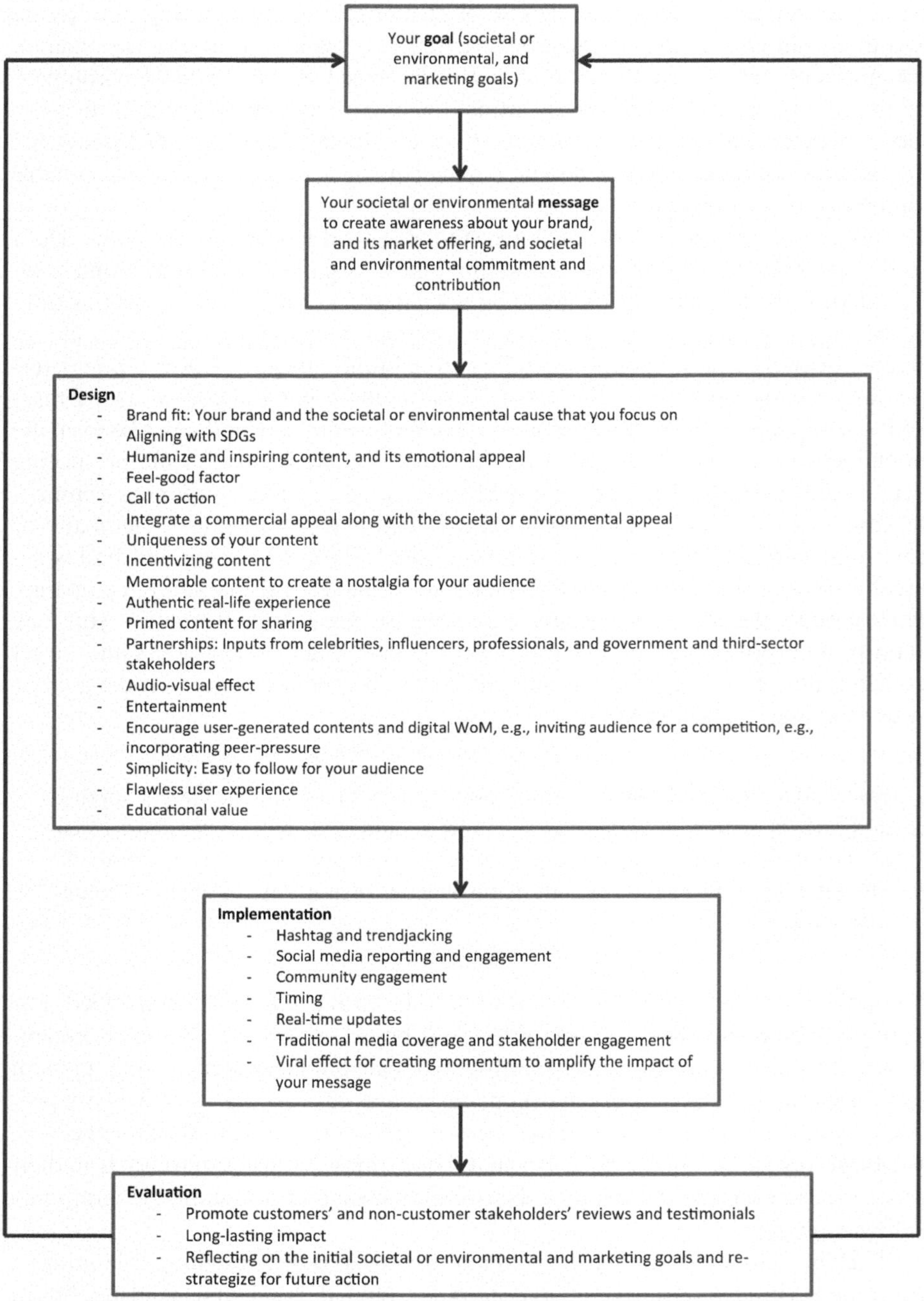

Fig. 6.5 Designing, implementing, and evaluating sustainable viral marketing campaign (adapted from PC Social, 2023)

you are causing cancer and now you are curing it! ► Section 1.1 discusses more about brand fit to carefully select a well-aligned societal and/or environmental cause to promote a brand.

- Aligning with Sustainable Development Goals (SDG): A sustainable viral marketing campaign should align with one or more SDGs too. It will help the campaign to contribute to our global pursuit of SDGs, while creating a socially and/or environmentally caring image for the brand.
- Feel good factor—Humanize and inspire content, and its emotional appeal: Using a humanistic compassionate tone to design your brand message, and promoting the message, based on inspiring and emotional content while personifying the content, would be instrumental to make the campaign attractive to your audience. For example, ► Manager's Corner 6.1 describes that McDonald's also contributes to the American Red Cross's Annual Disaster Giving Program to help Red Cross helping people during natural calamity. For a viral marketing campaign centered on a cause related to natural calamity, incorporating human appeal could attract the target audience's interest and engagement, e.g., to raise fund to help people who are affected by the calamity, as it gives your brand's audience a "feel good factor" to engage.
- Call to action: A sustainable viral marketing campaign should call its target audience for action instantly (e.g., right now) to contribute (event to a small scale) to address the societal and/or environmental concern. For example, if a sustainable viral marketing campaign is related to societal cause, such as young girl's mental health, the campaign should encourage its audience to act right now to help young girls, e.g., to plan for a non-detrimental digital life. It could be, for example, educating girls on how they can avoid digital bullying, how they can remain safe in the digital arena, etc.
- Incentivizing content—Integrate a materialistic benefit along with the societal or environmental appeal: Incorporating a materialistic benefit for your target audience while promoting a societal and/or environmental cause, could be instrumental to attract additional engagements from your audience. For example, offering a limited-time discount of your product or service, e.g., buy one get one free, while you run the campaign can create further interest in your audience, as it gives them an optional materialistic benefit, as well as good for your brand to attract some additional sales.
- Uniqueness of your content: Uniqueness of the content of your sustainable marketing campaign is important for its success. It is because if your audiences realize that the content is copied from a particular source (e.g., an old movie), it could reduce the excitement of the audience to engage.
- Memorable content to create nostalgia for your audience: A viral marketing campaign should help its audience to make memories, e.g., a digital photography competition that could make them nostalgic in future. It can be helpful for the audience to recall the societal and/or environmental appeal associated with your viral marketing campaign in the future, which would be instrumental to motivate the audience to contribute to the societal and/or environmental cause even in future.

- Authentic real-life experience: A viral marketing campaign should give its target audience an authentic real-life experience to encourage them to engage. For example, creating awareness and engagement (e.g., like, share, comments) through social media and traditional media, and simultaneously directing the audience on what to do, how to do, when to do to engage in real life (e.g., donating to a fund-raising campaign within a specific deadline, through a specific online and/or traditional payment-processing channel etc.), and then sharing the trustworthy information of how the funds that have been raised are used to address the societal- and/or environmental cause, can create a sense of reliability among the audience to engage with the call of your brand now and in future. Overall, authenticity and reliability are important to attract the audience's interest to engage them with the content of a viral marketing campaign.
- Partnerships—Inputs from celebrities, influencers, professionals, traditional media, and government and third-sector stakeholders: A digital sustainable viral marketing campaign can endorse a celebrity, partner with an influencer, professionals (e.g., ad agency), government department and third-sector stakeholders (e.g., climate activists), and traditional media, in order to promote the call and the associated societal and/or environmental cause further.
- Audio-visual effect and entertainment: The audio-visual effect of the digital content of a viral marketing campaign should be entertaining for its audience to attract and retain their interest to watch the content, entirely.
- Simplicity—easy to follow for your audience for a flawless user experience: The content should be simple while entertaining, so that the audience can follow and understand the prompt/s in the content easily, which is important for a flawless content-user experience for your audience.
- Primed content for sharing, and encouraging user-generated contents and digital WoM, e.g., inviting audience for a competition, e.g., incorporating peer pressure: The content should be primed properly, so that your audience can easily engage (e.g., like, share, comments, etc.) through social media to attract more digital WoM for the campaign. Also, the content could include a competition for your audience. For example, ▶ Sect. 6.1 describes briefly a successful sustainable viral marketing case of ALS, i.e., the ALS ICE Bucket Challenge campaign went viral, as the campaign invited participants to pour a bucket of ice water over their heads, and challenge others among family and friends to do the same, or to donate to ALS research. Such competitions among your target audiences can increase interest and engagement, especially, if the competition is designed based on harmless fun, like the ALS' Ice Bucket Challenge.
- Educational value: The content of a sustainable viral marketing campaign should have some educational value for its audience. For example, how the societal and/or environmental concern make our society or environment more difficult for sustainable living. What has been done so far, and what else needs to be done to address the challenge. What your brand has contributed so far? How the target audience of your brand can contribute? What is the short-, mid-, and long-term goal to achieve based on all collective contributions, in order to address the societal and/or environmental challenge? Incorporating this information into a sustainable viral marketing campaign could be instrumental to its success.

► Marketing in Action 6.2 presents two more brief case studies of sustainable viral marketing campaigns. Fourthly, after designing a sustainable viral marketing campaign, its implementation is a crucial stage. At this stage, marketers could consider using hashtag through social media and applying trendjacking technique to create initial awareness about your viral marketing campaign, its goal, message, and overall content. Trendjacking is an "art of inserting your brand into popular conversations by creating content around those topics or events to connect with your audience and draw attention to your products and services" (Arinze, 2024, np). For example, during a mega event, e.g., FIFA World Cup Football, Olympic or ICC Cricket World Cup, endorsing a celebrity athlete, a brand can promote a societal cause (e.g., helping lame people to increase their physical mobility) to draw attention of the audience of the brand to engage (e.g., to donate to a medical research for lame people), while running the relevant sustainable viral marketing campaign (e.g., fund-raising for the medical research) throughout the period of that mega event. Timing is an important success factor too. For example, considering whether a particular time of a year, e.g., summer months, would be relevant to run a campaign related to an environmental cause, e.g., addressing global warming.

Community engagement and real-time updates are another two crucial factors to consider during the implementation stage of viral marketing campaign. For example, upon creating initial awareness, for example to raise funds for a societal and/or environmental cause, fund-raising events at the community level can be organized to increase the fund-raising efforts. Also, real-time updates, e.g., every day how much amount has been raised for a month-long fund-raising campaign can be updated through different digital channels, which is an example of social media reporting that (e.g., the initially raised funds) can attract further interest and engagements from more audiences. Also, the momentum that could be created based on the initial viral effect can be amplified through traditional media coverage and stakeholder engagement (e.g., environmental activist groups).

Trendjacking is an art of inserting a brand message into popular conversations by creating content around those topics or events to connect with your audience and draw attention to your products and services.

Fifthly, evaluation of a sustainable viral marketing campaign is important to understand its impact and to determine future action plan. During and after a sustainable viral marketing, promotion of customers' and non-customer stakeholders' positive reviews and testimonials can prolong the impact of the campaign, in order to drive a long-lasting impact. Also, at the end of the campaign, analyzing the societal or environmental impact that has been generated by the campaign, reflecting on the initial campaign goal can be helpful to work for future actions to drive long-lasting impact.

Marketing in Action 6.2. Black Friday Donations and Planet or Plastics Viral Marketing Campaigns

Source: ▶ pexels.com/royalty-free-images/

The National Geographic

"The National Geographic Society was founded in 1888…Today, National Geographic continues to pursue its mission 'to increase and diffuse geographic knowledge while promoting the conversation of the world's cultural, historical, and natural resources'" (np). As part of their societal and environmental commitment and contribution, National Geographic creates awareness and calls for action, related to the "most pressing issues our planet and the animals and people on it have faced."

The tagline of one of their environmental campaigns in June 2018 is Planet or Plastic? The digital content of this campaign:

» Shows an image of what appears to be an iceberg but is actually a plastic bag. Part of the campaign is National Geographic's commitment to reduce their use of plastic by wrapping their magazine in paper instead of plastic for their UK, U.S., and Indian customers. This is estimated to save in excess of 2.5 million single-use plastic bags every single month. (np)

The National Geographic followers on social media went crazy for this campaign which attracted 881 million impressions across the social media platforms. The campaign was viral through digital channels and stimulated the National Geographic's community of 760 million people to take action. Certainly, the sustainable viral marketing campaign not only created the awareness about the environmental concern related to plastic and inspired people to take action, but also attracted the attention of National Geographic's target audience about its market offerings.

Patagonia

Patagonia is an American outdoor clothing brand, which was officially launched in 1973. Yvon Chouinard is its founder, who's "first test of environmental stewardship came years before he founded Patagonia." For example, in 1957, he learned how to make more environmentally friendly reusable pitons for rock climbing. His pitons were firmer and stronger than other existing ones and better suited for climbers for removal. Yvon Chouinard's passion for environment is reflected on Patagonia's operations. Among other socially and environmentally caring initiatives of Patagonia, their 2016 tweet for a Black Friday sell went viral. Patagonia tweeted that:

> This Black Friday we're donating 100% of sales to grassroots environment groups. #LoveOurPlanet. The result was phenomenal, with this tweet receiving 1000% more engagement than their other posts. Not only that, but the company ended up far surpassing their sales projection of $2 million. Instead, they achieved $10 million in sales on Black Friday that year.

The socially and/or environmentally caring sustainable viral marketing campaign helped Patagonia to not only establish a sustainable brand image, but also to underpin their commercial success. As a result, "Patagonia become (one of) the most reputable brand in the United States."

Information sources: Calamendrei, P. (2019). 3 companies that implemented viral marketing with their green initiatives. *Cargo Compass*. Retrieved May 12, 2024, from ► https://cargocompassworld.com/en/blog/3-companies-viral-marketing-green-initiatives/; National Geographic. (2024). Jan 13, 1888 CE: National Geographic Society founded. Retrieved May 12, 2024, from ► https://education.nationalgeographic.org/resource/national-geographic-society-founded/; Patagonia Brand Timeline. (2024). The story of one of the world's most active and responsible brands when it comes to caring for people and the planet. Retrieved May 12, 2024, from ► https://www.fatbuddhastore.com/patagonia-timeline-i278#; Sundheim, D. (2023). How Patagonia become the most reputable brand in the United States. *Forbes*. Retrieved May 12, 2024, from ► https://www.forbes.com/sites/dougsundheim/2023/12/12/how-patagonia-became-the-most-reputable-brand-in-the-united-states/.

Omnichannel Marketing Omnichannel marketing refers to an organization-wide effort across all departments of an organization to timely communicate with all internal and external stakeholders through all applicable traditional and contemporary channels from where stakeholders obtain information about the organization, so that the organization and its stakeholders both remain up-to-date about the organization's value proposition related to its marketing mix. For example, what are the product features, what its price is, from where (online or traditional shop or both) customers can purchase it, what is the core selling point (promotional message) to satisfy customers' needs, etc. Cui et al. (2021) defined omnichannel marketing as:

> Synergistic management of all customer touchpoints and channels both internal and external (and traditional and online touchpoints) to the firm to ensure that the customer experience across channels as well as firm-side marketing activity, including marketing mix and marketing communication (owned, paid, and earned), is optimized for both firms and their customers. (p. 104)

Here synergistic management is important, considering that a collaborative effort of all team members across different departments (e.g., marketing, sales, corporate office, HRM, finance, etc.) of an organization is a prerequisite to ensure that the organization is able to timely communicate with all its internal and external (customer and non-customer stakeholders) with its current value proposition, related to its marketing mix. Such communications occur via relevant traditional and contemporary media, such as newspapers, television, radio, email, website, social media, etc.

Omnichannel marketing refers to an organization-wide effort to timely communicate with all internal and external stakeholders through the traditional and contemporary channels, from where stakeholders obtain information about the organization, so that the organization and its stakeholders both remain up-to-date about the organization's value proposition related to its marketing mix.

The communication channels, whether it is a brand's website, social media platforms, newspaper ads, or its sales personnel, from where the brand's customers obtain information about the brand, are known as customer touchpoint. Customer journey, e.g., the entire process of customers to obtain information about a product, and to purchase and consume it can be mapped based on different customer touchpoints. Customers develop their experience of a brand through different customer touchpoints, which constitute a customer's overall experience of the brand. Therefore, planning, designing, and monitoring customer touchpoints are important to enrich customers' positive experience about a brand.

- **Customer Touchpoints**—The traditional and contemporary communication channels, whether it is a brand's website, social media platforms, newspaper ads or its sales personnel, from where the brand's customers obtain information about the brand, is known as customer touchpoint.
- **Customer Journey**—The entire process of customers to obtain information about a product, and to purchase and consume it can be mapped based on different customer touchpoints. The different customer touchpoints (e.g., a restaurant website, or having a dinner at the restaurant) that can be visible on the map collectively constitute the customer journey with the product.
- **Customer Experience**—Customers develop their experience about a brand through different customer touchpoints, which constitute a customer's overall experience of the brand.

► Section 6.3.3 discusses more about digital touchpoints and customer (digital) journey mapping to nurture customer experience.

6.3.3 Digital Touchpoints, Customer Journey mapping and Customer Experience

◘ Source: ▶ pexels.com/royalty-free-images/

Customers' digital journey can originate and progress through the digital touchpoints that customers would have while interacting with a brand through its digital channels to obtain information about the brand. The digital touchpoints would include:

- Digital presence of a brand
- Further influence of digital channels
- Digital transaction
- After-sales service through digital channels
- Digital marketing analytics to increase customers' share of wallet (loyalty)

In general, all aforementioned digital touchpoints collectively constitute customers' digital journey and influence customers' digital experience with a brand; these digital touchpoints can be considered to map customers' digital journey. ◘ Figure 6.6 illustrates customers' digital journey mapping to demonstrate how customers and marketers interact, and marketers collect and analyze customer data through different digital channels, with an aim to ensure a flawless customers' experience.

Digital Presence of a Brand In general, brands today convey their message about marketing mix (societal and environmental commitment and contribution, product features, price ranges, information about distribution channels, promotional appeal, etc.) through different digital channels. Such digital channels include websites, different other web contents (e.g., blog and vlog), email, social media, web commercials, digital apps, electronic billboards, among others. Customers obtain information about a brand through these digital touchpoints. Also, different web-based search engines, e.g., Google, Bing, Yahoo!, Yandex, etc., provide information to customers about competitive brands. Key priorities for sustainable marketers to design their digital platforms include (1) green web hosting to reduce their brand's and customers'

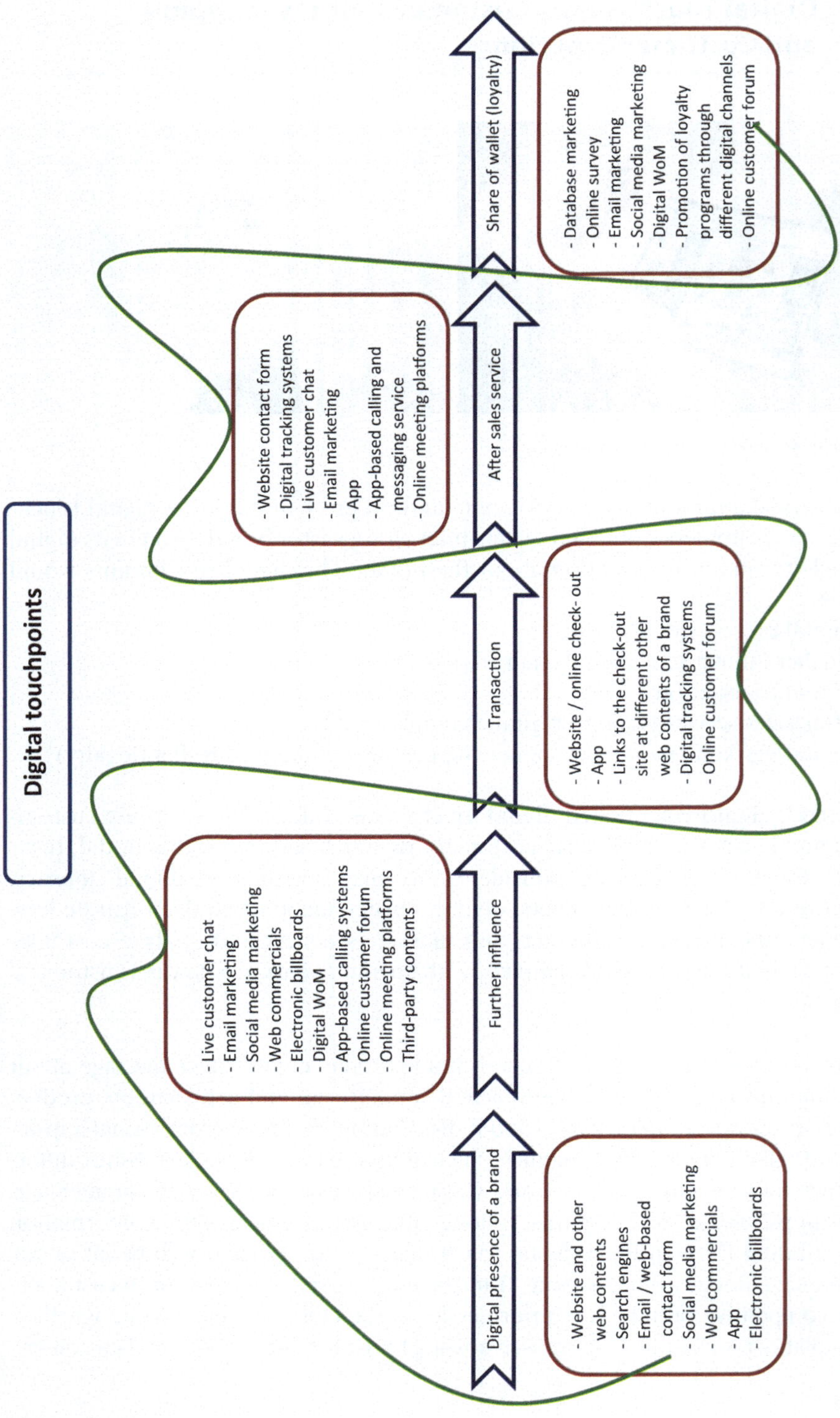

Fig. 6.6 Customers' digital journey mapping—Digital touchpoints of customers with a brand that influence customer experience

environmental footprint via digital channels, (2) designing the web contents that would be user-friendly, and (3) sharing information about the triple bottom line's stakeholders. For more discussion on green web hosting for a sustainable future, please, see the Opening Vignette of this chapter. In terms of user-friendliness of web platforms and contents, the web platforms and their contents should be easy for customers to download, browse, and navigate across different web links. In terms of sharing information about the triple bottom line's (People, Planet, and Profit) stakeholders, the web contents should provide information about the key selling points (e.g., why doing business with your brand would be more prolific for your customers, instead of making business with your competitive brands), and your brand's commitment and contributions to society and environment.

Further Influence of Digital Channels The digital touchpoints that are discussed under the above paragraph of "digital presence of a brand," these digital channels would have further consequences following an initial interaction between customers and brands. For example, an initial email from marketers as part of email marketing could lead to a series of emails over an extended period, especially for business-to-business (B2B) or industrial marketing, e.g., selling cleaning products in large quantity to a business buyer, e.g., a university. Similarly, social media marketing can attract auspicious or inauspicious digital word-of-mouth (WoM) for a brand, depending on its customers' extent of satisfaction and dissatisfaction with the brand's promise and its actual delivery to satisfy its customers' and non-customer stakeholders' needs.

In addition, different other web services could have further influence on customers' digital experience with a brand. For example, live chat option for customers is a web service to obtain instant support from a brand. Many brands offer this live chatting option to their customers; however, sometimes, we see that some brands do not reply to the live customer inquiries at all, or reply after a few days, and by that time, the customer may have already purchased from a competitive brand. However, the purpose of such live chatting option is to provide instant replies to the customers' inquiries. It means, if the resources of your brand do not permit to operate an effective live chatting option for your customers, it is better to not offer it as an ornamental option, considering that it can bring customers' dissatisfaction if it is not being operated properly.

Online customer forums (where customers share their satisfactory or dissatisfactory remarks and clarify why they are satisfied or dissatisfied) are another digital touchpoint for customers to obtain information about competitive brands. Such online customer forums could be owned by a brand, or it could be owned by a third party. For example, Airbnb runs its own online customer forums, which is a third-party digital platform for their suppliers, i.e., the hotels and other small accommodation service providers who offer their service through Airbnb. Examples of other third-party online customer forums include Reddit, Trustpilot, Quora, REVIEWS.io, Tripadvisor, etc., where customers share their positive and negative remarks about a brand. Such digital WoM as a digital touchpoint of customers also can influence customers' experience of a brand. Different app-based meeting, calling, and messaging service platforms, such as Microsoft Teams, Zoom, WhatsApp, or Messenger are used by different brands, especially for B2B marketing to communicate with customers. Therefore, these digital platforms also have implications for customers' digital experience. Other third-party contents, such as enthusiastic or unenthusiastic news about different brands (e.g., a brand's societal or environmental contribution, or a tech-giant's data breach scandal) that are published by different mainstream newspapers and online news portals through their digital channels, are also another digital source for a brand's customers to develop their digital experience about the brand.

Digital Transaction The digital transaction platforms (e.g., website or app) where customers place an order and pay to complete their purchase can influence customers' digital experience too. The features of such digital transaction platforms, such as placing order, specifications (type and quantity of product), shipment address, digital transaction option, link to the third-party digital tracking system for customers' order delivery, should be user-friendly. The terms and conditions for tracking an order and return policy should be clarified in a simplistic manner. The post-purchase (actually the post-delivery) option for customers to share their remarks about a product that was purchased online should be stated clearly to encourage customers to share their remarks about the product. The link of such digital transaction platforms may appear in other digital channels, e.g., a brand's social media, email or in digital messaging service. All these interrelated digital channels should be synchronized properly to offer a flawless experience to customers.

After-Sales Service Through Digital Channels All digital channels, which are discussed in this section are generally used to support customers further for after-sales service. For example, the digital tracking system should provide as accurate information about the progress of customer order delivery as possible. Similarly, web-based contact form, live chat option, email, app, app-based meeting, or messaging options can also be used to communicate for after-sales service. The key focus for after-sales service should be ensuring customers' peace of mind. Customers' peace of mind could be referred to as ensuring that customers will not have any concern to obtain and consume a product or service, exactly as per the specifications (e.g., product features and quality) and utilities (e.g., delivery options) as it was promised to the customers during the pre-purchase stage.

> **Peace of Mind**—Customers' peace of mind could be referred to as ensuring that customers will not have any concern to obtain and consume a product or service, exactly as per the specifications (e.g., product features and quality) and utilities (e.g., delivery options) as it was promised to the customers during the pre-purchase stage.

Digital Marketing Analytics to Increase Customers' Share of Wallet (Loyalty) The purpose of digital marketing analytics is to increase the share of wallet per customer through the digital channels. Customers' share of wallet refers to an individual customer's total spending to purchase a particular brand in her or his lifetime, in comparison to the customer's purchase of other competitive brands throughout the customer's entire life. Marketers primarily aim to increase the percentage of each customer's share of wallet, because, if share of wallet of customers could be increased, the total market share of a brand will increase, eventually.

> **Share of wallet** can be defined as the percentage of a particular customer's total spent to purchase a specific brand in their entire lifetime, in comparison to the total spending to purchase other competitive brands in the lifetime.

To increase customers' share of wallet, a popular strategy is database marketing. ▶ Section 6.4 discusses database marketing in more detail. Digital marketing research and digital marketing analytics aim to develop insights through digital channels on customers' and non-customer stakeholders' needs, wants, and expectations, in order to customize product and service to meet customers' expectations as accurately and competitively as possible. A communication tool for traditional marketing research is inviting customers via postal mail to respond to a survey questionnaire. Similarly, email marketing, social media marketing and other digital channels are used to communicate with customers to respond to an online customer survey. Furthermore, social media data, data from online customer forums, digital WoM are analyzed to understand customers' needs more accurately, in order to offer the best possible value proposition to them, with an aim to increase customers' loyalty to increase their share of wallet. Insights from such digital marketing analytics are also promoted through different digital channels to influence existing and potential customers further.

> **Digital Marketing Research and Digital Marketing Analytics** aim to develop insights through digital channels on customers' and non-customer stakeholders' needs, wants, and expectations, in order to customize product and service to meet customers' expectations as accurately and competitively as possible.

» With a major chunk of the world's population relying on the internet for their day-to-day operations and spending, digital marketing is rapidly gaining prominence… The integration of artificial intelligence and cloud-based platforms has also contributed to the market's growth. These platforms not only make operations easy and fast but also provide reliable recommendations (i.e., based on digital marketing analytics) to business owners about their marketing campaigns. (Digital Marketing Analytics Market, 2024, np)

Because of the effective application of digital marketing analytics, its global market has been growing steadily. The digital marketing analytics global market valued 6.31 billion in 2024, which is estimated to grow to 11.54 billion in 2029, and the largest market is North America; however, the fast-growing market of marketing analytics is Asia Pacific (Marketing Analytics Market Size, 2024). The major share of the global marketing analytics software comes from the consumer goods, media and communication, retail, industrial products, healthcare service, and the banking and financial industries (Market Size & Trends, 2024). ◘ Figure 6.7 shows the percentage of the global market share of the leading web analytics technologies in 2023.

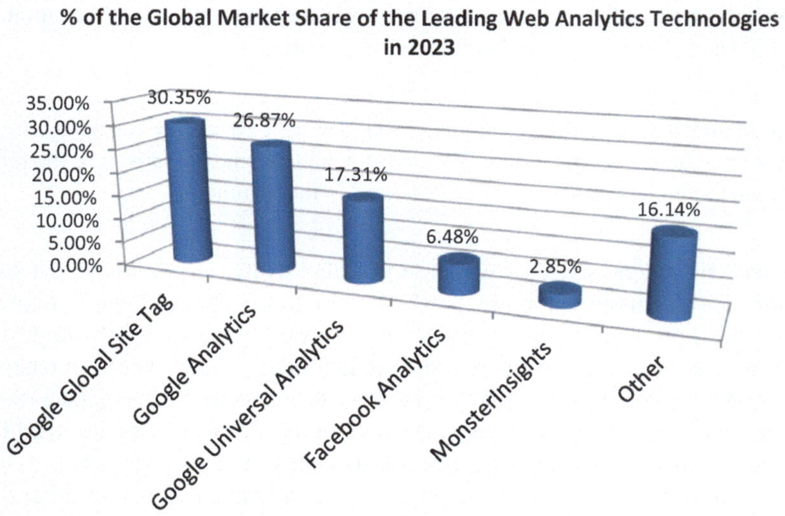

Fig. 6.7 Percentage of the global market share of the leading web analytics technologies in 2023 (Vailshery, 2023)

6.4 Database Marketing and Other Embryonic Technologies

Source: ► pexels.com/royalty-free-images/

Database Marketing The purpose of database marketing is to increase the share of wallet of each individual customer. We see a common database marketing practice in the supermarket industry. Database marketing develops insights into customers' spending and consumption habits to predict cross-selling and up-selling opportunities. For example, a customer regularly purchases from a particular supermarket and pays by an electronic channel, e.g., by credit or debit card at electronic funds transfer at point of sale (EFTPOS). The supermarket analyzes the customer's historical purchase record to understand the customer's total spending each time, the items the customer purchases regularly, the frequency and amount of purchase each time, etc. If the customer purchases pet food regularly, the supermarket would have a cross-selling opportunity to convince the customer to purchase toys or security devices for their pet. Also, based on analyzing the buying behavior, the supermarket would have an upselling opportunity if they are able to convince the customer to purchase more pet food than the customer used to purchase each time, e.g., by offering little discount or offering better quality food at the same price. In this context, cross-selling can be defined as selling another relevant item (e.g., car insurance), while selling the main item (e.g., the car). Up-selling can be defined as convincing customers to purchase the same product in more volume, in order to increase the product's overall sales. The **purposes of database marketing** would include:

- Trend analysis: To understand the current and embryonic changes in customers' preferences based on analyzing their purchase history.
- Forecasting: Based on monthly or yearly sales data, marketers try to forecast the product demand in upcoming months. Analyzing the ups and downs of sales figure in last 12 months can be helpful to predict when in future the sales will increase or decrease. For example, in the last 12 months, three international events, e.g., FIFA World Cup, or other smaller regional events took place during month 2, month 7, and month 11. As these international events attracted many international tourists, the sales during those months increased. Therefore, if any international event is scheduled in a forthcoming month, it would mean that the product demand would increase in that month in future, so that the production, sales, and delivery teams should also be alerted to prepare for this prospective increase in product demand in that particular month. Similarly, the demand for example, for cold beverages (e.g., ice-cold Coke or Pepsi) could increase during the summer months.
- Segmentation: Segmenting the total market (i.e., all customers) into different segments. For example based on different age groups, e.g., younger customers (1–18 years old), young customers (18–35 years old), middle-aged customers (36–55 years old), aged customers (56–75 years old) and older-age customers (76 years or older), marketers aim to understand different age groups' needs, wants and expectations more closely. Also, marketers further segment an already segmented market which is known as market niching or niche marketing, in order to understand the needs and preferences of each

smaller group of customers more accurately. For example, within the younger customer segment (1–18 years old), the need of a 6-year-old male or female child would be different than the need of a 12-year-old male or female child. Therefore, marketers subsegment one of the (or all of the) broader segments, for example, the younger customer segment further, e.g., into 1- to 5-year-old female subsegment or 6- to 10-year-old male subsegment; however, the entire market has segmented into five (younger, young, middle-aged, aged and older-age customer) broader segments by age-groups (Kotler et al., 2009; Kurtz, 2009; Shams, 2013). In this context, Kotler et al. (2009, p. 228) described a "niche is a more narrowly defined segment of a market or customer group seeking a distinctive mix of benefits" (as cited in Shams, 2013, p. 221).

- Information sharing: Based on analyzing customers' data and their historical buying patterns, marketers responsively share information with their team members across different departments, e.g., sales team, product design team, engineering team, and supply chain team, with an aim to ensure that all team members from their own contexts, can contribute to serving their customers reflecting on the customers' needs, wants, and preferences.
- Cross-selling and up-selling opportunities: Based on the supermarket example at the beginning of this current section, we already have discussed that exploring the cross-selling and up-selling opportunities based on analyzing customers' buying behavior is another purpose of database marketing.
- Targeted communication and personalization: Database marketing is also helpful for targeted communication with customers and personalizing product or service as per the specification of an individual customer. For example, upon analyzing the purchase history, customers could be alerted (e.g., via email, social media, in-store leaflet, and other channels) when a better-quality product or an additional service or a discount would be available for their desired products.
- Increasing the customers' share of wallet to drive loyalty: The eventual purpose of database marketing is keeping customers in an ongoing loop of customer-brand relationship dynamic, in a way that would be a win-win for both customers and a brand. For example, based on analyzing customers' buying behavior, if a brand is able to offer its target market the right product, based on the right price at the right time and deliver it through the right channel, and if customers would realize that the brand offers most competitive product, in this case, the customers would purchase frequently from that particular brand. It can ultimately increase each customer's share of wallet for the brand to drive customer loyalty.

Database marketing is defined by De Bock (2016, p. 1) as "the use of customer databases to enhance marketing productivity through more effective acquisition, retention and development of customers." This definition of De Bock clarifies the purpose of database marketing. Chaffey and Ellis-Chadwick (2019) also shared a definition of database marketing that highlights its function:

> The process of systematically collecting, in electronic or optical form, data about past, current and/or potential customers, maintaining the integrity of the data by continually monitoring customer purchases, by enquiring about changing status, and by using the data to formulate marketing strategy and foster personalized relationships with customers. (As cited in Ecommerce Forum, n.d., np)

Besides customers' data, a database marketing campaign can include data about other key stakeholders such as suppliers, employees, or the relevant societal or environmental activist groups to synchronize with the expectations of these key stakeholders. Another key consideration of database marketing is the responsible use of customer and other non-customer key-stakeholder data. In this context, we define database marketing is as responsively collecting, storing, and analyzing customers' and other key non-customer stakeholders' data to develop insights on their needs, wants, and expectations to customize market offerings for customers and other key stakeholders, in order to drive customer and stakeholder loyalty. More discussion on the responsible use of data is shared in ► Sect. 6.7 (digital marketing, marketing ethics, and regulations).

- **Database Marketing**—Responsively collecting, storing, and analyzing customers' and other key non-customer stakeholders' data to develop insights on their needs, wants, and expectations to customize market offerings for customers and other key stakeholders, in order to drive customer and stakeholder loyalty.
- **Cross-Selling**—Exploring opportunity to convince customers to purchase another relevant item (e.g., a car insurance) while selling the main item (e.g., the car).
- **Up-Selling**—Exploring opportunity to convince customers to purchase the same product in more volume, in order to increase the product's overall sales.
- **Niche** is a more narrowly defined segment of a market or customer group seeking a distinctive mix of benefits.

A database marketing campaign would have four operational **stages**:
- Database design: What data do we need? How can we get the data? How frequently we can collect the data? Where we can store data? How long we will store it? What are the analysis methods to retrieve value from data? Are we complying with data privacy ethics and law?
- Data collection: Is there any additional cost to collect the data?
- Data mining: What does the data tell us (i.e., the value of data)?
- Data cleansing: Is the data current and authentic?

Answering these questions is instrumental for effective database marketing. At these operational stages, considering the **5 Vs of data** is important (Yin & Kaynak, 2015; Bello-Orgaz et al., 2016; Rumsfeld et al., 2016; Shams & Solima, 2019):

- Data **Volume**: The amount of data each time we may be able to collect.
- Data **Variety**: The extent of variety in data each time we collect, e.g., customers' spending each time, types of purchases each time, when they purchase (e.g., at a particular month or at a particular day or at particular time of a day, etc.).
- Data **Velocity**: How frequently we would be able to collect new data.
- Data **Veracity**: Whether the source of data and the data itself are authentic or not.
- Data **Value**: Retrieving the value (for example, insights about customers' preferences) from raw data to influence customers based on customized offering relevant to customers' preferences.

◘ Table 6.3 describes the key considerations for database marketing.

Other Embryonic and Fuzzy Technologies Since the 1970s, the access to data and information has increased tremendously for mass people around the world, which is termed as knowledge democracy (Graham, 2003) to demonstrate individuals' rights and responsibilities to access and use data and information. The information revolution and the subsequent ease of using the World Wide Web, internet, and many other already existing and embryonic digital technologies and platforms enable the extraordinary inflow and outflow of data between data senders and receivers. These digital platforms and other embryonic digital technologies, such as big data, artificial intelligence (AI) and AI-supported robotics, virtual and augmented realities, cloud computing and advanced web analytics, blockchain, internet-of-things, among others, have implications, both benefits and challenges for marketing dynamics. For example, **big data analytics** help marketers to understand customers' and other stakeholders' needs, wants, and expectations, insights on market trends, and envisage the influence of different other societal, environmental, geopolitical, and legal issues on the changes in market dynamics.

Knowledge Democracy—Individuals' rights and responsibilities to access and use data and information.

The challenges of big data analytics would include exploring accurate data from authentic sources, and managing data velocity effectively to produce value out of raw data in a way that would be meaningful, even for non-expert data users. To simplify the big data management process, marketers should (Bikakis et al., 2018; Couture, 2018; Shams, 2024):

Table 6.3 Key considerations for database marketing

Goals for triple Ps' stakeholders	Strategies (Tapp, 2008)	Types of data	Few data sources	Information on databases	Examples of data	Analysis methods, ethics, and laws
Triple bottom line (i.e., People, Planet, and Profit): The type of insights that would be derived from a database to contribute to the triple bottom line, e.g., how a product can be developed in a way that would be more environmentally friendly by understanding customers' preferences for more socially caring and/or ecologically responsible products	- Data that is essential now - Data that can have value in future - Ensuring ease of the process of data collection and analysis - Costs must be offset by benefits - If you will not use it, don't collect it	- Behavioral data (e.g., frequency of purchase) - Volunteered data (e.g., customers' response on social media) - Profile data, e.g., linking insights from an existing database with other sources', e.g., customers' in-store response - Attributed data derived from market research	- List purchase - Electronic payments - Subscription - Questionnaire responses - In-store offer details - Requests for information - Responses at events and promotions (e.g., social media) - Other established channels (e.g., email, telephone, or purchasing raw data from third party)	- Customer contact and buying history - Buying behavior (e.g., frequency, items purchased, spent each time etc.) - Customer feedback - Market trend - Customers' reactions to campaigns and communications - Brand's reaction to customers' response - Customers' frequency of online access, e.g., a brand's website - Time spent online each time - Which time, what days and from where customers access to a brand's online platforms - What type of access, e.g., the type of information they search online on a brand's website - Profitability per customer (share of wallet)	- Customers' age, income, gender, ethnic groups and other ethnographic data, e.g., birthday (for personalized message), family unit, locations, interests and hobbies, etc., which are generally collated in groups, e.g., age groups or income level - Behavioral data, volunteered data, profile data, attributed data as described under the "types of data" column	- Qualitative, quantitative or mixed method - Complying with general ethical considerations to collect, store, use and abolish data - Complying with General Data Protection Act

- Use common business terminologies for data mining that could permit varied groups of stakeholders with diverse backgrounds to easily understand the meaning of big datasets. This is important, especially when big data and its original meaning are shared with cross-functional teams comprising non-IT experts.
- Streamline data inflow and outflow management to recognize the source of data and its relevance for different management levels and functional team members, such as front-desk customer service representatives, mid-management, and top management.
- Ensure the ease of interactive data exploration process for large and dynamic sets of volatile raw data. These processes should be focused on stable and simplified methods of information abstraction, sampling, and insight extraction, enabling non-expert data users to address current problems readily.
- Ensure that big data analysis software is user-friendly for different non-expert data users. It will confirm non-expert data users' access to the "right data, at the right time, in the right context" based on practical business and marketing needs.
- Empower different non-expert data users (e.g., customers, investors, customer service attendants, and other key stakeholders) to gain accurate insights and value from the ongoing big data flow as quickly as possible, minimizing reliance on IT experts.
- These patterns should be scalable and offer customized insights based on common business or management terms, minimizing the need for IT expertise and ultimately allowing the wider society to benefit from big data. This is particularly critical for the socioeconomic development of emerging economies, where the population may not be as tech-ready as in developed nations.

AI and AI-enhanced robotics, and virtual and augmented realities can influence brand loyalty. For example, in the case of AI use, specifically Siri, the extent of customers' trust on AI, its ease of use and interaction with users, perceived risk, and novelty in user experience can influence brand loyalty (Hasan et al., 2021). Similarly, **internet-of-things** (IoT) are used to produce optimized products or services. For example, a number of digital cameras capture photos from different corners and angles at a stadium that is full of 100,000 people, and IoT helps almost instantly to analyze all alternative photos to identify or produce the best possible photo to display at the stadium's giant screen. More discussion on responsible use of these embryonic and fuzzy technologies is shared in ► Sect. 6.7 (digital marketing, marketing ethics and regulations).

6.5 Internet-Savvy Firms Shorten Product Development and Marketing Cycle[2,3]

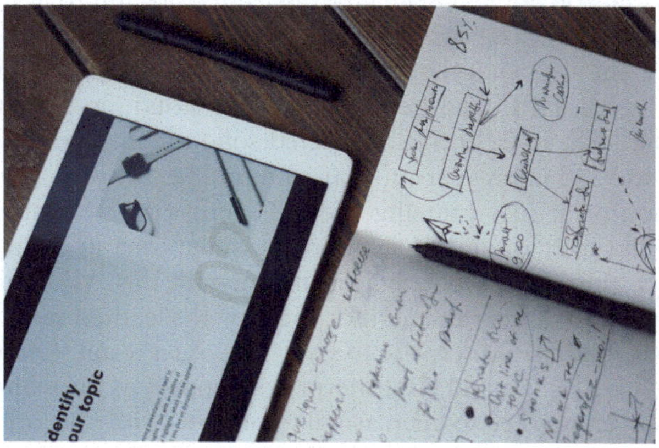

■ Source: ► pexels.com/royalty-free-images/

Marketing strategists often assume that marketers should first create a plan on the basis of a careful review of environmental and firm information and then execute that plan. However, in some cases, these two events occur simultaneously. In the article "The Convergence of Planning and Execution: Improvisation in New Product Development," published in the *Journal of Marketing*, authors Christine Moorman and Anne S. Miner define such convergences as "improvisation." They investigated the conditions in which improvisation is likely to occur and be effective.

The results show that organizational improvisation occurs moderately and that the frequency of improvised product development decreases as long as the com-

2 ► Section 6.5 (Internet-savvy firms shorten product development and marketing cycle) is sourced from Czinkota et al. (2021), and reprinted based on the permission of copyright owner, © Springer).

3 Information sources of ► Sect. 6.5: Moorman, C., & Miner, A.S. (1998). The Convergence of Planning and Execution: Improvisation in New Product Development. Journal of Marketing, 1–20; "Improvisation Can Help New Product Development," Marketing News (July 20, 1998): 42; Degenholtz, A. (1999, March 29). E-Commerce Fueling the Flame for New Product Development. Marketing News, p.18; Burrows, P. (2003, June 23). Innovation, Lego-Style. Business Week, p.74–76; Overgoor, G., Chica, M., Rand, W., & Weishampel, A. (2019). Letting the computers take over: Using AI to solve marketing problems. California Management Review, 61(4) 156–185; Brossard, M., Erntell, H., & Hepp, D. (2018). Accelerating product development: The tools you need now. McKinsey & Company. Retrieved June 29, 2020, from ► https://www.mckinsey.com/business-functions/operations/our-insights/accelerating-product-development-the-tools-you-need-now#; Quicke, S. (2020). Dell looking for server growth. MicroScope. Retrieved July 24, 2020, from ► https://www.computerweekly.com/microscope/news/252479324/Dell-looking-for-server-growth; Singh, P. (2019). Dell, HPE, Lenovo lead as cloud server market surges to a US$86 Billion in 2018. CounterPoint. Retrieved July 24, 2020, from ► https://www.counterpointresearch.com/dell-hpe-lenovo-lead-cloud-server-market-surges-us86-billion-2018/.

pany keeps good track of past product development experiences and increases as market uncertainties and turbulences increase. Although improvisation may reduce new product effectiveness, environmental and organizational factors can reduce negative effects and sometimes create a positive effect for improvisation. This suggests that, in some contexts, improvisation may be not only what organizations actually practice but also what they should practice in order to flourish.

For example, Dell, HPE, and Lenovo led as the cloud server market surged to US$86 billion in 2018. In 2019, Dell delivered solid numbers with fourth-quarter revenue up 1% to $24 billion year-on-year, and for the full year a 2%. In fact, the growing adoption of artificial intelligence (AI) across different applications is in parallel driving the need for greater server processing power and thus semiconductor capabilities within the servers. Synchronizing with this AI-dominated market trend analysis appears as instrumental for some of the leading server manufacturers like Dell EMC and HP Enterprise to grow fast. As a result, following the market trend enabled Dell EMC and HP Enterprise to appear as the major competitors in the global server market to increasingly compete with the other traditional major players such as IBM. In 2018, Dell EMC and HPE, both earned 16% market share in terms of total revenue. Electronic commerce puts forward-thinking companies, such as Dell EMC, on the leading edge of consumer need identification, helping them to improve new product development and marketing processes. Given the speed and ease of information gathering on the internet, market intelligence analysis and product development planning proceed increasingly in tandem. The internet has both upstream and downstream benefits in the corporate value chain. The two most important "downstream" benefits of online marketing are the ability to collect behavioral sales data rapidly and the consumer intimacy that stems from that. Behavioral data allow marketers to develop relevant marketing campaigns targeting their most valuable customers immediately. Consumer intimacy allows marketers to beat traditional manufacturers to market with successful new products. The important "upstream" benefit of online marketing is as the information tool that links an original equipment manufacturer (OEM) with suppliers, distributors, and contract manufacturers to reduce cost, eliminate waste, and speed up new product development.

In order to respond to the market competition, product developers have started to abridge the overall product development process with the help of internet-enhanced digital technologies. Firms across industries and sectors underpin their information technology architectures and capabilities to track consumer behavior more accurately, in order to adopt agile ways of serving consumers. Since competitors are not waiting; product developers need to accurately track consumer behavior and understand their target market's changing needs promptly, in order to accelerate the product innovation process, so that new product could be launched before the competitors launch a product to satisfy those changing needs of the target market. Utilizing artificial intelligence (AI), Amazon's market data mining forecasting method enable them "to anticipate product demand far enough in advance," which is an example of accurately and promptly exploring consumer behavior through data mining. "In the near future, AI is expected to make marketing more efficient by speeding up the decision-making process and providing marketing managers with information and insights that they could not develop in any other way."

The internet-enhanced digital platforms opened the gates for manufacturers to sell directly to consumers. Many argue that e-commerce is less intimate than the traditional retail, but it actually provides more targeted demographic and psychographic information. Manufacturers that traditionally sell through the retail channel benefit the most from e-commerce. Furthermore, customer information no longer is held hostage by the retail channel. Instead, marketers have access to robust behavioral data that allows them to segment best customers and identify trends and customer needs. Most importantly, the enormous flow of market data allows the development of relevant marketing messages aimed at important customers and initiating loyal relationships. Clearly, more interactive, improvised product development and marketing processes are increasingly becoming a mainstay marketing strategy, which further shortens the product life cycle.

6.6 Tech-readiness of Your Target Market and Projection of the Global Digital Market Growth

Source: ► pexels.com/royalty-free-images/

Tech-readiness could be referred to an individual's ability to safely use contemporary technologies, in order to be benefited by using the technology. An example of tech-readiness relevant to marketing dynamics would be to consider, whether a customer or a group of customers in a particular market or country can use the self-serviced vending machine to complete a purchase, e.g., to purchase a cola or to produce and purchase their own passport-size photos. Before entering into a market through digital channels, understanding the extent of tech-readiness of that market is important for marketers. Otherwise, a firm, for example, could invest a lot of resources to establish self-serviced vending machines throughout a country or communicate with its customers in a country mainly via digital channels, the vending machines remain unused or the brand message could be inaccessible by its target audience, if the target population is not tech-ready. For example, the countries where internet penetration rate is very low, it is highly likely that the digital channels would not be the most lucrative communication medium there for a brand to communicate with its customers.

Tech-readiness could be referred to an individual's ability to safely use contemporary technologies, in order to be benefited by using the technology.

◘ Figure 6.8 presents internet penetration rate data across different global regions that gives an indication of the current progress of tech-readiness of the global market. Similarly, analyzing the internet penetration rates of a particular country can be useful for projecting the overall digital market size and its potential growth. For example, Bangladesh has a population of 174.6 million people (Worldometers, n.d.). A May 2024 report demonstrates that in Bangladesh, internet subscribers

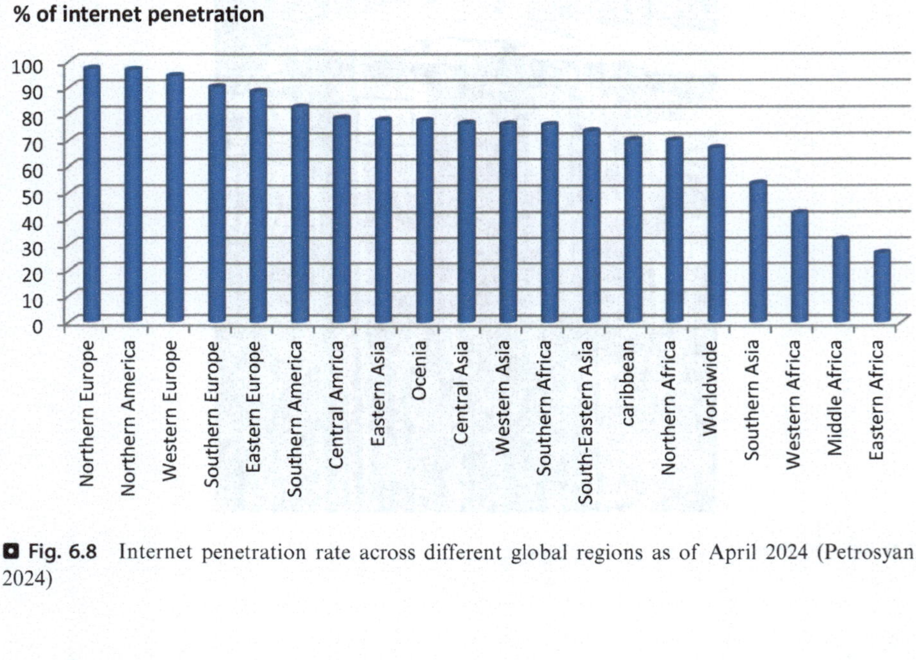

◘ **Fig. 6.8** Internet penetration rate across different global regions as of April 2024 (Petrosyan, 2024)

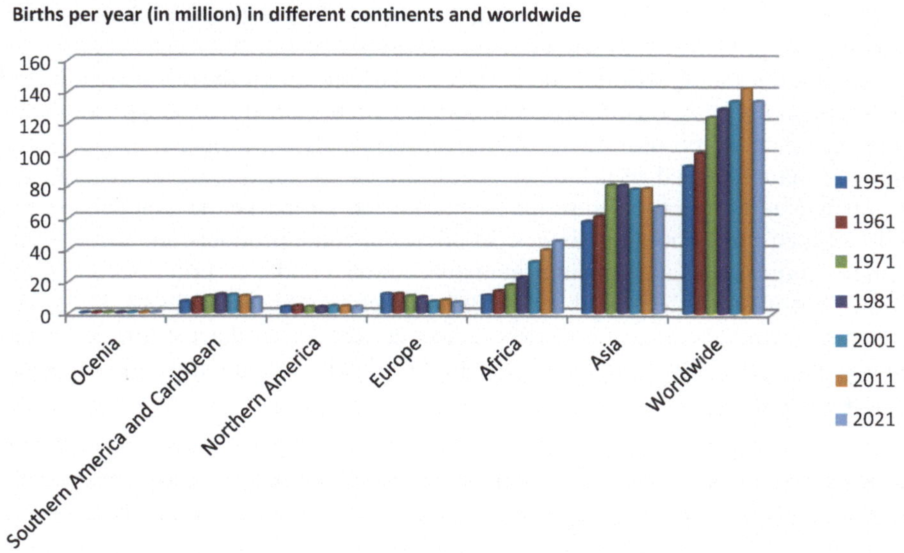

Fig. 6.9 Births per year (in millions) across different global regions (Our World in Data, n.d.)

have been increasing for two consecutive months. In April, the number of internet users increased by 4.3 million, and in February, the increase was 1.1 million internet subscribers. The total number of internet users in Bangladesh as of May 2024 is 134.7 million (Prothom Alo, 2024). This data indicates that the tech-readiness of the Bangladesh country market is increasing, with more than two-third of the population having internet access at least for basic purposes.

In general, teenagers are more tech-savvy than older adults. Now, if you are a marketer of an online streaming service provider, you would be interested to understand the potential market size of the teenage segment for your online streaming service, considering that the teenage segment would be one of the most lucrative segments.

Analyzing the births per year data across different global regions can be useful to predict the size of the teenage segment of different region-specific markets and the total global market for your online streaming service. For example, Fig. 6.9 shows that globally we had 142.13 million births in 2011 and 133.97 million births in 2021. It means an estimated number of 16-year-old people (or the 16-year-old market segment size) worldwide would be around 142.13 million in 2027, which would reduce to around 133.97 million in 2037.

6.7 Digital Marketing, Marketing Ethics, and Regulations

Marketing ethics can be defined as marketers' moral principles that guide them to plan, implement, and monitor marketing activities in a way that will not be unjust for individuals and society while influencing their target audiences. For example, payday loans are expensive as often the payday lenders charge high interest rate, and it can be predatory for customers to get trapped in a cycle of debt. As a result, it is not good for an individual and society as a whole. Similarly, other socially and environmentally irresponsible investments that can be harmful to society and environment, such as promoting gambling reprehensibly or promoting a product, which is manufactured in a factory that pollutes the local river, in general, are considered as immoral by sustainable marketers, as marketing ethics does not endorse such promotional activities. In addition, some products and services are legal to produce, sell, and consume in many countries under certain rules and regulations, but it would be arguable whether marketing ethics permits to do so. For example, it would be a matter of debate how ethical it is to market tobacco products by just saying that "smoking is harmful for health," as it has a serious impact on human health!

Marketing ethics can be defined as marketers' moral principles that guide them to plan, implement, and monitor marketing activities in a way that will not be unjust for individuals and society while influencing their target audiences.

Source: ▸ pexels.com/royalty-free-images/

It is evident that the contemporary innovations (i.e., AI) of big-tech brands are now applied in war and conflicts to apartheid mass people and obliterate societies (Kwet, 2024); however, it was not the purpose of these novel innovations, which are created to support the human civilization. Tech-giants are accused of breaching marketing ethics while promoting their digital products and services. For example, "the phone numbers and email addresses of 533 million Facebook users have been exposed in a data breach" (O'Flaherty, 2022, np). "Millions of Facebook and Google users hit by 2FA data leak. A SMS routing company used by Google, Facebook and TikTok is accused of leaking its own database" (Laird, 2024). Effectively protecting stakeholders' interests in such mega digital platforms, which are enhanced by AI and AI—supported virtual and augmented realities, while sincerely abiding with marketing ethics is a challenge. Schultz and Seele (2023) proposed a simple stakeholder-centric approach to address this challenge. Schultz and Seele (2023, p. 102) argue that:

> AI's are invisible and including AI many other digital technologies have unclear impacts on society raising ethical concern…Everyone who affects or is affected by products and services featuring AI (or other fuzzy technologies, all these stakeholders should be considered for a) … systematic approach to address stakeholders concerns and initiate dialogue, participation, and deliberation about AI (and other fuzzy technologies') impacts.

D'Cruz et al. (2022) argue that ethical concerns related to embryonic technologies should be recognized and addressed at three levels. For example, the types of ethical concerns related to the relevant stakeholders that would be predicted while designing and manufacturing a new technology, the designers should be able to predict the concerns to address them at the technology level, company level, and the wider society level. As laws help us to comply with our moral values, Trequattrini et al. (2016) discuss about a legal resistance-based proposition that can help marketers to comply with marketing ethics to protect stakeholders' privacy, safety, and security. Trequattrini et al. (2016) argue that:

- The existing legal rules, regulations and guidance should be followed strictly to govern a system, in order to ensure stakeholders' privacy, security, and safety.
- Transparency is the key to protect stakeholders' privacy, while ensuring to honor the stakeholders' "right to be left alone" (Warren & Brandeis, 1890; Reidenberg, 2000; Atzori et al., 2010).
- For stakeholders' safety and security on digital platforms, confidentiality of stakeholders' personal information, integrity of the overall digital dynamics, and availability of safety and security measures should be ensured (Mayer, 2009; Weber, 2013; Buttarelli, 2010; Dutton, 2014).
- At the governance level (Weber & Weber, 2010), two issues should be considered: Digital infrastructure governance and general regulation.
- For digital infrastructure governance, interoperability or connectivity between human beings, computers, and networks should be as user-friendly as possible. "Right to access (needs to be ensured), in order to guarantee a fair and non-discriminatory use by all interested stakeholders and businesses and to avoid any increase to the digital divide (Norris, 2001) in the IoT environment" (Trequattrini et al., 2016, p. 407). And reliability and "robustness of the infrastructure (needs to be confirmed to deal)…with change or loss of functionality" (Trequattrini et al., 2016, p. 407).
- For general regulation, a multi-stakeholder approach should be adopted as such digital platforms can influence diverse groups of stakeholders. Legitimacy and inclusion, transparency, and accountability need to be confirmed to ensure stakeholders' privacy, safety, and security.

However, the aforementioned proposition of Trequattrini et al. (2016) has been discussed centered on IoT; it can be adapted from the contexts of other embryonic technologies, such as AI, AI-centered virtual and augmented realities, and other digital platforms, based on the context-specific concerns for stakeholders' privacy, safety, and security (Shams, 2016).

> Individuals, and especially those within the millennial generation, are more socially aware than ever, and thus, the 'ethical consumer' has come about" (Laroche, 2017, p. 3). The term "ethical consumer" is defined as the consumers who are socially responsible, and value the ethical perspectives in their buying decisions, which is related to the "feel good factor" (Bonetto, 2015), so that consumers feel better about their purchase decisions, as well as about themselves because of the ethical significance of their decision (Laroche, 2017) that contributes to any socio-economic cause at least to some extent (for example, not buying plastic bags, instead using biodegradable bags). (Czinkota et al., 2021, p. 100).

6

Because of this "feel good factor," the **ethical consumers**, who are increasing in number around the world, have placed higher value on sustainability, corporate social responsibility, and cause-related marketing or sustainability (Choi et al., 2016). This has led to a purified scope for sustainable marketing, as a business management functional area to enable marketers to compete, centered on cause-related marketing campaigns, in order to help addressing a societal or environmental cause, so that the societal and environmental contribution of a brand can enhance its commercial competitive advantage. In this context, cause-related marketing as a form of sustainable marketing can be defined as aligning a societal or environmental cause with a marketing program, with an aim to help addressing the cause, so that the societal and environmental contribution of a brand can enhance its commercial competitive advantage. For example, the Monterey Bay Aquarium's (2024) marketing team launched their cause-related marketing campaign, "Act for the Ocean" in 2020. As part of the campaign, they started posting through social channels a varied range of video clips and streaming services related to the life under ocean, which are fun and engaging for kids and other age groups. Highlighting the sustainability concerns for the animals under water, the digital campaign strived "to educate the public about the climate change, aquatic life affected, and ways to help" (np), while promoting their services based on an alternative form, i.e., the streaming services.

- **Cause-Related Marketing** as a form of sustainable marketing can be defined as aligning a societal or environmental cause with a marketing program, with an aim to help addressing the cause, so that the societal and environmental contribution of a brand can enhance its commercial competitive advantage.
- **Feel Good Factor**—When consumers feel better about their purchase decisions, as well as about themselves because of the ethical significance of their decision (Laroche, 2017) that contributes to any socioeconomic cause at least to some extent.

To satisfy the growing number of ethical consumers, and to delight them with a "feel good factor," sustainable digital marketers should prioritize their strategic focus centered on a social or an environmental cause. ◘ Figure 6.10 presents the sustainable

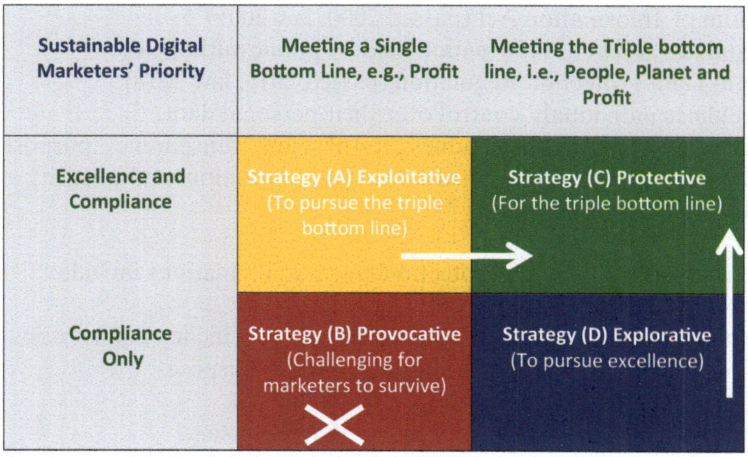

Fig. 6.10 Sustainable digital marketers' priority (adapted from Friedman & Miles, 2002; Shams, 2021)

digital marketers' priority matrix. The matrix highlights four alternative marketing strategies. Among the four strategies as described in Fig. 6.10, marketers' priority should be for strategy-C, as it meets the triple bottom line, while serving consumers' and other stakeholders' needs with a novel strategy, and complying with the applicable marketing ethics and business laws and regulations. Strategy-A as depicted in Fig. 6.10, is excellent in meeting consumers' and other stakeholders' needs while it complies with marketing ethics and business laws, but it does meet only a single bottom line. Therefore, marketers should exploit Strategy-A to explore how they can possibly meet the triple bottom line based on this strategy. Strategy-D however is good to meet the triple bottom line, it is not excellent to survive with competitive products, or competitors can easily copy this strategy to outplay its competitive advantage. Therefore, marketers should explore Strategy-D to understand how they can make this strategy more competitive. ▶ Section 7.4 (extending or possibly sustaining a competitive advantage) discusses in detail how marketers could extend or sustain their brands' competitive advantage. Strategy-B as depicted in Fig. 6.10 is provocative for marketers as it neither meets the triple bottom line, nor it is competitive enough for marketers to survive and prosper.

The requirements of **digital business laws and regulations** must be fulfilled by sustainable marketers to support their ethical behavior. Consumers and other stakeholders are willing to interact with brands by providing personal data (Kumar et al., 2014). But they are increasingly anxious over how their personal data is collated and utilized (Martin et al., 2017), especially, considering that we have already seen multiple high-profile data breaches (Kude et al., 2017) by big-techs. To protect individuals' privacy, laws are passed from time to time, for example (Legislation, n.d.; Local Government Association, 2018; Accessing Information, 2024):

- Data Protection Act (passed 1998, law 2000)
  - Regulates processing of data of any living person
  - Separate rules for health, sexuality, religion, disabilities

- Freedom of Information Act (passed 2000, law 2005)
  - Regulates access to information held by public authorities
- General Data Protection Regulation (passed 2016, law 2018)
  - Regulates individuals' control over their personal data
  - Build on 7 Principles: Lawfulness, fairness and transparency, purpose limitation, data minimization, accuracy, storage limitation, and integrity and confidentiality (security)

The key emphases of the data protection laws and regulations include:
- Data processing must be fair/lawful.
- Collected/obtained data must be used only for specific/legal purposes.
- Data must be adequate/relevant/not excessive to purpose.
- Data must be accurate.
- Personal data should not be kept longer than necessary.
- Personal data should be processed according to the subject's rights to (1) access, (2) prevent data being used for direct marketing, (3) prevent decisions being made on automated processing, and (4) prevent processing which may cause damage or distress.
- Data must be kept secure.
- Data must not be transferred outside approved countries without consent.

6.8 Digital Marketing for the Sustainable Development Goals

Source: ► pexels.com/royalty-free-images/

In order to satisfy the increasing number of ethical consumers, at consumers' digital touchpoints, which are discussed in ▶ Sect. 6.3.3 (digital touchpoints, customer journey mapping, and customer experience), sustainable marketers should attempt to provide an authentic experience, feeling, and impression to their consumers and other stakeholders that their brands are sincerely committed to the triple bottom line, People, Planet, and Profit. Throughout this chapter, a number of case examples are discussed in support of the discussed theoretical application that demonstrates how sustainable digital marketers have been contributing back to people and planet while underpinning their profit-making ventures, and contributing to our global pursuit of the United Nations' Sustainable Development Goals (SDGs).

◘ Table 6.4 summarizes these examples of sustainable digital marketing that help us pursue the SDGs. "Running the internet accounts for approximately 1.7% of total global energy demand. 40% of the energy used by data centres is for cooling…Carbon emissions associated with operating and accessing the internet are similar to the emissions released by the global airline industry" (Future Energy, 2024, np). Therefore, green web hosting as an environmentally friendly option to run a brand's website and other digital platforms contributes to operate a business with clean energy (SDG 7), responsible consumption and production (SDG 12), and climate action (SDG 13), as the digital operational process of the business produces less CO_2 emission. Green web hosting also helps develop industry infrastructure (web infrastructure) more environmentally friendly; thus, it helps contributing to the pursuit of SDG 9 (industry, innovation, and infrastructure). "Green web hosting companies aim to provide high-quality services while doing everything they can to limit their carbon emission" (Bustillos, 2024, np). The high-quality services of green web hosting companies contribute to a decent work environment (SDG 8). As the production of electricity is yet dependent mainly on fossil fuel, and running internet requires a lot of electricity; green web hosting helps protecting natural resources, people, and wildlife habitats that contribute to the life of future generations (Digital Sustainability, 2024). Thus, it helps us contributing to the pursuit of SDG 3 (good health and well-being), SDG 11 (sustainable cities and communities), and SDG 15 (life on land).

The next digital marketing campaign as presented in ◘ Table 6.4 is ALS viral marketing campaign. The campaign encourages community involvement, invites participants to perform physical activities, and encourages them to invite peers to join the competition, which contributes to good health and well-being (SDG 3) and sustainable communities (SDG 11). The campaign also partners with celebrities to raise funds for a cause, which contributes to SDG 17 (partnerships for the goals). McDonald's promotion of their environmentally cleaning items through digital channels encourages others to use eco-friendly cleaning products. The cleaning items that are used by McDonald's to clean french fry fat from their restaurant drains are made with natural ingredients instead of toxic dyes. Natural proteins are used to stone wash materials in the restaurant. It contributes to workers' good health and well-being (SDG 3), which eventually contributes to the sustainable community of McDonald's workers (SDG 11). The eco-friendly cleaning products contribute to improved sanitation system (SDG 6). The industry innovation (i.e., the cleaning items produced by bacteria instead of toxic dyes) contributes

Table 6.4 Sustainable digital marketing strategies for Sustainable Development Goals

Digital marketing projects	Section in this chapter, where it is discussed	Value creation for Sustainable Development Goals (SDGs)
Green web hosting	Preliminary Vignette	SDG 3 (good health and well-being), SDG 7 (affordable and clean energy), SDG 8 (decent work and economic growth), SDG 9 (industry, innovation and infrastructure), SDG 11 (sustainable cities and communities), SDG 12 (responsible consumption and production), SDG 13 (climate action), and SDG 15 (life on land)
ALS viral marketing campaign	► Sect. 6.1	SDG 3 (good health and well-being), SDG 11 (sustainable cities and communities), SDG 17 (partnerships for the goals)
Promotion of McDonald's environmentally friendly cleaning items through digital channels	► Sect. 6.2	SDG 3 (good health and well-being), SDG 6 (clean water and sanitation), SDG 9 (industry, innovation and infrastructure), SDG 11 (sustainable cities and communities), SDG 12 (responsible consumption and production), SDG 13 (climate action)
Promotion of McDonald's societal and environmental commitment and contribution through digital channels	► Manager's Corner 6.1	SDG 2 (zero hunger), SDG 3 (good health and well-being), SDG 4 (quality education), SDG 5 (gender equality), SDG 6 (clean water and sanitation), SDG 8 (decent work and economic growth), SDG 10 (reduced inequalities), SDG 11 (sustainable cities and communities), SDG 12 (responsible consumption and production), SDG 13 (climate action), SDG 15 (life on land), SDG 16 (peace, justice and strong institutions), and SDG 17 (partnerships for the goals)
British Government's celebrity endorsement and influencer marketing for a social cause	► Sect. 6.3.2	SDG 3 (good health and well-being), SDG 11 (sustainable cities and communities), and SDG 17 (partnerships for the goals)
Celebrity endorsement of Adidas for sustainable marketing	► Marketing in Action 6.1	SDG 9 (industry, innovation, and infrastructure), SDG 12 (responsible consumption and production), SDG 13 (climate action) SDG 14 (life below water), and SDG 17 (partnerships for the goals)

Digital marketing projects	Section in this chapter, where it is discussed	Value creation for Sustainable Development Goals (SDGs)
Planet or plastics viral marketing campaign of National Geographic	► Marketing in Action 6.2	SDG 3 (good health and well-being), SDG 6 (clean water and sanitation), SDG 12 (responsible consumption and production), SDG 13 (climate action), SDG 14 (life below water), SDG 15 (life on land)
Black Friday viral marketing campaign of Patagonia	► Marketing in Action 6.2	SDG 11 (sustainable cities and communities), SDG 13 (climate action), SDG 14 (life below water), SDG 15 (life on land)
Monterey Bay Aquarium's digital marketing campaign, "Act for the Ocean"	► Sect. 6.7	SDG 14 (life below water)

to SDG 9 (industry and innovation), responsible production (SDG 12), and climate action (SDG 13).

The promotion of McDonald's societal and environmental commitments and contributions through digital channels are discussed next in ■ Table 6.4. For example, their partnership (SDG 17) during the massive Turkey earthquake in 2023 and donations helped to feed (SDG 2) the people affected by the earthquake, which also was helpful for those people to maintain their good health and well-being (SDG 3). McDonald's contribution to the President Obama Center for Safety and Security for Communities, especially for marginalized people, and girls' empowerment and scholarships contribute to the pursuit of SDG 4 (quality education), SDG 5 (gender equality), SDG 10 (reduced inequalities) and SDG 11 (sustainable communities). Their Chicago-based investment of $3.5 million to 40 local nonprofits, and the partnership (SDG 17) to support local communities contribute to SDG 8 (decent work and economic growth) and SDG 10 (reduced inequalities). McDonald's employee assistance program and international relief support led by the Red Cross in Ukraine and across Europe contribute to SDG 2 (zero hunger, e.g., in conflict zones), SDG 3, i.e., good health and well-being of their employees and people in conflict zones. Also, their support to their employees and local communities in conflict zones contributes to the pursuit of SDG 16 (peace, justice, and strong institutional support).

McDonald's partnership (SDG 17) with the American Red Cross Annual Disaster Giving Program (This contribution ensures that McDonald's is supporting the needs of emerging disasters, even before they occur) contributes to SDG 11 (sustainable cities and communities). Their community grants and sponsorships contribute to SDG 4 (quality education) and sustainable communities (SDG 11). McDonald's contribution to the National Museum of the American Latino contributes to SDG 10 (reduced inequalities) and SDG 11 (sustainable communities). The Our Planet program (McDonald's climate action, responsible packaging and waste management, and activities for nature, forecasts, and water) contributes to SDG 6 (clean water and sanitation) and SDG 13 (climate action). Their efforts for food quality, food safety, responsible sourcing of food, ensuring nutrition, and responsible marketing practices contribute to SDG 3 (good health) and SDG 12 (responsible consumption and production). McDonald's support for animal health and welfare and responsible use of antibiotics contribute to SDG 15 (life on land). Employment creation; ensuring diversity, equity, and inclusion of people in McDonald's global family support for talents and benefits, human rights, safe and respectful workplace, and community connection contribute to SDG 8 (decent work and economic growth), SDG 10 (reduced inequalities), and SDG 11 (sustainable communities). Also, their community impact, philanthropies, and Ronald McDonald House Charities contribute to SDG 2 (zero hunger), SDG 3 (good health and well-being), and SDG 11 (sustainable communities).

During the COVID-19 pandemic, some people were reluctant to have the COVID-19 vaccine (National Library of Medicine, 2021), perhaps because of the unknown impact of the vaccine on the human body. But the governments across the world, on the one hand needed to protect the life of people, and on the other hand, needed to withdraw the lockdown and resume normal daily life across all sectors as soon as possible. Without vaccinating a large number of world populations, it was not possible to protect human life and to withdraw the lockdown. Therefore, the British Government partnered (SDG 17) with celebrities to influence Britons to get the COVID-19 vaccine. The government promoted the campaign through different traditional and contemporary digital channels, which contributed to protect human life (SDG 3: good health and well-being, and SDG 11: sustainable communities). The initiatives and partnership (SDG 17) of Adidas to recycle and reuse of raw materials to produce sports gears and dresses and its promotion through digital channels contribute to SDG 9 (industry and innovation), SDG 12 (responsible consumption and production), and SDG 13 (climate action).

The next sustainable digital marketing campaign discussed in ◘ Table 6.4 is National Geographic's "Planet or Plastics" viral marketing campaign. Because of excessive and irresponsible use of plastics and plastic products across the world, today human body contains micro-plastics, which we do not realize, but it is incredibly harmful for human body (Parker, 2023; Pinto-Rodrigues, 2023; Lee et al., 2023; Nature Medicine, 2024). If the human civilization can start using less plastics and plastic products, it can contribute to SDG 3 (good health and well-being), SDG 6 (clean water), SDG 12 (responsible consumption and production), SDG 13 (climate action), SDG 14 (life below water), SDG 15 (life on land). The Black Friday campaign of Patagonia (i.e., donating 100% of sales on Black Friday to grassroots environment groups) contributes to the pursuit of SDG 11 (sustainable communities) and SDG 13 (climate action). The Monterey Bay Aquarium's digital marketing campaign, "Act for the Ocean" also contributes to the pursuit of SDG 14 (life below water), and brings an additional promotional option for their original services in the aquarium.

Summary

The Preliminary Vignette (green web hosting for a sustainable future) introduces the core focus of this chapter that clarifies how sustainable digital marketers could contribute to society and environment, while underpinning their profit-making venture. Following the discussion on background, scope, and key definitions, the chapter first discusses about the digital environment and the digital marketing dynamics. Discussions here include the technological environment, and the impact of information technology on marketers and customers. Second, the chapter introduces the concept of sustainable digital marketing and its key considerations. Discussions in this part of the chapter include digital channels and its applications for sustainable digital marketing. Third, digital touchpoints, customer journey mapping, and customer experience are discussed. Fourth, database marketing and other embryonic digital technologies and their implications for sustainable marketing are discussed. Fifth, the discussions are evolved to demonstrate how tech-savvy firms shorten marketing cycle and can understand the tech-readiness of a target market. Sixth, digital marketing, marketing ethics, and regulations are discussed. Similar to all other chapters of this core sustainable marketing textbook, this digital marketing chapter also presents a number of practical cases throughout its different sections and subsections to demonstrate how sustainable digital marketers successfully pursue the triple bottom line, namely, People, Planet, and Profit, and contribute to our global pursuit of the United Nations' Sustainable Development Goals (SDGs). The last section of this chapter summarizes these practical examples of sustainable digital marketing that help us pursue the SDGs.

Review Questions

1. What is viral marketing? What is influencer marketing? Can you describe the links between influencer marketing and viral marketing?
2. Based on the discussion of ► Sect. 6.2 and your own research, please, describe the influence of information technology on sustainable marketers and ethical consumers.
3. What are the 7Ds of effective digital marketing? How could the 7Ds help sustainable digital marketers to plan for effective sustainable digital marketing?
4. What are the differences between digitization, digitalization, and digital transformation?
5. Please describe the purposes of database marketing. Why do you think the discussed purposes of database marketing are important to increase the customers' share of wallet?
6. What are the different types of influencers? How influencer marketing would be instrumental in effectively conveying a message related to a societal or environmental cause?
7. What is programmatic advertising? How the dynamics of programmatic advertising work?
8. What are the operational stages of database marketing? Why a digital marketer needs to consider the 5 Vs of data at these operational stages of database marketing?

Discussion Questions

1. Select a brand, any brand that has been operating for a minimum of two years, and has a publicly available website, and identify the customers' digital touchpoints for the brand. Make a list of customer and other non-customer stakeholders of the brand as its target audiences. How could the brand effectively utilize these digital touchpoints to influence its target audiences?
2. First, Google the term "trendjacking," and explore and discuss how different brands are employing trendjacking. Second, select an established brand from any industry, and describe your views on how the brand could apply effectively to promote a societal or environmental cause while promoting its commercial interest.
3. First, select a brand. Second, reflecting on ◘ Fig. 6.5 (designing, implementing and evaluating sustainable viral marketing campaign), design a viral marketing campaign for the brand. What would be the core sustainability message that you may want to communicate through your sustainable viral marketing cam-

paign? What are goals of this campaign for People and Profit or Planet and Profit? Describe the considerations to design and implement your sustainable viral marketing campaign for the brand.

4. Sustainable marketing, i.e., meeting the triple bottom line (People, Planet, and Profit) is applicable to organizations across public, private and third sectors. Browse the website of one organization/brand from each of these three sectors to recognize how each of these organizations across the three sectors communicates their societal and environmental messages through their websites. First make a list for the three websites to describe the messages that they covey related to the interests of society (people), environment (planet), and their profit-making initiatives. Second, share your views on why the design of one of these three websites is more attractive and user-friendly to communicate the sustainability message than the other two websites.
5. First, select a brand. Second, based on ■ Table 6.3 (key considerations for database marketing), prepare a database marketing strategy for the brand. What are the goals, you would set for the triple bottom line in your database marketing strategy? What are the strategies (described by Tapp, 2008) you would adopt for your database marketing campaign? What are the different data types you may be considering to collect, and what are the sources of these data? Applying the data protection principles, how would you plan to protect the privacy of customers and non-customer stakeholders of this brand?
6. First, Google the terms "data privacy scandals" and identify three data privacy scandals of three different digital brands. Second, based on the principles of data protection laws and regulations, describe your views on how these digital brands have violated the data protection principles. How you think that the data protection laws and regulations protect stakeholders' privacy?

Practice Quiz

1. There are two types of viral marketing, which are:
   (a) Organic viral marketing and amplified viral marketing
   (b) Organic viral marketing and exemplary viral marketing
   (c) Organic viral marketing and authenticated viral marketing
   (d) Exemplary viral marketing and amplified viral marketing
2. Organic viral marketing is a controlled viral marketing campaign.
   (a) True
   (b) False
3. The 7Ds of an effective digital marketing campaign include:
   (a) Digital goals, digital audiences, digital infrastructure, digital platforms, digital media, digital data, digital technology
   (b) Digital goals, digital audiences, digital devices, digital terms, digital media, digital data, digital technology
   (c) Digital goals, digital audiences, digital devices, digital platforms, digital media, digital data, digital technology
   (d) Digital goals, digital audiences, digital infrastructure, digital terms, digital media, digital data, digital technology

4. A micro-influencer should have 10,001 to 100,000 followers.
   (a) True
   (b) False
5. Exploring opportunity to convince customers to purchase the same product in more volume, in order to increase the product's overall sales, is a cross-selling strategy.
   (a) True
   (b) False
6. A database marketing campaign would have:
   (a) Four operational stages
   (b) Three operational stages
   (c) Eight operational stages
   (d) Six operational stages

7. 5Vs of data include:
   (a) Data volume, data variety, data velocity, data veniality, data value
   (b) Data volume, data venality, data velocity, data veracity, data value
   (c) Data volume, data variety, data velocity, data veracity, data value
   (d) Data volume, data variety, data visuality, data veracity, data value
8. The General Data Protection Regulation was passed in:
   (a) 2016
   (b) 2015
   (c) 2018
   (d) 2017
9. A digital marketing campaign would have three purposes, which are:
   (a) Identifying target audience, creating awareness, and forming a public opinion
   (b) Creating awareness, clarifying a brand's value proposition, and forming a public opinion
   (c) Identifying target audience, clarifying a brand's value proposition, and forming a public opinion
   (d) Identifying target audience, creating awareness, and forming a value proposition

(You can find the correct answers to these questions by taking the quiz and then comparing your answers with the answers that are provided at the end of the book, under the "Practice Quiz Answers" part of the book.)

References

Aberg, R. (2022). Car service M becomes Volvo On Demand will expand outside Sweden. FEBER. Retrieved May 5, 2024, from https://feber.se/bil/biltjansten-m-blir-volvo-on--demand/441928/.

Accessing Information. (2024). What is the Freedom of Information Act? *Metropolitan Police*. Retrieved June 1, 2024, from https://www.met.police.uk/foi-ai/af/accessing-information/what-is-the-freedom-of-information-act/#:~:text=The%20Freedom%20of%20Information%20Act%20was%20passed%20on%2030%20November,obligations%20on%20those%20public%20authorities.

Almukhtar, F., Mahmoodd, N., & Kareem, S. (2021). Search engine optimization: A review. *Applied Computer Science, 17*(1), 70–80.
Arinze, P. (2024). Trendjacking for a positive brand image: A guide to responsible participation. *Agility PR Solutions*. Retrieved May 12, 2024, from https://www.agilitypr.com/pr-news/public-relations/trendjacking-for-positive-brand-image-a-guide-to-responsible-participation/#:~:text=Trendjacking%20is%20the%20art%20of,around%20for%20a%20long%20time.
Atzori, L., Iera, A., & Morabito, G. (2010). The internet of things: A survey. *Computer Networks, 54*(15), 2787–2805.
Austin, D. (2023). 2023 internet minute infographic, by e-discovery today and LTMS!: eDiscovery trends. eDiscoveryToday. Retrieved May 4, 2024, from https://ediscoverytoday.com/2023/04/20/2023-internet-minute-infographic-by-ediscovery-today-and-ltmg-ediscovery-trends/.
BBC. (2014). The value of celebrity endorsements. BBC News. Retrieved May 11, 2024, from https://www.bbc.co.uk/news/business-30392829.
Bello-Orgaz, G., Jung, J. J., & Camachoa, D. (2016). Social big data: Recent achievements and new challenges. *Information Fusion, 28*(2), 45–59.
Berman, R., & Katona, Z. (2013). The role of search engine optimization in search marketing. *Marketing Science, 32*(4), 644–651.
Bikakis, N., Papastefanatos, G. and Papaemmanouil, O. (2018). Big data exploration, visualization and analytics. *Big Data Research*. Retrieved February 4, 2018, from https://www.journals.elsevier.com/big-data-research/call-for-papers/special-issue-on-big-data-exploration-visualization-and-anal.
Bonetto, L. (2015). *The ethical consumer—US.* Mintel Academic. Mintel Group Ltd.
British Government. (2021). Celebrities get back to the 'rhythm of life' in new film supporting Covid-19 vaccination programme. Press release. Retrieved May 11, 2024, from https://www.gov.uk/government/news/celebrities-get-back-to-the-rhythm-of-life-in-new-film-supporting-covid-19-vaccination-programme.
Bustillos, M. (2024). What is green web hosting? Host Papa. Retrieved May 5, 2024, from https://www.hostpapa.com/blog/web-hosting/what-is-green-web-hosting/.
Buttarelli, G. (2010). Internet of things: Ubiquitous monitoring in space and time. *European Privacy and Data Protection Commissioners' Conference,* Prague, Czech Republic, 29 April 2010.
Campbell, C., & Farrell, J. R. (2020). More than meets the eye: The functional components underlying influencer marketing. *Business Horizons, 64*(4), 469–479.
Chaffey, D., & Ellis-Chadwick, F. (2019). *Digital marketing*. Pearson.
Chaffey, D., & Smith, P. R. (2023). *Digital marketing excellence: Planning, optimizing and integrating online marketing* (6th ed.). Routledge.
Choi, J., Chang, Y. K., Li, Y. J., & Jang, M. G. (2016). Doing good in another neighbourhood: Attributions of CSR motives depend on corporate nationality and cultural orientation. *Journal of International Marketing, 24*(4), 82–102.
Clemons, E. K., & Row, M. C. (1992). Information, power, and control of the distribution channel: Preliminary results of a field study in the consumer packaged goods industry. *Semantic Scholar*. Retrieved July 2, 2019, from https://pdfs.semanticscholar.org/952c/8aa10ccdf4a48e4fb262f967eb7d5ac552d9.pdf.
Couture, N. (2018). Why data governance? Retrieved January 8, 2018, from https://www.cio.com/article/3245588/governance/why-data-governance.html.
Cui, T. H., Ghose, A., Halaburda, H., Iyengar, R., Pauwels, K., Sriram, S., Tucker, C., & Venkataraman, S. (2021). Informational challenges in Omnichannel marketing: Remedies and future research. *Journal of Marketing, 85*(1), 103–120.
Czinkota, M., Kotabe, M., Vrontis, D., & Shams, S. M. R. (2021). *Marketing management: Past, present and future* (4th ed.). Springer International Publishing.
D'Cruz, P., Du, S., Noronha, E., Parboteeah, K. P., Trittin-Ulbrich, H., & Whelan, G. (2022). Technology, megatrends and work: Thoughts on the future of business ethics. *Journal of Business Ethics, 180*, 879–902.
Day, K. (1995). Biotech's other benefits. *Washington Post*, H1.

De Bock, K. W. (2016). The brave new world of database marketing. In K. Coussement, K. W. De Bock, & S. A. Neslin (Eds.), *Advanced database marketing: Innovative methodologies and applications for managing customer relationships* (pp. 1–7). Routledge.

Dencheva, V. (2024). Influencer marketing market size worldwide from 2016 to 2024 (in billion U.S. dollars). Statista. Retrieved May 11, 2024, from https://www.statista.com/statistics/1092819/global-influencer-market-size/.

Digital Marketing Analytics Market. (2024). Industry trends in digital marketing analytics. *Future Marketing Insights*. Retrieved May 20, 2024, from https://www.futuremarketinsights.com/reports/digital-marketing-analytics-market.

Digital Sustainability. (2024). 6 reasons to choose green web hosting. *Solve*. Retrieved May 5, 2024, from https://solve.co.uk/ethical-business/reasons-to-choose-green-website-hosting/.

Dutton, W. H. (2014). Putting things to work: Social and policy challenges for the internet of things. *Info, 16*(3), 1–21.

Ecommerce Forum. (n.d.) Ecommerce skills development. Retrieved May 31, 2024, from https://ecomafrica.org/expertise/skills-development/.

6

Economist. (1995). Change at the check-out. p. R3.

European Commission. (n.d.) 2050 long-term strategy: Striving to become the world's first climate-neutral continent by 2050. Retrieved May 5, 2024, from https://climate.ec.europa.eu/eu-action/climate-strategies-targets/2050-long-term-strategy_en#:~:text=Striving%20to%20become%20the%20world's%20first%20climate%2Dneutral%20continent%20by%202050.&text=The%20EU%20aims%20to%20be,to%20the%20European%20Climate%20Law%20.

Fallon, J. (1994). U.K. is seen ahead in ECR, technology: Efficient consumer response. *Supermarket News*, 11.

Foster, R. N. (1988). *Innovation: The attacker's advantage* (p. 35). Summit Books.

Friedman, A. I., & Miles, S. (2002). Developing stakeholder theory. *Journal of Management Studies, 39*(1), 1–21.

Fukukawa, S. (1989). *The future of U.S.-Japan relationship and its contribution to new globalism* (pp. 10–11). Ministry of International Trade and Industry.

Future Energy. (2024). The carbon cost of the internet. Retrieved May 5, 2024, from https://ourfuture.energy/in-focus/the-carbon-cost-of-the-internet/#:~:text=Key%20Points%3A,network%20to%20access%20the%20internet.

Gastelu, G. (2022). Jeep saluting US military and veterans with $1,000 cash-back deal. Retrieved January 10, 2023, from https://www.foxnews.com/auto/jeep-saluting-u-s-military-veterans-cash-back-deal.

Gates, B. (2009). *Business @ the speed of thought: Succeeding in the digital economy*. Warner Books.

Graham, C. (2003). The transition from mode-1 to mode-2 society: The implications for education. Retrieved August 11, 2009, from http://www.futures1.net/docs/Mode-2%20Paper.doc.pdf.

Hasan, R., Shams, R., & Rahman, M. (2021). Consumer trust and perceived risk for voice-controlled artificial intelligence: The case of Siri. *Journal of Business Research, 131*, 591–597.

Heidbreder, L. M., Bablok, I., Drews, S., & Menzel, C. (2019). Tackling the plastic problem: A review on perceptions, behaviors, and interventions. *Science of the Total Environment, 668*, 1077–1093.

Jardine, P., & Thorn, B. (2023). How much CO_2 does a website produce? RootDesign. Retrieved May 4, 2024, from https://rootwebdesign.studio/articles/how-much-carbon-does-a-website-produce/#:~:text=According%20to%20figures%20from%20Google,105%20trees%20in%20a%20year.

Jeep. (2014). Drive for troops—Help support our soldiers. Retrieved January 10, 2023, from https://www.autoparkchryslerjeep.com/blogs/2638/lifestyle/drive-troops-help-support-soldiers/.

Kantar. (2023). The rise, fall, and rise again of eco-active consumers. Retrieved May 5, 2024, from https://www.kantar.com/company-news/the-rise-fall-and-rise-again-of-eco-active-consumers.

Kennedy, J. (2015). How digital disruption changed 8 industries forever. Retrieved October 18, 2016, from https://www.siliconrepublic.com/companies/digital-disruption-changed-8-industries--forever.

Kotler, P., Keller, K. L., Ang, S. H., Leong, S. M., & Tan, C. T. (2009). *Marketing management: An Asian perspective* (5th ed.). Prentice Hall.

Kovacevic, M. (2024). Top 6 sustainable digital marketing strategies. Retrieved May 4, 2024, from https://archseologist.com/blog/sustainable-digital-marketing-strategies/.

Kude, T., Hoehle, H., & Sykes, T. A. (2017). Big data breaches and customer compensation strategies: Personality traits and social influence as antecedents of perceived compensation. *International Journal of Operations & Production Management, 37*(1), 56–74.

Kumar, V., Zhang, X., & Luo, A. (2014). Modeling customer opt-in and opt-out in a permission-based marketing context. *Journal of Marketing Research, 51*(4), 403–419.

Kurtz, D. L. (2009). *Contemporary marketing*. South-Western Cengage Learning.

Kwet, M. (2024). How US big tech supports Israel's AI-powered genocide and apartheid. *Al Jazeera*. Retrieved June 1, 2024, from https://www.aljazeera.com/opinions/2024/5/12/how-us-big-tech-supports-israels-ai-powered-genocide-and-apartheid.

Laird, J. (2024). Millions of Facebook and Google users hit by 2FA data leak: A SMS routing company used by Google, Facebook and Tiktok is accused of leaking its own database. *Tech.Co*. Retrieved June 1, 2024, from https://tech.co/news/millions-facebook-google-2fa-data-leak.

Lamberton, C., & Stephen, A. T. (2016). A thematic exploration of digital, social media, and mobile marketing: Research evolution from 2000 to 2015 and an agenda for future inquiry. *Journal of Marketing, 80*(6), 146–172.

Laroche, S. (2017). *Cause-related marketing in five unique culture (honours dissertation)*. Texas Christian University.

Lee, Y., Cho, J., Sohn, J., & Kim, C. (2023). Health effects of microplastic exposures: Current issues and perspectives in South Korea. *Yonsei Med Journal, 64*(5), 301–308.

Legislation. (n.d.) Data Protection Act 1998. Retrieved June 1, 2024, from https://www.legislation.gov.uk/ukpga/1998/29.

Local Government Association. (2018). The General Data Protection Regulation (GDPR) Guidance for Members. Retrieved June 1, 2024, from https://www.local.gov.uk/sites/default/files/documents/The+General+Protection+Data+Regulation+(GDPR)+-+Guidance+for+Members.pdf.

Marketing Analytics Market Size. (2024). Report on industry size and market share analysis—Growth trends and forecasts (2024–2029). *Mordor Intelligence*. Retrieved May 20, 2024, from https://www.mordorintelligence.com/industry-reports/marketing-analytics-market.

Market Size & Trends. (2024). Marketing analytics software market size, share and trends analysis: Report by application (Solution, Service), by development, by organization size, by industry vertical, by region, and segment forecasts, 2023–2030. *Grand View Research*. Retrieved May 20, 2024, from https://www.grandviewresearch.com/industry-analysis/marketing-analytics-software-market.

Martin, K. D., Borah, A., & Palmatier, R. W. (2017). Data privacy: Effects on customer and firm performance. *Journal of Marketing, 81*(1), 36–58.

Mayer, C. P. (2009). Security and privacy challenges in the internet of things. *Electronic Communications of the EASST, 17*, 12.

Mazur, L. (1994). Evolution in the workplace. *Marketing*, 6.

Mironov, D. (2024). 12 top programmatic advertising platforms in 2024. Retrieved May 11, 2024, from https://improvado.io/blog/top-programmatic-platforms.

Monterey Bay Aquarium. (2024). 10 cause-related marketing campaign examples that inspire US. *HubSpot*. Retrieved June 1, 2024, from https://blog.hubspot.com/agency/5-tips-to-boost-your--next-cause-marketing-campaigns-reach.

Morin, R. (2002). When celebrity endorsers go bad. *Washington Post*. Retrieved May 11, 2024, from https://www.washingtonpost.com/archive/opinions/2002/02/03/when-celebrity-endorsers-go-bad/260776e6-d38c-4319-b683-eb466c499dce/.

Nace, T. (2019). Thailand supermarket ditches plastic packaging for banana leaves. *Forbes*. Retrieved June 29, 2019, from https://www.forbes.com/sites/trevornace/2019/03/25/thailand-supermarket-uses-banana-leaves-instead-of-plastic-packaging/#3aeb02c87102.

National Library of Medicine. (2021). Why are some people reluctant to be vaccinated for Covid-19? A cross-sectional survey among U.S. adults in May–June 2020. Retrieved June 2, 2024, from https://www.ncbi.nlm.nih.gov/pmc/articles/PMC8277541/.

Nature Medicine. (2024). Microplastics are everywhere—We need to understand how they affect human health. Retrieved June 2, 2024, from https://www.nature.com/articles/s41591-024-02968-x.

Net Zero (n.d.) Considerable changes need to take place before 2030, to reach net zero by 2050. *5D Net Zero*. Retrieved May 5, 2024, from https://www.5dnetzero.co.uk/net-zero/.

Net Zero Climate. (n.d.) What is net zero? Retrieved April 4, 2024, from https://netzeroclimate.org/what-is-net-zero-2/#:~:text=Net%20zero%20is%20the,remain%20consistent%20with%201.5C.

Norris, P. (2001). *Digital divide: Civic engagement, information poverty, and the internet worldwide*. Cambridge University Press.

O'Flaherty, K. (2022). Facebook data breach: Here's what to do now. *Forbes*. Retrieved June 1, 2024, from https://www.forbes.com/sites/kateoflahertyuk/2021/04/06/facebook-data-breach-heres-what-to-do-now/?sh=52f1c72b5708.

Our World in Data. (n.d.) Births per year, by world region (number of births over a given period. Refers to live births for annual civil calendar years from 1 January to 31 December). Retrieved May 31, 2024, from https://ourworldindata.org/grapher/annual-number-of-births-by-world-region.

Parker, L. (2023). Microplastics are in our bodies. How much do they harm us? The science is unsettled, but researchers say there is cause for concern. *National Geographic*. Retrieved June 2, 2024, from https://www.nationalgeographic.com/environment/article/microplastics-are-in-our-bodies--how-much-do-they-harm-us.

PC Social. (2023). The 10 most successful organic viral marketing campaigns. Retrieved May 4, 2024, from https://pcsocial.medium.com/the-10-most-successful-organic-viral-marketing-campaigns-f5983e1ed01c#:~:text=The%20ALS%20Ice%20Bucket%20Challenge,enormous%20impact%20on%20a%20cause.

Peerallya, J. A., Fuentes, C. D., & Figueiredo, P. N. (2019). Inclusive innovation and the role of technological capability-building: The social business Grameen Danone Foods Limited in Bangladesh. *Long Range Planning, 52*(6), 1–24.

Peterson, L. (2016). What are some famous or great examples of internal branding in a company? *Quora.* Retrieved June 12, 2019, from https://www.quora.com/What-are-some-famous-or-great-examples-of-internal-branding-in-a-company.

Petrosyan, A. (2024). Global internet penetration rate as of April 2024, by region. Statista. Retrieved May 31, 2024, from https://www.statista.com/statistics/269329/penetration-rate-of-the-internet--by-region/.

Pinto-Rodrigues, A. (2023). Microplastics are in our bodies. Here's why we don't know the health risks. There are big, open scientific questions about levels of exposure and toxicity. *Science News*. Retrieved June 2, 2024, from https://www.sciencenews.org/article/microplastics-human-bodies-health-risks#:~:text=In%20recent%20years%2C%20microplastics%20have,milk%20and%20in%20human%20blood.

Prothom Alo. (2024). Internet subscribers have increased by more than 43 lacks in one month. Retrieved May 31, 2024, from https://www.prothomalo.com/bangladesh/jsv5535ka7.

Rakic, B., & Rakic, M. (2013). Marketing challenges at the beginning of xxi century. *Actual Problems of Economics, 148*(10), 446–455.

Reidenberg, J. R. (2000). Resolving conflicting international data privacy rules in cyberspace. *Stanford Law Review, 52*, 1315–1371.

RFID Technology (2004). RFID technology is best suited to box and pallet. *Microscope*. p. 12.

Richards, J. I. (2022). *A history of advertising: The first 300,000 years*. Rowman & Littlefield.

Rosário, A. T., Lopes, P. R., & Rosário, F. S. (2023). The digital marketing for sustainable development. Retrieved May 10, 2024, from https://www.preprints.org/manuscript/202307.0330/v1.

Rumsfeld, J. S., Joynt, K. E., & Maddox, T. M. (2016). Big data analytics to improve cardiovascular care: Promise and challenges. *Nature Reviews Cardiology, 13*(6), 350–359.

Schultz, M. D., & Seele, P. (2023). Towards AI ethics' institutionalization: Knowledge bridges from business ethics to advance organizational AI ethics. *AI and Ethics, 3*, 99–111.

Shams, S. M. R. (2013). Implications of relationship marketing indicators to enable organizational growth: A stakeholder causal scope analysis. In R. Kaufmann & M. F. A. K. Panni (Eds.), *Customer-centric marketing strategies: Tools for building organizational performance* (pp. 214–244). IGI Global.

Shams, S. M. R. (2016). Branding destination image: A stakeholder causal scope analysis for internationalisation of destinations. *Tourism Planning & Development, 13*(2), 140–153.

Shams, S. M. R. (2019). Industry and managerial applications of internet marketing research. In A. Thrassou, D. Vrontis, Y. Weber, S. M. R. Shams, & E. Tsoukatos (Eds.), *The synergy of business theory and practice. Palgrave studies in cross-disciplinary business research, in association with EuroMed Academy of Business*. Palgrave Macmillan.

Shams, S. M. R. (2021). The triple bottom line (People, Planet and Profit) and marketers' strategic priority. Presented at the 53rd Academy of Marketing Annual Conference (Virtual Conference), UK.

Shams, S. M. R. (2024). Making big data useful to everyday users. The Business Standard. Retrieved May 31, 2024, from https://www.tbsnews.net/thoughts/making-big-data-useful-everyday--users-785946.

Shams, S. M. R., & Solima, L. (2019). Big data management: Implications of dynamic capabilities and data incubator. *Management Decision, 57*(8), 2113–2123.

Shankar, V., Grewal, D., Sunder, S., Fossen, B., Peters, K., & Agarwal, A. (2022). Digital marketing communication in global marketplaces: A review of extant research, future directions, and potential approaches. *International Journal of Research in Marketing, 39*(2), 541–565.

Shareef, M. A., Mukerjib, B., Dwivedic, Y. K., Ranac, N. R., & Islamd, R. (2019). Social media marketing: Comparative effect of advertisement sources. *Journal of Retailing and Consumer Services, 46*, 58–69.

Shiqun, Y., Chengjun, Z., & Yu, Z. (2021). The role and path of digital marketing in tourist souvenir brands. *The Proceedings of the E3S Web of Conferences, 251*, 1–4.

Statista. (2023a). Total greenhouse gas emissions released by Alphabet (Google) from financial year 2018 to 2022. Retrieved May 5, 2024, from https://www.statista.com/statistics/788517/ghg-emissions-released-by-google/.

Statista. (2023b). Number of social media users worldwide from 2017 to 2027 (in billions). Retrieved May 6, 2024, from https://www.statista.com/statistics/278414/number-of-worldwide-social-network-users/.

Stephen, A. T., Sciandra, M. R., & Inman, J. J. (2016). Is it what you say or how you say it? How content characteristics affect consumer engagement with brands on Facebook. Saïd Business School, University of Oxford, Research Papers. Retrieved May 8, 2024, from https://deliverypdf.ssrn.com/delivery.php?ID=398002017094087094124073022028115006024081090052024004069113068026080103004092127023016123121059011102019019107113001092121098114082027065044006117067078019086092065039008038118101082080096068065102086111004007084004066020127102096119087121071096083106&EXT=pdf&INDEX=TRUE.

Sustainability. (2020). Messi's boots today, recycled into yours tomorrow. Adidas. Retrieved May 11, 2024, from https://www.adidas-group.com/en/media/press-releases/messis-boots-today-recycled-into-yours-tomorrow.

Sustainability. (2022). Net-zero carbon. *Google*. Retrieved May 5, 2024, from https://sustainability.google/operating-sustainably/net-zero-carbon/.

Tapp, A. (2008). *Principles of direct and database marketing*. Prentice Hall.

The Independent. (2018). Agreement signed to produce jute-made 'Sonali Bag'. *The Independent*. Retrieved June 29, 2019, from http://www.theindependentbd.com/post/168851.

Townsend, S. (2018). 88% of consumers want you to help them make a difference. Forbes. Retrieved April 10, 2022, from https://www.forbes.com/sites/solitairetownsend/2018/11/21/consumers-want-you-to-help-them-make-a-difference/?sh=3dfef6866954.

Trequattrini, R., Shams, R., Lardo, A., & Lombardi, R. (2016). Risk of an epidemic impact when adopting the internet of things: The role of sector-based resistance. *Business Process Management Journal, 22*(2), 403–419.

United Nations. (2014). Leonardo DiCaprio (UN messenger of peace) at the opening of climate summit 2014. YouTube. Retrieved May 11, 2024, from https://www.youtube.com/watch?v=vTyLSr_VCcg.

United Nations. (n.d.) For a livable climate: Net-zero commitments must be backed by credible action. Retrieved May 5, 2024, from https://www.un.org/en/climatechange/net-zero-coalition.

Vailshery, L. S. (2023). Web analytics software market share worldwide 2023. *Statista*. Retrieved May 20, 2024, from https://www.statista.com/statistics/1258557/web-analytics-market-share-technology-worldwide/.

Verhoef, P. C., Broekhuizen, T., Bart, Y., Bhattacharya, A., Dong, J. Q., Fabian, N., & Haenlein, M. (2021). Digital transformation: A multidisciplinary reflection and research agenda. *Journal of Business Research, 122*, 889–901.

Volvo On Demand. (n.d.) A smarter way to have a car in Sickla with Volvo On Demand. Volvo Cars. Retrieved May 5, 2024, from https://www.volvocars.com/static/on-demand/media/AL_VolvoOnDemand_eng_Final.pdf.

Warren, S. D., & Brandeis, L. D. (1890). The right to privacy. *Harvard Law Review, 4*, 193–220.

Weber, R. H. (2013). Internet of things—Governance quo vadis? *Computer Law & Security Review, 29*(4), 341–347.

Weber, R. H., & Weber, R. (2010). *Internet of things*. Springer.

Worldometers. (n.d.) Bangladesh population (live). Retrieved May 31, 2024, from https://www.worldometers.info/world-population/bangladesh-population/.

Yin, S., & Kaynak, O. (2015). Big data for modern industry: Challenges and trends. *Proceedings of the IEEE, 103*(2), 143–146.

6

Further Reading

Ankita, A., & Khanna, A. (2024). Integrating environmental impact and digital marketing strategies for deteriorating items with preservation technology investment. *Process Integration and Optimization for Sustainability* (published online ahead of print). https://doi.org/10.1007/s41660-024-00419-1.

Bloom, P. N., Milne, G. R., & Adler, R. (1994). Avoiding misuse of new information technologies: Legal and societal considerations. *Journal of Marketing, 58*(1), 98–110.

Boegershausen, J., Datta, H., Borah, A., & Stephen, A. T. (2022). Fields of gold: Scraping web data for marketing insights. *Journal of Marketing, 86*(5), 1–20.

Bronnenberg, B. J., Dubé, J.-P., & Mela, C. F. (2010). Do digital video recorders influence sales? *Journal of Marketing Research, 47*(6), 998–1010.

Bruce, N. I., Murthi, B. P. S., & Rao, R. C. (2017). A dynamic model for digital advertising: The effects of creative format, message content, and targeting on engagement. *Journal of Marketing Research, 54*(2), 202–218.

Daviet, R., Nave, G., & Wind, J. (2022). Genetic data: Potential uses and misuses in marketing. *Journal of Marketing, 86*(1), 7–26.

Donath, J. (2021). Commentary: The ethical use of powerful words and persuasive machines. *Journal of Marketing, 85*(1), 160–162.

Dong, B., Zhuang, M., Fang, E., & Huang, M. (2024). Tales of two channels: Digital advertising performance between AI recommendation and user subscription channels. *Journal of Marketing, 88*(2), 141–162.

Dretsch, H. J., Kirmani, A., & Lundberg, J. (2024). Designing brand cocreation activities to increase digital consumer engagement. *Journal of Interactive Marketing, 59*(2), 151–166.

Du, R. Y., Netzer, O., Schweidel, D. A., & Mitra, D. (2021). Capturing marketing information to fuel growth. *Journal of Marketing, 85*(1), 163–183.

Eigenraam, A. W., Eelen, J., van Lin, A., & Verlegh, P. W. J. (2018). A consumer-based taxonomy of digital customer engagement practices. *Journal of Interactive Marketing, 44*(1), 102–121.

Elgaaied-Gambier, L., Bertrandias, L., & Bernard, Y. (2020). Cutting the internet's environmental footprint: An analysis of consumers' self-attribution of responsibility. *Journal of Interactive Marketing, 50*(1), 120–135.

Garvey, A. M., Kim, T., & Duhachek, A. (2023). Bad news? Send an AI. Good news? Send a human. *Journal of Marketing, 87*(1), 10–25.

Ghose, A., Lee, H. A., Nam, K., & Oh, W. (2024). The effects of pressure and self-assurance nudges on product purchases and returns in online retailing: Evidence from a randomized field experiment. *Journal of Marketing Research, 61*(3), 517–535.

Gustavo, J. U., Jr., Trento, L. R., de Souza, M., Pereira, G. M., de Sousa Jabbour, A. B. L., Ndubisi, N. O., Jabbour, C. J. C., Borchardt, M., & Zvirtes, L. (2021). Green marketing in supermarkets: Conventional and digitized marketing alternatives to reduce waste. *Journal of Cleaner Production, 296*, 1–14.

Hollebeek, L. D., & Macky, K. (2019). Digital content marketing's role in fostering consumer engagement, trust, and value: Framework, fundamental propositions, and implications. *Journal of Interactive Marketing, 45*(1), 27–41.
Kim, J., Yoon, Y., Choi, J., Dong, H., & Soman, D. (2024). Surprising consequences of innocuous mobile transaction reminders of credit card use. *Journal of Interactive Marketing, 59*(2), 135–150.
Krafft, M., Sajtos, L., & Haenlein, M. (2020). Challenges and opportunities for marketing scholars in times of the fourth industrial revolution. *Journal of Interactive Marketing, 51*(1), 1–8.
Kumar, V. (2018). Transformative marketing: The next 20 years. *Journal of Marketing, 82*(4), 1–12.
Kumar, A., Bezawada, R., Rishika, R., Janakiraman, R., & Kannan, P. K. (2016). From social to sale: The effects of firm-generated content in social media on customer behavior. *Journal of Marketing, 80*(1), 7–25.
Labrecque, L. I., vor dem Esche, J., Mathwick, C., Novak, T. P., & Hofacker, C. F. (2013). Consumer power: Evolution in the digital age. *Journal of Interactive Marketing, 27*(4), 257–269.
Lin, Y., Yao, D., & Chen, X. (2021). Happiness begets money: Emotion and engagement in live streaming. *Journal of Marketing Research, 58*(3), 417–438.
Luangrath, A. W., Peck, J., Hedgcock, W., & Xu, Y. (2022). Observing product touch: The vicarious haptic effect in digital marketing and virtual reality. *Journal of Marketing Research, 59*(2), 306–326.
Naylor, R. W., Lamberton, C. P., & West, P. M. (2012). Beyond the "like" button: The impact of mere virtual presence on brand evaluations and purchase intentions in social media settings. *Journal of Marketing, 76*(6), 105–120.
Petit, O., Velasco, C., & Spence, C. (2019). Digital sensory marketing: Integrating new technologies into multisensory online experience. *Journal of Interactive Marketing, 45*(1), 42–61.
Pittman, M., & Abell, A. (2021). More trust in fewer followers: Diverging effects of popularity metrics and green orientation social media influencers. *Journal of Interactive Marketing, 56*(1), 70–82.
Puntoni, S., Reczek, R. W., Giesler, M., & Botti, S. (2021). Consumers and artificial intelligence: An experiential perspective. *Journal of Marketing, 85*(1), 131–151.
Rangaswamy, A., Moch, N., Felten, C., Van Bruggen, G., Wieringa, J. E., & Wirtz, J. (2020). The role of marketing in digital business platforms. *Journal of Interactive Marketing, 51*(1), 72–90.
Shmargad, Y., & Watts, J. K. M. (2016). When online visibility deters social interaction: The case of digital gifts. *Journal of Interactive Marketing, 36*(1), 1–14.
Šikić, F. (2021). Using Instagram as a communication channel in green marketing digital mix: A case study of bio&bio organic food chain in Croatia. In M. Topić & G. Lodorfos (Eds.), *The sustainability debate (critical studies on corporate responsibility, governance and sustainability)* (Vol. 14, pp. 221–236). Emerald Publishing Limited.
Stallone, V., Wetzels, M., Mahr, D., & Klaas, M. (2024). Enhancing digital advertising with blockchain technology. *Journal of Interactive Marketing, 59*(1), 76–98.
Swaminathan, V., Sorescu, A., Steenkamp, J.-B. E. M., O'Guinn, T. C. G., & Schmitt, B. (2020). Branding in a hyperconnected world: Refocusing theories and rethinking boundaries. *Journal of Marketing, 84*(2), 24–46.
Tellis, G. J., MacInnis, D. J., Tirunillai, S., & Zhang, Y. (2019). What drives virality (sharing) of online digital content? The critical role of information, emotion, and brand prominence. *Journal of Marketing, 83*(4), 1–20.
Thangam, D., & Chavadi, C. (2023). Impact of digital marketing practices on energy consumption, climate change, and sustainability. *Climate and Energy, 39*(7), 11–19.
Venkatraman, V., Dimoka, A., Vo, K., & Pavlou, P. A. (2021). Relative effectiveness of print and digital advertising: A memory perspective. *Journal of Marketing Research, 58*(5), 827–844.
Yalcin, G., Lim, S., Puntoni, S., & van Osselaer, S. M. J. (2022). Thumbs up or down: Consumer reactions to decisions by algorithms versus humans. *Journal of Marketing Research, 59*(4), 696–717.
Yim, M. Y.-C., & Yoo, C. Y. (2020). Are digital menus really better than traditional menus? The mediating role of consumption visions and menu enjoyment. *Journal of Interactive Marketing, 50*(1), 65–80.

Strategic Marketing, Sustainable Competitive Advantage, and People, Planet, and Profit

Contents

S. M. R. Shams et al., *Sustainable Marketing*, Classroom Companion: Business,
https://doi.org/10.1007/978-3-031-79130-7_7

Preliminary Vignette: How Easily Competitors Can Outplay the Commercial Competitive Advantage, Even of a Triple Bottom Line-Focused Value Proposition

◼ Source: ► pexels.com/royalty-free-images/

We produce more food than the number of people we need to feed globally (The Guardian, 2022). However, an August 2023 World Vision report shows that across the world, "as many as 828 million people—about 10% of the global population—regularly go to bed hungry" (np). Different stages of food insecurity (i.e., cutting budget, changing diet, decreasing food consumption first by adults and then children) in households "are consistent in both developed and developing countries and across languages and cultures" (Smith & Meade, 2019, np). Unfortunately, the amount of food we produce, 56% in developed economic countries and 44% in developing economic countries, are wasted or lost at some point in our food value network, i.e., at retail, at restaurant or at household (Ishangulyyev et al., 2019).

What should we do with the leftover food (near to its expiry) at commercial kitchens? In response to this question, ◼ Fig. 7.1 shows the results of a 2022 consumer survey in the United States. Seventy-two percent consumers strongly agree to give this food to charities, 58% strongly suggest to give it to the restaurant staff members for free, 54% strongly agree to give it to the public for free, and 45% responders strongly suggest selling it at discounted price to consumers (Statista, 2023a). In ► Chap. 1, Marketing in Action ► 1.1 (A sustainable Marketing Dynamic to Deliver Value to Customers, Partners and Society at Large) describes about the "Look, Smell, Taste" app, based on what sustainable marketeers are reflecting on the suggestions of 45% consumers to sell it at discounted price, which not only helps the restaurants to reduce their operational loss and make some profit, but also helps to address food waste, feed more people in our society, and helps address environmental challenges related to food waste, considering that 38% of total energy produced are used in our global food system (United Nations, 2023).

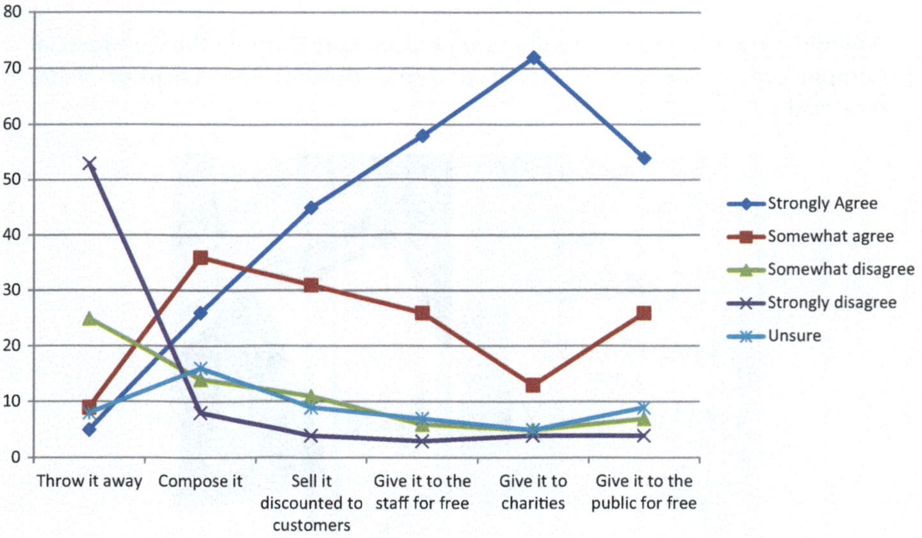

Fig. 7.1 Consumers' opinions on what to do with surplus food from businesses that are approaching expiration. (Information source: Statista, 2023a)

Upon seeing that people are searching in bins for food, and realizing the extent of food waste each day at their own restaurant, a restaurant owner in India placed a refrigerator outside her restaurant to store leftover food for homeless and hungry people to grab for free (The Independent, 2016; The Mirror, 2016). The popularity of this idea has attracted affluent people, as they have also started donating their leftover food to this refrigerator to help those who need it most. The campaign:

- Received huge mainstream and social media coverage for the restaurant.
- Created a positive brand position in the consumers' mind that helped the restaurant to establish a socially and environmentally responsible brand image.
- Increased the restaurant's profit margin, because, the popularity of this idea started bringing new consumers to the restaurant.

Considering the potential societal, environmental, and commercial value of this triple bottom line-focused value proposition, we have seen other giant restaurant businesses too started to implement this idea. For example, partnering with Feeding America, Starbucks announced that they "will give away 100 percent of its food waste to those in need" (Henderson, 2016, np).

Alongside the societal and environmental benefits, the idea of the Pappadavada restaurant owner in India, who named/branded her refrigerator as "nanma maram" (meaning "tree of goodness") to distribute leftover food for anyone to grab for free (The Independent, 2016) is also able to establish an initial commercial competitive advantage for her restaurant, and the restaurant owner could enjoy this commercial competitive advantage in its full extent, until the competitors would be able to copy and implement the same or similar idea, perhaps even with greater value proposition. It is likely that competitors would also prefer to increase their restaurants' popularity, based on this or similar other socially or environmentally caring campaigns.

Yes, as sustainable marketeers, our aim is to implement such socially and/or environmentally caring campaigns more and more, including among our competitors. But simultaneously, we need to protect our commercial interest, because, our commercial supremacy among our competitors also enables us to pay our employees better and contribute to our society and environment better. As discussed in ► Chap. 1, sustainable marketing is not philanthropy, but the necessity to satisfy the growing number of socially and environmentally conscious customers in our global marketing dynamics, based on planning, implementing and evaluating our triple bottom line—focused value proposition that would offer superior societal, environmental and commercial (SEC) value to our triple bottom line's stakeholders.

Since competitors can copy and implement such SEC-value-centered sustainable marketing strategies, and outplay a brand's commercial competitive advantage; let's have a look at a simulative example to understand how marketeers could possibly defend and prolong their commercial competitive advantage. Let's assume that you are marketing director of a refrigerator brand in your country. To secure a niche market at least for a considerable period of time, you are, for example, signing a contract with the health department or National Health Service (NHS) in your country. How many medical facilities, i.e., hospitals and other facilities the NHS may have throughout the country, and at all these country-wide medical facilities, how many refrigerators the NHS may need to purchase and repair over a period of 5 years. In this context, as a single buyer, NHS is one of the largest refrigerator buyers for your refrigerator brand and its competitors.

Let's assume that as per the contract, the NHS will purchase all their new refrigerators and repair the old ones only from your brand during the next 5-year period. And, you will put your refrigerators at common places in major cities/towns across the country, where restaurants or individuals would be able to donate their leftover food for anyone to grab for free. In this way, don't you think that you would be able to prolong the SEC-value-oriented short-lived commercial competitive advantage of your brand, at least for one of the largest refrigerator buyers in your country for the contracted period of 5 years? You would also be able to promote your brand, through socially and environmentally caring marketing messages based on this campaign, which would also be instrumental for you to influence further the other segments or niches in your country-wide refrigerator market, in order to develop an authentic socially and environmentally caring brand image for your refrigerator.

7.1 Introduction

The Preliminary Vignette of this chapter discusses how easy it would be for competitors to outplay a short-lived commercial competitive advantage of a brand, even for a value proposition that is focused on the triple bottom line. This is why, we see that competitive brands such as iPhone and Galaxy need to relaunch them every year. It is because that the short-lived competitive advantage that iPhone was able to establish based on their most recent version of smartphone, has been outplayed by Samsung Galaxy, based on their latest version (or vice-versa), considering that competitors closely monitor the features of competitive products and services, and persistently attempt to copy a same or add a superior feature in their own product or service. Such intense head-to-head competition among competitive brands requires a lot of resources, time and energy. But the question is how many brands would have as much resource as iPhone and Galaxy have? Therefore, marketeers need to consider how they could possibly sustain or at least prolong their short-lived commercial competitive advantage over a mid- or long-term period.

The last two paragraphs of the Preliminary Vignette of this chapter also discuss about a simulative case example to clarify how marketeers could possibly prolong their short-lived commercial competitive advantage. Recent strategic marketing research demonstrates that our existing marketing knowledge is inept to meaningfully sustain the commercial competitive advantage of a brand in practice, even only from a single bottom line context (Vrontis et al., 2021), i.e., focusing only on profit-making initiatives to ensure the brand's long-term commercial competitive advantage. Subsequently, the existing strategic marketing principles are further impractical to pragmatically support a brand's initiative to prolifically pursue the triple bottom line i.e., contributing to people and planet while making profit (Peerally et al., 2019; Möller et al., 2020; Robiady et al., 2021). To address this gap in existing marketing knowledge, this chapter will help prepare the future marketeers to apply different strategic cross-disciplinary and cross-functional business theories to support their sustainable marketing strategies, with an aim to possibly sustain or at least prolong the commercial competitive advantage of their triple bottom line-focused value propositions.

In ▶ Sect. 1.1 (Introduction) of ▶ Chap. 1, we referred that when we use the terms in this textbook, such as "sustainability" or "sustainable", we refer it not only for societal or environmental sustainability, but also for commercial sustainability (i.e., creating a business which is commercially sustainable and simultaneously contributes to its society and/or environment for societal and/or environmental sustainability). In this context, this current chapter discusses about how marketeers could equip themselves with cross-disciplinary and cross-functional knowledge, such as VRIN capabilities and dynamic capabilities theories from strategic management knowledge stream, and nudge theory from behavioral economics and psychology, in order to support their sustainable marketing initiatives, and sustainable value propositions, with an aim to sustain the overall operational (i.e., societal, environmental, and commercial) competitive advantage of their brands.

Following a preliminary discussion on competitive advantage (► Sect. 7.2), the remainder of this chapter discusses about firm resources and dynamic capabilities (► Sect. 7.3), how a short-lived competitive advantage could possibly be sustained or at least be prolonged (► Sect. 7.4), marketing capabilities for sustainable competitive advantage (► Sect. 7.5), embryonic changes in marketing dynamics and sustainable marketing (► Sect. 7.6), and finally, summarizing from the discussions of this chapter, how entrepreneurial marketeers have been applying strategic marketing to pursue the Sustainable Development Goals (SDGs) is discussed in ► Sect. 7.7.

7.2 Competitive Advantage

◘ Source: ► pexels.com/royalty-free-images/

► Section 1.2.5 of ► Chap. 1 discusses that "value" is the core concept of marketing. It is because, competitive brands compete among themselves, with an aim to create superior value, in order to address the same problem and need of their common customers. Eventually, the core purpose of creating value through marketing capabilities (e.g., product design, competitive pricing, convenient distribution channel, and appealing promotional message) is establishing a competitive advantage of our brand, over its competitive brands. In its simplistic sense, the competitive advantage of a brand could be defined as the superior value proposition of the brand, in comparison to its competitive brands. In general, marketeers attempt to create superior value through their marketing capabilities, in order to demonstrate to their target audiences why doing business with our brand would be more prolific for the target audiences, instead of doing business with our competitive brands, so that the brand's overall operation could survive and prosper. Therefore, operational competitive advantage of a brand could be defined as a brand's target audiences' perceived experience about why doing business with the brand would be more prolific for them, instead of its competitive brands.

- **Competitive advantage** is superior value proposition of a brand, in comparison to its competitive brands.
- **Operational competitive advantage** reflects through a brand's target audiences' perceived experience about why doing business with the brand would be more prolific for them, instead of doing business with its competitive brands, so that the brand's overall operation could survive and prosper.
- **Sustainable competitive advantage** is "a distinct reason of supremacy of a... brand that offers superior value for...(its) target market in association with all involved (triple bottom line's) stakeholders, comparing to any other existing or emerging substitute (brands)..., as well as share that value with the associated (triple bottom line's) stakeholders, with a long-term outlook, potentially for life-time" (Shams, 2013a, p. 244).

In terms of developing an operational competitive advantage, the goal of sustainable marketeers should be designing, delivering, and evaluating superior value propositions for their triple bottom line's societal, environmental, and commercial stakeholders, so that the involved stakeholders would remain motivated to continue their business with the brand, instead of doing business with its competitive brands. Since brands co-create value in association of its key stakeholders (see ▸ Fig. 1.8 in ▸ Chap. 1), such as suppliers, investors, business partners, employees, distributors, city council (e.g., collecting back the used packaging of your product, so that the product's production cost can be reduced while improving its environmental footprint), and other key stakeholders; sustainable marketeers should demonstrate to their stakeholder groups how creating value together with our brand would be most beneficial for them over a long-term period, instead of creating value with our competitive brands. In this context, sustainable competitive advantage could be defined as "a distinct reason of supremacy of a...brand that offers superior value for...(its) target market in association with all involved (triple bottom line's) stakeholders, comparing to any other existing or emerging substitute (brands)..., as well as share that value with the associated (triple bottom line's) stakeholders, with a long-term outlook, potentially for life-time" (Shams, 2013a, p. 244).

To motivate the associated triple bottom line's stakeholders first time, and keep that motivation intact over a long-term period, one of the options would be sharing dependency, business risk and uncertainty among the involved stakeholders. ▸ Section 1.6 in ▸ Chap. 1 presents more discussions and examples of sharing dependency, risk and uncertainty. Another option would be enabling the involved stakeholders feeling that the value of partnering with our brand, instead of the competitive brands, not only from the context of engineering/manufacturing quality of our product, but also the "feeling" quality of our product and its overall value proposition. Manager's Corner ▸ 7.1 presents a discussion on "feeling" quality of Japanese car brands.

Manager's Corner 7.1: Changing Concept of Product Quality: From Body to Soul as From "Engineering" Quality to "Feeling" Quality

Source: ▶ pexels.com/royalty-free-images/

Japanese companies, once were reeling from the dramatic appreciation of the yen, and thus were having tremendous difficulty pricing their products competitively in the US market. Consequently—and hastily, however—US business executives as well as MBA students began to believe in the end of the "Japan era" in global competition. Can we really relax? Uh oh, Detroit, watch out. Once again, something extraordinary happened in Japan. Just as US car makers improved their product quality, the Japanese were redefining and expanding the term. The new concept is called *miryokuteki hinshitsu*—making cars that are more than reliable; cars that fascinate, bewitch, and delight. In plain English, this expression translates into "things gone right". It signaled a campaign by the Japanese to engineer unprecedented measures of look, sound, and "feel" into everything from family sedans to luxury models, at the same time that they continued to improve reliability.

As Mark Twain once wrote, "if you stand still on the road, you will get run over." You can never relax in an era of global competition. When the Big Three—GM, Ford, and Chrysler—narrow their competitive gap vis-à-vis their Japanese rivals, they should move on relentlessly to look for a next generation of competitive advantage. That is the only known means of survival in a competitive world.

7

To pursue the next generation of competitive advantage, the Japanese companies intertwined their engineering capacities well with their global customers' needs and feelings. It is not just mechanically developing a car engine or its interior, but to reflect on the customers' preferences and feelings as the soul of the body of a car. "The soul leaves the body" is a common euphemism for death, but what about "when the soul leaves business"? This is increasingly the case, as business managers care more about profit than about decency and curative behavior. As a recent study explores in "Qualitative Marketing Research," business must change: the soul must come back home.

Wrestling for the soul of business is nothing new. Each year we are reminded about just how far some companies are falling short. In 2015 Volkswagen cheated emissions regulators. More recently, United Airlines bloodied a passenger while dragging him from a plane the company had overbooked. Most stunning to it is that not a single United employee—pilot, ground crew, or flight attendant—interceded to say, "This is wrong."

The traditional pillars of business—risk, competition, product or service decision, subsequent profit and ownership—are insufficient for the contemporary age. Companies focus too much on hedging risks, ousting their competitors, maximizing profits and expanding assets, and forget the cardinal rule of human decency: reverence for others. Product or service decision without soul takes on a sinister character, which charges exorbitant rates for on-flight pillows or hotel mini-bar snacks. In a true sense, business is meant to create value through its product or service decision, for itself and the customer, not leach off others.

There was a time when managers cared about all these things. Years ago, a Japanese firm was dissatisfied with five product failures in a 12,000-item shipment. The company demanded that the supplier conduct an investigation into its production process to prevent future product problems. Their supplier was dismissive, offering instead 20 free units as compensation. The company stood firm, demanding a commitment to flawless excellence or their contract would be canceled. The investigation took place, and the product continues to be superior. Perhaps *Profiles in Courage* the 1957 book by John F Kennedy, needs to be reread today.

Furthermore, the old pillars need a refurbishing, placing a seat on top of all decision-making related to product or service design and delivery to give the customers their rightful priority. New pillars should be Truthfulness, Authenticity, Simplicity, Expanded Participation, and Personal Responsibility. When problems arise, managers often obfuscate, develop complicated excuses, or disengage by claiming "I didn't know, I just work here." Managers must take responsibility for the effects on their businesses, consequences of the flaws of their product design and delivery decisions, and initiate curative action for past damage and to prevent future harm. Not all conditions can be presented in an employee handbook; however, the acid test question is: "Does your mother know and approve of what you do?"

How to instill these new pillars in business, and in particular product decisions? One solution is to promote mindfulness in corporate life. It is easy to lose sight of what is important, spending each day laboring in an office far from the impacts of work. Practicing mindfulness can help overcome this.

Another is to specifically focus on how to bring the soul back into business operation. For this, we can look to the tenets of the Jesuits, whose vigorous promotion of honor and service deserves our praise and emulation.

If businesses are to thrive in this era, they must rediscover their soul. We need companies like the old GM—a car company of car people. Passion and commitment to excellence must drive business and its product design, delivery and evaluation decisions. This is the soul of business that should be reflected in product/service decision-making, and just as the body dies without its soul, so does the business or its products/services as market offerings. Therefore, intertwining the technical aspects of product development with soft skills like understanding customers' and societal and environmental stakeholders' voices and feelings is imperative to bring the soul back into new product development, in order to deal with the competitive gap(s) of the product.

Information Sources:

A new era for auto quality (1990, October 22). *Business Week*. 83.

Bowonder, B., & Miyake, T. (1994). Creating and sustaining competitiveness: An analysis of the Japanese robotics industry. *International Journal of Technology Management*, 9, 575–601.

Unknown. (2003, June 12). Birthday Blues. *Economist*.

Czinkota, M. R. (2018). *In Search for the Soul of International Business.* Business Expert Press.

Czinkota, M. R., Kotabe, M., Vrontis, D., & Shams, S. M. R. (2021). *Marketing management: Past, present and future* (4th edn.). Springer International Publishing. (Reprinted a slightly revised version by permission of the copyright owner, © Springer).

7.3 Firm Resources and Dynamic Capabilities to Develop Competitive Advantage

In order to develop a short-lived competitive advantage for the first time, and to possibly sustain or at least prolong that short-lived competitive advantage, marketeers need to proactively deal with their existing and potential resources and capabilities. It is because; marketeers' existing and potential resources and capabilities are instrumental to create value, and integrate that value into the 4Ps or extended marketing mix elements. The current section, discusses about firms' (or marketeers') resources and capabilities that they need to employ, in order to create value for their target markets. Marketeers could source resources from three different sources of resources, namely:

- Market-based resources
- Resource-based resources
- Knowledge-based resources

> Analysts have tended to define assets too narrowly, identifying only those that can be measured, such as plant and equipment. Yet the intangible assets, such as a particular technology, accumulated customer information, brand name, reputation, and corporate culture, are invaluable to the firm's competitive power. In fact, these invisible assets are often only real source of competitive edge that can be sustained over time. (Hiroyuki Itami, 1991; Analyzing Resources and Capabilities, 2007, p. 123)

Analyzing a firm's these three (market-based resource, resource-based resource and knowledge-based resource) sources of tangible and intangible resources can help marketeers to explore and employ those resources, in order to develop a commercial competitive advantage for their brands. To explore all three sources of a firm's resources, market orientation is important to probe and understand triple bottom line stakeholders' major and minor needs, wants and expectations, in order to reflect on those needs while developing, delivering and monitoring a value-delivery process. ▶ Section 1.4.1 (Ongoing Market Orientation) in ▶ Chap. 1 discusses about market orientation.

7.3.1 Market-Based View of Resources

The market-based view of resources give emphasis to the role of market conditions, related to a firm's current and potential relationships and interactions with its internal and external stakeholders to develop a (marketing) strategy (McGee, 2015). Such internal and external stakeholders for market orientation to the stakeholders' needs, relevance, and influence on a firm's business would include, but not limited to (Miguel & Santiago, 2010; Shams, 2017):

- Orientation toward employees.
- Orientation toward buyers (both consumers and customers).

- Orientation toward partners and suppliers
- Orientation toward external agencies (e.g., media, community groups, cause-related activist groups)
- Orientation toward government agencies
- Orientation toward non-government organizations (e.g., leading policy and research firms/institutions)
- Orientation toward society
- Orientation toward environment
- Orientation toward competitors
- Orientation toward technological progress
- Orientation toward geopolitical issues
- Orientation toward economic conditions and changes
- Orientation toward legal conditions

McGee (2015) argues that such orientation could also be emphasized from the context of Porter's five competitive factors, in order to understand the influences of existing competitors, buyers' bargaining power, suppliers' bargaining power, threats of new entrants, and threats of substitute product or service. ► Section 1.4.2 (Competitive Forces in Your Market) in ► Chap. 1 discusses about Porter's five competitive forces and stakeholders' bargaining power. The purpose of such stakeholder-focused market orientation to explore stakeholder-specific market-based resources is twofold (Srivastava et al., 2001):

- Exploring stakeholder-specific relational resources
- Exploring stakeholder-specific intellectual resources

Source: ► pexels.com/royalty-free-images/

For example, you are planning for your next international marketing project targeting an overseas market. But you or your firm do not know much about the socioeconomic, ecological, and overall market conditions of that overseas market. In this case, for example, you have established a relationship with your new business partner (e.g., an investor or distribution channel partner) in that overseas market. The overseas market knowledge that you would be able to obtain through your relationships and interactions with your new overseas business partner is considered as market-based relational knowledge that a firm can explore and apply in association with their business partners.

Web Exercise 7.1

The Unilever's global web page, "Partner with Purpose" (▶ https://www.unilever.com/suppliers/partner-with-purpose/) presents information on their current and prospective future partners/suppliers across the globe. The web page provides the following information to attract new partners and keep motivated the existing partners in their global market:

- A brief introduction to the future of partnerships
- The growth ecosystem (for partners)
- How brands with a purpose, such as "clean future" or "force for good," can grow better
- More information about purpose-led collaboration
- How purpose-led people thrive and how purpose-led brands survive and prosper
- People-led and planet-led plans
- Partner promises and partner awards

Unilever's purpose-led partnership, such as Clean Future (2024) contributes to use less water and less energy in their value-delivery process (e.g., from product development to product delivery) that helps to pursue SDG-6 (clean water) and SDG-7 (i.e., affordable energy) in their global value chain. Browse Unilevers' and other global brands' partnership websites to explore how these brands underpin their commercial competitive advantage, based on their relational resources that they obtain from their overseas partners.

Market-based intellectual resources are those specific knowledge or skills of a firm (e.g., a firm's employees' knowledge or skills) that help the firm to develop a superior competitive advantage in a market. Such intellectual resources would be relevant to how to execute a particular task more cost-effectively, or executing a task in a way that offers superior value to its customers and/or non-customer stakeholders. For example, let's assume that there are ten Italian restaurants on a street. All ten restaurants offer the same cuisine, same price, and same level of customer service. However, one out of these ten restaurants could become much popular in comparison to its nine other competitors, because of the unique recipe prepared by its chef. The question is why someone else can make a rocket engine that can take us to the planet Mars, but why I can't? Because of such natural distinctive abilities that we see from person to person, there is a talent war among competitive firms to attract and retain such extraordinary employees (i.e., the chef in the case of the example discussed here), who offer superior value to their employers and its customers and non-customer-stakeholders. This talent war is applicable to any industry and market, as competitive brands try to underpin their competitive advantage based on their intellectual resources.

- **Relational resources** are those market-specific resources, e.g., market knowledge that a firm would be able to obtain from their market-specific stakeholder relationships and interactions.
- **Intellectual resources** are those specific knowledge or skills of a firm's employees that help the firm to develop a superior competitive advantage in a market.

Renewable energy and social business, both are well-recognized fields of practice that help us to pursue our Sustainable Development Goals. However, the application of the social business concept on the renewable energy sector is scarce. More specifically, considering the potential implications of social business to effectively manage the value dynamics of renewable energy for sustainable socioeconomic development, scholars argue that we need "empirical evidence on social entrepreneurship initiatives in the field of renewable energy" (Becker et al., 2017, p. 33). Experts emphasize on the potential application of the social business concept on the renewable energy sector, considering that their joint application could help our global community to address our socioeconomic and ecological concerns more meaningfully. The socioeconomic value dynamics of renewable energy could include but not limited to sustainable clean energy production, management, marketing, and consumption of renewable energy to support the socioeconomic progress of all (low, mid, and high income) population groups of a society, alongside monitoring the value dynamics. In this context, we have seen recent initiatives to apply the social business concept in the renewable energy sector. For example, in Malawi, a landlocked country in South-Eastern Africa, "energy services social enterprises are emerging as a sustainable response to the challenges of achieving SDG 7 (affordable and clean energy), but detailed case study insight is lacking to inform the sector" (Eales et al., 2020, p. 1). Think about how, as a sustainable marketeer, you could explore and apply your intellectual resources, e.g., by attracting talented individuals both from the social business and renewable energy sectors to form a project team, in order to contribute to people (e.g., more job creation) and planet (e.g., offering clean and affordable energy) while making profit through a social business-led renewable energy venture. Such partnerships between people or organizations from the social entrepreneurship and renewable energy sectors can help us pursue SDG 17 (partnerships for the goals).

7.3.2 Resource-Based View of Resources

The resource-based view of resources refers to those resources that a firm can obtain from its physical resources (Williamson, 1975, 1981), human resources (Becker, 1964) and organizational resources (Tomer, 1987; Rotheim, 1990; Altman,

2003). For example, buildings, furniture, machines all such physical objects that are owned or accessed by a firm to use them, are considered as physical resources. Human resources refer to the manpower that a firm could have, e.g., either through their regular (full-time, part-time, casual) staff members or through their outsourced agencies.

A firm's day-to-day and short-term, mid-term and long-term strategic organizational decision-making ability and process, and its efficiency, collectively are referred to the organizational resources of the firm. For example, McDonald's have 40,275 restaurants in their global value chain across the world (Statista, 2022). The McDonald's (2022) global headquarters is located in Chicago, Illinois, United States. It is normal that McDonald's Chairman or Chief Executive would not be able to know instantly the types of satisfactory and dissatisfactory remarks that they receive every day from their 40,275 global locations. The question is, how quickly the customers' remarks that are received by their front-line customer service attendants across the world, could be conveyed to the McDonald's top management at their global headquarters via middle management and via their local and regional offices or directly to headquarters, and how efficiently they are able to process this information to adjust their strategy to reflect on those customers' remarks? Answering this question would help us to understand how efficient the McDonald's organizational resources are from their day-to-day operational context. How efficiently McDonald's are able to conduct market research, in order to decide where to open a new restaurant, what would be the prospective demand in that particular location, what would be the location-specific (or culture-specific) promotional message, what would be the supply chain dynamics (e.g., inward supply of their burger's ingredients), and how effective McDonald's decision-making related to these questions to retrieve back their investment? Answering such questions would help us understand the efficiency of McDonald's organizational resources from their strategic context.

The efficiency of organizational resources (i.e., the ability to understand and reflect on customers' views promptly and accurately) could differ among competitive brands. For example, let's assume that five young social entrepreneurs received $10,000 business grant each from their local council. By the end of the first year, one entrepreneur was able to produce $12,000, one $11,000, one 10,500, one 10,200 and one $9000. However, the funds all five entrepreneurs had to launch their businesses were equal; they were able to make profit or incur loss in different extents. It was because all five entrepreneurs' capability to explore, apply, and monitor their organizational resources, based on both day-to-day and strategic contexts, was not equal. In order to establish and prolong or at least possibly sustain a short-lived commercial competitive advantage based on resource-based view of resources, the resources (i.e., physical, human and organizational resources) that are engaged to develop that competitive advantage, need to be heterogeneous (distinctive) and immobile among the competitors. For example, the employee who is very skilled to conduct market research, in order to understand customers' and non-customer stakeholders' needs accurately and promptly, that market research skill needs to be unique among the other market researchers of the competitive brands. Also, in order to prolong or sustain a competitive advantage, a brand should satisfy that unique employee, so that the employee will not switch job to join a competitive brand because of superior salary and benefits that would be offered by the competitive brand.

7.3.3 Knowledge-Based View of Resources

Source: ▶ pexels.com/royalty-free-images/

The knowledge-based theory of firm resources considers knowledge as a firm-specific distinctively unique resource, and considers "the firm as a dynamic, evolving, quasi autonomous system of knowledge production and utilization" (Spender, 1996, p. 59; as cited in Kearns & Sabherwal, 2006, p. 132). Its proponents argue that knowledge-based resources, which are unique to each firm, are usually difficult to imitate by competitors. Therefore, such heterogeneous and socially complex knowledge bases and capabilities among competitive firms are the major determinants of sustained competitive advantage and superior marketing performance. In this context, how a firm could influence its customers' and non-customer stakeholders' behavior through the firm-specific unique knowledge (i.e., Coca-Cola's unique recipe) is the key consideration in knowledge-based view of resources to establish, prolong, and possibly sustain a brand's competitive advantage (Sabherwal & Sabherwal, 2005; Kearns & Sabherwal, 2006). Sveiby (2001) suggested managers' nine questions to ask themselves, as answering these questions is instrumental to explore, apply, and monitor firm-specific unique knowledge-based resources:

1. How could we acquire knowledge from people within our organization?
2. How could our employees improve the knowledge and competence of our customers and non-customer stakeholders?
3. How could our customers and non-customer stakeholders improve our employees' knowledge and competence?
4. How could we improve our systems and tools based on the individually held knowledge within our organization?

5. How could we improve our employees' individual knowledge using our existing systems and tools?
6. How could we increase dialogue among our customers and non-customer stakeholders so that they can improve their knowledge about us?
7. How could the knowledge of our customer and non-customer stakeholders improve our systems, tools, and market offering?
8. How could our existing systems and tools help improve the knowledge of our customers and non-customer stakeholders about us?
9. How could our organizational systems and tools be effectively integrated so that the systems and tools can complement each other?

In ► Sect. 7.3.4, after discussing a firm's "dynamic capabilities," we have discussed these nine questions of Sveiby (2001) in ◘ Fig. 7.2 in detail with examples, considering that answering these nine questions to explore a firm's knowledge-based resources also can help the firm to improve its organizational dynamic capabilities.

7.3.4 Dynamic Capabilities

◘ Source: ► pexels.com/royalty-free-images/

Based on 3Ps: Process, Position, and Path (dependency) of a firm, organizational dynamic capabilities of the firm are denoted to the "coordination, integration, learning, reconfiguration and transformation of processes (of a firm to)…strategically allocate/reallocate its resources, based on the path/strategic direction it pursues, with an aim to prolifically dealing with the competitive business environment" (Teece et al., 1997, as cited in Shams, 2016a, p. 673). Organizational dynamic capabilities should focus on allocating or re-allocating a company's resources, aiming to establish a competitive advantage for its products or services in a way that can sustain the competitive advantage. This should be the case even in a very dynamic market or industry that encounters rapid changes or one that has industry-specific competitive factors (e.g., rapid technological innovation).

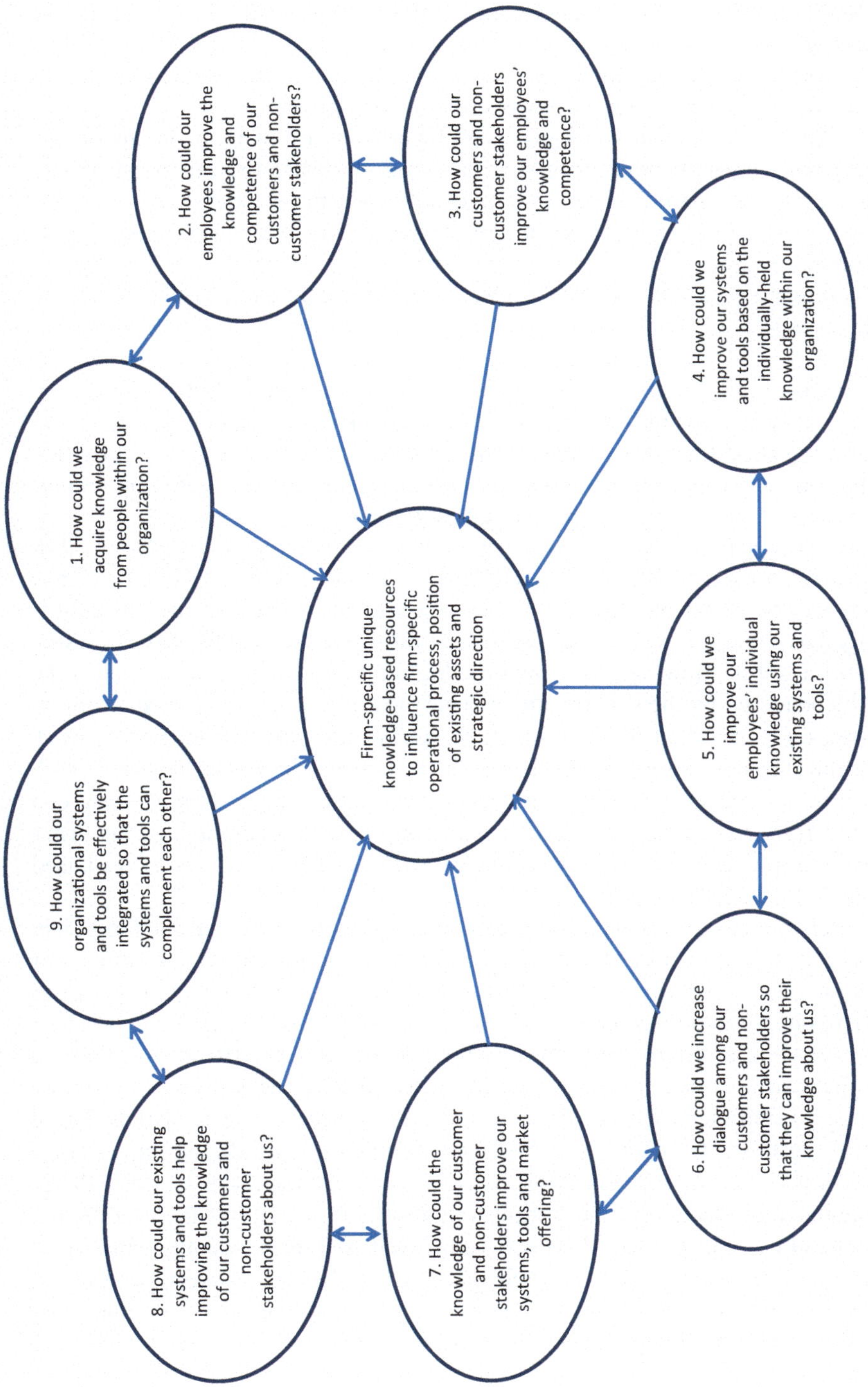

Fig. 7.2 Nine questions to explore firm-specific unique knowledge-based resources in order to influence a firm's dynamic capabilities. (Adapted from Teece et al., 1997; Sveiby, 2001)

In organizational dynamic capabilities, Process refers to an organization's processes that comprise its day-to-day operational and long-term strategic decisions. For example:

- How a service firm organizes its daily schedule or weekly work roster for its front-line customer service employees (day-to-day process)?
- How the front-line customer service employees develop insights on customers' perceptions through their customer interaction (so that the organization can reflect on those customers' perception for a long-term period)?
- How the front-line customer service employees convey such customers' insights to top management, in order to enable the firm to re-design its service offerings upon integrating those insights into its new or modified service offerings (long-term focus)?

Such organizational processes are applied to organize a firm's all different activities. For example, customer service, human resource management, inbound and outbound logistics, financial management, promotional and marketing activities, coordination between different cross-functional areas, e.g., in between finance department and marketing department, among others.

Position in organizational dynamic capabilities refers to the extent of accessibility of a firm to its available resources at the right time. For example, a firm may want to sell one of its buildings, but they cannot sell it, because the building is mortgaged to a bank. It means that the firm cannot have the benefit of the building (e.g., cannot sell it); however, they own the building. Similarly, a firm may not be able to be benefited by the skill of one of their most talented and efficient employees during the time when the firm needs the service of that particular employee, because the employee could be ill. Therefore, the concern here is to ensure a firm's resources' position, its availability and accessibility in a way, so that the firm can utilize the resource when they need it. Ensuring this accessibility to the firm's resources is applicable to all types of resources (e.g., market-based, resources-based and knowledge-based resources).

Path dependency in organizational dynamic capabilities refers to the long-term strategic goal that a firm has been pursuing so far or would pursue in future. For example, ► Sect. 1.5.1 (Sustainable Vision and Mission) in ► Chap. 1 discusses that Microsoft's initial mission in 1992 was to confirm "a computer on every desk and in every home" (Bae, 2015). By the turn of the century (from twentieth to twenty-first century), we have seen that Microsoft have achieved their initial mission (i.e., "a computer on every desk and in every home") at least at their home market in the USA. It means until Microsoft fulfilled its initial mission, "a computer on every desk and in every home" was its path dependency or strategic direction to manage its resources to accomplish that mission. By end of the first decade of this current century, once Microsoft (2024) felt they have accomplished their initial mission, they have redefined their mission, i.e., "empower every person and every organization on the planet to achieve more." Since then this redefined mission is Microsoft's path dependency or strategic direction to allocate or reallocate their resources to accomplish this mission. Manager's Corner ► 1.1. (Microsoft's Redefined Mission Statement to Empower the Triple Bottom Line's Stakeholders) in ► Chap. 1 discusses more about Microsoft's redefined mission.

Dynamic capabilities of an organization refer to its organizational processes, assets' position and path dependency (strategic direction), and how efficiently the organization implements and evaluates its processes to deploy its resources, in order to pursue its organizational goal.

A firm's vision, mission, strategic goals and objectives collectively construct the firm's path dependency or strategic direction within their organizational dynamic capabilities. ► Section 1.5.1 (Sustainable Vision and Mission) in ► Chap. 1 discusses about a firm's vision and mission in detail. Now, let's have a look at Sveiby's (2001) nine questions (in ◘ Fig. 7.2) that can help marketeers not only to explore their firm-specific unique knowledge-based resources but also can help to enrich their firm-specific dynamic capabilities.

? Question 1: How could we acquire knowledge from people within our organization?

It is normal that all employees within an organization will not have an equal level of knowledge, industry experience and skill (e.g., operational know-how). Thus transferring knowledge from the most knowledgeable and experienced employees to our early-career employees could not only ensure that our organizational process will not be interrupted when that most knowledgeable employee goes for annual leave or if that particular employee resigns from our organization, but also such knowledge transfer among our employees ensures that more employees within our organization have advanced skills (e.g., learning the use of an industry-specific specialized software) to apply their creativity, in order to contribute to our overall organizational dynamic capabilities, i.e., contributing to and improving our organizational process, planning for our resources position and contributing to our path dependency.

? Question 2: How could our employees improve the knowledge and competence of our customers and non-customer stakeholders?

When Uber, the taxi service provider, was first launched in 2009 (Blystone et al., 2024), many of their customers did not know how to use Uber taxi, using their App. Thus, to educate their customers, Uber (2024) demonstrates a step-by-step process (i.e., create an account, enter your destination, meet your driver, check your trip, sit back and relax, and rate your trip) to improve their customers' knowledge and competence on how to use Uber service. It helps Uber to establish and make popular a new process of using taxi service that aims to offer most competitive value to their users.

Question 3: How could our customers and non-customer stakeholders improve our employees' knowledge and competence?

It is just the opposite of question 2. The question 2 helps marketeers to make their customer and non-customer stakeholders knowledgeable about their market offerings. The focus of question 3 is how our employees' knowledge and competence to serve our target audiences could be improved based on the inputs of our customers and non-customer stakeholders. For example, our employees improve their industry knowledge, based on attending a training program organized by the local Chamber of Commerce (which is a non-customer stakeholder).

Question 4: How could we improve our systems and tools based on the individually held knowledge within our organization?

"After a period of rapid expansion, Starbucks competitor Luckin Coffee drummed up more revenue than Starbucks for the first time last quarter in China" (Fantozzi, 2023, np). Luckin Coffee and Starbucks are the top two competitors in the Chinese coffee chain shop industry. To retrieve back their lost position in the Chinese market, let's assume that Starbucks have appointed a new CEO for Starbucks China, who is a local Chinese executive and very knowledgeable about the Chinese coffee chain industry and Chinese coffee drinkers' needs, wants, and expectation. This new Chinese CEO may have introduced a new process in Starbucks China based on their individually held knowledge and experience about the Chinese market that could improve Starbucks Chinese operational systems and tools in a way that would be instrumental to delight the Chinese coffee drinkers, in order to win back their lost position. It applies to our existing employees, as well. For example, at a large organization, e.g., in McDonald's, one of their employees at one of their very globally remote locations may have applied a new way of customer service, e.g., organizing customers during busy periods based on a different queuing process. If McDonald's sense that this new queuing process can help them serve more customers in a better way, especially during the busy periods, they may consider adopting this new process in their other global locations to improve their customer service system.

Question 5: How could we improve our employees' individual knowledge using our existing systems and tools?

It is just the opposite of question 4. In this case, an organization provides training to their employees to improve their employees' individual skills. For example, training about a newer version of the organization's database marketing software, or training a new employee to use the existing database marketing software.

Question 6: How could we increase dialogue among our customers and non-customer stakeholders so that they can improve their knowledge about us?

Many brands have their own publicly available online customer forums. In these forums, customers share their experiences and ask questions about the brands' market offerings, its price ranges, customer service options, etc. Along with the brands' own moderators, other customers also answer those questions to help each other's knowledge about the market offerings of these brands. Nurturing customers' dialogue in such publicly available customer forums can help brands to improve their customers' knowledge about their overall market offerings. ► Web Exercise 7.2 shares a practical example of such customer forums.

Web Exercise 7.2

"BT (2024) Group plc is a British multinational telecommunications holding company headquartered in London, England" (np). BT runs a customer forum (► https://community.bt.com/), where BT's customers share their user experience and ask questions about BT's services. For example, questions about Connecting a Second TV (2024), Second Line and Split Account (2024), etc., are asked by some customers. These questions are answered and liked by other customers that attracted more engagements with other customers (e.g., views) as well. As a result, BT is able to make their customers more knowledgeable about their market offerings.

Have a look at BT's and other brands' Customer Community Forums to understand how these brands nurture dialogue among their customers to increase their customers' knowledge about their products and services.

Question 7: How could the knowledge of our customer and non-customer stakeholders improve our systems, tools, and market offering?

From time to time, the knowledge of our customer and non-customer stakeholders can be instrumental to improve our organizational processes, systems, tools, and overall market offering. Based on the hospitality industry, an example is shared in ► Web Exercise 7.3.

Web Exercise 7.3

► https://www.booking.com/ is a large provider of information on accommodation service. For each of their affiliated accommodation provider, they have an option for their customers to share their customers' satisfactory and dissatisfactory remarks and suggestions. For example, a Glamping Service Provider (2023) received an overall "very good" remark, with a customer rating of 8 out of 10 for a particular service. However, alongside this overall satisfactory remark, the customer shared a suggestion to improve the service provider's service delivery process, for example keeping a "sign at the entrance would have been useful" (np). Learning from this particular customer's experience can help the service provider to improve their overall service delivery process for their future customers.

Look at the website of ► https://www.booking.com/ and other brands' websites in the hospitality industry to explore how customers' suggestions from their experience and knowledge can help those brands improve their service delivery process.

Question 8: How could our existing systems and tools help improving the knowledge of our customers and non-customer stakeholders about us?

Stock market business could be a complex task for many micro investors who do not have specialized knowledge about this business, for example, to understand when to invest, how to invest, how much to invest, how to obtain up-to-date information about the changes in the stock market, etc. Let's assume that a Stock Market brand has launched a new software that not only offers up-to-date information and tips and tricks for investors, but also this new software as a business tool is very user-friendly even for non-expert users. In this way, the Stock Market brand, based on this new software could help improve the micro investors' knowledge about their service, in order to attract and engage with more micro investors.

Question 9: How could our organizational systems and tools be effectively integrated so that the systems and tools can complement each other?

Let's assume that you are attending a lecture of your "Sustainable Marketing" course, and your lecturer is about to start the lecture. But, the projector of your classroom's PowerPoint presentation system is not working, even after a few attempts by your lecturer to rectify the issue. After spending a few minutes to rectify the issue, once it remains unresolved, your lecturer called the university's IT team to come promptly to that particular classroom, in order to resolve the issue, so that the lecture can commence without much interruption before you lose a lot time for that 1-h lecture. An IT staff member has come promptly, resolved the issue, and your lecture has commenced without losing much time. Imagine, if the IT staff member had taken a long time to come to the classroom, the delivery of that particular lecture on that day may needed to be canceled and rescheduled for another day! Yes, your classroom is well-equipped with contemporary technology and your university also has a good IT support department. But lack of integration or coordination between different departments of your university can devalue your learning experience. Therefore, effectively integrating organizational tools and systems together can help a brand to improve their operational process.

7.4 Extending or Possibly Sustaining a Competitive Advantage

7.4.1 Why a Competitive Advantage Would Be Vulnerable

◘ Source: ► pexels.com/royalty-free-images/

In ► Sect. 1.2.2 (Strategic Marketing and Sustainable Marketing) of ► Chap. 1, we have discussed an example of competitive advantage of two coffee shops. For example, two coffee shops are located next to each other, and offer same price, same taste of coffee and same customer service; overall both coffee shops offer the same extent of materialistic benefits to their customers. However, one of these two coffee shops, for example, advertises a sign at their front door, saying that a percentage of the price that you pay for each cup of coffee, we donate to a particular social or ecological cause. In this case, from where, out of these two coffee shops, would you be purchasing your coffee, even though both offer the same materialistic benefits? Asking this question to a number of postgraduate marketing students across different student cohorts that consist of different nationalities, they answered that all of them would be purchasing their coffee from the shop that contributes to a social or ecological cause. It indicates a latent need or aspiration of contemporary customers and non-customer stakeholders that they prefer that organizations/brands should help them avoid harming the society and environment, but to protect it. This is why, sustainable marketing is not philanthrophy, but a necessity to survive and prosper in the contemporary marketplace.

Based on that donation, the coffee shop is able to establish a competitive advantage over its next-door rival; however, both coffee shops offer same materialistic benefits to their common customers. But, the problem is that the same idea of donating a percentage of selling price to a social or ecological cause can be imitated (even from a better context) by the next-door coffee shop as well. As soon as the next-door coffee shop imitates the idea, the competitive advantage of the first coffee shop turns into competitive parity among the two competitive coffee shops. This is why competitors, e.g., iPhone and Galaxy need to relaunch (the newer version of their smartphone) every year, considering that the short-lived competitive advantage that was established by their smartphone's previous version has already been outplayed by competitors' newer version. But not too many companies would have as many resources as owned by Apple or Samsung to relaunch them every year. So, the question is how a brand could possibly make their competitors unable or reluctant to imitate them, as making competitors reluctant or unable to imitate could help a brand to sustain or at least prolong their short-lived competitive advantage to a mid- or long-term scale. Examples are discussed in ► Sects. 7.4.2 and 7.4.3.

7.4.2 Strategic Management for Strategic Marketing: VRIN Capabilities for Sustainable Competitive Advantage

The commonalities and substitutability features of a product, and its competitors' capability to imitate the product features, turn the product's short-lived competitive advantage into competitive parity (Eisenhardt & Martin, 2000; Peteraf et al., 2013). For example, the resources (e.g., launching an advanced software to make a product delivery and usage more customer-friendly) and/or capabilities (e.g., a unique recipe of a chef in the hospitality industry) which are used to develop a new product and establish its initial competitive advantage, those resources (e.g., the software or the recipe) could be re-developed by competitors or could be substituted (e.g., a competitive restaurant could hire another chef who cooks food that tastes same or even better). As a result, the short-lived competitive advantage that was established by that new product can be outplayed by competitors. Therefore, marketeers need to consider how they can make their product or service and/or its features rare, inimitable and non-substitutable by competitors (Barney, 1991), so that the initial short-lived competitive advantage could be prolonged over a mid-term period or sustained over a long-term period.

An example to make competitors unable to outplay a product's competitive advantage is Coca-Cola's unique recipe. Although Coke and Pepsi both taste almost similar, the loyal customers of Coke feel that Coke as a soda tastes better (Staples-Otis, 2022). It is, however, an enduring rivalry between Coke and Pepsi, and an individual customer's loyalty over the both brands; Coke loyalists are frustrated when they are offered Pepsi. "While there's other sodas out there, like store brands or Pepsi, Coca-Cola drinkers are most loyal to their brand. In fact, the internet is full of memes about customers being frustrated with being asked, 'is Pepsi okay?' when they order a Coke" (Staples-Otis, 2022, np). It means, based on their unique recipe, Coke is able to establish a global base of loyal customers, who always prefer Coke over rival brands. Because of Coke's unique taste and a long-established brand image, it becomes difficult for the rivals to attract those customers who are really loyal to Coke. As a consequence, Coke is able to sustain their competitive advantage over a long period of time.

Following the discussion thus far in this current section, it can be argued that making competitors unable to imitate our product or its features is a precondition to sustain or at least prolong the short-lived competitive advantage of our product. Barney (1991) and Barney et al. (2001, 2011) postulate a theory, which is known as VRIN competency of a product that can enable the product to make its competitors unable to imitate the product or its features, so that the product can prolong or sustain its commercial competitive advantage:

- V = **Valuable** to satisfy customers' needs
- R = **Rare** among competitors
- I = **Inimitable** by competitors
- N = **Non-substitutable** by competitors

Valuable, here refers to a product's utility (value) to satisfy customers' needs. However, offering a mere utility is not enough to preclude competitors to copy that utility. For example, a university offers an academic program that has a value to satisfy students' need to develop in-depth theoretical and practical knowledge in that particular subject area, so that students can apply that knowledge in real life once they graduate. The challenge is that other competitive universities can also offer the same academic program, which brings a competitive parity among the competitive universities to attract students, instead of enabling a particular university to sustain or at least prolong their competitive advantage. Therefore, making the value of that academic program, e.g., the program features rare among the competitive universities will be essential to prolong or sustain the competitive advantage. In this case, human resource management would play a crucial role. For example, if a university employs a number of world-class researchers, including few Nobel laureates in that academic subject area (and other competitive universities in that region do not have that many globally elite scholars) to teach students in that subject area, in this case, the university would be able to sustain their competitive advantage for that academic program, because, their academic standing in the subject is "rare" among the competitors.

VRIN competency of a brand enables its market offering to proffer value to satisfy customers' needs in a way that is rare among the competitive brands, and make that value inimitable and non-substitutable by the competitive brands, in order sustain the brand's competitive advantage.

Source: ▶ pexels.com/royalty-free-images/

Also, making a product and/or its features valuable and rare would not be sufficient to sustain the product's competitive advantage. Developing product or product features in a way that would be inimitable by competitors is therefore important to sustain or at least prolong the commercial competitive advantage of the product. For example, iPhone has a unique design among all competitive smartphone brands. Don't you think Samsung Galaxy or other competitive smartphone brands can copy the iPhone's design? Surely, the engineers at Galaxy or at other competitors can do, but the competitors do not want to imitate iPhone's design. It is because iPhone has already established a very well-known brand image based on their unique design, and if the competitive brands imitate that design, customers will have a perception that it's not iPhone, but it looks like iPhone as the design has just been copied. It means, even customers will recall their iPhone experience when they will use a competitive smartphone that is designed based on the iPhone's design. As a result, the competitive brands not only devalue their own brand value but also will help increase the iPhone's brand image, if they copy iPhone's design. Similarly, don't you think McDonalds, Hungry Jacks, Subway, Burger King, or other established fast-food brands (not the non-branded next-door restaurants) can fry chicken in a way that is fried by KFC? Yes, they can, but they do not want to fry chicken in the way KFC does, because of the same reason that is discussed here in the case of iPhone. In this way, iPhone, KFC, and other brands try to make their product inimitable by their competitors.

Source: ► pexels.com/royalty-free-images/

Non-substitutability is another condition to sustain the competitive advantage of a product. For example, there are almost countless sea beaches around the world. But there are only very few sea beaches in the world from where we can observe both sunrise and sunset from the same sea beach, and because of the natural location, we can observe only either sunrise or sunset (but not both) from the rest of the sea beaches on our planet. Let's assume that you are planning to have your next holiday on a sea beach in the Mediterranean region in next summer. If just one sea beach in the Mediterranean region out of all other sea beaches in this global location offers the opportunity to observe both sunrise and sunset from their beach, don't you think that particular sea beach won't have a competitive advantage among the international tourists over the other competitive sea beaches in the Mediterranean region? Yes, observing sunrise and sunset is not the only feature tourists consider when they evaluate the tourism destination features. But, if most other features, e.g., hospitality, price, safety, security, etc., appear equal among the competitive sea beaches, in this case, the sea beach from where tourists can observe both sunrise and sunset will have a sustained competitive advantage among the competitive sea beaches. It is because the experience of observing sunrise and sunset from the same sea beach is non-substitutable by the competitive sea beaches, considering that the competitors' natural location do not allow such a unique experience to observe.

However, it would be really difficult to make a product and its utility completely non-substitutable by competitors. If it is not possible to make a product entirely non-substitutable; try to add some features in your product or promote your product based on a feature that would be non-substitutable by competitors. Let's have a look at two examples, where marketeers tried to make their market offering non-substitutable, based on promoting a particular feature of their offering, because it is really difficult to make the entire offering non-substitutable. The first example is based on the international education industry (IEI). In the global IEI, the top five host countries of international students include USA, the UK, Canada, France, and Australia, who are the major competitors to attract most international students (Statista, 2023b).

Fig. 7.3 Major home countries of international students from where (marked in green in the map) students move to another country to obtain their international education. (Information source: British Council, 2013; Shams & Hasan, 2020. The world map is adapted from ▶ pexels.com/royalty-free-images/)

Figure 7.3 shows the major home countries of international students from where (marked in green in the map) students move to another country to obtain their international education. You can see in Fig. 7.3 that South-East Asia and Far-East Asia together are one of the bigger geographical segments in the global education market from where students move to another country to obtain international education. Let's assume that a prospective international student from the South-East or the Far-East Asian market segment in the global IEI is currently researching among the top five host countries to decide where the student would move next year to obtain an international degree. In general, international students consider education quality, tuition fees, living cost, postgraduate employability, part-time work opportunity, among others, to decide where to go for their international education. If all these major factors appear quite equal among the major international education destinations, then the prospective international students also consider other factors such as weather, proximity to their home country, social life, etc. (Shams, 2016a). If international students find the major factors (i.e., education quality, tuition fees, living cost, postgraduate employability, part-time work opportunity) quite equal among the top five host countries (USA, UK, Canada, France, and Australia), then the other factors would have a key influence on their decision-making to choose their international education host country. From this context, the Australian universities further promote their market offering (international education) based on the features that would be non-substitutable by the other four major host countries. Such additional features include "warm weather" and "proximity to home country," because these two features, especially "proximity" for the South-East and Far-East countries is non-substitutable by the other four major competitive host countries (see the global map in Fig. 7.3 for "proximity" between Australia and South-East and Far-East Asia, in comparison to USA, UK, Canada, and France). Why you have chosen Australia as your international education destination? To answer this question, international students answered that "'Australia is a good place to study [...] considering its wonderful climate' (Chen, 2015, np). 'I had heard so many good things about Brisbane! It has a warmer climate'" (Mackinnon, 2015, np) (as cited in Shams, 2016a, p. 685).

> When international students from Asia as the major market of the global international education industry (UNESCO, 2014; International Students Australia, 2015) consider the proximity of their international education's host country (Phang, 2013; Singh et al., 2013), naturally, Australia comes first, compared to their main global competitors, such as the UK and the North American countries. (Shams, 2016a, p. 685)

Consequently, another international student answered that "one of the reasons influencing my decision to study in Australia is its proximity to Singapore" (Tan, 2014, np; as cited in Shams, 2016a, p. 685). Since making the entire market offering (i.e., the education service) fully VRIN-competent is not possible for Australia in comparison to the other major international education destinations; Australia promotes their market offering further based on such additional features. Targeting the particular segments in the global education market, "proximity" and "warm weather" as additional features enhance Australia's VRIN competency over the four other major competitors for particular segments, i.e., "proximity" for the South-East and Far-East Asian segment, and "warm weather" for the international student segment that prefers a lifestyle that offers more sunshine in the natural environment.

Celebrity endorsement is a common example to add VRIN feature in a product, especially when it would not be possible to make the entire product VRIN-competent. Lionel Messi's brand ambassador role for Adidas and Cristiano Ronaldo's brand ambassador role for Nike is a prime example here (Shams, 2016a). "Although he (Messi) did wear Nike boots for a few years in his youth, even featuring in a "Remember My Name" Nike ad, the Argentinian has been an Adidas athlete since 2006. He has probably been the brand's longest-serving number one "face" of all time due to his sustained excellence at the highest level of the sport" (Asher, 2022, np). "Messi's deal with Adidas is now (in 2023) worth a humongous $25m per year. While there are footballers and athletes whose yearly deal may exceed this huge figure, the World Cup winner will be entitled to this yearly fund for the rest of his life" (Austen-Hardy, 2023, np). On the rival front, "Ronaldo has been contracted to Nike since the early stages of his career in 2003 and is on a £147 million, 10-year deal with the American giants" (Rashid, 2023, np). However, there is a speculation that Ronaldo's deal with Nike would be in trouble, because of a recent counter deal of Adidas with Ronaldo's club, Al Nassr (Millar, 2023). Celebrities/influencers have huge fan base. In general, the fans consciously and/or subconsciously follow their celebrities. Thus, until Messi promotes Adidas, don't you think Adidas would have an additional VRIN feature to sustain their competitive advantage in the Argentinian market and in the global market among Messi's followers? Similarly, don't you think the same will be applicable in the case of Ronaldo and Nike until Ronaldo promotes the brand?

7

Marketing in Action 7.1: iPhone' Unique Design and Its Commitment on Sustainability

Source: ► pexels.com/royalty-free-images/

Design Focus

Why iPhone looks different than other smartphones? Why iPhone's design is standalone among the competitive smartphone brands? Let's have a look at an article published in *The Independent* in 2022. Based on the views of Apple executives, the article presents inside story on the key factors that are considered in designing iPhone. "Apple's Craig Federighi, senior vice president of software engineering, and Alan Dye, vice president of design talked exclusively to the Independent about…Apple's approach to design in general" (Phelan, 2022, np). "Collective design" is a key factor that influences Apple's product design, where the design for software and hardware is considered simultaneously. Referring to the Android operating system that is used in different hardware made by different smartphone brands, the executives describe that it is difficult to establish and sustain a "unique ecosystem" to synchronize between hardware and software. As Apple manufactures both hardware and software for iPhone, it helps them to optimize the performance of their product, considering that hardware design, human interface design, and software development are conceived together.

Apple designs its own software to use in iPhone's hardware. However, it is not the case for Android, which is designed by Google. It gives Apple the opportunity to have a better control on the changes in software to update their hardware in a more synchronized manner, which is another factor that helps making unique design for iPhone and sustain their competitive advantage. Intrinsic relationship in between Apps and how these Apps can be conveniently used by iPhone users is the next factor that helps Apple to sustain their competitive advantage in designing their smartphone. Long-term thinking and envisaging the evolving nature of smartphone usage enables Apple to offer unique experience to iPhone users, while they use different features, e.g., monitoring live activities. For example:

» You've ordered a food delivery, the Dynamic Island will automatically show it's arrived without you having to wait for a notification or open the app, and stays on screen even if you're switching apps, say. Or if you're tracking a sports game, the score will update in real time. This has been in the planning for a long time. (Phelan, 2022, np)

Aiming for magic in design thinking is the next factor in the success of iPhone design. For example, "it's like watching a magician (the software) performing a trick where you can't quite see the card up their sleeve. Sleek animations govern the notifications and other interactions, designed to delight the eye. And that's the point, Mr. Federighi says" (Phelan, 2022, np). "Making the features inevitable" for a unique and valuable smartphone usage experience enables Apple to make their users' experience more engaging. The inputs from Apple's cross-disciplinary team in designing the iPhone also play a significant role to offer a unique user experience.

Sustainability Focus
Apple's Product Environmental Report (2023) describes that aluminum, rare earth elements, copper, plastic, tungsten, cobalt, and tin are used as raw materials to produce Apple products. The report also confirms that all these raw materials (excluding plastic), which are used by Apple are recycled 100%. For plastic, the report describes that "we're transitioning from fossil fuel-based plastics to renewable or recycled alternatives. For iPhone 15 Pro and iPhone 15 Pro Max, we use 35% or more recycled plastic in multiple components" (p. 5). In the manufacturing process, Apple also follows a responsible approach (Apple's Product Environmental Report, 2023, p. 4):

- Sourcing raw materials: "Maximize recycled content in products and responsibly source primary materials."
- Manufacturing Apple products: "Make products responsibly, upholding high standards of labor and human rights and minimizing our environmental impact."
- Packaging and shipping: "Eliminate plastic from our packaging and minimize emissions from transportation."
- User experience: "Minimize emissions from product use, enhance product longevity, and seek to avoid exposure to harmful chemicals."
- Recovery: "Reuse products through Apple Trade In and maximize material recovery at end of life."

Think about Apple's approach in product design and their sustainability focus to apply in designing products in other industries. For example, how the concepts like collective design, unique ecosystem, control on the different aspects of product design, benefits of a cross-disciplinary product team, etc., would be applicable to design a new product, for example, in the car industry. Similarly, consider how the sustainability focus of Apple in their product manufacturing process could be adapted in the car industry to ensure a responsible approach in manufacturing cars.

Information Sources:

Apple's Product Environmental Report. (2023). iPhone 15 Pro and iPhone 15 Pro Max. Apple. Retrieved April 13, 2024, from ▶ https://www.apple.com/environment/pdf/products/iphone/iPhone_15_Pro_and_iPhone_15_Pro_Max_Sept2023.pdf

Phelan, D. (2022). Eight secrets of Apple's design success: iPhone executives reveal how its devices are really created. *The Independent*. Retrieved April 13, 2024, from ▶ https://www.independent.co.uk/tech/apple-design-secrets-success-iphone-b2201215.html

Web Exercise 7.4

Google "Coke's VRIN Competency", and Google will provide you few examples of Coke's VRIN Competency.

- Review those examples of Coke.
- Then consider developing VRIN examples of other renowned brands.
- Try to develop VRIN examples of two major competitive brands in a particular industry, e.g., iPhone and Galaxy, Coke and Pepsi, or FC Barcelona and Real Madrid CF.
- Compare the VRIN competency examples you have developed for the competitive brands to determine the brand that you think has more competitive advantage over its rival.

Another option to add VRIN features into a product or service to sustain or at least prolong its competitive advantage is establishing, maintaining, and monitoring a win-win value network of trusted key stakeholders who help create value for your product or service to offer that value to your customers at a particular market for a long time. As a consequence, because of such trusted partnerships with the key stakeholders at a particular local/regional market, it becomes difficult for competitors to outplay the competitive advantage of your market offering in that particular market. It is because the key stakeholders in that market have a long-standing win-win business relationship with you, which makes it difficult for competitors to make new partnerships with those key stakeholders to influence that particular market further. For example, the Opening Vignette of ▶ Chap. 1 (The Value Chain of a For-Profit Firm for its People, Planet, and Profit) describes the case of Aspuna (a British small- and medium-sized enterprise (SME) in the food processing industry) that how they have established a win-win key stakeholder value network in their African market. Aspuna recognized that the African farmers don't receive the right price for their crops. Thus, Aspuna established a cooperative of African farmers and also established food processing plants in Africa

(Aspuna, n.d.). Aspuna purchases the crops of African farmers and offers them the right price for the crops. The farmers also work at Aspuna's food processing plants in Africa to process food to market it in the local and overseas market, which creates additional employments for local people in Africa. Furthermore, Aspuna (n.d.) partnered with local micro-entrepreneurs in Africa to market the processed food in local market. This partnership of Aspuna with the key stakeholders in local African market enables them to source their raw materials (raw crops) in competitive price, while also enabling them to contribute back to the local marginalized population to improve their life, based on creating additional employments for small farmers and micro-entrepreneurs in Africa. Consequently, it will be difficult for Aspuna's competitors to outplay Aspuna's competitive advantage in Africa (i.e., partnering with those local farmers to source raw materials in competitive price from Africa, and partnering with those African micro-entrepreneurs to conveniently market competitive products in local African market), until Aspuna keeps up their trusted partnership with these local key stakeholders.

7.4.3 Behavioral Economics for Marketeers: Implications for Sustainable Competitive Advantage

Background of Behavioral Economics

◘ Source: ► pexels.com/royalty-free-images/

Behavioral economics has started evolving since early 1950s, and has had a persistent growth as a sub-discipline of economics throughout the last twentieth century (Svorenčík & Truc, 2022; Buttenheim et al., 2023; Baddeley, 2024). The growth of behavioral economics as an academic discipline and a field of practice has reached a new height since Professor Richard H. Thaler received the Nobel Prize in Economics in 2017, because of his contribution to behavioral economics based on nudge theory. However, the progress of behavioral economics was not without its criticism. Since its initial inception, some mainstream economists have not considered it as a sub-discipline of economics, as their view is that psychology is the root of behavioral economics, which is true too. According to Professor Thaler (2016), "in recent years there has been growing interest in the mixture of psychology and economics that has come to be known as 'behavioral economics'" (p. 1577). Along the progress of its growth, because of its significant contribution to the field of economics, behavioral economics today is widely considered as a sub-discipline of the broader field of economics.

In the same vein, some of the cross-disciplinary and cross-functional theories and concepts, such as "resource-based view of resources, market-based view of resources, knowledge-based view of resources, VRIN capabilities, dynamic capabilities, behavioral economics and nudge theory", which are discussed in this current chapter of a marketing textbook are not discussed in many other marketing textbooks or at marketing classrooms, considering that these theories and concepts are mainly from the field of strategic management and behavioral economics. Once you complete reading this chapter, we hope you understand the influence and contribution of these cross-disciplinary and cross-functional theories and concepts to reinforce your knowledge on marketing, in order to design your marketing and sustainable marketing strategies more prolifically.

Considering that behavioral economics aims to analyze people's psychological state for economic decision-making, "Camerer (1999) defines behavioral economics as a research program aimed at reunifying psychology and economics" (as cited in Berg, 2010, p. 1). "Behavioral economists, however, suggest that our preferences are not always consistent or well defined (Camerer & Loewenstein, 2004); choices often depend on their contexts or on how they are framed" (as cited in Jabbar, 2011, p. 446). The theoretical development of behavioral economics is influenced by different theories and concepts, predominantly the prospect theory, framing effects, status quo bias, paradox of choice, and intrinsic motivation. The prospect theory assumes that someone would act differently than others while deciding on prospective gains in comparison to prospective losses when people have less, no or an uncertain extent of control over a particular situation (Kahneman & Tversky, 1979; Bruhin et al., 2010). Also, the presentation and the circumstances related to alternative choice options can influence people's decision-making, which is discussed in framing effects (Chang et al., 1987; Camerer et al., 1997; Kahneman, 2003). Behavioral economists argue that many individuals generally wish to maintain status quo, even when the individual would have more attractive alternatives to choice (Madrian & Shea, 2000; O'Donoghue & Rabin, 1999, 2001; Benartzi & Thaler, 2000; Thaler & Benartzi, 2004; DellaVigna, 2007; Choi et al., 2011). A key reason for such behaviors is that many people would have less extent of adaptive willingness and/or capability with a changing situation than others.

- **Prospect theory** assumes that someone would act differently than others while deciding on prospective gains in comparison to prospective losses when people have less, no or an uncertain extent of control over a particular situation.
- **Framing effects** refer to the presentation and the circumstances related to alternative choice options that can influence people's decision-making.
- **Status quo** refers to the existing state of affairs, especially regarding social or political issues.

The paradox of choice claims that when people have more options to select one, people may behave strangely (Bertrand et al., 2004; Iyengar & Kamenica, 2010). For example, alongside many benefits of the enormous flow of information in today's global society, one of the disadvantages of enormous flow of information, e.g., about many alternative choices of competitive products can cause customers confused; thus they may have difficulty selecting the most appropriate product related to their needs, while analyzing a lot of information about the many competitive products and their alternative value propositions. The concept of intrinsic motivation argues that someone's self-interest can motivate her or him, in spite of any extrinsic or external rewards (Frey, 1997; Kreps, 1997; Camerer et al., 2004; Heyman & Ariely, 2004). For example, religious belief can influence someone even to pay more for a food, e.g., pay more or drive a longer distance to purchase halal or kosher food. Similarly, health-conscious customers, because of their intrinsic motivation tend to pay more for organic food, while they have an option to pay less for non-organic food. Based on this background of people's psychology and its impact on their decision-making, behavioral economics analyses the alternative aspects of human behavior to develop insights on how people (e.g., different customers and non-customer stakeholders) make decisions (Thaler, 2017). Such decision-making may be related to people's personal and professional decision-making, e.g., to purchase a window cleaner to use personally at home, or to purchase that same window cleaner in bulk volume to clean windows at an office building.

Behavioral economics analyses the alternative aspects of human behavior to develop insights on how people (e.g., different customers and non-customer stakeholders) make decision.

- **Paradox of choice** claims that when people have more options to select one, people may behave strangely.
- **Intrinsic motivation** argues that someone's self-interest can motivate her or him, in spite of any extrinsic or external rewards.

Behavioral Economics and Marketing

Let us discuss the intersection of behavioral economics and some marketing principles (i.e., individual's environment for decision-making, evaluation of alternative products, enormous flow of information, individual's needs, marketeers' presentation, and promotion) to influence a brand's target audience, so that marketeers can correlate the implications of behavioral economics to understand the psychology and behavior of their target audience (i.e., customers and non-customer stakeholders) better to influence them, responsively. Both the principles of marketing to influence a brand's target audience and behavioral economics have the same origin, i.e., psychology. ◘ Figure 7.4 describes the implications of behavioral economics for marketing principles to understand how individuals (e.g., customers and non-customer stakeholders) behave and make decisions.

Culture has a significant influence on how an individual behaves and makes decisions. Culture is considered as the combination of specific values, beliefs, and assumptions that an individual inherits from her or his early childhood environment (Hofstede et al., 2010). Research shows that early childhood environment has a long-lasting impact on forming people's personalities (Understanding How Childhood Shapes Who We Become, 2023). Consequently, these cultural philosophes (i.e., the values, beliefs, assumptions, and way of seeing life) that are ingrained in people's mind since their childhood, reflected through their regular decision-making process, and appear as vibrant dimensions of differences in human behavior from culture to culture (Steenkamp, 2001). Following this mindset of a particular cultural philosophy and differences between different cultures, customers evaluate which market offering among the competitive offerings will be most

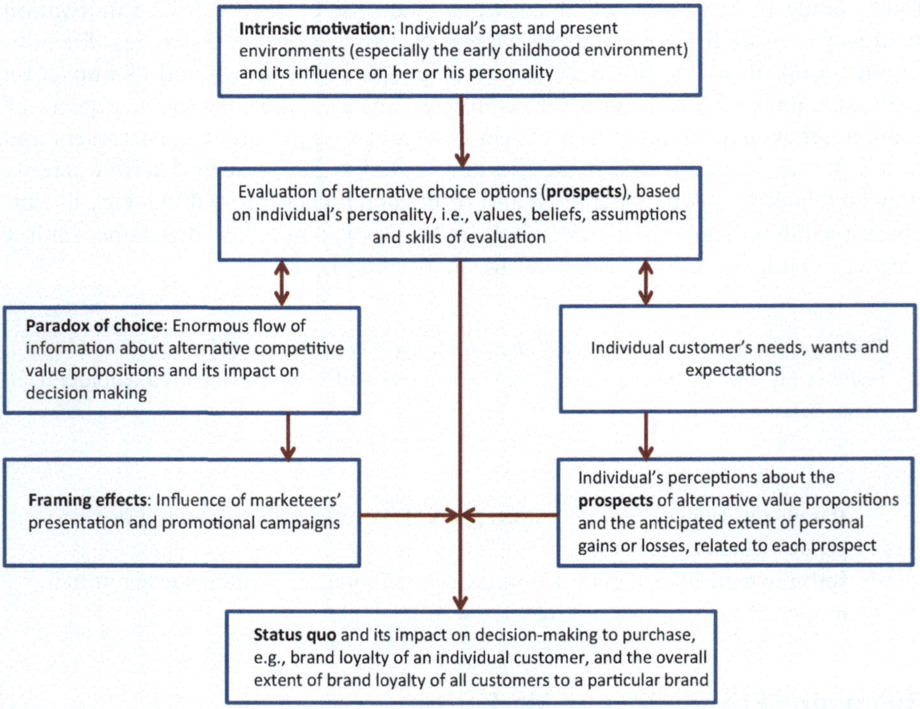

◘ **Fig. 7.4** Behavioral economics and principles of marketing to map customers' decision-making process. (Adapted from Steenkamp, 2001; Berner & Tonder, 2003; Kotler, 2003; Hofstede et al., 2010; Jabbar, 2011)

relevant to their needs, wants and expectation (Kotler, 2003). Because of this influence of culture on customers' decision-making, when a South-Asian customer goes to purchase beef from a supermarket in London or New York, they look for beef in curry piece, whereas a native customer look for the same beef in stake piece. The top arrow sign of ◘ Fig. 7.4 indicates this *intrinsic motivation* of each individual customer (that they inherit from their own environment) and its influence on their evaluation of alternative competitive value propositions (*prospects*) based on their individual needs, e.g., a South-Asian customer may prefer to travel longer to purchase beef in curry piece if her or his local supermarket does not offer it. As a result, many supermarkets today offer cross-cultural products and services, especially in locations where they cannot ignore the number of cross-cultural customers (marketeers' framing effects).

Customers are encircled with enormous flow of information about alternative value propositions of competitive products (Berner & Tonder, 2003), which is a *paradox of choice* for customers. Following this knowledge (information of competitive value propositions) and their preferences (income, mobility and search costs, personal preferences, word-of-mouth, etc.), customers form and reform an expectation (value anticipation) about their estimated product or service value and act on it, i.e., *status quo* to make their purchase decision that reflects through the extent of their brand loyalty (Kotler, 2003).

Value anticipation is customers' own interpretation about a particular product's utility before they purchase and consume it, and how they think of what they may need to sacrifice and what utility they may obtain in return of their sacrifice after they consume it.

Nudge Theory for Sustainable Competitive Advantage

> In 2008, the behavioral economist Richard Thaler and the legal scholar Cass Sunstein published a book (Nudge—Improving Decisions About Health, Wealth and Happiness) in which they advocated a novel approach to public policy based on the notion of a "nudge." Roughly speaking, a nudge is an intervention in the decisional context that steers people's decisions by acting on their cognitive biases. (Congiu & Moscati, 2022, p. 188)

In this context, the purpose of nudging is to help people to mindfully evaluate the alternative options that they have for a decision-making, with an aim to enabling them to take the most appropriate decision that would be valuable for them individually and/or collectively. Marketing in Action ▶ 7.2 describes an example of how British government nudged the British employees to help them taking a decision for pension scheme in a way that is not only beneficial for an individual employee but also for the collective British society.

> A nudge, as we will use the term, is any aspect of the choice architecture that alters people's behavior in a predictable way without forbidding any options or significantly changing their economic incentives. To count as a mere nudge, the intervention must be cheap and easy to avoid. Nudges are not mandates. Putting fruit at eye

> level [hoping that people then choose fruit over unhealthy alternatives] counts as a nudge. Banning junk food does not. (Thaler & Sustein, 2009, p. 6; as cited in Kosters & der Heijden, 2015, pp. 278–279)

Here we see that the advocates of nudge theory suggest that we should not make a nudge compulsory for our target audiences, but should keep options for them to avoid the nudge, as well, and then the target audience will take decision among the alternative options that would be better for them. Thaler (2017) "reasons that the concept of 'nudging' means helping people have more self-control to take decisions" (Santander, n.d., np). One of the underlying concepts of nudge theory is libertarian paternalism that "involves the application of behavioral techniques to nudge people to choose options that are in their own best interests, while keeping any cost of resisting a nudge very low" (Holland, 2014, p. 331). The critics of nudge theory argue that nudge can be applied in a way that may keep no additional option for people, but to comply with the nudge. This is where marketing ethics or responsible business has a role to play to decide whether marketeers should design a nudge to influence their target audiences, based on:

- Keeping alternative options (indirect nudge or pull-effect) for nudging,
- Or not keeping an option (direct nudge or push-effect) for nudging.

Keeping alternative options for people's decision-making is ethical while designing a nudge effect, as it does not impact negatively on individual's freedom to evaluate alternative options (e.g., among healthy food and junk food) to take a decision that would be most relevant for their interest. On the contrary, not keeping alternative options for people's decision-making may not be unethical while designing a nudge effect, if the purpose of a particular nudge is to increasing collective societal benefit, i.e., for the best interest of an individual and society (see an example in Marketing in Action ► 7.2).

Nudge is predicting how people behave (act and/or react) based on alternative options for making a decision (e.g., a purchase decision), and influencing their decision based on the prediction, while helping people to take a better decision.

Direct nudge refers to designing a nudge based on not keeping alternative options for people's decision-making.

Indirect nudge refers to designing a nudge based on keeping alternative options for people's decision-making or keeping any cost of resisting a nudge very low (e.g., if you prefer paper bills for your utility, you need to pay an additional $4.00 per year as a low cost to resisting the nudge. The purpose here could be nudging customers to opt in for paperless digital billing system, as it can increase responsible behavior to our environment by reducing printing).

Marketing in Action 7.2: Financial Security After Retirement: A Direct Nudge for Societal Benefit

Until 2012, it was not compulsory, but optional for employees both in the private and public sectors in the UK to contribute a percentage of their monthly income into their own pension scheme. As most employees decided not to contribute to their pension scheme from their monthly income, the rate of pension scheme memberships and savings in the national pension schemes in the UK were decreasing alarmingly from 2008 to 2012 (see ◻ Fig. 7.5). As a result, the British government was anxious thinking of how they will feed people once they retire!

To encourage British employees to save for their own retirement, the British government applied a simple direct nudge (i.e., the push-effect of nudging) in 2012, based on making it compulsory that all British employees, either in the public or private sector, must need to contribute a small percentage of their monthly salary into their own pension scheme. As a result, we see in ◻ Fig. 7.5 that the rate of pension scheme membership and the savings in the national pension scheme both have started increasing steadily since 2012 in the private and public sectors, which makes the British government happy that they can now ensure people's financial security in the UK after they retire.

You see that the British government did not keep any option for employees to not to contribute to their pension scheme from monthly income, as all employees must need to contribute a small percentage of monthly salary to their individual pension scheme since 2012, if they want to work in the UK. Designing such a direct nudge to push the target audience (in this case, the British employees) without keeping any alternative option for their decision-making could be okay, if it aims to support greater societal interest (i.e., feeding Briton after they retire).

Let's assume you are a marketeer of a gas and electricity company, and think how you could nudge your customers based on a direct nudge effect to act in a way that is more environmentally responsible. An option would be nudging your customers by keeping no paper billing option, as printing papers can have an impact on our environment. In this case, you may be a little lenient to the segment of your older-age customers to keep the paper billing option only for them, as they may be less proficient in using digital bills, while educating them on how to use digital bills.

Information Sources:

Morrison, B. (2013). How a small nudge is helping people save for their retirement. *Civil Service Quarterly*. Retrieved April 21, 2024, from ► https://quarterly.blog.gov.uk/2013/10/22/how-a-small-nudge-is-helping-people-save-for-their-retirement/

Cibb, J., & Emmerson, C. (2016). *What happens when employers are obliged to nudge? Automatic enrollment and pension saving in the UK*. Institute for Fiscal Studies. Retrieved April 21, 2024, from ► https://ifs.org.uk/sites/default/files/output_url_files/wp1619.pdf

Office for National Statistics. (2021). Occupational pension schemes survey, UK: 2018. *Census*. Retrieved April 21, 2024, from ► https://www.ons.gov.uk/peoplepopulationandcommunity/personalandhouseholdfinances/pensionssavingsandinvestments/bulletins/occupationalpensionschemessurvey/2018#:~:text=Active%20membership%20of%20occupational%20pension,2017%20levels%20(7.7%20million)

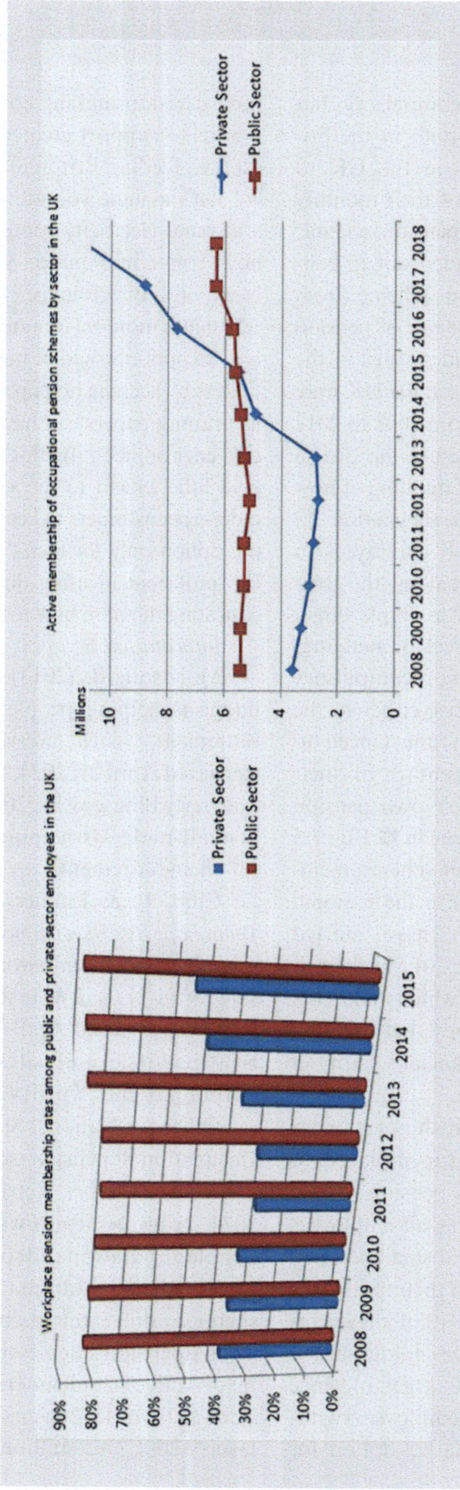

Fig. 7.5 How a direct nudge (applied in 2012) has protected the British national pension scheme. (Information source: Cibb & Emmerson, 2016; Office for National Statistics, 2021)

Marketing in Action ► 7.2 describes the push-effect of a direct nudge by not keeping an option for the target audience for their decision-making, where the purpose of this particular nudge for the British employees is to protect the well-being of Britons after they retire. However, the focus of nudge theory is helping target audience for their decision-making (e.g., purchasing your environmentally friendly product) while keeping other options available (e.g., the competitive products that may not all be environmentally friendly), which we refer to as indirect nudge or the pull-effect of nudging. For example, clarifying to your target audiences why doing business with you would be more beneficial (e.g., environmentally friendly product, competitive pricing options for all income segments etc.) for your target audiences instead of doing business with your competitors, in order to pull them toward your market offering. In this case, a nudge keeps options for your customers for decision-making without forbidding any alternative options (e.g., alternative competitive products), as the core marketing message of such indirect nudges should be focusing on the superior value proposition of your product related to People (e.g., competitive pricing), Planet (environmentally friendly product), and Profit (e.g., establishing a loyal customer group who are more socially and environmentally concerned customers), so that customers can decide whether to be loyal to you or your competitive brands. ◘ Table 7.1 describes the examples of indirect nudges.

The Preliminary Vignette of this chapter clarifies how easily competitors can outplay a short-lived commercial competitive advantage, even of a triple bottom line-focused (People-, Planet, and Profit-focused) value proposition, based on copying the underlying idea of the proposition. In the same vein, competitors can imitate the nudge that you have designed to influence your target audiences, in order to outplay the short-lived competitive advantage of your nudge effect. ► Section 7.4.2 (Strategic Management for Strategic Marketing: VRIN Capabilities for Sustainable Competitive Advantage) of this chapter discusses how marketeers can possibly sustain or at least prolong the short-lived competitive advantage of their market offerings to a mid- or long-term period, based on applying VRIN theory. In this context, a nudge that you may design to influence your target audiences may or may not have VRIN competency. See the left column of ◘ Table 7.1 that lists some nudges that have VRIN competency and some other nudges that may not have the VRIN competency.

As seen in ◘ Table 7.1 that a "unique cooking style" could be a nudge effect of a restaurant to attract customers. For example, how KFC fries chicken in a crunchier way. McDonald's and other competitive fast-food brands also can fry chicken in a way that is fried by KFC; but the competitors do not want to imitate the KFC style of fried chicken, as KFC have already developed a long-established global reputation to consistently fry chicken in that crunchier way. In this case, if competitors imitate the KFC style, customers will recall their KFC experience even when they purchase from KFC's competitors who fry chicken in the KFC's way of frying it, e.g., when we purchase fried chicken from non-branded local next-door fast-food restaurants that follow the KFC's style of frying chicken; it automatically comes to our mind that KFC fries chicken in this way! Therefore, KFC's fried chicken as a "unique recipe" is a VRIN-competent nudge effect to influence KFC's

Table 7.1 An exercise of indirect nudge effects that may (or may not) have VRIN competency

Indirect nudge effects: Try to recognize the indirect nudge effects here that you think have and not have VRIN competency	Correlate an individual nudge effect from the left column with a particular brand of this (right) column
– A unique formula or flavor	– ALDI
– Value for money	– Walmart
– User friendliness or easy to use	– MAC operating system
– A unique cuisine or cooking style	– Coca-Cola
– Cost-effectiveness	– Emirates Airlines
– Peace of mind	– Toyota
– Elegance	– iPhone
– A unique reputation	– McDonald's
– A unique design	– Microsoft Windows
– Competitive pricing	– Alibaba
	– Uber
	– Samsung Galaxy
	– Mercedes Benz
	– Unilever
	– Amazon
	– Hyundai
	– KFC
	– Ford
	– Jet Airways

7

target audiences, considering that KFC's major global competitors do not want to imitate it.

For "User-friendliness or easy to use" as seen in Table 7.1 as another indirect nudge effect to influence target audiences, some customers (e.g., especially the tech-savvy generations) may think that Uber is more user-friendly than the traditional taxi service providers. However, some other customers (e.g., the older-adults) may think that the service of the traditional taxi service providers is more easy to use, in comparison to the service of Uber, as these customers have more favorable experience from the traditional taxi service providers, in relation to their specific needs, wants, and expectations. Thus, whether a particular product is user-friendly or not would depend on individual customer's personal experience with that particular product and its competitive products. Also, let's assume that most customers perceive that Uber is more user-friendly than traditional taxi service providers. But in this case, the traditional taxi service providers can also upgrade their app and service in a way that would be even more user-friendly. It means, "user-friendliness" as an indirect nudge may not have VRIN competency as it can be imitated by competitors. Let's first, recognize the indirect nudge effects from the left column of ◘ Table 7.1 that you think have and may not have the VRIN competency. Second, try to correlate each of the indirect nudges from the left-hand column with the brands listed in the right-hand column of ◘ Table 7.1 that you think a particular nudge effect here is most relevant with a specific brand listed on this table. Remember that your experience with a particular brand can significantly differ from the experience of your friends. So, there is no right or wrong answer, but it is important how you justify your answer, based on your personal experience with that particular brand that you are thinking of.

Designing a nudge, either for a direct or an indirect nudge effect, is a challenge, as it requires critical evaluation of people's psychology across different backgrounds, e.g., ethnicity, gender, income level, age, etc., and misunderstanding here can lead to marketing myopia. Therefore, careful data collection, analysis, and synthesis are prerequisite to understanding customers' needs, wants, and expectations, in order to not only design and deliver a product in a way that would be expected and accepted by customers, but also to avoid losing market share by offering a product that is not that relevant or less relevant to customers' needs, in comparison to competitive products.

> **Marketing myopia** is a short-sightedness in understanding target audiences' needs, wants, and expectations, and designing a product or service based on that wrong understanding.

To understand customers' needs, wants, and expectations, marketeers need to understand customers' intrinsic motivation, their status quo, paradox of choice among alternative (e.g., the competitive products) as a prospect to satisfy their need, and framing effects (e.g., presentation of alternative value propositions by competitive marketeers). Intrinsic motivation, i.e., self-interest about a particular product feature may arise from an individual's environmental influence, specifically the childhood environment that helps form people's perception (Understanding How Childhood Shapes Who We Become, 2023). An example of customers' status quo from the context of consumer goods could be a customer may prefer to use a 2005 Nokia model, instead of a contemporary smartphone, but it is not necessarily that the customer cannot purchase a contemporary smartphone, but she or he feels that the old model is more user-friendly. Also, a social or political issue, e.g., the US sanctions on Chinese technology companies may motivate a US customer to avoid purchasing a Chinese-brand of smartphone. Prospect refers to people's idiosyncratic behavior in decision-making, an individual could act differently than others while deciding on prospective gains in comparison to prospective losses when people have less, no or an uncertain extent of control over a particular situation (Kahneman & Tversky, 1979; Bruhin et al., 2010). Why customers may behave in such dissimilar ways across different market segments? Let's have a look at Thaler's (2017) proposition based on rational perception, fairness perception, and lack of self-control (or collective control of a group of people) to consider how such variables could influence or be influenced by intrinsic motivation, status quo, paradox of choice, prospects, and framing effects. ◘ Figure 7.6 illustrates these dynamics of people's psychology. Analyzing these dynamics for a particular target market segment would be instrumental to design a (direct or indirect) nudge to influence that segment.

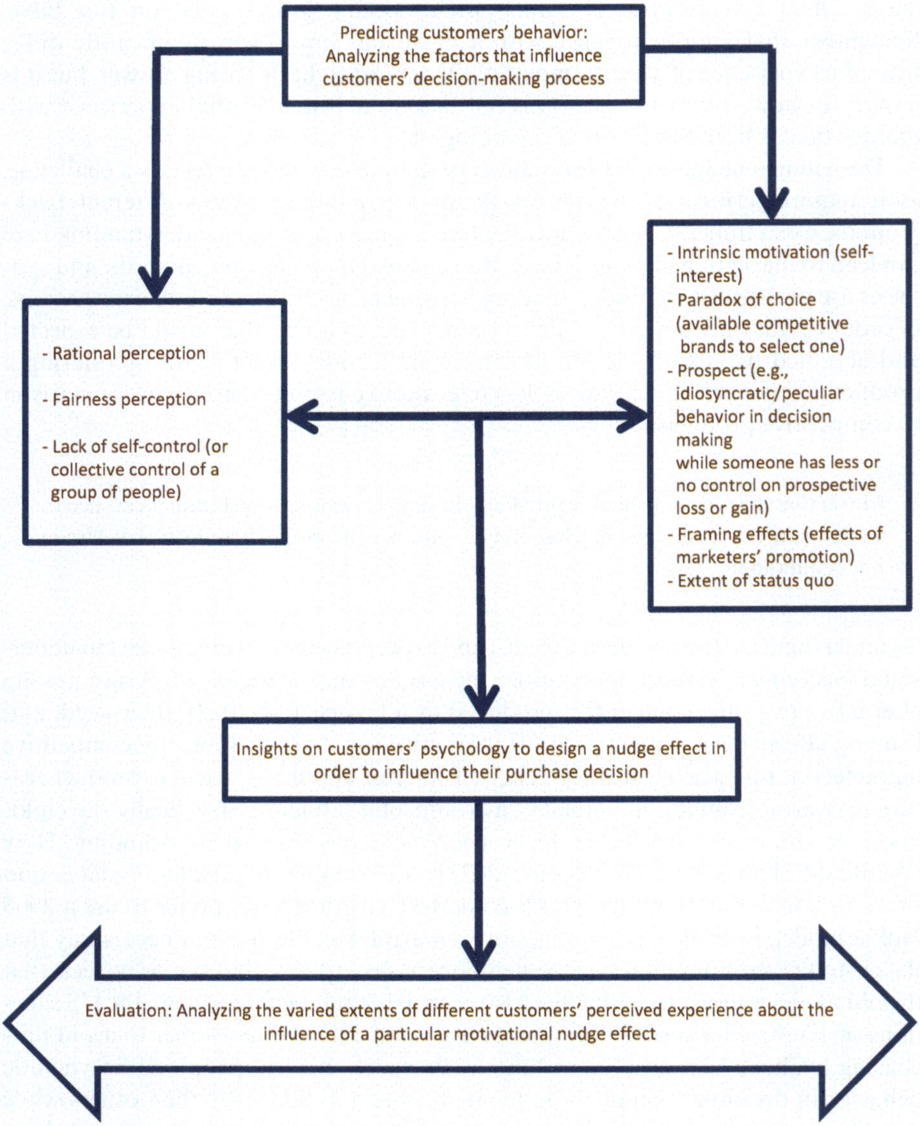

◘ **Fig. 7.6** Predicting customers' behavior to design and evaluate a nudge effect in order to influence customers' purchase decision. (Adapted from Thaler & Sunstein, 2008; Jabbar, 2011; Gundlach & Neville, 2012; Shams, 2016b; Thaler, 2017)

◘ Figure 7.6 presents the dynamics of two sets of factors that can influence people's decision-making. The first set includes rational perception, fairness perception and lack of self-control (or collective control of a group of people). The second set includes intrinsic motivation, status quo, paradox of choice, prospects, and framing effects. Analyzing these factors could help us predicting how customers' act and/or react while making a purchase decision. Rational perception refers to an individual's own calculation about prospective loss or gain from a particular decision. For example, someone in general will purchase the most relevant product among the alternative competitive products that is most relevant to their needs, budget, and preferences. Fairness perception refers to an individual's perception on how their purchase decision, i.e., how purchasing a particular product or service could impact on people and planet, and how an individual considers such impacts while deciding to purchase a particular product or service. For example, the socially and environmentally caring customers would prefer to not purchasing a garment brand that has scandal in sourcing their garments items, e.g., the brand may not pay well the workers who sew the garments or may be the workers' working environment is not good for health and safety. Similarly, a customer may prefer not purchasing a brand if it has high negative impact on environment. Since the number of such socially and environmentally caring customers are increasing around the world (Townsend, 2018); fairness perception of customers is also a very important consideration for responsible marketeers while designing a nudge. Sometimes, because of lack of self-control on our purchase decision, we as customer may not be able to purchase a particular product or service that we would have purchased in general based on our rational perception. For example, you are planning to spend a holiday in the Mediterranean region during next summer. But, most of your family and/or friends prefer to spend the holiday in Africa this year. In this case, you may have to decide to go for an African safari; however, a Mediterranean holiday could have been more cost-effective for you and more relevant to your interest! Therefore, lack of self-control in decision-making can be defined as peers' influence on an individual's purchase decision

Rational perception refers to an individual's own calculation about prospective loss or gain from a particular decision, and the individual will, in general, will take the decision among the alternative options that would be most prolific for their needs and budget.

Fairness perception refers to an individual's perception on how their purchase decision, i.e., how purchasing a particular product or service could impact on People and Planet, and how an individual considers such impacts while deciding to purchase a particular product or service.

Lack of self-control in decision-making can be defined as peers' influence on an individual's purchase decision.

The second set of factors in ◘ Fig. 7.6 are discussed in section "Background of Behavioral Economics." The first and second sets of factors would have influence on each other while an individual evaluates her or his alternative options to make a purchase decision (the thinner double-headed arrow sign of ◘ Fig. 7.6 indicates this influence). For example, an individual would be confused about recognizing the most relevant value proposition (rationale perception) because of paradox of choice (too many information to process about competitive brands to select one), because of framing effects (e.g., confused and influenced by marketeers' promotion), and thus could take an idiosyncratic decision (e.g., prospect). Also, an individual's intrinsic motivation and status quo may influence her or him to make a decision to pay more for a halal or kosher or an organic food against rational perception (e.g., financial loss and gain). Similarly, because of the fairness perception and lack of self-control customers could take a decision avoiding the influences of their intrinsic motivation, status quo, framing effects, or ignoring their paradox of choice and prospect, e.g., traveling to the destination-A for a holiday instead of destination-B because of peers' influence.

Based on analyzing such factors, the British government was able to predict the British employees' prospective behavior (action or reaction) while designing the small nudge to make it compulsory in 2012 that all British employees must have to contribute a small percentage of their monthly income to their own pension scheme. For example, by making a direct nudge (i.e., not keeping alternative options for employees to protect their job) and promoting the greater societal interest (i.e., people's well-being after they retire), the British government neutralized the rational perception of employees, e.g., some may think that it is my income and it is better for me to consume it all right now, instead of saving for unknown future. The government also promoted a fairness perception here, e.g., well-being of older adults in the British society, and ensured that employees will not have any control on their decision, i.e., they must have to contribute regularly to their pension scheme. Similarly, this particular direct nudge (i.e., not keeping alternative options for decision-making) neutralized employees' intrinsic motivation (self-interest) and status quo to consume all monthly incomes by the end of each month, while promoting the message clearly (i.e., you just need to contribute to your pension scheme), the government did not keep any option for employees for being biased by paradox of choice or framing effects.

After applying a nudge, especially an indirect nudge where customers have alternative options for decision-making, a brand must collect customer data periodically to understand the actual effect of the nudge and its progress over a period of time to influence customers' decision-making. For a single nudge effect, different customers would have different views based on their individual perceived experience with the brand. For example, "everyday low price" as the promotional tag of ALDI as they promote them as the "most affordable supermarket" (which is a nudge to influence customers to purchase from ALDI), may have different impact on different customers' extent of loyalty depending what ALDI promises and what is perceived by its customers. Some customers may feel that ALDI is really the most affordable supermarket, and some others may feel opposite based on their individual experience. For example, a North American ALDI customer describes that "I usually shop at ALDI and get a few items here and there at wegmans. My bill at ALDI usually comes to $70–80. I usually get enough to fill 3–4 bags. This week, I went to wegmans instead. 3 bags of my usual stuff came to $115" (Dizzy, 2022, np). On the contrary, a North-European ALDI customer describes that "having been an ALDI shopper for 5 years, I've seen them gone from undercutting every supermarket to suddenly taking advantage of the current situation (i.e., increasing living costs) and hiking their prices up at a rate that means they are no longer the cheapest anymore" (Montana, 2022, np). In this context, for example, out of 100 ALDI customers, 20 customers would perceive that ALDI is neither cheaper nor expensive than its competitors, 20 customers would think that ALDI is cheaper in 75% cases and 20 other customers may think that ALDI is cheaper in 100% cases. On contrary, 20 customers may think that ALDI is cheaper in 25% cases, and the other 20 customers think that ALDI is expensive (not cheaper) in 100% cases. The thicker double-headed arrow of ◘ Fig. 7.6 indicates such diverse customers' perceptions about a particular nudge effect, e.g., cheaper price of ALDI as a nudge effect for the case of this discussion. Collecting and analyzing customer data periodically is instrumental to understand the impact of a nudge and its progress to influence customers' decision-making.

Nudging for Sustainable Competitive Advantage Based on the discussions in this chapter thus far, we understand that developing a nudge (e.g., designing a product based on nudge effect, or adding nudge effect in product features or its service provision), making the product VRIN-competent or adding VRIN features to the product are prerequisite to sustain or prolong a short-lived commercial competitive advantage of a brand. Based on direct nudge and indirect nudge effects, let's have a look at two other examples of a product or product features that could have VRIN competency to a considerable extent, in order to sustain or prolong its competitive advantage. The first example is based on direct nudge (i.e., not keeping any option for customers to purchase an alternative product). In this case, let's consider the intense on-the-field and off-the-field rivalry of two sports teams, e.g., Manchester United and Manchester City football clubs. Both clubs enjoy a long-established loyalty from their ardent fans in their local and global markets. It would be very difficult for Manchester United to motivate the City's fans to purchase United-branded products, e.g., a souvenir and vice-versa. Let's assume that Manchester United, in association with a pharmaceutical company, have invested a lot of money and other resources to launch a life-saving drug or vaccine, e.g., a Corona vaccine. The vaccine comes with a Manchester United logo on its packaging and its promotion. The vaccine has been certified as very effective to prevent the virus with zero side effects, and it is also offered at very competitive price in comparison to other major vaccines. In this case, don't we think that Manchester United would be able to successfully nudge the ardent Manchester City fans to accept the United-branded vaccine, as it is a life-saving medicine, and there is no other vaccine available that is close to the number of features of this vaccine, such as its effectivity to protect human health, zero side effect, very competitive price etc.? In this case, Manchester United would be able to sustain its competitive advantage at least until any other vaccine could outplay the United's vaccine, in terms of its effectivity, zero side effects, pricing, etc.

The second example is based on the indirect nudge effect (i.e., keeping alternative options for customers' decision-making to purchase). Let's go back to ◘ Table 7.1, where we have listed "unique reputation of a brand" as a nudge effect. In general, Amazon as a long-established digital superstore that offers a huge range of products to purchase online has been pioneering in the global online shopping industry since its inception. In general, it could be argued that over a considerable period of time, they are able to establish a "peace of mind." In their customers' mind (which is another nudge effect), based on their transparent return policy, protecting customers' and vendors' interests, and offering a truly wide selection of alternative products, etc. "Buying from Amazon can be safe if you take certain precautions. Amazon is a reputable online marketplace that strives to provide a secure shopping experience for its customers" (np), and it has a reliable reputation, good trustworthiness, transparent customer reviews and ratings, comprehensive options for guarantee, and a wide selection of products (Emery, 2023). In this case, based on a "unique reputation of reliability" as an indirect nudge, Amazon would be able to sustain its commercial competitive advantage over its competitors, as it takes long time, resources, and efforts to establish such a unique reputation for a brand.

7.4.4 Criticisms of VRIN Capability and Nudge Theory, and Implications for Sustainable Marketing

Like most other social science theories and concepts, resource-based view (RBV) and VRIN capability, and nudge theory have criticisms. ◘ Table 7.2 summarizes a few **criticisms of RBV and VRIN capability** to sustain or prolong a short-lived competitive advantage, and our response to these criticisms.

The **criticism of nudge theory** includes the misuse of the nudge theory to influence public interest for individual's or a small group of people's interest, e.g., in politics. But, it is not a fault of nudge theory. For example, we know that dynamite was invented initially to progress the human civilization, e.g., to construct roads and highways by slightly abolishing hills and mountains; however, the same technology and its upgraded versions have been applied to develop weapons and used in war and conflict! Similarly, sustainable marketeers have the responsibility to ensure that nudge theory will not be misused to influence a brand's target audience. A prime example of misusing nudge theory include applying nudges inappropriately during election to influence voters. "Digital political marketing tools are used for deceptive and coercive ends. Of course, deception is a long-observed tactic deployed during election campaign in democracies" (Herbst, 2016; Perloff, 2018; as cited in Bakir, 2020, p. 4).

Table 7.2 Criticisms of RBV and VRIN capability to sustain or prolong a short-lived competitive advantage, and our response to these criticisms

Criticisms of RBV and VRIN capability	Our response
Because of the lack of empirical evidence from diverse contexts of business environments across the world, in their 2011 paper, Barney et al. themselves argue that "RBT [resource-based theory in 2011] lies at a critical juncture, one that will be followed either by revitalization of the theory or by its decline" (p. 1299)	It can be argued that the empirical evidences on VRIN capability have been progressed considerably in the recent decade. As a consequence, we are able to share a number of empirical examples of VRIN capability in this chapter. Also, we agree that it is really difficult to make a new product or converting an existing product to 100% VRIN-competent. In this case, we suggest adding additional VRIN attributes to a product or its service or promotional provisions. Examples are shared in this chapter
"Using meta-analytic techniques, we find that versatile resources (e.g., multipurpose utilities or adaptive capacities or reusable capacities) are associated with higher levels of growth (or sustainability), whereas VRIN resources are not" (Nason & Wiklund, 2018, p. 32)	We argue that developing a product or designing a nudge, based on versatile resources or capabilities (e.g., designing and launching a new academic program, led by an experienced academic who has not only the superior subject knowledge but also has depth understanding about the progress of market/student demand of that subject area) can address this issue
Another inadequacy in RBV is it is difficult to ensure sufficient competitive advantage through tradable or readily transferable resources, while employing the resource-based approach (Arndt & Pierce, 2018), e.g., a superior brand image could bring an advantage for the brand in long-run, but would be difficult to immediately transform the value of a superior brand image into liquid assets	We agree with this criticism that RBV and VRIN capability could bring an advantage for a brand in mid- or long run, but would be difficult to immediately transform the value of a superior brand image into liquid assets. But, it is not a core purpose of establishing a sustainable commercial competitive advantage of a brand to earn immediate liquid assets. But, the purpose of sustaining a brand's commercial competitive advantage is developing more fixed and variable resources, and tangible and intangible resources for the brand over a period of time, based on sustaining or at least prolonging its initial short-term competitive advantage over a mid- or long-term period

- Brad Parscale, (the) digital director of Trump's 2016 U.S. presidential election campaign used Facebook's advertising platform to automatically expand the number of people the campaign could target on Facebook by identifying voters who were not Trump supporters, but had "common qualities" that "look like" known Trump supporters on Facebook (Green and Issenberg, 2016). Similarly, since 2015 U.K. digital campaigning has seen increasing use of data analytics and data management approaches in order to profile and identify target audience, including "persuadables (nudges)" and swing voters. (The Electoral Commission, 2018, p. 4; as cited in Bakir, 2020, p. 4)

In India, the application of nudge theory also attracts criticism. For example:

- India Prime Minister Narendra Modi tried with his unexpected announcement Nov. 8 (in 2016) that all 500-rupee ($7.5) and 1,000-rupee notes would be banned. When the Dec. 30 deadline passed to exchange the notes, about 23 billion now-worthless bills had been switched for new ones or converted into electronic deposits. Modi says his goal is to combat counterfeit cash and unaccounted-for money and to nudge a cash-dominated society in a digital direction. (Nag & Beniwal, 2016, np)

It attracted huge criticisms from the Indian civil society, as it was perceived that the nudge for a digital society is used here to pursue a bigger political goal of the government. In some societies, bribery is not a secret issue. For example, "in the purview of the Nudge theory, we see that monetary incentives and benefits (bribery as a nudge) to policemen indirectly prevent them from misusing 'lathi' (stick)—this falls under the philosophical idea of 'Libertarian Paternalism' in Nudge Theory" (Krisnamurthy, 2021, np).

Implications for Sustainable Marketing Because of the effective application of nudge theory to influence people's behavior and decision-making, nudge theory has become very popular around the world since its inception in 2008. However, alongside its benefit, the nudge theory has attracted criticisms mainly for its misuse, especially for direct nudges (i.e., not keeping additional options for target audience to make decision) when it perceived as applied for pursuing an interest of individual or a small group of people. On the contrary, we have seen the responsible application of nudge to bring a positive change in societies. For example, the application of nudge by the British government for British employees to make their monthly contribution compulsory to their own pension scheme (Morrison, 2013). During the COVID-19 pandemic, once the COVID-19 vaccine was first introduced, we have seen that there was some reluctance among people in some countries to take the COVID-19 vaccine, which was a concern for governments to protect people's health during the pandemic. For example, the health chiefs (in the UK) are particularly "worried about the number of people who are still undecided, and about vaccine scepticism among NHS (National Health Service) staff" (Campbell, 2020, np). To address this issue, the UK government decided for celebrity endorsement (as a nudge), considering that celebrities have huge national and international fan base, and in general, the fans maintain a relationship with their celebrities (e.g., following the celebrities' lifestyle). In this context, with the hope to favorably influence people's decision to take COVID-19 vaccine, the British government drawn up a list of "very sensible" popular celebrities, with the aim that the celebrities' advice and promotion to get immunized would be trusted widely by their nation-wide followers (Campbell, 2020).

Similarly, in India, we have seen another positive application of nudge theory for a greater societal benefit. For example, "the Government of India's use of nudge theory in the first three months of the (COVID-19) pandemic helped to tackle the virus on numerous fronts, a new study suggests" (University of Cambridge, 2024, np). During the Covid-19 pandemic in India:

> Policy nudges were focused on evaluating the risk of incoming travellers from China and extending surveillance at international airports. But the narrative (of nudges) soon shifted to address other pressing concerns. (For example,) nudges sought to impose new restrictions on travel, discouraging people from visiting crowded and public spaces, and strict social distancing. Modi (the Indian prime minister) told the nation that "21 days is critical to breaking the infection cycle…or else the country and your family could be set back 21 years". (This "21 days and 21 years" nudge helped to let the people understand better the necessity for lockdown in an overly crowded-country, like India, as) The next day, the country entered phase 1 of lockdown. (University of Cambridge, 2024, np)

In general, there are positive and negative applications of nudge theory. However, there are more positives than negatives. For example, "most newspaper quotes on nudges were positive, but there were substantial differences between categories and between themes (of nudges)" (Tummers, 2023, p. 1019). Tummers' (2023) research based on 19 different categories and themes of nudges recognizes that 65% of newspaper reports on nudge are positive, while 35% accounts for negative application of nudge theory to influence people's decision. So, business ethics and responsible marketing are important factors sustainable marketeers should consider, while designing, applying and evaluating a nudge effect. For example, sustainable marketeers should be guided by the moral principles of nudges aiming to underpin benefit for greater society and environment, instead of applying nudges merely for personal gain.

7.5 Marketing Capabilities for Sustainable Competitive Advantage

► Figure 1.7 in ► Chap. 1 presents the 4Ps' marketing mix (i.e., product, price, place and promotion) along with the additional 18Ps as extended marketing mix. These 22Ps in the extended marketing mix, e.g., designing, developing, and evaluating a product properly reflecting on customers' needs or a promotional campaign that would be appealing to its target audiences are specific capabilities of marketeers to influence their customers and non-customer stakeholders. Therefore, based on these 22 marketing capabilities, ◘ Table 7.3 discusses the core focus of sustainable marketing related to each of the 22 marketing capabilities.

Table 7.3 Focus on marketing capabilities for sustainable marketing

Marketing capabilities	Focus for sustainable marketing
Probing for marketing information and intelligence system (Baumgartner, 1991; Vorhies & Morgan, 2005)	Ongoing market orientation focusing on the triple bottom lines' (People/Societal, Planet/Environmental, Profit/Commercial) stakeholders to understand their current and embryonic needs and distresses (e.g., increasing living cost)
Product/service (McCarthy, 1960; Borden, 1964) design	Designing products and services as an integrated value package for the triple bottom lines' stakeholders
Competitive pricing strategy (McCarthy, 1960; Borden, 1964)	Ensuring equity, diversity, and inclusion (EDI) of customers across different segments, including the low-income consumer groups. For example, Unilever offers different prices for their Lux soap for different global consumer segments. In North Europe, a single piece of Lux soap would cost £0.60 (Amazon, 2024a), whereas it costs Indian Rupee 36.00 in India, roughly equivalent to £0.35 (Amazon, 2024b). It is because that the price that Unilever offers in North Europe for Lux, if they offer the same price in India, it will not be affordable to Indian consumers. To ensure that Unilever can make profit in India, while offering an affordable price to the Indian consumers, they produce Lux in India and promote it using local celebrities; it enables them to reduce the production cost to offer it at a lower price. Similarly, marketeers should consider the affordability of their different target segments to ensure EDI
Convenient distribution channel for customers (McCarthy, 1960; Borden, 1964)	A consumer may prefer to purchase just a 150 ML Coke at 3.00 AM. The issue is whether Coca-Cola has placed a self-serving vending machine nearby to this consumer, e.g., at a town center, so that the consumer can reach to Coke to purchase it after 5–10-min walk. If Coca-Cola does not have this convenient distribution channel, but Pepsi Cola has, in that case, Pepsi will have a competitive advantage over Coke. Similarly, sustainable marketeers should consider the convenience of their different market segments, including older adults, disabled customers, and other disadvantaged groups while planning for distribution channel
Branding, positioning, communication and promotional message (McCarthy, 1960; Borden, 1964)	Since the number of socially and environmentally caring customers is increasing across the globe and consumers are helping businesses drive sustainability (Close, 2021; Deloitte, 2024) at a growing rate; sustainable marketing, responsible product and service development, and its branding and promotion appealing to its contribution to society, environment, and profit (e.g., competitive pricing) is not charity but a necessity for a brand today to survive and prosper

(continued)

7

Table 7.3 (continued)

Marketing capabilities	Focus for sustainable marketing
Contribution to people (Judd, 1987)	The value that a brand can contribute to society, that societal contribution can enhance the brand's commercial competitive advantage
Contribution to planet (Looney & Seiler, 2003)	The value that a brand can contribute to environment, that environmental contribution can enhance the brand's commercial competitive advantage
Making profit (Baumgartner, 1991)	Making profit is a key bottom line for any business. However, sustainable marketeers should be transparent to make profit responsibly. For example, making profit by harming the interests of People and Planet should be avoided at any cost
Convenient process for stakeholders (Booms & Bitner, 1981)	Making business process easier for stakeholders is a key priority to proactively engage stakeholders in the business process. For example, a North American consumer research (Lebow, 2021) shows that the popularity of paperless billing has increased from 62% in 2019 to 75% in 2021 across all generations, as it is cost-effective, quick, and environmentally friendly. Similarly, making business process People- and Planet-friendly (e.g., easily accessible by stakeholders, zero carbon emission, etc.) should be a prime priority of sustainable marketeers
Stakeholders' participation (Booms & Bitner, 1981)	Stakeholder consultation is a prerequisite for co-creation of a product or service and establishing a business process that would be stakeholder-friendly. For example, the triple bottom line's stakeholders should be consulted to understand and integrate their value expectation into product features, pricing and distribution options, promotional messages that would be desirable by these stakeholder groups, in order to ensure the triple bottom line stakeholders' participation in product development, pricing, delivery, and promotion
Distinctive physical evidence (Booms & Bitner, 1981)	Physical evidence of a brand refers to the physical attributes, e.g., color, design, message, etc., of the brand's logo, physical signage, and other physical features, such as interior design of an outlet. For example, when we visit a McDonalds or a KFC, the color combination of its interior design and its overall presentation enable us to distinguish between these two competitive brands. Such physical attributes of a brand, e.g., a banner, a hoarding, a digital signage, etc., should be designed to convey the message of the brand's devotion to society and environment, which will not only be instrumental in creating a socially and environmentally caring brand image, but also be helpful for customers and non-customer stakeholders as a reminder about our societal and environmental concerns
Adaptation to stakeholders' preference (Looney & Seiler, 2003; Shams, 2013b)	Unless marketeers appreciate the preferences of their triple bottom line stakeholders' preferences; it would be really difficult to create value for their societal, environmental, and commercial stakeholders

(continued)

Table 7.3 (continued)

Marketing capabilities	Focus for sustainable marketing
Forming/influencing public opinion (Kotler, 1984)	What comes to your mind first when you compare the core value offered by ALDI with its competitive supermarkets? In response to this question, a number of postgraduate marketing students across different cohorts answered in between mid-2019 and mid-2024 that ALDI is more affordable (slightly cheaper) than its competitors, which is exactly what (as a public opinion) ALDI's marketeers wanted their customers to think about them. Sustainable marketeers should create a socially and environmentally caring brand image in a way that when their customers think about them, customers should have a perceived feeling that, "yes," the brand truly cares about society and/or environment. To create a public opinion, a key prerequisite is what a brand promises to contribute and what the brand actually contributes, in relation to their promises over a period of time
Public relations (Baumgartner, 1991)	Public relation activities, e.g., press release, press conference, road shows, social and traditional media presence, of a responsible brand should be focused on communicating what they promise to their triple bottom line's stakeholders and what they contribute
Political power (Kotler, 1984)	Sustainable marketeers should take advantage of their proximity to political power for the interests of the triple bottom line, i.e., contributing to people and planet while making profit
Proving advantage through stakeholders' experience (Kurtz, 2009; Shams, 2013b)	"All of Coca-Cola (nd) European Partners (CCEP) bottles and cans have been 100% recyclable" (np). In response to Coke's sustainability initiatives, their "EuroPacific Partners' Vice President, Customer Service and Supply Chain, Javier Sanchez Gandarias, discusses the manufacturer's latest supply chain innovations and they will help to reduce GHG emissions by 30% by 2030 and reach net zero by 2040" (Bush, 2022, np). Here, Coca-Cola promotes the advantage of partnering with them for environmentally friendly initiatives through the experience of their stakeholdersh
Partition (Baumgartner, 1991) for stakeholders' convenience	Partition refers to the two version of a product, i.e., a generic version that offers the core utility with a cheaper price (e.g., budget airlines), and a value-added version of the same product or service (e.g., luxury airlines, or economy class and business class at a flight) that is offered with premium price. Sustainable marketeers should provide equal importance for each version of their product, while designing it to contribute to the triple bottom line's stakeholders

(continued)

Table 7.3 (continued)

Marketing capabilities	Focus for sustainable marketing
Prioritizing (Baumgartner, 1991) and balancing	On the one hand, in general, it is more challenging for sustainable marketeers to satisfy their stakeholders, in comparison to the single bottom line-oriented (profit-only) marketeers. It is because sustainable marketeers need to deal with diverse groups of stakeholders, i.e., societal, environmental, and commercial, whereas profit-only marketeers need to deal mainly with their commercial stakeholders (Ballesteros-Sola, 2014; Haigh et al., 2015; Gregory-Smith et al., 2017; Apostolidis et al., 2022; Shams et al., 2022; Shams et al., 2024). On the other hand, it is not possible for any brand to equally contribute to everyone. Therefore, it is imperative to prioritizing and balancing between different stakeholders' priorities, while ensuring to contribute to all three (societal, environmental and commercial) stakeholder groups to a realistic extent
Sales management (Vorhies & Morgan, 2005)	A 2020 McKinsey Consumer Sentiment Survey (2024) reveals that "more than 60 percent of respondents said they'd pay more for a product with sustainable packaging" (np). The same study recognized that "products making ESG-related (Environmental and Societal Governance-related) claims averaged 28 percent cumulative growth over the past five-year period, versus 20 percent for products that made no such claims" (np), and the study concludes that "consumers care about sustainability—and back it up with their wallets" (np). Consequently, from the phases of product design and development to delivery (sales) management and after-sales-service phases, sustainable marketeers and sales personnel should clarify and promote with evidence how their brand contributes back to society and environment, in order to increase its sales
Positive implementation (Baumgartner, 1991)	Positive implementation could be referred to a brand's commitment on what the brand promises to its triple bottom line's stakeholders and what it delivers. Therefore, promises and claims of contribution should be realistic
Overall planning, performing (Baumgartner, 1991) and evaluating based on a balanced integration of relevant marketing capabilities	This table lists a number of marketing capabilities. But, a single marketing or sustainable marketing project may always not need all capabilities. Therefore, focusing on the right capabilities instrumental for a specific sustainable marketing project, and all those capabilities' balanced integration through the sustainable marketing project's design, implementation and evaluation phases is important to deliver the promised value to the triple bottom line's stakeholders

(continued)

Table 7.3 (continued)

Marketing capabilities	Focus for sustainable marketing
Position (Baumgartner, 1991)	Position of a brand in its target audience or target market, e.g., occupying targeted market share in terms of sales volume is an important consideration for the brand and its growth. Along with targeting a market share, sustainable marketeers should also target a share to occupy in its customers and non-customer stakeholders' mind as a socially and/or environmentally caring brand, among the competitive brands. As the number of socially and/or environmentally caring customers are increasing; once the share/percentage of a considerable number of customers will be considering a brand as socially and/or environmentally caring brand; the brand's position in terms of its sales and market share will increase, eventually

7.6 Embryonic Changes in Marketing Dynamics and Sustainable Marketing

► Section 1.3.1 and ► Table 1.2 in ► Chap. 1 discuss the evolvement of marketing dynamics in different eras since the early 1900s to post-2020s, and the market influences and marketing's response to the changes over time. This section discusses the ongoing marketing environment and its embryonic nature, and its implications for sustainable marketing.

Increasing Societal and Environmental Concerns and Marketing The Belgian chemist, Leo Baekeland, received the patent for inventing synthetic plastic in 1907 (Science Museum, 2019). Before 1907, the world did not have threat from plastic. Today, "millions of tons of plastic waste enter the ocean each year" (Darrah, 2023, np). "It's been estimated that plastic pollution kills 100,000 marine mammals every year. 81 out of 123 marine mammal species are known to have eaten or been entangled in plastic, and all seven sea turtle species are affected" (WWF-Australia, 2023, np). "According to a report by the United Nations, there will be more plastic in the ocean than fish by 2050" (Darrah, 2023, np). Because of the easy availability of plastic and plastic products, and its irresponsible usage around the world, along with animals on land and ocean, human life is under significant threat. For example, "some scientists have estimated the average person might eat 5 grams of microplastics in a week—about the weight of a ► credit card. Another study breaks that down to up to ► 52,000 particles annually from various food sources" (Francis, 2022, np). In addition to these significant societal and environmental concerns because of plastic, the human society and our ecology are also under-threat because of increased pollutions in different industries across the world and increased CO_2 emissions. To address this grand challenge, sustainable marketeers should introduce socially and environmentally caring alternative services. For example, "the Rimping Supermarket chain in Chiang Mai (Thailand) has gotten rid of plastic packaging for vegetables and is now instead using banana leaves to wrap them. Their initiative has gained global recognition" (NDTV, 2023, np). In a bid to decrease plastic waste, Bangladeshi scientist Mubarak Ahmad Khan invented a plastic bag alternative with the help of a local natural fiber" (CNN, 2023, np). The local natural fiber used to invent this 100% biodegradable plastic bag is jute, a natural fiber produced in Bangladesh. This decomposable jute-made plastic bag is branded as "Sonali Bag." The meaning of sonali in Bangla is golden. Since the processed jute looks golden, the jute-made fully biodegradable plastic bag is branded as "Sonali Bag," which has been adopted by UNDP (2023) to combat the global plastic pollution.

Equity, Diversity, and Inclusion (ED) and the Marketing Mix Rising inflation rate and increasing living costs "are widening health inequalities, weakening food security and diminishing health and social care" (Kilfoyle, 2023, np). The cost-of-living crisis raises major challenges for food, housing, clothing, health service and so on, especially for marginalized communities around the world. Sustainable marketeers should endorse EDI in their marketing mix, e.g., offering competitive price, offering multiple versions of a product in different price ranges that would be affordable by different market segments (including low-income consumer group), and promoting products while endorsing EDI for an inclusive society.

SDG-16—Peace, Justice and Strong Institution, and the Role of Marketing The war and conflicts severely impact on businesses and subsequently on the global value chain that helps to ensure the inflow and outflow of goods and services across national borders. Because of the Ukraine-Russia war that started in early 2022 (and is still continuing in February 2025 while reading the proof of this textbook), western brands needed to close their business in Russia. As a result of closing 850 Russian outlets (Onita, 2022), McDonald's "overall revenue dipped 3% to $5.8 billion in the second quarter largely because McDonald's closed its company-owned eateries in Russia and Ukraine after the war broke out. Profits fell by nearly half due to a $1.2 billion charge related to the sale of the Russian business" (Fickenscher, 2022, np). Also, the aftermath of war and conflict drives boycotts for brands from consumers. For example, following the Israel-Palestine war that started in late 2023, "McDonald's has missed a key sales target, partly due to customers boycotting the firm…(consequently)…The fast food chain reported its first quarterly sales miss in nearly four years due to weak growth in its international business division" (BBC, 2024, np).

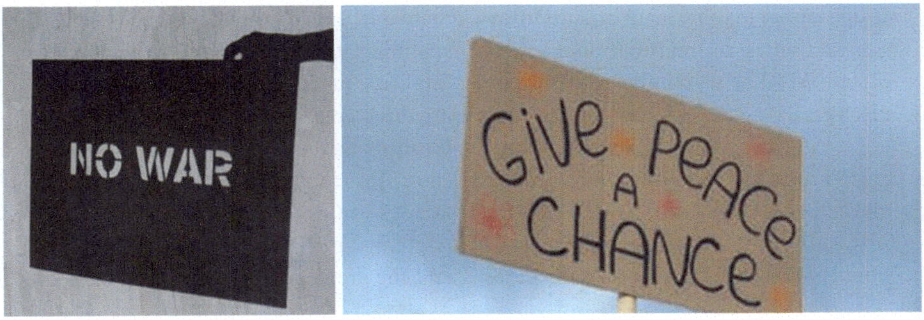

■ Source: ► pexels.com/royalty-free-images/

Sustainable marketeers have been promoting different social and environmental cause, such as recycling, e.g., Coke's Sustainability TVC (2019); or mental health, e.g., Saving Lives with Music—Logic, A Hashtag for More Mental Health—JanSport, Excuses to Escape Online Fatigue—TBWA, Using the Power of Shared Laughter to Build More Resilience—Maltesers, or A Platform to Combat Anxiety and Depression—Maybelline New York (Heintz, 2022), while promoting their brands. However, it is rare to see that marketeers promote peace while promoting their brands. It is high time for marketeers to work for peace, justice and strong institution while promoting their brands (e.g., creating awareness through branding message for peace), as peace can not only ensure that brands will be able to run their business smoothly across the world, but also will be able to protect our global value chain, which is essential for a sustainable global economy.

Marketing in Conflict Regions It is not expected that war and conflict will continue; however, marketeers need to do marketing even in conflict regions. Manager's Corner ► 7.2 discusses an approach for marketeers to plan for marketing in conflict regions.

Manager's Corner 7.2: Marketing in Conflict Regions

Conflict refers to the risk or actual encounter of violent acts designed to cause fear and intimidation. Despite posing a critical threat to internationally active firms, there is a paucity of knowledge that could address the distinctive challenges that conflict across the world brings for marketeers today, especially in the case of international marketing activities of firms, while marketing in the conflict regions.

Here first, a theoretical background on conflict and its effects on international marketing is discussed. Then operational costs, marketing planning, supply chain management and distribution activities of multinational enterprises (MNEs) in conflict regions are discussed.

It is recognized that significant costs are involved in the international marketing budget of MNEs. Firms with substantial resources and international experience appear to have more alternatives, which allow them to cope better with the effects of conflicts than their less-endowed peers.

Conflicts or terrorism (whatever the name we give it) is a salient threat to organizational competitiveness in international marketing. It is the premeditated use or threat to use violence by individuals or sub-national groups to obtain a political or social objective through the intimidation of a large audience beyond that of the immediate noncombatant victims.

For terrorists, perception matters! Terrorist attacks around the world have increased greatly in the past decades, spanning 92 countries and over 28,000 fatalities in 2015 alone. Most attacks are directed at civilians, businesses, and business-related infrastructure. The five countries most exposed to terrorist attacks in recent years are Iraq, Afghanistan, Pakistan, India, Nigeria and the greater Middle-Eastern region.

Emerging markets are particularly affected by such conflicts since their businesses and citizens have comparatively less opportunity to protect themselves. Among the possible environmental contingencies that can affect marketing organizations are weak economic conditions, rising energy prices, and financial crises. However, conflict itself is identified as potentially the most serious threat.

Since terrorists select their targets with high flexibility, intensity and precision, international firms seek a competitive advantage through the expansion of production, distribution and marketing of products and services across multiple national boundaries. Conflicts sharply reduce corporate enthusiasm to expand. Measures to counter conflicts in turn are based on restricted freedom of movement and increased government regulation, both of which impair the global economy. The border-crossing effect of conflicts creates slowdowns for international transactions reaching 2.5% of merchandise value, which is comparable to the average level of global tariffs.

International trade depends on the efficiency and cost-effectiveness of global transportation systems. Conflicts increase the transaction costs of international commerce and delay global supply chains and distribution channels. Terrorism's main impact reaches far beyond its immediate and direct effects. What is key is the long-term results from the indirect effects that occur in national and global economies. These include widespread anxiety and uncertainty that affect buyer demand, shifts or interruptions in the supply of needed inputs, new government regulations and procedures enacted to deal with conflicts, and longer-term perceptions that alter patterns of global trade and investment. Conflicts can also affect managerial attitudes toward risk, shift the risk absorption capacity of firms, and reduce the likelihood of embarking on international ventures or new investments abroad.

A Google search of the NGram viewer system analyzed the extent of terrorism-related writings, and checked for correlations with the key terms "trade", "investment" and "risk." The results indicate a rapid increase in concern about terrorism since 1998. This development serves as an indicator of the growing preoccupation (in the English-speaking literature at least) with conflicts. Concurrently, and as expected in terms of theory-based postulations, actual risk increased while trade and investment interests declined.

It is unfortunately, assumed that conflicts will continue to be a significant factor in international marketing for decades to come. The rise of conflicts signals a new type of threat relevant to both developed and emerging economies. As governments increase security of public facilities, the likelihood of attacks against the softer targets of firms' international operations is likely to increase. Emerging economies need to find ways to increase their security in order to retain their attractiveness for foreign sourcing and investments. Corporate preparedness for the unexpected is a vital task. Innovative marketeers develop appropriate resources, and undertake planning and strategies to accommodate dislocations and sudden shocks. Conflicts represent an organizational crisis whose ultimate effects may be unexpected and unknown, posing a significant threat to the survival or performance of the firm.

Conflicts present the firm with a dilemma that requires new decision-making and behaviors that will result in organizational change. Firms that neglect to devote resources and capabilities to respond flexibly to terrorist-triggered disruptions and risk sudden or sometimes even total loss of competitive advantage.

We follow the thinking of former US Secretary of Defense Donald Rumsfeld who stated: "There are known knowns, which are things we know that we know. There are known unknowns; that is to say, there are things that we now know we don't know. But there are also unknown unknowns, these are things we don't know we don't know." The goal should be to analyze the role of conflicts under all these three conditions, in order to plan for your marketing mix in conflict regions.

Information Source:

Czinkota, M. R. (2018). *In search for the soul of international business.* Business Expert Press. (Reprinted a slightly revised version by permission of the copyright owner, © Springer).

Geopolitical Tensions and Marketing Even during peacetime, the rising geopolitical tensions among nations impact severely on businesses. Sanctions, anti-sanctions and its aftermath create a more complex environment for marketing. For example, you are a marketing director of an international brand. Following the existing success of your brand in international markets, perhaps, you have invested time, money and efforts in last two years to expand the brand's international market further. After two years of dedicated efforts and investments, you are now about to launch the operation of your brand in a new international location in next month. But, just before you launch your brand, a new sanction has been imposed by your government. As a result, the government of your targeted international location has increased the tariff for products and services that would be exported from your country. It can make your efforts futile that you invested since last two years to expand your business in this new international location. Therefore, marketeers need to consider plan-B or C, based on anticipating the changing market environment.

Embryonic Technologies and the Combination of Marketplace, Marketspace, and Simulated Market We have our traditional marketplace, i.e., the physical shop, which will remain relevant in the distribution of products and services in future. Since the turn to the twenty-first century, we have digital marketspace, such as the online superstores, social media platforms to promote products and services, and other digital services, such as electronic funds transfer at point of sale (EFTPOS) payment option and other online payment option, e.g., PayPal. In addition, we have an embryonic simulated market, which is more virtual than our contemporary marketspace, as it attempts to offer a virtually simulated experience to buyers and sellers that would feel similar to the real-life in-person business experience. For example, Facebook Meta and a few other tech giants are trying to launch an online platform where business stakeholders will meet virtually from distance, but they will feel that they have participated in a traditional in-person face-to-face business meeting. Applying the traditional marketing principles (i.e., the marketing mix, branding, sales management etc.) and the contemporary and emerging thoughts in marketing discipline, marketeers today need to deal with a market that is formed in combination of the traditional marketplace, the contemporary marketspace and the embryonic simulated market.

◘ Source: ► pexels.com/royalty-free-images/

Artificial intelligence, robotics, augmented reality, virtual reality, big data and advanced analytics, blockchain, internet-of-things, and other embryonic technologies will have a significant influence on future marketing dynamics. We already have seen that robots are serving foods in restaurants, or virtual assistants are helping us in our personal and professional life. Marketeers should ensure that not only they comply with the applicable business law, but they also comply with marketing ethics to protect their and their customer and non-customer stakeholders' rights while applying contemporary and embryonic technologies.

◘ Source: ► pexels.com/royalty-free-images/

Data Analytics and General Data Protection Regulation (GDPR) We have seen customers' data breach by big brands. For example, "50 million Facebook profiles harvested for Cambridge Analytica in major data breach" (Cadwalladr & Graham-Harrison, 2018, np). "Meta fined $276 million over Facebook data leak involving more than 533 million users" (Roth, 2022, np). More recently, "millions of Facebook and Google users hit by 2FA data leak" (Laird, 2024, np). Marketeers need to prioritize the protection of their customers' privacy and safety over profit. Product managers should ensure that products or services are designed and delivered in a way that ensures the principles of "responsible production and consumption" (i.e., SDG-12). It is important considering that irresponsible production not only influences society negatively but also forms a product image that is damaging for a brand.

■ Source: ► pexels.com/royalty-free-images/

Interplanetary Space and Marketing In 2024, it takes around 23 h for a direct commercial flight to arrive at Sydney from London (British Airways, n.d.). Students who are reading this textbook, perhaps, you would be marketing a commercial flight that would take passengers to Sydney from London in 4 h! Space tourism is already in existence, "since the flight of the world's first space tourist, American businessman Dennis Tito, on April 28, 2001, space tourism has gained new prominence as more suborbital and orbital tourism opportunities have become available" (Seedhouse,

2024, np). Space tourism is "the recreational travel to space. The 'space' area is pegged around 50 to 600 miles above ground; the International Space Station, for example, orbits at about 200 miles above the ground" (Zero-G, 2022, np). Since Dennis Tito as the first-ever space tourist, a few other people have already self-funded their space tourism that include (Zero-G, 2022) the South African Croesus Mark Shutteleworth in 2022, American businessman Gregory Olsen in 2005, Iranian-American entrepreneur Anousheh Ansari in 2006 who is the first female space tourist, American billionaire Charles Simonyi who traveled to the space twice in 2007 and 2009, the video game developer Richard Garriott in 2008, Canadian entrepreneur Guy Laliberté in 2009, and the Japanese entrepreneur Maezawa Yusaku and his production assistant Yozo Hirano in 2021. SpaceX, Blue Origin, Virgin Atlantic and a few other government, and non-government sectors' organizations have been trying to make space tourism more affordable and accessible for people. Consequently, many of you who are reading this textbook today as an undergraduate or postgraduate student will market space tourism at some stage soon in your marketing career.

Is Paperless Society the Future? Dubai in 2021 introduced the world's first paperless government (Heric, 2022). Similarly, some areas of our personal and professional lives have become paperless today, which include but not limited to paperless meeting, paperless payment, paperless business, paperless bookings for travel and tourism. It would take some time for the contemporary digital payment systems and cryptocurrency to replace the traditional paper-based bank notes; however, the human civilization has been moving to the state of a society, where perhaps the usage of paper would be very minimal.

Long-Term Impact of the COVID-19 Pandemic The COVID-19 pandemic propelled the half of the world's population to stay home under lockdown by April 2020 (Sandford, 2020). The global community struggled severely because of the COVID-19 pandemic since the beginning of 2020. "The Covid-19 outbreak is not only about health care, but also businesses, livelihoods, wellbeing, innovation and resilience" (Krishna & Kummitha, 2020, p. 9). By mid-May 2020, our established management system in all sectors including public health, education, business, communication, transportation, public administration, among others across the world had been impacted seriously because of this COVID-19 pandemic. Indicating to the unknown long-term impact of the pandemic, researchers (i.e., Xie et al., 2020) argue that "while this is still 'an emerging, rapidly evolving situation' (Centers for Disease Control and Prevention, 2020), COVID-19 has already exposed numerous serious problems. Its long-term impact may not be fully realized until years later" (p. 1). Also, the pandemic impacted different sectors or industries to different ways and extents. Thus, it would take few decades or more to realize the actual impact of the pandemic on world population. Marketeers need to be vigilant about the latent aspects of the impact of the COVID-19 pandemic in their respective industry.

Crisis Management and Increased Importance on Agile Performance In addition to the big crises, like the COVID-19 pandemic, businesses can be impacted by different other ongoing political, environmental, societal, technological, legal, or environmental (PESTLE) issues. Businesses need to be agile to cope up with the nascent business dynamics, in order to ensure that their value-delivery network can proactively deal with any small or big crisis in future. "Strategic agility is a fuzzy concept that may be counter intuitive as well as confounding to some scholars in terms of the agile strategies' contextual issues. At the same time, the need to be agile is crucial for firms" (Shams et al., 2021, p. 1). "Agility clearly extends the concept of flexibility and is associated with speedily sensing and responding to opportunities and threats in the business environment" (Mao et al., 2015, p. 361). Information technology offers considerable tools to evaluate and understand the emerging nature of prospective crises, e.g., from tracking historical customer complaints to understand the emerging nature of PESTLE dynamics, in order to timely recognize the triggers of crisis and apply the agile strategies (e.g., plan-B or plan-C) to protect a firm's value-delivery network.

◘ Source: ► pexels.com/royalty-free-images/

Increased Significance of Cross-Cultural Marketing It, in general, can be argued that there are more cross-cultural customers in a city today than ever before. As a result, we see more cross-cultural products in our supermarkets today than ever before. Alongside international marketing, cross-cultural marketing therefore has a role to play in marketing in your home market, as well. Since it would be difficult to ignore the growing number of cross-cultural customers' demand for a particular product or services; it brings an opportunity for marketeers to support EDI by bringing more people across different race in the consumption dynamics.

Consumption dynamics refer to the overall inflow and outflow of goods and services in a society, and the extents of different customer segments' (i.e., high-, mid-, and low-income segments') ability to purchase and consume fundamental and luxury goods and services.

Increased Buying Power and New International Markets China did a tremendous job in between mid-twentieth and the end of the twentieth century to bring out a huge number of Chinese populations from poverty (BBC News, 2019). For example:

> In 1990 there were more than 750 million people in China living below the international poverty line—about two-thirds of the population. By 2012, that had fallen to fewer than 90 million, and by 2016—the most recent year for which World Bank figures are available—it had fallen to 7.2 million people (0.5% of the population)... This suggests that overall, 745 million fewer people were living in extreme poverty in China than were 30 years ago. (BBC Reality Check, 2021, np).

What China did in the last century and especially in last three decades, many other emerging economic countries have been doing it more recently, i.e., bringing out population from poverty level. Some Latin, South-Asian, and East-Asian countries have been performing well to improve the lifestyle of their people. For example, one of the co-authors of this textbook is from this region of the world. During the childhood of the co-author, having at least one holiday a year even within the country, traveling via airplane, purchasing a car or home was luxury for middle-class people in his society. Today, all these items which were once luxury have become a regular part of the life of middle-income population. It brings opportunities for marketeers, as the affordability (purchasing power) of customers across the globe has been increasing to purchase more. The increased purchasing power is good for making profit; however, sustainable marketeers should promote responsible consumption to pursue the SDG-12 (Responsible Consumption and Production).

Intergenerational Marketing Marketeers need to influence customers from different generations. For example, baby boomers who were born in between 1946 and 1964, Generation X who were born in between 1965 and 1980, Generation Y was born in between 1981 and 1994, Generation Z who were born in between 1995 and 2009, and Generation A or Alpha who were born in between 2010 and 2024, who follow the Generation B or Beta who will born in between 2025 and 2039 (Trinet, 2024; MCCRINDLE, 2024). Designing a mass-standardized product that would meet the needs of all generations is not easy.

> Demographics suggest that Americans over the age of 60 will soon become a larger consumer group in terms of income than 20-to-34-year-olds, yet companies still position their brands as if the bulk of the market is people under the age of 40. The answer, however, is not to shift marketing strategies from one generation to another. Research has suggested that boundaries between generations are fuzzy, if not arbitrary. Plus there's a bewildering degree of variability across individuals within the same generation. Instead, marketeers should aim to emphasize common values across different age groups and build "post-generational" brands. Guillén, 2024, np)

Understanding the common values, norms, and ways of seeing life across different generations would be instrumental to design a mass-standardized product; however, if resources permit, customized and niche marketing would be more relevant to satisfy customers across different generations more profoundly. Both mass-standardized and customized products have their advantages and disadvantages for marketeers. For example, the monitoring of the value-delivery network of a mass-standardized product would be more convenient and less resource-intensive, in comparison to monitoring the multiple value-delivery networks of different customized products. On the contrary, customized products or niche marketing would be more instrumental to influence the target customer groups more appealingly. Therefore, marketeers should consider their available resources to decide whether mass-standardization or customization would be more relevant for them.

The Debate of Globalization Vs. Localization (This subsection is sourced from Czinkota et al., 2021, and reprinted by permission of copyright owner, © Springer)

In recent decades there is a debate on globalization vs. localization. This subsection discusses that globalization and freedom of a local area (e.g., job security of local communities) are simultaneously possible.

Globalization, trade, and investment deserve our "Thank You" for their achievements. Yes, in Europe and the United States, popular discontent is forcefully expressed. An introvert trend has emerged, fed by nationalism, populism, xenophobia and anti-globalization rhetoric. Globalization is not new; it has existed for centuries. What is different today is the speed of globalizing the world, made possible by new technologies, transportation networks, media, and marketing across borders. Many claim that never before in history has there been so much evidence of strong opposition to globalization. However, any comparison with the past is highly inaccurate. Only a few records of resistance to globalization have been preserved for us today.

One main feature of globalization in our age is the conveyance of freedom, which in turn becomes the foundation of the good life. There are three dimensions to this freedom. For one, the opportunity to collaborate with others, to streamline processes and procedures and to increase the variety of products and services cause great freedom of choice. There is the freedom conveyed by sovereignty, where nations and their governments maintain and exercise preferences for their citizens. Individual countries will always seek to have sovereign freedom. Then there is the freedom of principles and options. Such freedom of choice is primarily expressed by the recourse to deeply rooted principles which provide the boundaries of the playing field. If there is no limitation and no alternative, then there is no freedom. A true alternative requires a decision and the exercise of virtue. If there are no principles to which one can seek recourse, there is no freedom.

In reciprocal causality, freedom both causes and facilitates international marketing, while international marketing is a key support of the cause of freedom. The focus and aim of international marketing is on crossing borders, with the goal of providing more choices for consumers and letting them be the ones to maximize their satisfaction. International marketing does so in all corners of the globe, the glamorous ones as well as in the small and remote ones where its efforts are not seen by others. When the long-standing rivalry between socialism and market economy came to an end, market forces directly improved human rights and the extent of freedom. Exchanges across national markets have clearly demonstrated greater efficiency and effectiveness in their ability to satisfy the needs of people. Now they must demonstrate their ability to persist through continuity and balance.

In spite of complaints about slowness of change, biases in wealth distribution, and economic disparities and inequalities, a majority of the participants in market-oriented conditions are now better off than they were before. World poverty has been reduced significantly. However, one of the main discontents of globalization is low incomes in formerly successful regions, a high level of inequality within and between countries, and between the world's economic classes. People at large will not tolerate such glaring inequality anymore. One key dimension of freedom is to allow people to think outside the box. National borders are where business, government and people usually find their limits. Freedom knows no international boundaries. It thrives on how to successfully cross national borders, on coping with local differences once the crossing is done, and on profitably reconciling any conflicts. Domestic borders limit economic expansion, while international approaches contain the freedom of much wider opportunities.

The anti-globalization narrative in Western European and the United States is caused by imbalances in both people and goods, as well as an intransigent enforcement of rules from disconnected centers. Migration and refugee pressures force people to move from rural homes into urban areas, from their developing countries into industrialized ones, and to take any measures to be relieved of wartime conditions. For both sides, little freedom is involved here. Most individuals who do the moving would much rather stay home but cannot afford to do so. The recipient countries might not want to welcome the refugees and migrants but have to do so in response to humanitarian pressures.

Globalization may have been part of what triggered some of these shifts and migrations, but it also can be instrumental in restoring the freedom to stem the tide. Globalization can provide economic opportunities and alternatives for individuals in their home countries, free from pressures to shift locations, so they can become productive economic contributors. To make globalization work, governmental, managerial, and corporate virtue, vision, and veracity are required. Trading and investment partners need to understand the current context and the longer-term effect of policy and business actions taken. Joint actions based on mutual interests but also accepting divergent interests will make better outcomes not just possible but likely.

Fragility of the Capitalist Economic Concept and the Role of Marketing The core focus of the capitalist economy is more industrialization, more production, and more consumption. We have seen that increased industrialization and production derived societal and environmental concerns, including pollution, increased CO_2 emissions and subsequent climate change. It attracts significant criticisms for the capitalist economic concept from different societal and environmental activist groups. In addition, over a sustained period of the capitalist economic concept's contribution to the world economy, we have also witnessed the fragility of this economic concept. For example, a Harvard Business Review (HBR) article summarized the reasons behind the fragility of the capitalist economic concept: The forces of disruptions; the fragility of the capitalist-concept-led global financial system; breakdowns in global trade; inequality and populism; migration; environmental degradation; failure of the rule of law; decline of public health and education; rise of state capitalism; radical movements, terrorism and war; evolution of pandemics; inadequacy of institutions (Bower et al., 2011).

As a result, "we are entering the postcapitalist era. At the heart of further change to come is information technology, new ways of working and the sharing economy. The old ways will take a long while to disappear, but it's time to be utopian" (Mason, 2015, np). How businesses and marketeers could deal with the challenges in the postcapitalist era? To answer this question, the two participant groups out of four of the research published in the HBR suggest:

> The third group, which we labeled "business as innovator," saw business as better able than government to address serious challenges but thought that business would do so not by influencing policy but through innovations in products, services, strategies, and business models. The fourth group, which we dubbed "business as activist," argued that business can and must become more engaged in shaping public policy, spurring government (which they believed could not on its own solve major problems) toward policies that would strengthen the market system. (Bower et al., 2011, np)

Another option for marketeers to address the challenges in the postcapitalist era is looking back to move forward. For example, "scholars acknowledge that determining the direction of where a field (e.g., marketing) is headed often requires a reflection of its founding principles, the transformation it has endured, and the driving forces that shape its existence" (Czinkota et al., 2021, p. 789). In such reflections on the past economies and societies, the pioneering marketeers borrowed concepts from the past societies, analyzed how people in the past applied those concepts to address their challenges during primitive, medieval or in most recent eras, and adapted some of those concepts in a way that can be applied in the contemporary global markets for a win-win outcome for customers and non-customer stakeholders. For instance, the micro-credit or the social business concept has been popularized in our lifetime by Professor Yunus, and the concept has become popular across the world both in academia and in industry once professor Yunus won the Nobel Prize in 2006 because of his contribution to the life of marginalized communities based on this concept. However, Seibel (2005) reported that:

> The birth of microfinance in Europe dates back to tremendous increases in poverty since the 16th and 17th century (p. 1). The case of India shows that the origins of microfinance predate those reported above in Ireland and Germany by more than two and perhaps even three millenniums. (p. 6; as cited in Shams & Kaufmann, 2016, p. 1255).

The microfinance and the social business concepts and sustainable marketing are closely related concepts, considering that all these concepts are focused on to addressing societal and environmental challenges through a profit-making venture. In fact, these socially and environmentally caring business concepts, i.e., social business, social marketing, sustainable marketing are introduced to deal with the capitalist economic concept's fragility. Therefore, sustainable marketeers should research the business practices of the preceding societies for such concepts that could be adapted to address the contemporary societal and environmental challenges, with an aim to deal with issues in the postcapitalist era.

Increased Literacy and Increased Life Expectancy Around the world in many countries, the rate of literacy and the rate of average age of people are increasing. ◘ Figure 7.7 represents the increasing global literacy rate and ◘ Fig. 7.8 represents the increasing life expectancy.

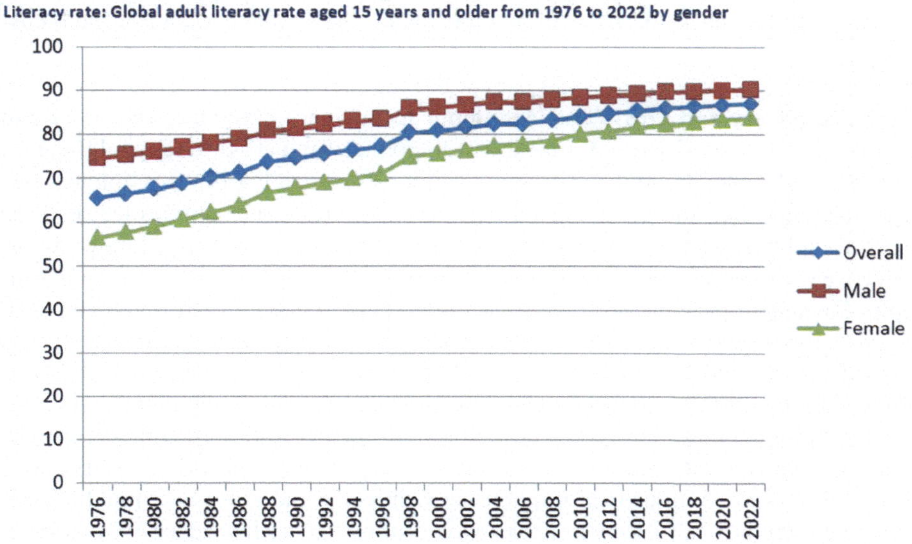

◘ **Fig. 7.7** The increasing literacy rate of global population. (Information source: O'Neill, 2024)

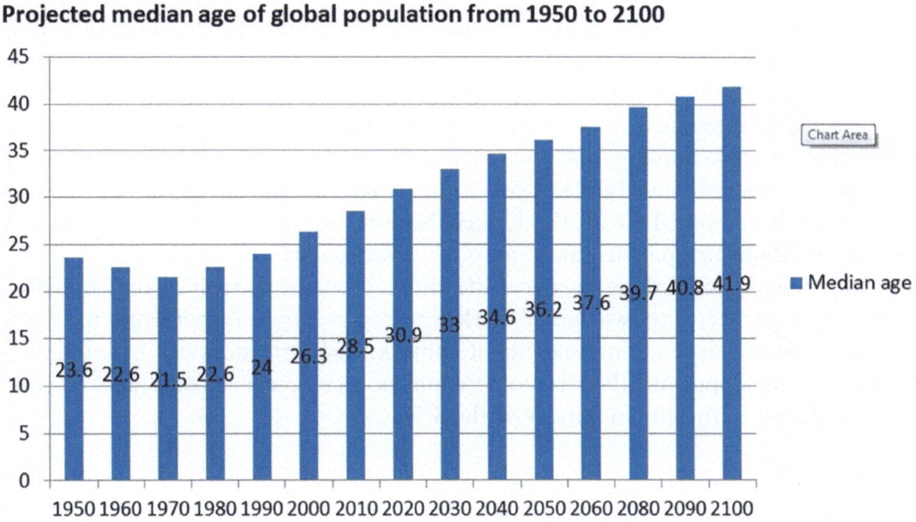

◘ **Fig. 7.8** The increasing life expectancy of global population. (Information source: Dyvik, 2024)

The increasing literacy and life expectancy as illustrated in ◘ Figs. 7.7 and ◘ 7.8 have at least two implications for marketeers. Firstly, increased literacy means marketeers need to be more transparent to clarify to their customers and non-customer stakeholders on why doing business with us is better for you instead of doing business with our competitors, and transparency is good for sustainable marketing. Secondly, increasing life expectancy of global population brings the opportunity for marketeers to aim for increasing the share of wallet of each customer.

Cross-Disciplinary and Cross-Functional Input to Influence Marketing Dynamics and Vice-Versa We have discussed some cross-disciplinary and cross-functional theories, concepts and principles in this marketing textbook, especially from psychology, strategic management, and economics. Hope, you appreciate the significance of these cross-disciplinary and cross-functional inputs to enrich your marketing knowledge to influence your sustainable marketing career further. In the same vein, marketing principles can help improving the knowledge and practice in other cross-disciplinary and/or cross-functional areas. For example, branding principles from marketing are applied for internal branding as part of human resource management (HRM) to attract and retain talented employees. This practice of cross-disciplinary and cross-functional inputs into marketing knowledge and practice and marketing's contribution to other subject areas will continue to supplement each other. Marketeers, therefore, should continue exploring cross-disciplinary and cross-functional knowledge to enrich their marketing practice, while aiming to enrich practice in other business functional areas, such as HRM, leadership, strategy and so on based on their marketing knowledge, in order to complement the overall business functions of a firm.

SDG-17—Partnerships for the Goals The number 17 Sustainable Development Goal (SDG), "Partnerships for the Goals" is nothing but marketing. More specifically it is relationship marketing. On the one hand, one of the definitions describes "marketing that reflected an integrative activity involving personnel from across the organisation, with emphasis on facilitating, building and maintaining a range of relationships (with internal and external stakeholders) over time" (Brodie et al., 2008, p. 85). On the other hand, based on SDG-17, the United Nations advised us all to establish and nurture relationships/partnerships among stakeholders in government, non-government and the third sector across the globe to pursue the rest of the 16 SDGs. Therefore, marketeers' knowledge of marketing brings a huge opportunity for them to apply relationship-building and maintaining knowledge and skills to build and nurture relationships with their customers and non-customer stakeholders to contribute together to the global pursuit of the SDGs.

Mission of Sustainable Marketeers and Nudging for People, Planet, and Profit—Marketeers As a Change Agent Last but most important consideration for sustainable marketeers would be to remember their professional mission: "Applying marketing principles to create value for People, Planet, and Profit, while inspiring customers and non-customer stakeholders, eying on reinforcing equity, diversity and inclusion of our global community for a shared sustainable development." Working to pursue such a professional mission of sustainable marketing to nudge customers and non-customer stakeholders for People, Planet, and Profit could also be instrumental for marketeers to differentiate them, in order make their contribution standalone from the crowds of ordinary marketeers (who work for a single bottom line, e.g., profit at any cost), as an effort of personal branding to accelerate in their marketing career. Entrepreneurial marketeers are change agent, because they envisage and introduce a change in our lifestyle. For example, "Bill Gates, Steve Jobs, Mark Zuckerberg and other innovative entrepreneurial marketeers all were able to understand the changing environmental patterns and the future needs of their customers" (Czinkota et al., 2021, p. 726). Accordingly, they introduced a new lifestyle in our global society.

7.7 Be a Marketeer (Not a Marketeer) to Apply Strategic Marketing for the Sustainable Development Goals

The words "marketeer" and "marketeer" could be synonymous to some extent; however, there is a fundamental difference in between marketeer and marketeer. Marketeers primarily focus on selling a product or service. On the contrary, marketeers aim to offer a unique product and/or service experience while selling it. For example, to design, distribute, sell, and evaluate a product and its performance related to the triple bottom line (People, Planet, and Profit), marketeers' prime consideration is to ensure how convenient and pleasing would be for customers to find their product, consume it and responsively recycle it after its consumption, in comparison to the competitive products. In this context, all marketeers are marketeer too (as beside providing a unique experience they sell a product too), but not all marketeers are marketeers (as they do not aim to provide a unique experience but only to sell a product). Sustainable marketeers should aim to provide a unique experience to their societal, environmental and commercial stakeholders, in return to their business relationships and interactions, with an aim to pursue the global SDGs. Summarizing from the discussion of this chapter, ◘ Table 7.4 describes how marketeers have been creating and contributing value, or can create and contribute value to our global pursuit of SDGs, based on the sustainable marketing principles discussed in this chapter.

Table 7.4 Sustainable marketing for creating value for the Sustainable Development Goals

Marketing management projects	Section in this chapter, where it is discussed	Value creation for the Sustainable Development Goals (SDGs)
– Look, Smell, Taste app – Managing food waste – The partnership of a refrigerator brand with NHS	Preliminary Vignette	SDG 2 (zero hunger), SDG 3 (good health and well-being), SDG 7 (affordable and clean energy), SDG 10 (reduced inequalities), SDG 11 (sustainable cities and communities), SDG 12 (responsible consumption and production), SDG 13 (climate action), and SDG 17 (partnerships for the goals)
Product quality—from body to soul as from "engineering" quality to "feeling" quality	Manager's Corner ► 7.1	SDG 9 (industry, innovation and infrastructure) and SDG 12 (responsible consumption and production)
Unsilvers' purpose-led partnership	► Web Exercise 7.1	SDG 6 (clean water and sanitation), SDG 7 (affordable and clean energy)
Social enterprises in Malawi	► Section 7.3.1: Market-Based View of Resources	SDG 7 (affordable and clean energy)
Recycling and reusing raw materials: iPhone's unique design and its commitment on sustainability	Marketing in Action ► 7.1	SDG 8 (decent work and economic growth), SDG 9 (industry, innovation and infrastructure) and SDG 12 (responsible consumption and production), SDG 13 (climate action), SDG 14 (life below water), SDG 15 (life on land)
Financial security after retirement—a direct nudge for societal benefit	Marketing in Action ► 7.2	SDG 2 (zero hunger), SDG 3 (good health and well-being), SDG 11 (sustainable cities and communities) and SDG 12 (responsible consumption and production)
'Everyday low price'—ALDI's nudge effect	Section Nudge Theory for Sustainable Competitive Advantage	SDG 2 (zero hunger), SDG 3 (good health and well-being) and SDG 10 (reduced inequalities)
Coke's recycling campaigns	◘ Table 7.3	SDG 9 (industry, innovation and infrastructure), SDG 11 (sustainable cities and communities), SDG 12 (responsible consumption and production), SDG 13 (climate action), SDG 14 (life below water), SDG 15 (life on land)

Marketing management projects	Section in this chapter, where it is discussed	Value creation for the Sustainable Development Goals (SDGs)
Biodegradable packaging project	► Section 7.6: Embryonic Changes in Marketing Dynamics and Sustainable Marketing	SDG 3 (good health and well-being), SDG 11 (sustainable cities and communities), SDG 12 (responsible consumption and production), SDG 13 (climate action), SDG 14 (life below water), SDG 15 (life on land)
EDI and marketing mix	► Section 7.6: Embryonic Changes in Marketing Dynamics and Sustainable Marketing	SDG 10 (reduced inequalities)
Sustainable marketing and endorsing peace	► Section 7.6: Embryonic changes in marketing dynamics and sustainable marketing	SDG 16 (peace, justice and strong institution)
Embryonic technologies, data analytics and GDPR	► Section 7.6: Embryonic Changes in Marketing Dynamics and Sustainable Marketing	SDG 12 (responsible consumption and production)
Paperless businesses	► Section 7.6: Embryonic Changes in Marketing Dynamics and Sustainable Marketing	SDG 12 (responsible consumption and production), SDG 13 (climate action)
Cross-cultural marketing and EDI	► Section 7.6: Embryonic Changes in Marketing Dynamics and Sustainable Marketing	SDG 10 (reduced inequalities)
Sustainable marketing for social business	► Section 7.6: Embryonic Changes in Marketing Dynamics and Sustainable Marketing	SDG 1 (no poverty), SDG 2 (zero hunger), SDG 3 (good health and well-being), SDG 11 (sustainable cities and communities), SDG 12 (responsible consumption and production), SDG 13 (climate action), SDG 14 (life below water), SDG 15 (life on land)

◘ Source: ► pexels.com/royalty-free-images/

The Preliminary Vignette of this chapter describes a case of addressing food waste in restaurants. Yes, the described case of addressing food waste can contribute to society and environment, while attracting more customers to increase profit. However, competitive restaurants can also copy the strategy to outplay the short-lived competitive advantage of the restaurant, who first implemented it. To sustain or at least prolong this short-live commercial competitive advantage, the vignette also discusses another simulative case example of addressing food waste, based on a partnership between a refrigerator brand and one of its largest single customer (e.g., the National Health Service who need to purchase and maintain a lot of refrigerators throughout their country-wide medical establishments). These cases demonstrate that the strategies employed in these two case studies (one is real-life and the other one is simulative) can contribute to SDG 2 (zero hunger), SDG 3 (good health and well-being), SDG 10 (reduced inequalities, e.g., feeding more people in a society) by addressing food waste that can help contributing to the progress of sustainable cities and communities (SDG 11). Addressing food waste can also be instrumental for responsible production, e.g., food production (SDG 12). Once there is less food waste, it can help contributing to more affordable energy (SDG 7) as a climate action (SDG 13). Also, we have seen a case example in the Preliminary Vignette that describes a partnership (SDG 17) that aims to pursue the other SDGs by addressing food waste. After indicating these case studies from the Preliminary Vignette of this chapter, ◘ Table 7.4 describes how the strategy described in Manager's Corner ► 7.1 to pursue SDG 9 and SDG 12. Manager's Corner ► 7.1 describes a case of responsible design of car, based on "feeling" quality of a product (i.e., car in this case). Truthfulness, authenticity, simplicity, expanded participation of customers to contribute to product design, and personal responsibility are discussed as instrumental to ensure "feeling" quality in product design. Once such features (especially, personal responsibility) as "feeling" quality are considered meaningfully in product design, it contributes not only to enrich innovation (SDG 9) to design a more user-friendly and socially and environmentally caring product, but also can contribute to responsible production and consumption (SDG 12).

The next two strategies presented in ◘ Table 7.4 are Unilever's purpose-led partnership (► Web Exercise 7.1) and social enterprises in Malawi (► Sect. 7.3.1). Unilever's purpose-led partnership, such as Clean Future (2024), contributes to use less water and less energy in their value-delivery process (e.g., from product development to product delivery) that helps to pursue SDG-6 (clean water) and SDG-7 (i.e., affordable energy) in their global value chain. In Malawi, a landlocked country in SouthEastern Africa, "energy services social enterprises are emerging as a sustainable response to the challenges of achieving SDG 7 (affordable and clean energy)" (Eales et al., 2020, p. 1). Based on iPhone's unique design and its commitment on sustainability, Marketing in Action ► 7.1 describes a recycling and reusing strategy of raw materials, which is listed as the next socially and/or environmentally caring strategy in ◘ Table 7.4. Apple's Product Environmental Report (2023) describes that aluminum, rare earth elements, copper, plastic, tungsten, cobalt, and tin are used as raw materials to produce Apple products. The report also confirms that all these raw materials (excluding plastic), which are used by Apple, are recycled 100%. For plastic, the report describes that "we're transitioning from fossil fuel–based plastics to renewable or recycled alternatives. For iPhone 15 Pro and iPhone 15 Pro Max, we use 35 percent or more recycled plastic in multiple components" (p. 5). Once such raw materials which are limited in nature, will be recycled and reuse, it will contribute to economic growth (SDG 8) by minimizing production cost as part of industry innovation (SDG 9) to improve production efficiency. The increased recycling and reusing of raw materials also contribute to responsible production (SDG 12) as part of climate action (SDG 13), which can contribute significantly to pursue SGD 14 (life below water) and SDG 15 (life on land). For example, once plastics will be recycled more or more renewable plastics will be used, it can protect the life below land and life on land from plastic pollution.

◘ Table 7.4 next describes the British government's initiative to ensure financial security of Britons after they retire based on a nudge affect (see Marketing in Action ► 7.2). Once financial security of older adults is confirmed, it can contribute to our pursuit of SDG 2 (zero hunger), SDG 3 (good health and well-being) of older adults in our society, which can also contribute to the sustainability of our cities and communities (SDG 12). To ensure the financial security of older adults in the UK, the nudge that is applied by the British government ensures that British employees must contribute a percentage of their monthly salary to their own pension scheme, instead of consuming the entire salary within the month. It also contributes to responsible consumption (SDG 12). Next, the ALDI's "everyday low price" is listed in ◘ Table 7.4. Once the basic food items (i.e., bread, butter, vegetable, milk, egg, meat, fish, etc.) are more affordable, especially for low-income consumer groups, it can contribute to reduce inequalities (SDG 10) by helping those low-income consumer groups to ensure their nutrition needs. It helps ensuring the inclusion of people from different segments in a society to be part of the society's overall consumption dynamics that can help us to contribute to our pursuit of SDG 2 (zero hunger) and SDG 3 (good health and well-being). From ◘ Table 7.3, the next strategy that is discussed in ◘ Table 7.4 is Coke's recycling campaigns. Similar to iPhone's recycle and reuse of raw materials as discussed in this current section, Coke's recycling efforts can also contribute to industry innovation (SDG 9) to reduce production cost by recycling and reusing production materials. Such recycling and reusing of raw materials can also help us contribute to our pursuit of responsible production (SDG 12), as part of our climate action (SDG 13) to protect the life below water (SDG 14) and life on land (SDG 15). Consequently, such sustainability initiatives to recycle and reuse of raw materials can be instrumental to the progress of sustainable cities and communities (SDG 11).

Seven more sustainable strategies are discussed in ◘ Table 7.4 from ► Sect. 7.6. The first strategy here is the biodegradable packaging projects in Bangladesh (i.e., the 100% biodegradable plastic bag made by natural fiber) and in Thailand (using banana leaves for packaging in supermarkets) that contribute to responsible consumption and production (SDG 12). As part of climate action (SDG 13), such 100% biodegradable packaging can contribute to life below water (SDG 14) and life on land (SDG 15), in order to ensure good health and well-being (SDG 3). Collectively, such sustainability initiatives can be instrumental to progress of sustainable cities and communities (SDG 11) by reducing pollution from plastic and other non-biodegradable (e.g., synthetic) packaging materials. Next, emphasizing equity, diversity, and inclusion (EDI) of different customer segments while designing a product's marketing mix are discussed. EDI in marketing mix can help reduce inequalities (SDG 10) in our global consumption dynamics by ensuring the inclusion of different consumer groups, especially the low-income consumers. Sustainable marketing for endorsing peace is discussed as the next responsible marketing strategy, which can contribute to our global pursuit of SDG 16 (peace, justice and strong institution). Supporting general data protection regulation (GDPR) is also discussed as an important element to protect customers' and non-customer stakeholders' data and privacy, especially when marketeers use embryonic technologies and data analytics tools. Ensuring GDPR can help contribute to our pursuit of responsible production (SDG 12) especially in the digital platforms.

Next, paperless business is discussed, which is instrumental to reduce paper-based wastes. Eventually, the paperless business goal can help us contributing to SDG 12 (responsible production and consumption) and SDG 13 (climate action). Next, the significance of endorsing EDI in cross-cultural marketing is highlighted in ◘ Table 7.4 from ► Sect. 7.6, which can be useful for reduced inequalities (SDG 10) across different market segments. Finally, ◘ Table 7.4 highlights sustainable marketing for social business from the discussion in ► Sect. 7.6. A social business that is commercially viable can contribute to a number of SDGs. For example, it creates employment for marginalized community that helps us to pursue SDG 1 (no poverty), SDG 2 (zero hunger) and SDG 3 (good health and well-being) across different segments in a society. In general, social businesses follow the responsible consumption and production (SDG 12) principle more than profit-only businesses, and responsible consumption and production can contribute to our pursuit of SDG 13 (climate action), SDG 14 (life below water), SDDG 15 (life on land), in order to contribute to the progress of sustainable cities and communities (SDG 11).

We have now reached the end of the last chapter of this core sustainable marketing textbook. We authors, together with a team of reviewers, researchers, assistants, editors, designers, and printing and publishing executives, have produced this textbook to educate, inform, and delight you. We hope the text has offered you a delightful learning experience, and you have enjoyed reading it. We also hope that you find the different marketing principles and cross-disciplinary and cross-functional business theories that are discussed in this core textbook of "sustainable marketing," instrumental to equip with, in order to contribute back to your society and environment while enriching the commercial competitive advantage of your brand to make profit.

Summary

Strategic marketing scholars argue that our existing marketing knowledge is inept to meaningfully sustain the competitive advantage of a brand, even only from a single bottom line context, i.e., focusing only on profit-making initiative to ensure the brand's long-term commercial competitive advantage. Subsequently, the existing strategic marketing principles are further impractical to rationally support a brand's initiative to prolifically pursue the triple bottom line, i.e., contributing to people and planet while making profit. To address this gap in existing marketing pedagogy, this chapter will help preparing the future marketeers to apply different strategic cross-disciplinary and cross-functional business theories to support their strategic marketing plan, with an aim to more meaningfully pursue the triple bottom line, and sustain or extend the short-lived commercial competitive of a brand.

At the intersections of strategic management, psychology, economics, cognitive science, and marketing principles, this chapter:

- First, discusses the resources and capabilities that can help marketeers to establish an initial competitive advantage.
- Second, the chapter, clarifies why a brand's competitive advantage would be vulnerable in a mid- or long-term period.
- Third, how marketeers could equip with different cross-disciplinary and cross-functional theories to underpin their strategic marketing plans, in order to prolong the commercial competitive advantage of their brand (e.g., possibly extending a brand-specific short-term competitive advantage to a mid- or longer-term period).
- Fourth, the chapter discusses how an enriched strategic marketing plan in support of different cross-disciplinary and cross-functional business theories could meaningfully contribute to people and planet while making profit.
- Fifth, the embryonic changes in marketing dynamics and the role of sustainable marketeers are discussed.
- Sixth, summarizing the discussions across different sections, the last section of this chapter clarifies how the discussed strategic marketing plans help marketeers to pursue the Sustainable Development Goals.

Similar to all other chapters in the book, this chapter also presents discussions based on real-life marketing schemes, socioeconomic, and ecological case examples, web exercises, manager's corners and sustainable marketing-in-actions, in order to clarify the application of complex cross-disciplinary and cross-functional arguments in a simplified manner. The futuristic evolutionary aspects of strategic marketing education and practice, and the subsequent propositions that are discussed in this chapter, will help tomorrow's marketeers to understand the practical significance of the propositions to meaningfully contribute back to their society and ecology, while sustaining (or at least extending) the commercial competitive advantage of their brands.

Review Questions

1. What are the relational resources and intellectual resources of a brand?
2. Define dynamic capabilities of an organization.
3. Firstly, following the discussion in ► Sect. 7.3.1 (Market-Based View of Resources), select a brand of your choice to identify the internal and external stakeholders of that brand that you think you, as a marketing manager of the selected brand should analyze the needs, wants and expectations of these stakeholders, in order to develop a market orientation. Secondly, consider how you could distinguish the prospective market-based resources (e.g., a supplier's knowledge in an overseas market's supply chain system) that you could obtain from these identified stakeholder/s of the selected brand.
4. Please share your views to clarify why an initial competitive advantage of a brand would be vulnerable to sustain over a longer-period of time.
5. Describe the background of behavioral economics.
6. What is a nudge?
7. Describe the criticisms of nudge theory. How, as a sustainable marketeer, you would design a nudge to address the criticisms of the nudge theory?

Discussion Questions

1. Select an established brand from any industry and analyze the extent of the brand's VRIN competency. It could be any brand from any industry or market that has been in their business for at least for a few years, and has established a loyal group of customers and non-customer stakeholders. For example, it could be KFC from the fast-food industry, Walmart from the supermarket industry, Arsenal football club from the sports industry, Dolce & Gabbana from the fashion industry, your own university from the education sector, or any other (less-known) brand. There is no right or wrong answer in your analysis of the selected brand's VRIN competency, but it is important how you justify your answer, related to each of the components of VRIN theory, i.e., valuable, rare, inimitability, and non-substitutability from the context of the selected brand.
2. Based on the discussion related to ◘ Fig. 7.2, first select a brand of your choice, and then analyze the nine questions to explore the brand-specific unique knowledge-based resources that you think would be instrumental to establish and initial competitive advantage for the brand.

3. Let's assume that the chairman of your organization has assigned you to plan, implement, and monitor a new sustainable marketing campaign to create a public awareness about your organization's societal and environmental initiatives and contributions. Based on the 4Ps' traditional marketing mix and extended marketing mix, ◘ Table 7.3 describes a number of marketing capabilities and their focus for sustainable marketing. As marketing director of your organization, which marketing capabilities from ◘ Table 7.3 you would be selecting to plan for the awareness creation campaign, and why you would be selecting those particular marketing capabilities? Justify your answer.
4. It is really difficult to make a product 100% VRIN-competent and it takes a long time too in many cases. Consequently, marketeers also try to add additional VRIN feature(s) into the product. For example, celebrity endorsement or partnership with a large buyer etc. Select a brand of your choice, and think of how you would be able to add an additional VRIN feature to this brand.
5. Select a brand, and design a nudge to influence the brand's customers to be more environmentally responsible. To design the nudge, try to correlate the context of the environmental responsibility with the business of the selected brand. For example, Coke encourages their customers to recycle plastic bottles of Coke. Will you be designing a direct or an indirect nudge?

Practice Quiz

1. Strategic management scholars have three sources of firm resources. What are they?
   (a) Resource-based view of resources, firm-based view of resources, knowledge-based view of resources
   (b) Resource-based view of resources, firm-based view of resources, market-based view of resources
   (c) Resource-based view of resources, market-based view of resources, knowledge-based view of resources
   (d) Resource-based view of resources, market-based view of resources, industry-based view of resources
2. Market-based view of resources would include:
   (a) Relational resources and intellectual resources
   (b) Relational resources and interactive resources
   (c) Relational resources and organizational resources
   (d) Organizational resources and interactive resources
3. Sveiby (2001) suggested how many questions to managers to answer to explore firm-specific unique knowledge-based resources?
   (a) 7
   (b) 9
   (c) 5
   (d) 11
4. Organizational dynamic capabilities include 3Ps. What are they?
   (a) Process, Position, Power
   (b) Process, Power, Provision

   (c) Position, Provision, Power
   (d) Process, Position, Path
5. VRIN stands for:
   (a) Valuable, Rare, Inimitable, Non-comparable
   (b) Valuable, Rare, Inseparable, Non-comparable
   (c) Valuable, Rare, Inimitable, Non-substitutable
   (d) Valid, Rare, Inimitable, Non-substitutable
6. The theoretical development of behavioral economics is influenced predominantly by:
   (a) Prospect theory, framing effects, status quo bias, paradox of choice, and intrinsic motivation
   (b) Prospect theory, framing effects, status quo bias, paradox of choice, and extrinsic motivation
   (c) Prospect theory, relational effects, status quo bias, paradox of choice, and intrinsic motivation
   (d) Prospect theory, framing effects, status quo bias, paradox of identity, and intrinsic motivation
7. Two types of nudges are discussed in this book. What are they?
   (a) Single nudge and diverse nudge
   (b) Direct nudge and diverse nudge
   (c) Direct nudge and indirect nudge
   (d) Singular nudge and plural nudge
8. What are the three factors that influence an individual's (e.g., a customer's) decision-making?
   (a) Intellectual perception, fairness perception, and lack of self-control
   (b) Rational perception, fairness perception, and lack of self-control
   (c) Rational perception, fairness perception, and lack of information
   (d) Relational perception, fairness perception, and lack of information
9. When Professor Thaler received the Nobel Prize because of his contribution based on the nudge theory?
   (a) 2017
   (b) 2018
   (c) 2019
   (d) 2015
10. All marketeers are marketeers, but not all marketeers are marketeers.
   (a) True
   (b) False

(You can find the correct answers to these questions by taking the quiz and then comparing your answers with the answers that are provided at the end of the book, under the "Practice Quiz Answers" part of the book.)

References

Altman, M. (2003). Reviewed work: Human firm: A socio-economic analysis of behavior and potential in a new economic age by John Tomer. *The Review of Social Economy, 61*(1), 124–128.

Amazon. (2024a). 12 x Lux soft touch for soft fragrant soap bars with French Rose & Almond Oil 80g (Pack of 12). Retrieved April 22, 2024, from https://www.amazon.co.uk/Soft-Touch-Fragrant-french-Almond/dp/B00B4RA09E/ref=sr_1_5?dib=eyJ2IjoiMSJ9.BZx86A8Ro5_r416BHbXPv8EPmfEYDLPqjuD1mpUnxMAt6IWtxTCkKk8fxSWKxEM3FuO-JmNE5X_hjz52JPb9bR99ahWne1EJzrj0UfqNFWtyvKR3dCUZOCuIzzDvN6OCd0IhqJfYy-VifFaJDKMh00Ir8ARnl02H81EGTEt9o5iiXkhChOi7b28KmdE3W-3PQYVQWrNpDtLqr8NmlA1Mq4fnTdIo-CAnufkgISz3SB-TSRAuGlyrP6V4FQfj2shfH70ywDmZuJKMIup9bgSe2W3O8I-QWjkVc2BwosPnhlc.SS5jOMiifKd4L9srWXieasBl6Edr7BdSbOH42fwLoj8&dib_tag=se&keywords=lux+soap&qid=1713789861&sr=8-5

Amazon. (2024b). LUX rose & vitamin E soap, bathing soap with floral fragrance, soft, moisturized and glowing skin, 100 g. Retrieved April 24, 2024, from https://www.amazon.in/Vitamin-Bathing-Fragrance-Moisturized-Glowing/dp/B00U1CAPZC?th=1

Analyzing Resources and Capabilities. (2007). Retrieved February 22, 2024, from https://www.blackwellpublishing.com/content/GrantContemporaryStrategyAnalysis/6th_Edition/CSAC05.pdf

Apostolidis, C., Shams, R., Gregory-Smith, D., Vrontis, D., Bian, X., Belyaeva, Z., & Papgiannidis, S. (2022). Technology as a catalyst for sustainable social business: Advancing the research agenda. *Technological Forecasting and Social Change, 183*, 1–6.

Arndt, F., & Pierce, L. (2018). The behavioral and evolutionary roots of dynamic capabilities. *Industrial and Corporate Change, 27*(2), 413–424.

Asher, G. (2022). Messi spotted wearing Nike trainers?! Problems with Adidas? Retrieved April 13, 2024, from https://m.allfootballapp.com/news/All/Messi-spotted-wearing-Nike-tainers-Problems-with-Adidas/2778351#:~:text=Although%20he%20did%20wear%20Nike,highest%20level%20of%20the%20sport

Aspuna. (n.d.). Who we are. Retrieved December 3, 2023, from https://www.aspuna.com/about/

Austen-Hardy, P. (2023). Inside Lionel Messi's bumper lifetime Adidas deal that pays MLS star a fortune. *The Daily Star*. Retrieved April 13, 2024, from https://www.dailystar.co.uk/sport/football/lionel-messi-adidas-lifetime-deal-30843244

Baddeley, M. (2024). Behavioral economics: Past, present and future. Retrieved April 14, 2024, from https://www.bbvaopenmind.com/en/articles/behavioral-economics-past-present-and-future/

Bae, H. (2015). Bill Gates' 40th anniversary email: Goal was 'a computer on every desk'.

Bakir, V. (2020). Psychological operations in digital political campaigns: Assessing Cambridge Analytica's psychographic profiling and targeting. *Frontiers in Communication, 5*, 1–16.

Ballesteros-Sola, M. (2014). Social business planning: Management and financial issues. In A. Grove & G. Berg (Eds.), *Social business*. Springer.

Barney, J. (1991). Firm resources and sustained competitive advantage. *Journal of Management, 17*(1), 99–120.

Barney, J., Ketchen, D. J., Jr., & Wright, M. (2011). The future of resource-based theory: Revitalization or decline. *Journal of Management, 37*(5), 1299–1315.

Barney, J., Wright, M., & Ketchen, D. J., Jr. (2001). The resource-based view of the firm. Ten years after 1991. *Journal of Management, 27*, 625–641.

Baumgartner, J. (1991). Nonmarketing professionals need more than 4Ps. *Marketing News, 22*, 28.

BBC. (2024). McDonald's sales dented by Israel-Gaza boycotts. Retrieved April 25, 2024, from https://www.bbc.co.uk/news/business-68209085#:~:text=The%20fast%20food%20chain%20reported,about%204%25%20after%20the%20announcement

BBC News. (2019). China anniversary. How the country became the world's 'economic miracle'. *BBC*. Retrieved April 26, 2024, from https://www.bbc.co.uk/news/business-49806247

BBC Reality Check. (2021). Has China lifted 100 million people out of poverty. *BBC*. Retrieved April 26, 2024, from https://www.bbc.co.uk/news/56213271#:~:text=China's%20rapid%20reduction%20in%20poverty,on%20the%20poorest%20rural%20areas.&text=The%20government%20has%20relocated%20millions,remote%20villages%20into%20apartment%20complexes

Becker, G. S. (1964). *Human capital: A theoretical and empirical analysis, with special reference to education*. University of Chicago Press.

Becker, S., Kunze, C., & Vancea, M. (2017). Community energy and social entrepreneurship: Addressing purpose, organisation and embeddedness of renewable energy projects. *Journal of Cleaner Production, 147*, 25–36.

Benartzi, S., & Thaler, R. H. (2000). Myopic loss aversion and the equity premium puzzle. In D. Kahneman & A. Tversky (Eds.), *Choices, values, and frames* (pp. 301–316). Cambridge University Press and Russell Sage Foundation.

Berg, N. (2010). Behavioral economics. MPRA Paper No. 26587. Available at: https://mpra.ub.unimuenchen.de/26587/1/Berg_Behavioral_Economics_for_Sage_v9afterreviewsreceived.pdf#:~:text=Similarly%2C%20Camerer%20(1999)%20defines%20behavioral%20economics%20as,of%20economic% (accessed 15 February 2010).

Berner, A., & Tonder, C. L. V. (2003). The postmodern consumer: Implications of changing customer expectations for organisational development in service organisation. *The Journal of Individual Psychology, 29*(3), 1–10.

Bertrand, M., Mullainathan, S., & Shafir, E. (2004). A behavioral economics view of poverty. *American Economic Review, 94*(2), 419–423.

Blystone, D., James, M., & Jackson, A. (2024). The history of Uber: How the controversial ride-sharing company came to dominate its market worldwide. *Investopedia*. Retrieved March 10, 2024, from https://www.investopedia.com/articles/personal-finance/111015/story-uber.asp#:~:text=Key%20Takeaways,emerge%20since%20the%20late%202000s

Booms, B. H., & Bitner, M. J. (1981). Marketing strategies and organization structures for service firms. In J. H. Donnelly & W. R. George (Eds.), *Marketing of services* (pp. 47–51). American Marketing Association.

Borden, N. H. (1964). The concept of the marketing mix. *Journal of Advertising Research, 1964*, 2–7.

Bower, J. L., Leonard H. B., & Paine, L. S. (2011). Global capitalism at risk: What are you doing about it? *The Harvard Business Review*. Retrieved April 26, 2024, from https://hbr.org/2011/09/global-capitalism-at-risk-what-are-you-doing-about-it

British Airways. (n.d.). Flights to Sydney. Retrieved April 26, 2024, from https://www.britishairways.com/en-gb/destinations/sydney/flights-to-sydney#:~:text=Sydney%20flight%20FAQs&text=How%20long%20is%20a%20flight,hours%2050%20minutes%20in%20total

British Council. (2013). *The shape of things to come: The evolution of transnational education: Data, definitions, opportunities and impact analysis*. Retrieved August 28, 2015, from www.britishcouncil.org/sites/britishcouncil.uk2/files/the_shape_of_things_to_come_2.Pdf

Brodie, R. F., Coviello, N. E., & Winklhofer, H. (2008). Contemporary marketing practices research program: A review of the first decade. *Journal of Business & Industrial Marketing, 23*(2), 84–94.

Bruhin, A., Fehr-Duda, H., & Epper, T. (2010). Risk and rationality: Uncovering heterogeneity in probability distortion. *Econometrica, 78*(4), 1375–1412.

BT. (2024). Telecom TV. Retrieved March 10, 2024, from https://www.telecomtv.com/content/bt/#:~:text=BT%20Group%20plc%20is%20a,subscription%20television%20and%20IT%20services

Bush, J. (2022). Coke zero: Coca-Cola Eruopacific partners have ambitious net zero targets. *The Manufacturer*. Retrieved April 22, 2024, from https://www.themanufacturer.com/articles/coke-zero-coca-cola-europacific-partners-has-ambitious-net-zero-targets/

Buttenheim, A., Moffitt, R., & Beatty, A. (2023). *Behavioral economics: Policy impact and future directions*. The National Academies Press.

Cadwalladr, C. and Graham-Harrison, E. (2018). Revealed: 50 million Facebook profiles harvested for Cambridge Analytica in major data breach. *The Guardian*. Retrieved April 26, 2024, from https://www.theguardian.com/news/2018/mar/17/cambridge-analytica-facebook-influence-us-election

Camerer, C., Babcock, L., Loewenstein, G., & Thaler, R. (1997). Labor supply of New York City cabdrivers: One day at a time. *Quarterly Journal of Economics, 112*(2), 407–441.

Camerer, C. F., & Loewenstein, G. (2004). Behavioral economics: Past, present, future. In C. F. Camerer, G. Loewenstein, & M. Rabin (Eds.), *Advances in behavioral economics* (pp. 3–51). Russell Sage Foundation and Princeton University Press.

Camerer, C. F., Loewenstein, G., & Rabin, M. (2004). *Advances in behavioral economics*. Russell Sage Foundation and Princeton University Press.

Campbell, D. (2020). NHS to enlist 'sensible celebrities to persuade people to take coronavirus vaccine. *The Guardian*. Retrieved April 21, 2024, from https://www.theguardian.com/society/2020/nov/29/nhs-enlist-sensible-celebrities-coronavirus-vaccine-take-up#:~:text=Ministers%20and%20NHS%20England%20are,vaccine%20scepticism%20among%20NHS%20staff

Centers for Disease Control and Prevention. (2020). *Coronavirus disease 2019 (COVID-19) situation summary*. Retrieved September 26, 2020, from https://www.cdc.gov/coronavirus/2019-ncov/summary.html

Chang, O. H., Nichols, D. R., & Schultz, J. J. (1987). Taxpayer attitudes toward tax audit risk. *Journal of Economic Psychology, 8*(3), 299–309.

Chen, G. (2015). Bachelor of International Hotel and Tourism Management from China at the University of Queensland. Retrieved August 24, 2015, from www.uq.edu.au/international-students/garmenchen

Choi, J. J., Laibson, D., & Madrian, B. C. (2011). $100 bills on the sidewalk: Suboptimal investment in 401(k) plans. *Review of Economics and Statistics, 93*(3), 748–763.

Cibb, J., & Emmerson, C. (2016). *What happens when employers are obliged to nudge? Automatic enrolment and pension saving in the UK*. Institute for Fiscal Studies. Retrieved April 21, 2024, from https://ifs.org.uk/sites/default/files/output_url_files/wp1619.pdf

Clean Future. (2024). *Unmissably superior products that are sustainable and great value*. Unilever. Retrieved February 23, 2024, from https://www.unilever.com/brands/home-care/clean-future/

Close, C. (2021). *The global eco-wakening: How consumers are driving sustainability*. World Economic Forum. Retrieved April 22, 2024, from https://www.weforum.org/agenda/2021/05/eco-wakening-consumers-driving-sustainability/

CNN. (2023). *Fighting plastic pollution with a biodegradable bag*. Retrieved November 20, 2023, from https://edition.cnn.com/videos/tv/2022/04/01/sonali-bag-spc-intl.cnn

Coke's Sustainability TVC. (2019). *AdNews Australia*. Retrieved April 25, 2024, from https://www.youtube.com/watch?v=0Crpab4WWic

Congiu, L., & Moscati, I. (2022). A review of nudges: Definitions, justifications, effectiveness. *Journal of Economic Surveys, 36*, 188–213.

Connecting a Second TV. (2024). *BT community*. Retrieved March 10, 2024, from https://community.bt.com/t5/EE-TV-YouView/Connecting-a-second-tv/m-p/2352592#M77210

Czinkota, M., Kotabe, M., Vrontis, D., & Shams, S. M. R. (2021). *Marketing management: Past, present and Future* (4th ed.). Springer International Publishing.

Darrah, P. (2023). How many marine animals die from plastic pollution? *GVI Planet*. Retrieved April 25, 2024, from https://www.gvi.co.uk/blog/smb-how-many-marine-animals-die-from-plastic-pollution/

DellaVigna, S. (2007). *Psychology and economics: Evidence from the field* (NBER Working Paper No. 13420). National Bureau of Economic Research.

Deloitte. (2024). The sustainable consumer. Retrieved April 22, 2024, from https://www2.deloitte.com/uk/en/pages/consumer-business/articles/sustainable-consumer.html

Dizzy, T. (2022). Is ALDI even cheaper anymore? *Reddit Customer Forum*. Retrieved April 21, 2024, from https://www.reddit.com/r/aldi/comments/10clx5j/is_aldi_even_cheaper_anymore/

Dyvik, E. H. (2024). Projected global median age from 1950 to 2100. *Statista*. Retrieved April 26, 2024, from https://www.statista.com/statistics/672669/projected-global-median-age/#:~:text=This%20statistic%20shows%20the%20median,be%2041.9%20years%20of%20age

Eales, A., Frame, D., Coley, W., Bayani, E., & Galloway, S. (2020). Sustainable delivery models for achieving SDG7: Lessons from an energy services social enterprise in Malawi. Paper presented at the 10th IEEE Global Humanitarian Technology Conference (29 October–1 November, Online).

Eisenhardt, K. M., & Martin, J. A. (2000). Dynamic capabilities: What are they? *Strategic Management Journal, 21*, 1105–1121.

Emery, E. (2023). Is buying from Amazon safe? What type of things should not be bought from Amazon? *Quora Customer Forum.* Retrieved April 21, 2024, from https://www.quora.com/Is-buying-from-Amazon-safe-What-type-of-things-should-not-be-bought-from-Amazon#:~:text=Amazon%20is%20a%20reputable%20online,online%20marketplaces%20in%20the%20world

Fantozzi. (2023). Luckin overtakes Starbucks as top coffee chain in China. *Nation's Restaurant News.* Retrieved March 10, 2024, from https://www.nrn.com/news/luckin-overtakes-starbucks-top-coffee-chain-china

Fickenscher, L. (2022). McDonald's revenue takes hit after Russia exit. *The New York Post*. Retrieved April 25, 2024, from https://nypost.com/2022/07/26/mcdonalds-revenue-takes-hit-after-russia-exit/

Francis, A. (2022). *How much microplastic am I eating? And is there any way to avoid it? Here's what you need to know about microsplastics in our food chain.* Retrieved April 25, 2024, from https://www.bonappetit.com/story/microplastics-food#:~:text=Some%20scientists%20have%20estimated%20the,annually%20from%20various%20food%20sources

Frey, B. S. (1997). On the relationship between intrinsic and extrinsic work motivation. *International Journal of Industrial Organization, 15*(4), 427–439.

7

Glamping Service Provider. (2023). *Booking.com.* Retrieved March 10, 2024, from https://www.booking.com/hotel/gb/islay-westcote-glamping.en-gb.html?label=en-gb-booking-desktop-LtZCxJDUEbB5qbLEgydvTwS652804039385%3Apl%3Ata%3Ap1%3Ap2%3Aac%3Aap%3Aneg%3Afi%3Atikwd-24091000324%3Alp9046776%3Ali%3Adec%3Adm&gclid=EAIaIQobChMIleqY_fDphAMVuGdBAh0rzAikEAMYASAAEgKZsPD_BwE&aid=2336990&ucfs=1&arphpl=1#tab-reviews

Gregory-Smith, D., Manika, D., & Demirel, P. (2017). Green intentions under the blue flag: Exploring differences in EU consumers' willingness to pay more for environmentally-friendly products. *Business Ethics: The Environment & Responsibility, 26*(3), 205–222.

Guillén, M. F. (2024). The future of marketing is intergenerational. *The Harvard Business Review*. Retrieved April 26, 2024, from https://hbr.org/2024/02/the-future-of-marketing-is-intergenerational

Gundlach, H., & Neville, B. (2012). Authenticity: Further theoretical and practical development. *Journal of Brand Management, 19*, 484–499.

Haigh, N., Kennedy, E. D., & Walker, J. (2015). Hybrid organizations as shape-shifters: Altering legal structure for strategic gain. *California Management Review, 57*(3), 59–82.

Heintz, E. (2022). *5 groundbreaking campaigns promoting mental health.* Retrieved April 25, 2024, from https://dmexco.com/stories/5-groundbreaking-campaigns-promoting-mental-health/#:~:text=%232%20A%20hashtag%20for%20more%20mental%20health%20%E2%80%93%20JanSport&text=With%20the%20motto%20%E2%80%9CStrength%20through,to%20normalize%20conversations%20about%20it

Henderson, E. (2016). Starbucks will give away 100 per cent of its food waste to those in need. *The Independent*. Retrieved December 28, 2023, from https://www.independent.co.uk/news/world/americas/starbucks-will-be-giving-away-100-per-cent-of-its-food-waste-to-those-in-need-a6947491.html

Herbst, S. (2016). The history and meaning of public opinion. In A. J. Berinsky (Ed.), *New directions in public opinion*. Routledge.

Heric, S. (2022). Moving toward a paperless society by 2024. *KM World*. Retrieved April 26, 2024, from https://www.kmworld.com/Articles/Editorial/ViewPoints/Moving-toward-a-paperless-society-by-2024-151281.aspx#:~:text=In%202021%2C%20for%20example%2C%20the,before%20travelling%2C%20avoiding%20paper%20forms

Heyman, J., & Ariely, D. (2004). Effort for payment: A tale of two markets. *Psychological Science, 15*(11), 787–793.

Hofstede, G., Hofstede, G. J., & Minkov, M. (2010). *Cultures and organizations: Software of the mind* (3rd ed.). McGraw-Hill.

Holland, S. (2014). Libertarian paternalism and public health nudges. In Michael Freeman, Sarah Hawkes, and Belinda Bennett (Eds), Law and global health: current legal issues 16, 331–353.

International Students Australia. (2015). Nationalities: Where do international students come from? Retrieved August 19, 2015, from www.internationalstudents.com.au/nationalities/

Ishangulyyev, R., Kim, S., & Lee, S. H. (2019). *Understanding food loss and waste—Why are we losing and wasting food?* National Library of Medicine: National Center for Biotechnology Information. Retrieved December 28, 2023, from https://www.ncbi.nlm.nih.gov/pmc/articles/PMC6723314/#B6-foods-08-00297

Itami, H. (1991). *Mobilizing invisible assets*. Harvard University Press.

Iyengar, S. S., & Kamenica, E. (2010). Choice proliferation, simplicity seeking, and asset allocation. *Journal of Public Economics, 94*(7–8), 530–539.

Jabbar, H. (2011). The behavioral economics of education: New directions for research. *Educational Researcher, 40*(9), 446–453.

Judd, V. C. (1987). Differentiate with the 5th P: People. *Industrial Marketing Management, 16*(4), 241–247.

Kahneman, D. (2003). Maps of bounded rationality: Psychology for behavioral economics. *American Economic Review, 93*(5), 1449–1475.

Kahneman, D., & Tversky, A. (1979). Prospect theory: An analysis of decision under risk. *Econometrica, 47*(2), 263–291.

Kearns, G. S., & Sabherwal, R. (2006). Strategic alignment between business and information technology: A knowledge-based view of behaviors, outcome, and consequences. *Journal of Management Information Systems, 23*(3), 129–162.

Kilfoyle, M. (2023). How is the cost of living crisis affecting public health? *Economics Observatory*. Retrieved April 25, 2024, from https://www.economicsobservatory.com/how-is-the-cost-of-living-crisis-affecting-public-health

Kosters, M., & der Heijden, J. V. (2015). From mechanism to virtue: Evaluating nudge theory. *Evaluation, 21*(2), 276–291.

Kotler, P. (1984). *Marketing management: Analysis, planning and control* (5th ed.). Prentice-Hall.

Kotler, P. (2003). *Marketing management, International edn*. Prentice Hall.

Kreps, D. M. (1997). Intrinsic motivation and extrinsic incentives. *American Economic Review, 87*(2), 359–364.

Krishna, R., & Kummitha, R. (2020). Smart technologies for fighting pandemics: The techno- and human-driven approached in controlling the virus transmission. *Government Information Quarterly, 37*, 1–11.

Krisnamurthy, A. (2021). Police brutality and nudge theory. *Deccan Herald*. Retrieved April 21, 2024, from https://www.deccanherald.com/opinion/police-brutality-and-nudge-theory-993517.html

Kurtz, D. L. (2009). *Contemporary marketing*. South-Western Cengage Learning.

Laird, J. (2024). Millions of Facebook and Google users hit by 2FA data leak: A SMS routing company used by Google, Facebook and TikTok is accused of leaking its own database. *TECH*. Retrieved April 26, 2024, from https://tech.co/news/millions-facebook-google-2fa-data-leak

Lebow, S. (2021). Paperless bill payments are increasing across all generations. *EMARKETER*. Retrieved April 22, 2024, from https://www.emarketer.com/content/paperless-bill-payments-increasing-across-all-generations

Looney, H. F., & Seiler, M. (2003). Invitation marketing: Using customer preferences to overcome Ad avoidance. *Accenture*. Retrieved May 26, 2010, from http://www.accenture.com/Global/Research_and_Insights/Outlook/By_Alphabet/InvitationAvoidance.htm

Mackinnon, M. (2015). *Study abroad*. Retrieved August 24, 2015, from www.uq.edu.au/studyabroad/what-ourstudents-are-saying

Madrian, B., & Shea, D. (2000). The power of suggestion: Inertia in 401(k) participation and savings behavior. (NBER Working Paper No. 7682). National Bureau of Economic Research.

Mao, H., Liu, S., & Zhang, J. (2015). How the effects of IT and knowledge capability on organizational agility are contingent on environmental uncertainty and information intensity. *Information Development, 31*(4), 358–382.

Mason, P. (2015). The end of capitalism has begun. *The Guardian*. Retrieved April 26, 2024, from https://www.theguardian.com/books/2015/jul/17/postcapitalism-end-of-capitalism-begun

McCarthy, E. J. (1960). *Basic marketing: A managerial approach*. Irwin.

MCCRINDLE. (2024). *Understanding generation Alpha*. Retrieved April 26, 2024, from https://mccrindle.com.au/article/topic/generation-alpha/generation-alpha-defined/#:~:text=That%20is%20why%20the%20generations,born%20from%202025%20to%202039

McDonald's. (2022). MHQ & office locations. Retrieved February 23, 2024, from https://careers.mcdonalds.com/mhq-and-field-offices#:~:text=At%20our%20global%20headquarters%20in,Chicago%20in%20the%20West%20Loop

McGee, J. (2015). Market-based view. In C. L. Cooper (Ed.), *Wiley encyclopedia of management* (Vol. P. 1). Wiley Publications.

McKinsey Consumer Sentiment Survey. (2024). Consumers care about sustainability—And back it up with their wallets. Retrieved April 22, 2024, from https://www.mckinsey.com/industries/consumer-packaged-goods/our-insights/consumers-care-about-sustainability-and-back-it-up-with-their-wallets

Microsoft. (2024). *About Microsoft*. Retrieved March 10, 2024, from https://www.microsoft.com/en-us/about

Miguel, B. C., & Santiago, G. B. (2010). Application of the total quality management approach in a Spanish retailer: The case of Mercadona. *Total Quality Management & Business Excellence, 21*(12), 1365–1381.

Millar, C. (2023). Cristiano Ronaldo sparks 'contract breach' speculation as Al Nassr thrashed 5-0. *The Mirror*. Retrieved April 13, 2024, from https://www.mirror.co.uk/sport/football/news/cristiano-ronaldo-nike-contract-adidas-30513092

Möller, K., Nenonen, S., & Storbacka, K. (2020). Networks, ecosystems, fields, market systems? Making sense of the business environment. *Industrial Marketing Management, 90*, 380–399.

Montana. 2022). ALDI no longer the cheapest supermarket! *Reddit Customer Forum*. Retrieved April 21, 2024, from https://www.reddit.com/r/UKPersonalFinance/comments/x34qgx/aldi_no_longer_the_cheapest_supermarket/

Morrison, B. (2013). How a small nudge is helping people save for their retirement. *Civil Service Quarterly*. Retrieved April 21, 2024, from https://quarterly.blog.gov.uk/2013/10/22/how-a-small-nudge-is-helping-people-save-for-their-retirement/

Nag, A., & Beniwal, V. (2016). *India's scramble to switch 23 billion banknotes: Quick take Q&A*. Bloomberg, UK. Retrieved April 21, 2024, from https://www.bloomberg.com/news/articles/2016-11-15/india-s-scramble-to-switch-23-billion-banknotes-quicktake-q-a

Nason, R. S., & Wiklund, J. (2018). An assessment of resource-based theorizing on firm growth and suggestions for the future. *Journal of Management, 44*(1), 32–60.

NDTV. (2023). *Supermarket praised for replacing plastic packaging with banana leaves*. Retrieved April 25, 2024, from https://www.ndtv.com/offbeat/thailand-supermarket-praised-for-replacing-plastic-packaging-with-banana-leaves-2021417

O'Donoghue, T., & Rabin, M. (1999). Doing it now or later. *American Economic Review, 89*(1), 103–124.

O'Donoghue, T., & Rabin, M. (2001). Choice and procrastination. *Quarterly Journal of Economics, 116*(1), 121–160.

O'Neill, A. (2024). Global adult literacy rate aged 25 years and older from 1976 to 2022 by gender. *Statista*. Retrieved April 26, 2024, from https://www.statista.com/statistics/997360/global-adult-and-youth-literacy/#:~:text=In%20the%20past%20five%20decades,to%2087.01%20percent%20in%202022

Office for National Statistics. (2021). Occupational pension schemes survey, UK: 2018. *Census*. Retrieved April 21, 2024, from https://www.ons.gov.uk/peoplepopulationandcommunity/personalandhouseholdfinances/pensionssavingsandinvestments/bulletins/occupationalpensionschemessurvey/2018#:~:text=Active%20membership%20of%20occupational%20pension,2017%20levels%20(7.7%20million)

Onita, L. (2022). McDonalds takes $1.4bn hit from closing its 850 Russian outlets: Fast food giant to pull out of Russia amid fallout from Ukraine war. *The Telegraph*. Retrieved April 25, 2024, from https://www.telegraph.co.uk/business/2022/05/16/mcdonalds-in-russia-closes-restaurants-ukraine-war-losing-investment/

Peerally, J. A., Fuentes, C. D., & Figueiredo, P. N. (2019). Inclusive innovation and the role of technological capability building: The social business Grameen Danone Foods Limited in Bangladesh. *Long Range Planning, 52*, 101843. https://doi.org/10.1016/j.lrp.2018.04.005

Phelan, D. (2022). Eight secrets of Apple's design success: iPhone executives reveal how its devices are really created. Apple's Craig Federighi and Alan Dye explain the decision behind the iPhone 14's iconic dynamic island. Available at: https://www.independent.co.uk/tech/apple-design-secrets-success-iphone-b2201215.html (accessed 15 February 2025).

Perloff, R. M. (2018). *The dynamics of political communication: Media and politics in a digital age*. Routledge.

Peteraf, M., Stefano, G. D., & Verona, G. (2013). The elephant in the room of dynamic capabilities: Bringing two diverging conversations together. *Strategic Management Journal, 34*(2), 1389–1410.

Phang, S.L. (2013). *Factors influencing international students' study destination decision abroad*. Masters dissertation, University of Gothenburg.

Rashid, A. (2023). Cristiano Ronaldo risks Nike contract breach after appearing to wear Adidas shin pads during Al Nassr's 5-0 loss to Celta Vigo. *The Daily Mail*. Retrieved April 13, 2024, from https://www.dailymail.co.uk/sport/football/article-12319389/Cristiano-Ronaldo-risks-Nike-contract-breach-appearing-wear-Adidas-shin-pads-Al-Nassrs-5-0-loss-Celta-Vigo-despite-147million-deal-rivals.html

Robiady, N. D., Windasari, N. A., & Nita, A. (2021). Customer engagement in online social crowdfunding: The influence of storytelling technique on donation performance. *International Journal of Research in Marketing, 38*(2), 492–500.

Roth, E. (2022). Meta fined $276 million over Facebook data leak involving more than 533 million users. *The Verge*. Retrieved April 26, 2024, from https://www.theverge.com/2022/11/28/23481786/meta-fine-facebook-data-leak-ireland-dpc-gdpr

Rotheim, R. J. (1990). Review: Review essay: "Organizational capital". *The Review of Social Economy, 48*(1), 70–83.

Sabherwal, R., & Sabherwal, S. (2005). Knowledge management using information technology: Determinants of short-term impact on firm value. *Decision Sciences, 36*(4), 531–567.

Sandford, A. (2020). Coronavirus: Half of humanity now on lockdown as 90 countries call fro confinement. *Euro News*. Retrieved April 26, 2024, from https://www.euronews.com/2020/04/02/coronavirus-in-europe-spain-s-death-toll-hits-10-000-after-record-950-new-deaths-in-24-hou

Santander. (n.d.). *Nudge theory and why it led to a Nobel prize?* Retrieved April 21, 2024, from https://www.santander.com/en/stories/nudge-theory-and-why-it-led-to-a-nobel-prize#:~:text=This%20theory%20was%20developed%20in,in%20regard%20to%20their%20finances

Science Museum. (2019). *The age of plastic: From parkesine to pollution*. Retrieved April 25, 2024, from https://www.sciencemuseum.org.uk/objects-and-stories/chemistry/age-plastic-parkesine-pollution#:~:text=Belgian%20chemist%20and%20clever%20marketeer,phenol%2C%20under%20heat%20and%20pressure

Second Line and Split Account. (2024). *BT Community*. Retrieved March 10, 2024, from https://community.bt.com/t5/Bills-Packages/2nd-line-and-Split-account-query/m-p/2354648#M166348

Seedhouse, E. (2024). *Space tourism*. Britannica. Retrieved April 26, 2024, from https://www.britannica.com/topic/space-tourism

Seibel, H.D. (2005). *Does history matter? The old and the new world of microfinance in Europe and Asia*. Working Paper No. 2005, 10. Development Research Center, University of Cologne.

Shams, R., Sohag, K., Islam, M. M., Vrontis, D., Kotabe, M., & Kumar, V. (2024). B2B marketing for industrial value addition: How do geopolitical tension and economic policy uncertainty affect sustainable development? *Industrial Marketing Management, 117*, 253–274.

Shams, S. M. R. (2013a). Implications of relationship marketing indicators to enable organisational growth: A stakeholder causal scope analysis. In H. Kaufmann & M. F. A. K. Panni (Eds.), *Customer centric marketing strategies: Tools for building organizational performance* (pp. 214–244). Business Science Reference.

Shams, S. M. R. (2013b). Stakeholder causal scope centric market positioning: Implications of relationship marketing indicators. In H. Kaufmann & M. F. Panni (Eds.), *Customer-centric marketing strategies: Tools for building organizational performance* (pp. 245–263). Business Science Reference.

Shams, S. M. R. (2016a). Capacity building for sustained competitive advantage: A conceptual framework. *Marketing Intelligence & Planning, 34*(5), 671–691.

Shams, S. M. R. (2016b). Branding destination image: A stakeholder causal scope analysis for internationalisation of destinations. *Tourism Planning & Development, 13*(2), 140–153.

Shams, S. M. R. (2017). Transnational education and total quality management: A stakeholder-centred model. *Journal of Management Development, 36*(3), 376–389.

Shams, S. M. R., & Hasan, R. (2020). Capacity building for transnationalisation of higher education: Knowledge management for organisational efficacy. *European Business Review, 32*(3), 459–484.

Shams, S. M. R., & Kaufmann, H. R. (2016). Entrepreneurial co-creation: A research vision to be materialised. *Management Decision, 54*(6), 1250–1268.

Shams, S. M. R., Vrontis, D., Belyaeva, Z., Ferraris, A., & Czinkota, M. (2021). Strategic agility in international business: A conceptual framework for "agile" multinationals. *Journal of International Management, 27*(1), 1–10.

Shams, S. M. R., Vrontis, D., & Christofi, M. (2022). Stakeholder causal scope analysis-centered big data management for sustainable tourism. *Journal of Hospitality & Tourism Research, 46*(5), 972–978.

Singh, N., Kaur, J., Schapper, J., & Jack, G. (2013). The importance of place for international students' choice of university: A Malaysian case-study. In *Paper presented at the HERDSA Annual International Conference, Auckland, 1–4 July*.

Smith, M. D., & Meade, B. (2019). *Who are the world's food insecure? Identifying the risk factors of food insecurity around the world*. Economic Research Service: U.S. Department of Agriculture. Retrieved December 28, 2023, from https://www.ers.usda.gov/amber-waves/2019/june/who-are-the-world-s-food-insecure-identifying-the-risk-factors-of-food-insecurity-around-the-world/

Spender, J. C. (1996). Making knowledge the basis of a dynamic theory of the firm. *Strategic Management Journal, 17*, 45–63.

Srivastava, R. K., Fahey, L., & Christensen, H. K. (2001). The resource-based view and marketing: The role of market-based assets in gaining competitive advantage. *Journal of Management, 27*(6), 777–802.

Staples-Otis, L. (2022). *Brand loyalty examples: 3 brands that have the most loyal customers*. Retrieved April 9, 2024, from https://www.bazaarvoice.com/blog/brand-loyalty-success-stories-3-brands-that-have-the-most-loyal-customers-and-how-they-do-it/

Statista. (2022). *Number of McDonald's restaurants worldwide from 2005 to 2022*. Retrieved February 23, 2024, from https://www.statista.com/statistics/219454/mcdonalds-restaurants-worldwide/#:~:text=Number%20of%20McDonald's%20restaurants%20worldwide%20 2005%2D2022&text=Global%20foodservice%20retailer%20McDonald's%20operated,from%20 40%2C031%20restaurants%20in%202021

Statista. (2023a). *Consumers opinions on what to do with surplus food from businesses that is approaching expiration in the United States (U.S.) in 2022*. Retrieved December 28, 2023, from https://www.statista.com/statistics/1362778/consumers-opinions-on-what-to-do-with-extra-food-in-the-us/#:~:text=Consumers%20opinions%20on%20what%20to,in%20the%20U.S.%20in%20 2022&text=In%202022%2C%20approximately%2072%20percent,the%20day%20to%20charity%20organizations

Statista. (2023b). *Top host destination of international students worldwide in 2022, by number of students*. Retrieved April 13, 2024, from https://www.statista.com/statistics/297132/top-host-destination-of-international-students-worldwide/

Steenkamp, J. B. E. M. (2001). The role of national culture in international marketing research. *International Marketing Review, 18*(1), 30–44.

Sveiby, K. (2001). A knowledge-based theory of the firm to guide in strategy formulation. *Journal of Intellectual Capital, 2*(4), 344–358.

Svorenčík, A., & Truc, A. (2022). *A history of behavioral economics and its applications: What we know and future research directions*. The National Academies Press.

Tan, W. (2014). *Undergraduate*. Retrieved May 10, 2014, from www.uq.edu.au/international-students/international-undergraduate-student-testimonials

Teece, D. J., Pisano, G., & Shuen, A. (1997). Dynamic capabilities and strategic management. *Strategic Management Journal, 18*(7), 509–533.

Thaler, R. H. (2016). Behavioral economics: Past, present, and future. *American Economic Review, 106*(7), 1577–1600.

Thaler, R. H. (2017). The prize in economic sciences 2017. *The Nobel Prize.* Retrieved April 14, 2024, from https://www.nobelprize.org/prizes/economic-sciences/2017/popular-information/

Thaler, R. H., & Benartzi, S. (2004). Save More TomorrowTM: Using behavioral economics to increase employee saving. *Journal of Political Economy, 112*(S1), S164–S187.

Thaler, R. H., & Sunstein, C. R. (2008). *Nudge: Improving decisions about health, wealth and hapiness.* Penguin.

Thaler, R. H., & Sustein, C. R. (2009). *Nudge (revised edition).* Penguin.

The Electoral Commission. (2018). *Digital campaigning: Increasing transparency for voters.* Retrieved March 26, 2020, from https://www.electoralcommission.org.uk/sites/default/files/pdffile/Digital-campaigning-improving-transparency-for-voters.pdf

The Guardian. (2022). Can the world feed 8bn people sustainably? Retrieved December 28, 2023, from https://www.theguardian.com/global-development/2022/nov/15/can-the-world-feed-8bn-people-sustainably

The Independent. (2016). Restaurant owner installs fridge outside to feed the homeless. Retrieved December 28, 2023, from https://www.independent.co.uk/news/world/asia/restaurant-owner-installs-fridge-outside-to-feed-the-homeless-a6963861.html

The Mirror. (2016). This restaurant is tackling the problem of hunger and food waste in a brilliant way. Retrieved December 28, 2023, from https://www.mirror.co.uk/news/world-news/restaurant-tackling-problem-hunger-food-7669392

Tomer, J. F. (1987). *Organizational capital—The path to higher productivity and wellbeing.* Praeger Publishers.

Townsend, S. (2018). 88% of consumers want you to help them make a difference. *Forbes.* Retrieved April 10, 2022, from https://www.forbes.com/sites/solitairetownsend/2018/11/21/consumers-want-you-to-help-them-make-a-difference/?sh=3dfef6866954

Trinet. (2024). Generational differences in the workplace: Boomers, Gen X, Gen Y, and Gen Z explained. Retrieved April 26, 2024, from https://www.trinet.com/insights/generations-in-the-workplace-boomers-gen-x-gen-y-and-gen-z-explained

Tummers, L. (2023). Nudge in the news: Ethics, effects, and support of nudges. *Public Administration Review, 83*, 1015–1036.

Uber. (2024). *A guide to using Uber.* Retrieved March 10, 2024, from https://www.uber.com/gb/en/ride/how-it-works/

Understanding How Childhood Shapes Who We Become. (2023). The role of early childhood experiences in shaping personality. *Medium.* Retrieved April 19, 2024, from https://medium.com/@virtued.reception/the-role-of-early-childhood-experiences-in-shaping-personality-b134dd946451

UNDP. (2023). *UNDP to raise awareness on plastic pollution by introducing jute-polymer bags.* Retrieved November 20, 2023, from https://www.undp.org/bangladesh/press-releases/undp-raise-awareness-plastic-pollution-introducing-jute-polymer-bags

UNESCO. (2014). *Global flow of tertiary-level students.* Retrieved August 23, 2015, from www.uis.unesco.org/Education/Pages/international-student-flow-viz.aspx

United Nations. (2023). International day of awareness on food loss and waste reduction, 29 September. Retrieved November 19, 2023, from https://www.un.org/en/observances/end-food-waste-day#:~:text=the%20consumption%20level.-,An%20estimated%2017%20percent%20of%20total%20global%20food%20production%20is,in%20the%20global%20food%20system

University of Cambridge. (2024). *Extent of India's Covid nudge campaign revealed.* Retrieved April 21, 2024, from https://www.cam.ac.uk/research/news/extent-of-indias-covid-nudge-campaign-revealed

Vorhies, D. W., & Morgan, N. A. (2005). Benchmarking marketing capabilities for sustainable competitive advantage. *Journal of Marketing, 69*, 80–94.

Vrontis, D., Shams, S. M. R., Thrassou, A., Pereira, V., & Siachou, E. (2021). Strategic marketing and the realities and illusions of sustained competitive advantage—Revising the resource-based view. *Journal of Strategic Marketing*. Retrieved December 15, 2021, from https://think.taylorandfrancis.com/special_issues/strategic-marketing-sustained-competitive-advantage/?utm_source=TFO&utm_medium=cms&utm_campaign=JPG15743&fbclid=IwAR1nzLdv2rU8_xWh3Pu4EF59cMWIAHOPrTDH3vNeobhtduR34clz9XkAyk0#?utm_source=CPB&utm_medium=cms&utm_campaign=JPG15743

Williamson, O. E. (1975). *Markets and hierarchies: Analysis and antitrust implications*. The Free Press.

Williamson, O. E. (1981). The economics of organization: The transaction cost approach. *The American Journal of Sociology, 87*(3), 548–577.

World Vision. (2023). *10 world hunger facts you need to know*. Retrieved December 28, 2023, from https://www.worldvision.org/hunger-news-stories/world-hunger-facts

WWF-Australia. (2023). *Plastic in our oceans is killing marine mammals*. Retrieved April 25, 2024, from https://wwf.org.au/blogs/plastic-in-our-oceans-is-killing-marine-mammals/

Zero-G. (2022). A brief history of space tourism. Retrieved April 26, 2024, from https://www.gozerog.com/article/a-brief-history-of-space-tourism/

Further Reading

7

Baker, W. E., & Sinkula, J. M. (2005). Environmental marketing strategy and firm performance: Effects on new product performance and market share. *Journal of the Academy of Marketing Science, 33*(4), 461–475.

Bell, S. J., Whitwell, G. J., & Lukas, B. A. (2002). Schools of thought in organizational learning. *Journal of the Academy of Marketing Science, 30*(1), 70–86.

Chandrashekaran, M., Mehta, R., Chandrashekaran, R., & Grewal, R. (1999). Market motives, distinctive capabilities, and domestic inertia: A hybrid model of innovation generation. *Journal of Marketing Research, 36*(1), 95–112.

Day, G. S. (2011). Closing the marketing capabilities gap. *Journal of Marketing, 75*(4), 183–195.

Dickson, P. R., Farris, P. W., & Verbeke, W. J. M. I. (2001). Dynamic strategic thinking. *Journal of the Academy of Marketing Science, 29*(3), 216–237.

Fang, E., Palmatier, R. W., & Grewal, R. (2011). Effects of customer and innovation asset configuration strategies on firm performance. *Journal of Marketing Research, 48*(3), 587–602.

Ghose, A., Lee, H. A., Nam, K., & Oh, W. (2023). The effects of pressure and self-assurance nudges on product purchases and returns in online retailing: Evidence from a randomized field experiment. *Journal of Marketing Research, 61*, 517. https://doi.org/10.1177/00222437231180494

Goswami, I., & Urminsky, O. (2016). When should the ask be a nudge? The effect of default amounts on charitable donations. *Journal of Marketing Research, 53*(5), 829–846.

Hogreve, P. J., Matta, S., Hettich, A. S., & Reczek, R. W. (2021). How do social norms influence parents' food choices for their children? The role of social comparison and implicit self-theories. *Journal of Retailing, 97*(2), 173–190.

Huang, S., Maimaran, M., & Kupor, D. (2024). EXPRESS: Using price promotions to drive children's healthy choices in a developing economy. *Journal of Marketing Research*. https://doi.org/10.1177/00222437241237483

Hunt, S. D. (1999). The strategic imperative and sustainable competitive advantage: Public policy implications of resource-advantage theory. *Journal of the Academy of Marketing Science, 27*(2), 144–159.

Johnson, J. L., Lee, R. P.-W., Saini, A., & Grohmann, B. (2003). Market-focused strategic flexibility: Conceptual advances and an integrative model. *Journal of the Academy of Marketing Science, 31*(1), 74–89.

Laran, J., Janiszewski, C., & Salerno, A. (2019). Nonconscious nudges: Encouraging sustained goal pursuit. *Journal of Consumer Research, 46*(2), 307–329.

Luo, X., Sivakumar, K., & Liu, S. S. (2005). Globalization, marketing resources, and performance: Evidence from China. *Journal of the Academy of Marketing Science, 33*(1), 50–65.

Menguc, B., & Auh, S. (2006). Creating a firm-level dynamic capability through capitalizing on market orientation and innovativeness. *Journal of the Academy of Marketing Science, 34*(1), 63–73.

Mishra, S., & Modi, S. B. (2016). Corporate social responsibility and shareholder wealth: The role of marketing capability. *Journal of Marketing, 80*(1), 26–46.

Mrkva, K., Posner, N. A., Reeck, C., & Johnson, E. J. (2021). Do nudges reduce disparities? Choice architecture compensates for low consumer knowledge. *Journal of Marketing, 85*(4), 67–84.

Palazzolo, M., & Pattabhiramaiah, A. (2021). The minimum wage and consumer nutrition. *Journal of Marketing Research, 58*(5), 845–869.

Polman, E., & Maglio, S. J. (2023). Nudges increase choosing but decrease consuming: Longitudinal studies of the decoy, default, and compromise effects. *Journal of Consumer Research, 51*, 542. https://doi.org/10.1093/jcr/ucad081

Reeck, C., Posner, N. A., Mrkva, K., & Johnson, E. J. (2023). Nudging app adoption: Choice architecture facilitates consumer uptake of mobile apps. *Journal of Marketing, 87*(4), 510–527.

Robitaille, N., Mazar, N., Tsai, C. I., Haviv, A. M., & Hardy, E. (2021). Increasing organ donor registrations with behavioral interventions: A field experiment. *Journal of Marketing, 85*(3), 168–183.

Romero, M., & Biswas, D. (2016). Healthy-left, unhealthy-right: Can displaying healthy items to the left (versus right) of unhealthy items nudge healthier choices? *Journal of Consumer Research, 43*(1), 103–112.

Ruiz-Fernández, L., Rienda, L., & Marco-Lajara, B. (2024). Hotel chains and sustainable development: Degree of internationalization, SDGs and dynamic capabilities as drivers of successful performance. *Environ Dev Sustain*. https://doi.org/10.1007/s10668-024-04721-3

Singh, R. (2022). Dynamic capabilities and performance of the firm: The mediating role of VRIN resources. In Rajagopal & R. Behl (Eds.), *Inclusive businesses in developing economies. Palgrave studies in democracy, innovation, and entrepreneurship for growth*. Palgrave Macmillan. https://doi.org/10.1007/978-3-031-12217-0_19

Slotegraaf, R. J., & Dickson, P. R. (2004). The paradox of a marketing planning capability. *Journal of the Academy of Marketing Science, 32*(4), 371–385.

Tran, T. K. (2024). Can sustainable entrepreneurship be achieved through green knowledge sharing, green dynamic capabilities, and green service innovation? *Environmental Science and Pollution Research, 31*, 3060–3075.

Varadarajan, R. (2018). A commentary on "Transformative marketing: The next 20 years". *Journal of Marketing, 82*(4), 15–18.

Xie, B., He, D., Mercer, T., Wang, Y., Wu, D., Fleischmann, K.R., Zhang, Y., Yoder, L.H., Stephens, K.K., Macker, M., Lee, M.K., 2020. Global health crises are also information crises: a call to action. J. Assoc. Inform. Syst., 71, 1419–1423 (https://doi.org/10.1002/asi.24357).

Wansink, B. (2017). Healthy profits: An interdisciplinary retail framework that increases the sales of healthy foods. *Journal of Retailing, 93*(1), 65–78.

Supplementary Information

S. M. R. Shams et al., *Sustainable Marketing*, Classroom Companion: Business,
https://doi.org/10.1007/978-3-031-79130-7

Practice Quiz Answer

Chapter 1

The correct answer is highlighted in ***bold*** and ***italic*** font.

1. For a brand, sustainable marketing is
   (a) A notional strategy
   (b) A charity
   (c) An optional strategy
   (d) ***A necessity***
2. The word "society" was included in the American Marketing Association's definition of marketing in
   (a) 1985
   (b) 1935
   (c) ***2007***
   (d) 2004
3. Target market of a brand includes
   (a) All customers and shareholders
   (b) ***All customers and non-customer stakeholders***
   (c) All customers
   (d) All shareholders and suppliers
4. The core purpose of overall marketing activities includes
   (a) Need management, industry management, and relationship management
   (b) ***Need management, differentiation management, and relationship management***
   (c) Industry management, relationship management, and target audience management
   (d) Need management, differentiation management, and industry management
5. For marketers, strategic marketing is a
   (a) ***Dual task***
   (b) Three-way task
   (c) Single task
   (d) Four-way task
6. The triple bottom line (three Ps) of sustainable marketers, include
   (a) People, planet, and promise keeping
   (b) ***People, planet, and profit***
   (c) Price, payment, and profit
   (d) Price, profit, and people
7. The simplest definition of value is
   (a) What we produce, in return of what we receive
   (b) ***What we receive, in return of what we sacrifice***
   (c) What we donate, in return of what we receive
   (d) What our memory forgets, in return of what we sacrifice

8. In general, customers can develop an anticipated value
   (a) ***Before they consume a product or service***
   (b) Instantly after they complete consumption of a product or service
   (c) Both before and after they consume a product or service
   (d) After a long time, they complete consumption of a product or service
9. In marketing, the market includes only
   (a) Current customers
   (b) Both current and potential non-customer stakeholders
   (c) ***Both current and potential customers and non-customer stakeholders***
   (d) Potential customers and non-customer stakeholders
10. The symbolic form of value started to influence customers from
    (a) Early 2010s
    (b) ***Late 1960s***
    (c) Mid-1940s
    (d) Early 1990s
11. Marketing is
    (a) Science and technology
    (b) Science of market analysis
    (c) Art of promise
    (d) ***Art and science***
12. The United Nations introduced the Sustainable Development Goals in
    (a) 2014
    (b) 2013
    (c) 2012
    (d) ***2015***
13. The Rs of the circular economy include
    (a) Refuse, rethink, reduce, reuse, refurbish, remanufacture, repurpose, recycle, recover
    (b) Refuse, rethink, reduce, redirect, return, refurbish, remanufacture, repurpose, recycle, recover
    (c) Refuse, rethink, reduce, redirect, repair, refurbish, remanufacture, repurpose, recycle, recover
    (d) ***Refuse, rethink, reduce, reuse, repair, refurbish, remanufacture, repurpose, recycle, recover***
14. How many Ps are discussed in the expanded marketing mix of this chapter?
    (a) 25
    (b) 16
    (c) ***22***
    (d) 21
15. Homburg et al. (2011) discussed how many steps in market orientation?
    (a) 16
    (b) 7
    (c) ***5***
    (d) 11

16. Porter's five competitive market forces include
    (a) New customers, substitute products, future competitors, buyers, and suppliers
    (b) New entrants, non-substitute products, future competitors, buyers, and suppliers
    (c) New entrants, substitute products, future competitors, buyers, and suppliers
    (d) ***New entrants, substitute products, current competitors, buyers, and suppliers***
17. Stakeholders' bargaining power analysis include
    (a) Number and size of buyers and suppliers, importance of specific buyer and supplier, is the item vital, and ability to snatch each other's employees
    (b) Number and size of government and non-government stakeholders, importance of specific government and non-government stakeholder, is the item vital, and ability to enter into each other's industry
    (c) ***Number and size of buyers and suppliers, importance of specific buyer and supplier, is the item vital, and ability to enter into each other's industry***
    (d) Number and size of government and non-government stakeholders, importance of specific buyer and supplier, is the item vital, and ability to enter into each other's industry
18. Drummond and Ensor (2001) describe how many characteristics of a feasible mission statement?
    (a) 4
    (b) 10
    (c) ***6***
    (d) 8
19. Four essentials to pursue the triple bottom line (people, planet, and profit) include
    (a) Equity in orientation, triple Ps' stakeholder-led co-creation, visible change for the society, and co-evaluation
    (b) ***Equity in orientation, triple Ps' stakeholder-led co-creation, visible change for the triple Ps' stakeholders, and co-evaluation***
    (c) Equity in orientation, triple Ps' stakeholder-led co-creation, visible change for the environment, and co-evaluation
    (d) Equity in orientation, triple Ps' stakeholder-led co-creation, visible change for the triple Ps' stakeholders, and co-destruction

Chapter 2

The correct answer is highlighted in ***bold*** and ***italic*** font.

1. How did the era of mass production contribute to the evolution of product development processes?
    (a) It emphasized personalized and artisanal production.
    (b) ***It led to the standardization of products and assembly line techniques.***
    (c) It had no impact on product development.

2. Explain the role of capitalism in shaping product development strategies and priorities.
   (a) Capitalism has no influence on product development.
   (b) Capitalism encourages sustainable and eco-friendly product development.
   (c) ***Capitalism influences strategies to maximize profit and market share.***
3. Discuss the impact of consumer demand on the mass production era and the development of standardized products.
   (a) Consumer demand had no influence on mass production.
   (b) Consumer demand led to customized and unique products.
   (c) ***Consumer demand led to standardized products and assembly line production.***
4. How does the concept of planned obsolescence relate to product development in a capitalist market?
   (a) It encourages long-lasting and durable products.
   (b) ***It promotes continuous innovation and frequent product upgrades.***
   (c) It has no relevance to product development.
5. Explore the motivations behind companies engaging in premiumization and how it influences product development.
   (a) Premiumization has no impact on product development.
   (b) Companies engage in premiumization to offer affordable products.
   (c) ***Premiumization aims to create high-end, exclusive products for higher profit margins.***
6. Analyze the ethical considerations associated with planned obsolescence in the context of consumer behavior.
   (a) Planned obsolescence is an ethical business practice.
   (b) ***It raises ethical concerns as it encourages wasteful consumer behavior.***
   (c) Consumers prefer planned obsolescence for constant product upgrades.
7. Evaluate the role of innovation in driving product development strategies during the mass production era.
   (a) Innovation had no role in the mass production era.
   (b) ***Innovation was crucial for introducing new products and production techniques.***
   (c) The mass production era discouraged innovation.
8. Discuss how capitalism has influenced the marketing of products, including tactics like upselling and cross-selling.
   (a) Capitalism has no impact on marketing strategies.
   (b) Capitalism encourages transparent marketing practices.
   (c) ***Capitalism promotes upselling and cross-selling to maximize revenue.***
9. Explain the challenges and opportunities for sustainable product development within a capitalist framework.
   (a) Sustainable product development aligns perfectly with capitalism.
   (b) ***Capitalism poses challenges but also opportunities for sustainable product development.***
   (c) Sustainable product development is not feasible in a capitalist system.

10. Provide examples of companies that have successfully implemented premiumization strategies in their product lines.
    (a) All companies avoid premiumization strategies.
    (b) Examples are not available as premiumization is an unsuccessful strategy.
    (c) ***Companies like Apple and Rolex have successfully implemented premiumization.***
11. Discuss the role of advertising and branding in promoting premium products and influencing consumer perceptions.
    (a) Advertising and branding have no impact on consumer perceptions.
    (b) ***They play a crucial role in promoting premium products and shaping consumer perceptions.***
    (c) Premium products do not rely on advertising and branding.
12. Explore the potential consequences of overconsumption in a capitalist-driven market.
    (a) Overconsumption has no consequences.
    (b) ***It leads to resource depletion, environmental degradation, and societal issues.***
    (c) Overconsumption is encouraged for economic growth.
13. How can companies balance the pursuit of profit with responsible and sustainable product development practices?
    (a) Profit and sustainability are incompatible.
    (b) By prioritizing profit over sustainability.
    (c) ***By integrating responsible and sustainable practices into their business models.***
14. Assess the impact of globalization on product development strategies in a capitalist economy.
    (a) Globalization has no impact on product development.
    (b) It encourages localized and region-specific product development.
    (c) ***It influences product development strategies to cater to a global market.***
15. In which phase of the product lifecycle do sales typically stabilize, and the market becomes saturated?
    (a) Growth Phase.
    (b) ***Maturity Phase.***
    (c) Decline Phase.
16. What does a "cash cow" represent in the Boston matrix?
    (a) Products with high market share in a high-growth market.
    (b) Products with low market share in a high-growth market.
    (c) ***Products with high market share in a low-growth market.***

Chapter 3

The correct answer is highlighted in ***bold*** and ***italic*** font.

1. Which of these has a relatively "inelastic" demand?
    (a) Mortgages
    (b) Package holidays
    (c) Champagne
    (d) ***Salt***

2. If everyone in a country must buy all of their water from the government's water authority, what sort of market is this?
   (a) ***Pure monopoly***
   (b) Monopolistic competition
   (c) Pure competition
   (d) Oligopolistic competition
3. What is the calculation used to ascertain the break-even price of a product?
   (a) Fixed cost + (variable cost / unit sales)
   (b) ***Variable cost + (fixed cost / unit sales)***
   (c) Total sales / (fixed cost + variable cost)
   (d) Fixed cost / (price – variable cost)
4. Which pricing strategy is most likely to maximize profit early in the life of a technologically advanced product?
   (a) Product-line pricing
   (b) Customary pricing
   (c) ***Price skimming***
   (d) Second-market discounting
5. What are the three main categories of pricing policy?
   (a) Needs-, wants-, and demands-based
   (b) Market-, manufacture-, and means-based
   (c) Contraction-, consolidation-, and conquest-based
   (d) ***Cost-, customer-, and competitor-based***
6. Which of these is an attractive offer that gains attention, generates customer footfall, and therefore enables sales of more profitable items?
   (a) ***Loss leader***
   (b) False footer
   (c) Hidden header
   (d) Dollar drainer
7. Which pricing technique is used by Duolingo and other apps to encourage customer trials before increasing their profitability?
   (a) Honey trap pricing
   (b) ***Freemium pricing***
   (c) Sprat-and-mackerel pricing
   (d) Ads-or-your-money pricing
8. If a store buys a suit for £200 and sells it for £400, what is the mark-up and the margin, expressed as percentages?
   (a) 200% mark-up and 200% margin
   (b) 100% mark-up and 100% margin
   (c) 50% mark-up and 100% margin
   (d) ***100% mark-up and 50% margin***
9. Which pricing strategy is most closely associated with Walmart or its UK subsidiary, Asda?
   (a) Value-Added Pricing
   (b) Customary Pricing
   (c) ***Everyday Low Pricing (EDLP)***
   (d) Second-market discounting

10. Which websites collect content—including offers and prices—from other websites, enabling easier customer comparisons?
    (a) ***Aggregators***
    (b) Consolidators
    (c) Reconcilers
    (d) Unifiers
11. What is the term for the drop in costs associated with accumulated production experience?
    (a) The Law of Many
    (b) ***The Experience Curve***
    (c) The Parallax Error
    (d) The Volume Reward
12. What is usually the first stage of the price-setting process?
    (a) Development of profit targets
    (b) Market segmentation
    (c) Ideation
    (d) ***Development of pricing objectives***
13. A McDonald's Happy Meal is an example of which of these?
    (a) Menu pricing
    (b) By-product pricing
    (c) ***Product-bundle pricing***
    (d) Captive-product pricing
14. A pizza restaurant that offers a 30% discount to students at lunchtime is using which of these pricing approaches?
    (a) ***Second-market discounting***
    (b) Penetration pricing
    (c) Seasonal pricing
    (d) Odd-even pricing

Chapter 4

The correct answer is highlighted in **bold** and *italic* font.

1. What term refers to the unique elements that make a brand easily recognizable?
    (a) Sustainability
    (b) Differentiation
    (c) ***Distinctive assets***
    (d) Monopoly Power
2. How does "brand storytelling" contribute to a brand's identity?
    (a) By avoiding consumer engagement
    (b) By establishing a brand as a commodity
    (c) ***By sharing a narrative that resonates with consumers***
    (d) By encouraging unethical practices
3. What is the significance of "sustainability" in brand management?
    (a) It has no impact on consumer preferences
    (b) ***It is essential for building brand equity and loyalty***

   (c) It leads to greenwashing
   (d) It promotes monopoly power
4. What does "monopoly power" refer to in the context of brands?
   (a) Healthy competition
   (b) Unique brand storytelling
   (c) ***Dominance and control in the market***
   (d) Ethical branding practices
5. How do "consumer habits" influence brand success?
   (a) They have no impact on brands
   (b) ***They shape purchasing decisions and brand preferences***
   (c) They discourage brand loyalty
   (d) They lead to supply chain transparency
6. What is the role of "big tech" in shaping modern branding practices?
   (a) It has no influence on branding
   (b) ***To date, unethical branding practices have been witnessed***
   (c) It contributes to sustainable practices
   (d) It significantly shapes and influences brand strategies
7. How does "AI and branding" impact consumer experiences?
   (a) It has no impact on consumer experiences
   (b) ***It enhances personalization and engagement***
   (c) It discourages brand loyalty
   (d) It promotes greenwashing
8. What term refers to the emotional connection and commitment that consumers have toward a brand?
   (a) Monopoly power
   (b) Brand awareness
   (c) ***Brand loyalty***
   (d) Triple bottom line
9. Which of the following is associated with the responsibility of managing and enhancing a brand's reputation?
   (a) Supply chain transparency
   (b) ***Brand management***
   (c) Greenwashing
   (d) Unethical concerns
10. What does the "triple bottom line" consider in evaluating business success?
   (a) Profitability only
   (b) ***Social and environmental impact, in addition to profitability***
   (c) Competition and monopoly power
   (d) Cultural sensitivity
11. How does "greenwashing" relate to branding?
   (a) It promotes environmental sustainability
   (b) It discourages sustainable practices
   (c) ***It involves false or exaggerated sustainability claims***
   (d) It encourages ethical branding

12. What term refers to the dominance of a brand or a few brands in a specific market?
    (a) Brand equity
    (b) Status anxiety
    (c) ***Monopoly power***
    (d) Consumer activism
13. What concept involves involving various interested parties in decision-making processes and brand initiatives?
    (a) Ethical branding
    (b) Cultural sensitivity
    (c) ***Stakeholder engagement***
    (d) Brand equity
14. What term refers to the unique elements that make a brand easily recognizable?
    (a) ***Logo***
    (b) Status anxiety
    (c) Mental availability
    (d) Business models
15. What is the significance of "brand equity" in evaluating a brand's strength?
    (a) It measures the social and environmental impact.
    (b) It assesses the brand's profitability only.
    (c) ***It evaluates the brand's perceived value and strength.***
    (d) It considers monopoly power only.

Chapter 5

The correct answer is highlighted in ***bold*** and ***italic*** font.

1. Ill feelings and behavior between channel partners is known as what?
    (a) Channel dissent
    (b) Channel friction
    (c) ***Channel conflict***
    (d) Channel force
2. Which of these distribution strategies is likely to have the largest number of retailers?
    (a) Exclusive
    (b) ***Intensive***
    (c) Selective
    (d) Proactive
3. Which channel partner deals most capably with storage, breaking bulk, and offering variety?
    (a) Retailer
    (b) Dealer
    (c) Broker
    (d) ***Wholesaler***
4. What is the term for investments that are made in one specific channel partnership?
    (a) ***Idiosyncratic investments***

   (b) Bespoke investments
   (c) Unique investments
   (d) Reciprocal investments
5. Which of these is usually charged with finding supply for the customer, rather than finding custom for the supplier?
   (a) Retailer
   (b) Dealer
   (c) ***Agent***
   (d) Broker
6. Which of these strategies may entail making products available in-store, on the firm's website, and on third-party websites?
   (a) Optichannel
   (b) ***Omnichannel***
   (c) Interchannel
   (d) Megachannel
7. What is footfall?
   (a) ***Prospects and customers who enter sales premises***
   (b) b. A manufacturer's retailers within a given district
   (c) The average distance a customer is prepared to travel to a store
   (d) The speed at which a store serves its customers
8. What is the name of the phenomenon in which two channel members disagree about whether a task is their responsibility or that of their channel partner?
   (a) Cognitive dissonance
   (b) Buyer's remorse
   (c) ***Domain dissensus***
   (d) Caveat emptor
9. What sort of integration occurs when a manufacturer acquires its downstream channel partners?
   (a) Diagonal
   (b) Horizontal
   (c) ***Vertical***
   (d) Tangential
10. Which kind of power comes from legal and contractual requirements?
   (a) ***Coercive***
   (b) Referent
   (c) Expert
   (d) Reward
11. What do we call the goods held by a channel member?
   (a) Stock initiative
   (b) Stock innovation
   (c) ***Stock inventory***
   (d) Stock iteration
12. Which Key Performance Indicator expresses how many times in a given period a channel member sells and replaces their entire stock?
   (a) ***Stock turn***
   (b) Stock churn

   (c) Stock burn
   (d) Stock learn
13. Which of the following is a contractual agreement in which the downstream channel partner agrees not to represent its supplier's rival brands?
   (a) Monogamous retailing agreement
   (b) Loyal selling agreement
   (c) Universal supply agreement
   (d) ***Exclusive dealing agreement***
14. Which of these entails selling to customers without using one's downstream channel partners?
   (a) Deselection
   (b) ***Disintermediation***
   (c) Antidisestablishmentarianism
   (d) Indiscretion selling

Chapter 6

The correct answer is highlighted in ***bold*** and ***italic*** font.

1. There are two types of viral marketing, which are:
   (a) ***Organic viral marketing and amplified viral marketing***
   (b) Organic viral marketing and exemplary viral marketing
   (c) Organic viral marketing and authenticated viral marketing
   (d) Exemplary viral marketing and amplified viral marketing
2. Organic viral marketing is a controlled viral marketing campaign.
   (a) True
   (b) ***False***
3. The 7Ds of an effective digital marketing campaign include:
   (a) Digital goals, digital audiences, digital infrastructure, digital platforms, digital media, digital data, digital technology
   (b) Digital goals, digital audiences, digital devices, digital terms, digital media, digital data, digital technology
   (c) ***Digital goals, digital audiences, digital devices, digital platforms, digital media, digital data, digital technology***
   (d) Digital goals, digital audiences, digital infrastructure, digital terms, digital media, digital data, digital technology
4. A micro influencer should have 10,001 to 100,000 followers.
   (a) ***True***
   (b) False
5. Exploring the opportunity to convince customers to purchase the same product in more volume in order to increase the product's overall sales is a cross-selling strategy.
   (a) True
   (b) ***False***
6. A database marketing campaign would have:
   (a) ***Four operational stages***
   (b) Three operational stages

   (c) Eight operational stages
   (d) Six operational stages
7. 5Vs of data include:
   (a) Data volume, data variety, data velocity, data veniality, data value
   (b) Data volume, data venality, data velocity, data veracity, data value
   (c) ***Data volume, data variety, data velocity, data veracity, data value***
   (d) Data volume, data variety, data visuality, data veracity, data value
8. The General Data Protection Regulation was passed in:
   (a) ***2016***
   (b) 2015
   (c) 2018
   (d) 2017
9. A digital marketing campaign would have three purposes, which are:
   (a) Identifying target audience, creating awareness, and forming a public opinion
   (b) ***Creating awareness, clarifying a brand's value proposition, and forming a public opinion***
   (c) Identifying target audience, clarifying a brand's value proposition, and forming a public opinion
   (d) Identifying target audience, creating awareness, and forming a value proposition

Chapter 7

The correct answer is highlighted in ***bold*** and ***italic*** font.

1. Strategic management scholars have three sources of firm resources. What are they?
   (a) Resource-based view of resources, firm-based view of resources, knowledge-based view of resources
   (b) Resource-based view of resources, firm-based view of resources, market-based view of resources
   (c) ***Resource-based view of resources, market-based view of resources, knowledge-based view of resources***
   (d) Resource-based view of resources, market-based view of resources, industry-based view of resources
2. Market-based view of resources would include:
   (a) ***Relational resources and intellectual resources***
   (b) Relational resources and interactive resources
   (c) Relational resources and organizational resources
   (d) Organizational resources and interactive resources
3. Sveivy (2001) suggested how many questions managers to answer to explore firm-specific unique knowledge-based resources?
   (a) 7
   (b) ***9***
   (c) 5
   (d) 11

4. Organizational dynamic capabilities include 3 Ps. What are they?
   (a) Process, Position, Power
   (b) Process, Power, Provision
   (c) Position, Provision, Power
   (d) ***Process, Position, Path.***
5. VRIN stands for:
   (a) Valuable, Rare, Inimitable, Non-comparable
   (b) Valuable, Rare, Inseparable, Non-comparable
   (c) ***Valuable, Rare, Inimitable, Non-substitutable***
   (d) Valid, Rare, Inimitable, Non-substitutable
6. The theoretical development of behavioral economics is influenced predominantly by:
   (a) ***Prospect theory, framing effects, status quo bias, paradox of choice, and intrinsic motivation***
   (b) Prospect theory, framing effects, status quo bias, paradox of choice, and extrinsic motivation
   (c) Prospect theory, relational effects, status quo bias, paradox of choice, and intrinsic motivation
   (d) Prospect theory, framing effects, status quo bias, paradox of identity, and intrinsic motivation
7. Two types of nudges are discussed in this book. What are they?
   (a) Single nudge and diverse nudge
   (b) Direct nudge and diverse nudge
   (c) ***Direct nudge and indirect nudge***
   (d) Singular nudge and plural nudge
8. What are the three factors that influence an individual's (e.g., a customer's) decision-making?
   (a) Intellectual perception, fairness perception, and lack of self-control
   (b) ***Rational perception, fairness perception, and lack of self-control***
   (c) Rational perception, fairness perception, and lack of information
   (d) Relational perception, fairness perception, and lack of information
9. When did Professor Thaler receive the Nobel prize because of his contribution based on the nudge theory?
   (a) ***2017***
   (b) 2018
   (c) 2019
   (d) 2015
10. All marketers are marketers, but not all marketers are marketers.
    (a) ***True***
    (b) False

Glossary

A

Advertising Advertising is a form of marketing communication aimed at promoting products, services, or ideas to target audiences through various channels such as print, broadcast, digital, and social media. Objectives include persuading and influencing consumer behavior, driving sales, brand awareness, and customer engagement.

Agents and brokers Intermediaries between a product/service provider and the end user. Agents usually help customers by "shopping around" for a good deal from a number of potential suppliers, while brokers usually find customers for the originator firm and sell on their behalf.

Algorithms Algorithms are step-by-step procedures or formulas used by computers to solve problems or perform specific tasks, often relying on data inputs and rules to make automated decisions and predictions in various domains, such as machine learning, optimization, and data analysis.

Algorithmic amplification Algorithmic amplification refers to the process by which digital algorithms prioritize, promote, or magnify certain content, often based on user engagement metrics or predefined criteria, influencing the visibility and reach of information or media on online platforms, potentially shaping choices about brands.

Amplified viral marketing Amplified viral marketing is a controlled digital marketing campaign where marketers have a more active role in shaping the campaign goals and methods of transmitting the viral message.

Anticipated value Anticipated value is customers' own interpretation about a particular product's utility before they purchase and consume it, and how they think of what they may need to sacrifice and what utility they may obtain in return for their sacrifice after they consume it.

Antitrust Antitrust refers to laws and regulations aimed at promoting competition and preventing monopolistic practices and unfair business tactics, ensuring a level playing field for brands and protecting consumer welfare.

Artificial Intelligence (AI) Artificial Intelligence (AI) refers to the simulation of human intelligence processes by machines, encompassing tasks such as learning, reasoning, problem-solving, and perception, enabling computers to perform tasks that typically require human intelligence.

Art of marketing Art of marketing is influencing or forming the perception of a brand's target audience about the brand's core value and its promises.

Assembly line Assembly line is a manufacturing process where products are sequentially assembled by workers along a conveyor belt, often associated with mass production, volume, and linear manner.

Attention economy Attention economy refers to the economic system where the scarce resource is not goods or services but rather individuals' attention, driving competition among businesses and platforms to capture and monetize users' time and engagement through compelling content, advertising, and digital experiences.

Automation Automation is the use of machinery and technology to perform tasks without direct human involvement, often a time-saving resource.

Automation in marketing Automation in marketing involves the use of software and technology to streamline repetitive tasks, such as email campaigns, social media posting, and data analysis, increasing efficiency and allowing brands to focus on strategy and creativity.

B

Basepoint pricing/basing point pricing Incorporates transportation costs from a regional distribution point rather than the place of manufacture.

Behavioral economics Behavioral economics analyzes the alternative aspects of human behavior to develop insights on how people (e.g., different customers and non-customer stakeholders) make decisions.

Big tech Big tech refers to large, influential technology companies that dominate various sectors (not a single industry) of the digital economy, wielding significant power in areas such as online search, e-commerce, social media, smartphone manufacturing, and cloud computing.

Bounded rationality Bounded rationality posits that individuals make decisions based on limited information, cognitive limitations, and time constraints, leading to satisficing rather than optimizing behavior, influencing consumer choices and brand perceptions by shaping decision-making processes in marketing contexts.

Brand awareness Brand awareness is the level of familiarity and recognition that consumers have with a particular brand, reflecting its visibility, presence, and recall in the minds of target audiences, which is crucial for driving consumer consideration and purchase decisions.

Brand equity Brand equity refers to the intangible value associated with a brand, including factors such as brand awareness, perceived quality, customer loyalty, and the overall strength of the brand's image and reputation.

Brands A brand's value, or equity, comes from consumers' perceptions about its features, performance, name, and the company behind it. This distinguishes it from generic products by creating a unique image and connection with consumers.

Branding Branding is the process of strategically shaping and managing the perceptions, emotions, and associations that consumers have with a company, product, or service. It involves creating a distinct identity and conveying its value proposition to target audiences to build recognition, loyalty, and trust.

Brand loyalty Brand loyalty is the emotional attachment and repeat purchasing behavior exhibited by consumers toward a specific brand, often resulting from positive experiences, perceived value, and trust built over time.

Brand management Brand management involves overseeing and strategizing the development, maintenance, and enhancement of a brand's identity, perception, and equity across all brand touch points to ensure consistency, relevance, and resonance with target audiences.

Brand storytelling Brand storytelling is the art of using narratives, emotions, and experiences to convey a brand's values, mission, and personality in a compelling and memorable way, forging deeper connections with consumers. Storytelling establishes trust by humanizing the brand. It allows consumers to relate to the brand on a personal level, fostering a sense of authenticity.

Brand tone of voice Brand tone of voice refers to the consistent style, personality, and language used by a brand to communicate with its audience across various channels and touchpoints, reflecting its values, identity, and relationship with consumers.

Break-even price The price above which a sold item will make a profit but below which it will produce a loss.

Break-even volume The volume of sales above which a sold item will make a profit at a predetermined price but below which it will produce a loss.

Business models Business models refer to the way a company creates, delivers, and captures value, usually through revenue streams. Business models provide strategic direction for the key components and relationships between customers, products or services, channels, revenue streams, and cost structures to achieve sustainable and profitable operations.

By-product pricing Pricing a by-product in such a way that it supports the competitiveness of the primary product from whose manufacturer it was derived.

C

Capitalism Capitalism is an economic system used for producing the greatest volume and diversity of goods and services. For it to succeed, however, there must be enough buyers for all the goods and services that capitalism is capable of producing.

Captive product pricing Pricing a product keenly in the knowledge that it must consume other, more profitable products to function as desired.

Carbon footprint Carbon footprint is the total amount of greenhouse gases (including carbon dioxide and methane) that are generated by our actions.

Cash-back deals/cash rebates Refunding money to customers upon completion of a sale, often for "big ticket" items, which will be paid for by consumers through monthly repayments.

Category entry point Category entry point refers to the strategic positioning or introduction of a new product or brand into a specific market segment or niche, aiming to capitalize on unmet consumer needs, preferences, or emerging trends to gain competitive advantage.

Cause-related marketing Cause-related marketing is a form of sustainable marketing that can be defined as aligning a societal or environmental cause with a marketing program, with an aim to help address the cause so that the societal and environmental contribution of a brand can enhance its commercial competitive advantage.

Celebrity-endorsement Promoting a brand relying on a celebrity's popularity, based on a formal business relationship with the celebrity.

Celebrity-endorsement for sustainable marketing Promoting a brand's societal, environmental, and commercial commitment and contribution, relying on a celebrity's popularity, based on a formal business relationship with the celebrity.

Channel conflict Disagreements and underlying hostility and negativity between channel partners.

Circular economy (definition 1) Circular economy (definition 1) is a responsible process of production and consumption that focuses on sharing, recycling, refurbishing, and reusing limited current resources (e.g., a product) as much as possible so that a product's lifecycle can be prolonged, and more people could be benefited by that product.

Circular economy (definition 2) The circular economy is an economic model that aims to minimize waste and maximize resource efficiency by promoting the continual use, recycling, and repurposing of materials throughout their life cycles.

Climate change Climate change is a real and undeniable threat to our entire civilization. The effects are already visible and will be catastrophic unless we act now. Climate action refers to taking urgent action to combat climate change and its impacts.

Co-creation Co-creation is the process of collaborating with all stakeholders involved to produce value.

Coercive power Power derived from the use of force.

Commercialization Commercialization refers to the process of introducing a new product, service, or technology into the market and, ultimately, make available for consumption to generate revenue and achieve market success.

Commodity Commodity is a basic good or raw material that is interchangeable with other goods of the same type; commodities are uniform, and they typically possess few distinguishing features or qualities that set them apart from competitors' offerings.

Competitive advantage Competitive advantage is the superior value proposition of a brand in comparison to its competitive brands.

Community engagement Community engagement encompasses the active involvement of businesses with local communities, considering their needs, fostering positive relationships, and contributing to social development.

Competition Competition refers to the rivalry between brands operating in the same industry or market, vying for customers, market share, and profitability.

Competitor-based pricing Pitching prices against competitors' prices.

Concept development and market testing At these stages, ideas are refined into more detailed concepts, followed by testing and confirmation with target consumers to ensure market acceptance and feasibility.

Conceptualization phase Conceptualization phase is the initial stage where ideas are generated and explored to form the basis for a new product or service.

Conscious consumer identity Conscious consumer identity reflects an individual's intentional and informed choices in purchasing and consumption, prioritizing ethical, sustainable, and socially responsible products and practices.

Conspicuous consumption Conspicuous consumption refers to the public display and consumption of luxury goods and services as a means of signaling social status, wealth, and prestige, influencing consumer behavior and brand preferences by associating brands with status symbols and exclusivity.

Consumer activism Consumer activism refers to the collective actions taken by individuals or groups of consumers to advocate for change in business practices, policies, or products, often aiming to promote ethical standards, environmental sustainability, or social justice.

Consumer habits Consumer habits, shaped by the workings of the brain and influenced by cognitive biases, represent the ingrained patterns of behavior and preferences that individuals rely on when making purchasing decisions, influenced by factors such as lifestyle, culture, convenience, and brand loyalty.

Consumption culture Consumption culture is the extent of availability of basic and luxurious products and services among different consumer groups in a society.

Consumption dynamics Consumption dynamics refer to the overall inflow and outflow of goods and services in society and the extent of different customer segments (i.e., high-, mid-, and low-income segments) ability to purchase and consume fundamental and luxury goods and services.

Consumer identities Consumer identities encompass the multifaceted ways individuals construct and express their sense of self, often leveraging brands as symbolic props to convey desired identities, aspirations, and social status to others.

Content marketing Creating web contents, e.g., text, audio, video, and images, to communicate with a brand's target audiences, mostly using the digital marketing channels is content marketing.

Corporate social responsibility Corporate social responsibility is a management concept whereby companies integrate social and environmental concerns into their business operations.

Cost-based pricing Setting prices relative to the costs of producing the product or service.

Cost reduction Cost reduction is the process where companies reduce their costs (e.g., labor) and increase their profits.

Cost-plus pricing Covering the cost of producing the service or product, then adding a predetermined profit.

Cross-Selling Exploring opportunities to convince customers to purchase another relevant item, e.g., car insurance, while selling the main item, e.g., the car.

Cultural sensitivity Cultural sensitivity is the awareness, understanding, and respect for the cultural differences, norms, values, and practices of diverse individuals and communities, essential for effective communication, collaboration, and relationship-building across cultural boundaries.

Customary pricing Striving to continue charging customers at an established price point.

Customer-based pricing (or Value-based pricing) Shaping the firm's prices according to customer demands and needs.

Customer empowerment Customer empowerment is when brands provide customers with the information and the tools that they need to make a decision while also attempting to avoid manipulation/exploitation.

Customer experience Customers develop their experience of a brand through different customer touchpoints, which constitute a customer's overall experience of the brand.

Customer journey (Definition-1) The entire process of customers obtaining information about a product and purchasing and consuming it can be mapped based on different customer touchpoints. The different customer touchpoints (e.g., a restaurant website or having dinner at the restaurant) that can be visible on the map collectively constitute the customer journey with the product.

Customer journey (Definition-2) Customer journey (Definition-2) is the overall experience and interactions a customer has with a company, from initial awareness through to purchase and post-sales support. The customer can entail many touchpoints with a business, and a business needs to manage a seamless experience between these touchpoints.

Customer touchpoints The traditional and contemporary communication channels, whether it is a brand's website, social media platforms, newspaper ads, or sales personnel, from where the brand's customers obtain information about the brand, are known as customer touchpoints.

D

Dark patterns of design Dark patterns of design involve intentionally deceptive or manipulative user interface design techniques aimed at tricking users into taking actions they might not otherwise choose, often at the expense of their autonomy, comprehension, or well-being, highlighting ethical concerns about transparency, trust, and user empowerment in digital environments.

Database marketing Responsively collecting, storing, and analyzing customers' and other key non-customer stakeholders' data to develop insights on their needs, wants, and expectations to customize market offerings for customers and other key stakeholders in order to drive customer and stakeholder loyalty.

Data value Retrieving the value (e.g., insights about customers' preferences) from raw data to influence customers based on customized offerings relevant to customers' preferences.

Data variety The extent of variety in data each time we collect, e.g., customers' spending each time, types of purchases each time, when they purchase (e.g., at a particular month or at a particular day or at a particular time of a day, etc.).

Data velocity How frequently we would be able to collect new data.

Data veracity Whether the source of data and the data itself are authentic or not.

Data volume The amount of data we may be able to collect each time.

Demand pricing (aka dynamic pricing or surge pricing) Frequently adjusting prices to reflect the level of customer demand at any given point in time.

Differentiation It is the process of distinguishing a product or brand from its competitors by highlighting unique features, benefits, or qualities that set it apart in the minds of consumers.

Digital age "The current era in which all kinds of information–numbers, text, sound, and video–can be put into digital form such that it can be stored, processed, and forwarded by any computer."

Digital marketing Digital marketing can be defined as applying the traditional, contemporary, and embryonic marketing principles to influence a brand's target audience through different digital platforms.

Digital marketing research and digital marketing analytics Digital marketing research and digital marketing analytics aim to develop insights through digital channels on customers' and non-customer stakeholders' needs, wants, and expectations in order to customize products and services to meet customers' expectations as accurately and competitively as possible.

Digitalization Digitalization refers to converting a traditional (e.g., paper-based) business system into a technology-enhanced system.

Digital transformation Digital transformation refers to a situation when digitalization takes place to a large extent and leads to a major company-wide or a specific business-wide transformation, ultimately based on a fully digitalized new business model to communicate with all internal and external stakeholders.

Digitization Digitization refers to a system that transforms analog information into a digital format as a prerequisite for the subsequent stages of consumption.

Direct nudge Direct nudge refers to designing a nudge based on not keeping alternative options for people's decision-making.

Distinctive assets Are unique visual or auditory elements associated with a brand, such as logos, jingles, or packaging designs, that help to create strong brand recognition and recall among consumers.

Disposable culture Disposable culture is a societal trend characterized by the frequent disposal of products, contributing to environmental concerns and waste.

Distress a brand To undermine customer confidence in a brand inadvertently, and therefore, the firm's ability to charge a high price through the use of poorly conceived marketing initiatives, usually focused on deep discounts.

Downstream channel partner/channel intermediary A channel partner located between a firm and its customers.

Dynamic capabilities Dynamic capabilities of an organization refer to its organizational processes, assets' position and path dependency (strategic direction), and how efficiently the organization implements and evaluates its processes to deploy its resources in order to pursue its organizational goal.

E

Early adopters Customers who buy products (or a category of product) early in the product life cycle, during its growth stage.

Early majority Customers who buy products (or a category of product) in the early part of the maturity stage of the product life cycle.

Echo chambers Echo chambers are online environments or social networks where individuals are exposed predominantly to information, opinions, and perspectives that reinforce their existing beliefs with limited exposure to diverse viewpoints.

Eco-friendly materials Eco-friendly materials refer to materials that are produced and used in ways that minimize their environmental impact and carbon footprint, often through sustainable sourcing, reduced energy consumption, and recyclability.

Economies of scale (definition-1) The extent of production and production costs, which increases savings from production costs by increasing the production capacity.

Economies of scale (definition-2) Economies of scale (definition-2) refers to a firm's better use of its resources and capability to reduce its overall production and operational costs and increase its production to sell more products based on less production cost per unit product.

Electronic data interchange An electronic data entry system that allows manufacturers, logisticians, and vendors to communicate to maximize efficiency in production, delivery, and sales.

Ethical branding Ethical branding entails aligning a brand's values, actions, and messaging with ethical principles and practices, prioritizing honesty, integrity, and responsibility in all aspects of business operations and interactions with stakeholders.

Ethical co-design Ethical co-design is a collaborative and inclusive design process that prioritizes fairness, respect, and consideration for diverse perspectives, ensuring the ethical treatment of stakeholders and fostering equitable outcomes in the creation of products, services, or solutions.

Ethical sourcing Ethical sourcing involves the procurement of goods and services in a socially and environmentally responsible manner, considering factors such as fair labor practices, environmental sustainability, and ethical treatment of workers throughout the supply chain.

Everyday low pricing (EDLP) Using permanently low prices instead of sporadic discounts.

Exclusive dealing agreement A contractual agreement in which the downstream channel partner agrees not to represent its supplier's rival brands.

Exclusive distribution A strategy in which the originator brand or firm appoints a very small number of carefully selected downstream channel partners concentrated in key locations.

Exclusive network A very small number of channel partners chosen to represent a brand only in high-priority locations.

Expert power Power derived from the trust and respect that others have in a person or firm's expertise and knowledge.

F

Fair trade Fair trade is designed to help producers in developing countries achieve sustainable and equitable trade relationships.

Fairness perception Fairness perception refers to an individual's perception on how their purchase decision, i.e., how purchasing a particular product or service could impact people and the planet, and how an individual considers such impacts while deciding to purchase a particular product or service.

Feasibility analysis Feasibility analysis involves assessing the practicality and possibility of the suggested concept, considering factors such as technical, economic, and operational feasibility.

Feel good factor When consumers feel better about their purchase decisions, as well as about themselves, because of the ethical significance of their decision that contributes to any socioeconomic cause, at least to some extent.

Fixed costs Costs that do not change according to the number of products sold, appertaining to overheads like factory rent, machinery, and management wages.

FOB-origin pricing Passes responsibility for transportation costs to the customer.

Footfall The number of potential customers entering sales premises.

Framing Framing involves presenting information in a particular way to influence individuals' perceptions, judgments, and decisions, often shaping consumer attitudes and behaviors toward brands through the context and emphasis placed on different aspects of a message or experience.

Framing Effects Framing Effects refer to the presentation and the circumstances related to alternative choice options that can influence people's decision-making.

Franchisees Third-party organizations that agree to represent a brand by working on its behalf, usually in a sales and marketing capacity. The originator brand in such an agreement is a **franchisor**, and the contract tying the parties together is a **franchise agreement.**

Freight-absorption pricing Pricing that absorbs the cost of transportation into the amount to be charged, regardless of where the product is to be transported, so that the seller remains competitive in distant markets.

G

Generative AI Generative AI refers to a subset of artificial intelligence techniques that involve training models to generate new, original content, such as text, images, or music, based on patterns and examples from existing data.

Geographical pricing Charging different prices according to where the product is sold.

Good-value pricing Charging a price that is noticeably attractive to customers, representing perceived good value, often through the use of budget sub-brands.

Greenwashing Greenwashing refers to the deceptive practice of misleading consumers by falsely portraying a brand, product, or service as environmentally friendly or sustainable, often exploiting growing environmental concerns without making meaningful changes.

H

Horizontal integration The process of a firm acquiring rival firms that undertake the same function within a channel.

I

Idea generation Idea generation is the process of generating a wide range of potential concepts or solutions to address a particular problem of customers or an opportunity for them.

Idea screening Idea screening involves assessing and filtering the generated ideas to identify those with the most potential for further development.

Inclusive design Inclusive design is based on the simple principle that designing for the widest range of people creates better designs and benefits everyone.

Indirect nudge Indirect nudge refers to designing a nudge based on keeping alternative options for people's decision-making or keeping any cost of resisting a nudge very low (e.g., if you prefer paper bills for your utility, you need to pay an additional $4.00 per year as a low cost to resisting the nudge. The purpose here could be nudging customers to opt in for a paperless digital billing system, as it can increase responsible behavior to our environment by reducing printing).

Influencer marketing Promoting a brand relying on an influencer's influence on their followers, based on a formal business relationship with the influencer.

Industrialization Industrialization is the development and implementation of industrial methods (e.g., the assembly line) of production on a large scale.

Innovators Customers who buy products (or a category of product) very early in the product life cycle, typically at launch, through pre-launch orders, or immediately afterward.

Intensive network A very large network of retailers providing blanket coverage of a market territory.

Intellectual resources Intellectual resources are those specific knowledge or skills of a firm that help the firm to develop a superior competitive advantage in a market.

Intermediary An organization that deals with customers, end users, or other channel partners on behalf of its supplier/s.

Intrinsic motivation Intrinsic motivation argues that someone's self-interest can motivate her or him in spite of any extrinsic or external rewards.

K

Knowledge democracy Individuals' rights and responsibilities to access and use data and information.

L

Lack of self-control Lack of self-control in decision-making can be defined as peers' influence on an individual's purchase decision.

Labor exploitation Labor exploitation refers to the unethical practice of taking advantage of workers' vulnerabilities by subjecting them to poor working conditions, low wages, long hours, and other forms of mistreatment, violating basic human rights and ethical standards within the workplace.

Laggards Customers who buy products (or a category of product) very late in their product life cycle.

Late majority Customers who buy products (or a category of product) in the late part of the maturity stage and in the earlier part of the decline stage of the product life cycle.

Latent need Latent need could be defined as accepting a new utility or value as a solution of a customer problem with delightful satisfaction and without any predetermined plan to consume the utility.

Legitimate power Power derived from the law or a contractual agreement.

Life-cycle assessment (LCA) Life-cycle assessment (LCA) is a process of evaluating the effects that a product has on the environment over the entire period of its life, thereby increasing resource-use efficiency and reducing impact on the planet.

Logo Logo is a visual symbol or graphic representation that serves as a distinctive identifier for a brand, product, or organization, often embodying its values, identity, and purpose in a simple yet memorable design.

Location and time pricing Setting prices according to where and when the product is sold.

Loss leaders Loss-making items are heavily promoted or positioned prominently to attract customer footfall, which ultimately brings sales of more profitable items.

M

Margin The amount of profit on an item, expressed as a percentage of the selling price.

Market Market is the combination of people who are current and potential customers and non-customer stakeholders of a brand.

Market orientation Market orientation is closely recognizing and analyzing the needs, wants, and expectations of a brand's target audiences, i.e., customers and key non-customer stakeholders, to create competitive value reflecting on the target audiences' needs.

Marketing Marketing is a dynamic of designing, delivering, and evaluating a utility or value in the form of a product or service to meet the target markets' needs to establish a competitive advantage for the product or service.

Marketing ethics Marketing ethics can be defined as marketers' moral principles that guide them to plan, implement, and monitor marketing activities in a way that will not be unjust for individuals and society while influencing their target audiences.

Marketing myopia Marketing myopia is a short-sightedness in understanding target audiences' needs, wants, and expectations and designing a product or service based on that wrong understanding.

Marketplace Marketplace is a traditional place, e.g., a physical shop where two parties meet to exchange value.

Marketspace Marketspace is the internet-enhanced digital businesses, e.g., digital superstore, where buyers, sellers, and other third parties (such as suppliers) supply, sell, and purchase goods and services.

Market research Market research refers to the gathering of data, insight, and skills used to understand a market and the consumers within that market. Often used to roll out a new product development or a new service.

Mark-up pricing Adding a predetermined percentage onto the bought-in cost of a product before selling it for profit.

Mass manufacturing or mass production Mass manufacturing or mass production is the production of goods on a large scale using standardized processes and machinery.

Materialism Materialism refers to how consumers use brands as props to create a desired identity with an emphasis on acquiring and consuming material goods as a measure of success or happiness. The concept of consumerism is linked to the concept of materialism.

Memory structures and brands Memory structures and brands are intertwined as brands aim to create strong associations and memorable connections in consumers' minds through consistent messaging, experiences, and emotional resonance, fostering brand recall and preference.

Mental availability Mental availability refers to the likelihood of a brand coming to consumers' minds in relevant purchase situations, influenced by factors such as brand awareness, salience, and associations, crucial for steering purchase decisions.

Mental shortcuts, or heuristics Mental shortcuts, or heuristics, are cognitive strategies used by individuals to make quick and efficient decisions based on simplified rules of thumb, often influencing consumer behavior and brand perception through the unconscious processing of information.

Metaphorical marketspace Metaphorical marketspace is symbolic of marketers' and customers' interactions from a distance, based on different contemporary technologies, such as social media, mobile platforms, emails, digital ads, etc., and also based on the embryonic aspects of fuzzy technologies, such as AI, robotics, data analytics, meta (among others) that aim to offer a more symbolic experience to customers, e.g., client meeting can take place online from a distance, but both clients and marketers will have a feeling of in-person interactions.

The mission The mission is the combination of a plan of activities and actual activities that we strategize, implement, and evaluate to pursue our vision.

Mission statement Mission statement summarizes the core focus of a brand's activities to convey the message of how the brand's overall activities collectively help the brand to contribute value to its different stakeholder groups in order to pursue the brand vision.

Mixed-zone market Mixed-zone market consists of the traditional marketplace (i.e., the physical shops), a synchronized market (i.e., the mix of customers from different generations), and a metaphorical space (i.e., the technology-enhanced virtual space).

Mobile marketing Mobile marketing is using different wireless and portable devices, e.g., smartphones, tablets, or smartwatches, as a medium to convey a marketing message in order to influence a brand's target audience.

Monopoly power A monopoly is a situation where certain brands have a significant amount of market share, leading to a brand controlling much of an industry.

Multi-channel retailing Selling through more than one channel (e.g., retail stores, own website, and comparison/aggregator website).

N

Niche Niche is a more narrowly defined segment of a market or customer group seeking a distinctive mix of benefits.

Neural pathways Neural pathways are interconnected networks of neurons that transmit signals and information throughout the brain, shaping cognitive processes, learning, and memory formation, including the associations and perceptions individuals develop toward brands and experiences.

Nudge Nudge is predicting how people behave (act and/or react) based on alternative options for making a decision (e.g., a purchase decision) and influencing their decision based on the prediction while helping people to make a better decision.

Nudge theory Nudge theory proposes that subtle changes in the presentation of choices can influence individuals' decisions and behaviors, often steering them toward more beneficial or desirable outcomes, shaping consumer choices and brand preferences through behavioral economics principles.

O

Oligopoly A market in which a very limited number of suppliers compete with each other.

Omnichannel marketing Omnichannel marketing refers to an organization-wide effort to timely communicate with all internal and external stakeholders through traditional and contemporary channels, from where stakeholders obtain information about the organization so that the organization and its stakeholders both remain up-to-date about the organization's value proposition related to its marketing mix.

Omni-channel retailing Selling in a coordinated manner through different channels according to the immediate needs of the customer/prospect, often combining several channels within the sales process.

Operational competitive advantage Operational competitive advantage is a brand's target audiences' perceived experience about why doing business with the brand would be more prolific for them, instead of its competitive brands, so that the brand's overall operation could survive and prosper.

Optional-product pricing Pricing non-compulsory items, products, and accessories to tempt customers who are making a larger purchase.

Organic viral marketing Organic viral marketing is a type of digital marketing that spreads quickly and naturally through user-generated content and word-of-mouth, with little or no involvement or control from the brand.

P

Paradox of choice Paradox of choice claims that when people have more options to select one, people may behave strangely.

Peace of mind Customers' peace of mind could be referred to as ensuring that customers will not have any concern in obtaining and consuming a product or service exactly as per the specifications (e.g., product features and quality) and utilities (e.g., delivery options) as it was promised to the customers during the pre-purchase stage.

Penetration pricing Dropping prices by a small percentage to gain market share by leveraging on customers' price sensitivity.

Perceived value Perceived value is the difference between a brand's promises to its customers and what it delivers to the customers.

Persuasive technologies Persuasive technologies are digital tools and platforms designed to influence users' attitudes, behaviors, and decision-making processes through techniques such as persuasive design, nudges, and behavior change interventions, raising ethical considerations regarding user autonomy, manipulation, and privacy.

Predatory pricing Adopting highly aggressive prices with the intention of damaging a competitor and perhaps even eliminating them from the market.

Price skimming Charging very high prices in the earlier stages of the product life cycle, then gradually lowering them to sell to progressively more price-conscious customer segments.

Product bundle pricing Offering several products at a lower price than would be charged for them individually.

Product development Product development is the process of conceiving, designing, prototyping, and bringing a new product or service to market, involving various stages such as ideation, research, design, testing, and refinement.

Product lifecycle management (PLM) Product lifecycle management (PLM) is a holistic approach that involves the strategic management of a product or service from its initial concept and design through manufacturing/production, distribution, and use to its eventual disposal or recycling.

Production efficiency Production efficiency is the effectiveness and optimization of the manufacturing process to minimize waste and maximize output.

Product-line pricing Basing products' prices upon their relationships with other products in the seller's product range.

Product-mix pricing A range of pricing approaches that consider the firm's products in the context of each other.

Programmatic advertising Programmatic advertising is an automated algorithmic process to understand the real-time demand of digital ads based on recognizing the active digital target audiences at a given time in order to enable advertisers to reach their digital target audiences, relying on real-time data of active digital audiences.

Promotion Promotion involves the use of various marketing tactics to communicate and persuade target audiences about the benefits of a product, service, or idea. It encompasses activities such as advertising, sales promotions, public relations, and personal selling, all aimed at increasing brand visibility, sales, and customer loyalty.

Promotional pricing Short-term price adjustments to stimulate customer awareness, footfall, and demand.

Proof of concept Proof of concept is the stage where a prototype is developed to confirm and demonstrate the technical feasibility and potential success of the concept.

Prospect theory (Definition-1) Prospect theory (Definition-1) suggests that individuals' decisions are influenced by the potential gains and losses relative to a reference point rather than by absolute outcomes, providing insights into how consumer choices and preferences are shaped by perceived risks and rewards about certain brands in marketing contexts.

Prospect theory (Definition-2) Prospect theory (Definition-2) assumes that someone would act differently than others while deciding on prospective gains in comparison to prospective losses when people have less, no, or an uncertain extent of control over a particular situation.

Prosumption Prosumption describes the phenomenon where consumers not only consume products or services but also actively participate in their creation, customization, or dissemination, blurring the lines between producers and consumers.

Physical availability Physical availability refers to the accessibility and presence of a product or service in physical retail spaces or distribution channels, ensuring convenient access for consumers and increasing the likelihood of purchase. Brands must be easily available when a consumer decides to make a purchase.

Psychological pricing A range of pricing tactics aimed at manipulating customer perceptions of the brand and the competitiveness of its pricing.

Pure competition Markets or product categories with many buyers and sellers, none of whom are dominant.

Pure monopoly Markets or product categories in which customers can only buy from one supplier.

Purpose of marketing Purpose of marketing activities is primarily threefold: Need management, differentiation management, and relationship management to establish a brand's competitive advantage.

R

Rational perception Rational perception refers to an individual's own calculation about prospective loss or gain from a particular decision, and the individual will, in general, will make the decision among the alternative options that would be most prolific for their needs and budget.

Recommendation algorithms Recommendation algorithms are computational techniques that analyze user data and behavior to suggest personalized content, products, or services, enhancing user experience and engagement on digital platforms such as e-commerce websites, streaming services, and social media platforms.

Referent power Power derived from one partner's reputation and standing and other partners' willingness to emulate them or allow them to lead due to their reputation or standing.

Relational resources Relational resources are those market-specific resources, e.g., market knowledge, that a firm would be able to obtain from their market-specific stakeholder relationships and interactions.

Renewable energy Renewable energy is energy derived from natural sources that are replenished at a higher rate than they are consumed.

Representation in marketing Representation in marketing involves authentically showcasing diverse individuals and experiences, serving as positive role models and breaking down barriers, as when people see themselves reflected in various positions (i.e., in advertising and promotional materials), it inspires belief in the possibility of achieving similar success.

Responsible consumption Responsible consumption involves making mindful and ethical choices in purchasing and using products or services to minimize environmental and social impacts, promote sustainability, and contribute to a more equitable and responsible global economy.

Retailers Firms or individuals that sell to end customers, typically on-premises, designed to capture passing trade and to facilitate the storage, display, and selling of a variety of goods and services.

Return on Capital Employed (ROCE) The amount of money that is realized from an investment.

Return on Sales (ROS) A measure of the profitability of a company's sales.

Reward power Power derived from incentivizing and rewarding others.

S

Satisfaction Satisfaction is the extent to which customers' expectations and needs are met or exceeded by a product or service.

Scalability Scalability is the ability of a business to handle increased demands or growth without compromising performance or efficiency.

Science of marketing Science of marketing is collecting market data to analyze it to re-strategize marketing plans reflecting on customer segment-specific perceived experience about a brand.

Seasonal discounts Discounts offered to stimulate sales during periods of reduced customer demand.

Search engine optimization and search engine marketing Search engine optimization and search engine marketing refer to making a brand's website and other web content appealing and compatible with the requirements of online search engines in an attempt to bring the website and web content to the forefront of an online search result.

Second market discounting Charging a lower price to a secondary market segment for the same product or service.

Segmented pricing Applying different pricing approaches according to market segment or product type.

Selective network A network of downstream intermediaries, usually retailers, which provide widespread but not blanket coverage and are located in the most significant locations according to local market potential.

Service quality Service quality refers to the level of excellence or satisfaction experienced by customers about the services provided by a business.

Share of wallet Share of wallet can be defined as the percentage of a particular customer's total spent to purchase a specific brand in their entire lifetime, in comparison to the total spending to purchase other competitive brands in the lifetime.

Shareholder value Shareholder value refers to the financial worth or returns that a company generates for its shareholders, where the focus has often been on maximizing value for its investors.

Share of wallet Share of wallet can be defined as the percentage of a particular customer's total spent to purchase a specific brand in their entire lifetime, in comparison to the total spending to purchase other competitive brands in the lifetime.

Single-use products Single-use products are products designed for one-time use, often contributing to disposable culture and environmental issues.

Slogan Slogan is a brief and memorable phrase or tagline used in advertising and marketing campaigns to encapsulate a brand's message, values, or unique selling proposition, aiming to resonate with consumers and enhance brand recall.

Social impact Social impact refers to the impacts an organization's activities, products, or services have on the well-being, development, and overall welfare of individuals and communities, comprising both positive and negative consequences.

Social media marketing Social media marketing refers to using a social media platform to convey a marketing message in order to influence a brand's target audience.

Social responsibility Social responsibility is an ethical framework in which a person or a business must work and cooperate with other people and organizations for the benefit of the community.

Special-event pricing Applying pricing initiatives that are intended to support a special promotional event.

Stakeholder engagement Stakeholder engagement is the systematic identification, analysis, planning, and implementation of actions designed to influence stakeholders; an organization involves people who may be affected by the decisions it makes.

Standardization Standardization is the process of establishing and adhering to uniform specifications or criteria in the manufacturing and service delivery process of a product.

Status anxiety Status anxiety is the discomfort individuals feel about their perceived social standing, often influenced by comparisons with others and exacerbated by consumer goods, which can serve as visible markers of status and contribute to feelings of inadequacy or insecurity.

Status Quo Status Quo refers to the existing state of affairs, especially regarding social or political issues.

Stereotyping Stereotyping involves categorizing individuals or groups based on oversimplified and often inaccurate assumptions about their characteristics or behaviors, and using it during segmentation can lead to biased targeting, exclusion, and misrepresentation.

Stock turn The number of times annually that a firm sells each item in its stock.

Strategic marketing Strategic marketing is exploiting the current competitive advantage of a brand while exploring value-creation opportunities to pursue its future competitive advantage.

Stock inventory The products held by a channel member, and often a list of these.

Supply chain transparency Supply chain transparency involves accountability and the practice of openly disclosing information about the origins, processes, and impacts of products or services throughout their production and distribution.

Sustainability Sustainability entails managing brand operations in a manner that preserves the well-being of the environment, society, and future generations while also ensuring the long-term viability and success of the brand.

Sustainability reporting Sustainability reporting involves the systematic communication of a company's economic, environmental, and social performance, providing stakeholders with insights into its sustainability efforts and impacts.

Sustainable competitive advantage Sustainable competitive advantage is "a distinct reason of supremacy of a brand that offers superior value for (its) target market in association with all involved (triple bottom line's) stakeholders, comparing to any other existing or emerging substitute (brands), as well as share that value with the associated (triple bottom line's) stakeholders, with a long-term outlook, potentially for lifetime."

Sustainable consumption Sustainable consumption is consuming a product or service (e.g., at least related to basic human needs) in a way that ensures equity, diversity, and inclusion of all current and future consumer groups in society to fulfill consumers' basic needs to survive and prosper.

Sustainable development Sustainable development is a complete approach that seeks to meet the needs of the present without compromising the ability of future generations to meet their own needs, incorporating economic, social, and environmental considerations.

Sustainable digital marketing Sustainable digital marketing balances achieving marketing goals through making a positive impact on our planet and society.

Sustainable Influencer Marketing Promoting a brand's societal, environmental, and commercial commitments and contributions, relying on an influencer's influence on their followers, based on a formal business relationship with the influencer.

Sustainable innovation Sustainable innovation refers to the development and implementation of new ideas, processes, or products/services that address environmental, social, and economic challenges while minimizing negative impacts and promoting long-term ecological and societal well-being.

Sustainable marketing Sustainable marketing is contributing back to the society or the natural environment of a market where a brand operates its business so that the societal and/or environmental contribution of the brand could underpin its commercial competitive advantage.

Sustainable supply chain Sustainable supply chain integrates environmentally and socially responsible practices across sourcing, production, distribution, and consumption to minimize adverse impacts and promote ethical standards throughout the entire supply chain.

Sustainable viral marketing Sustainable viral marketing is creating awareness about a brand, its market offering, and societal and/or environmental commitment and contribution based on a well-aligned societal and/or environmental brand message to cost-effectively attract the interest and engagement of an enormous number of populations, mainly through digital platforms.

Sweatshops Sweatshops are workplaces characterized by poor working conditions, low wages, long hours, and often exploitation of labor, with the term "sweat" reflecting the harsh and oppressive environment where workers typically toil under strenuous conditions for minimal pay.

Synchronized space Synchronized space is the mix of customers from different generations, some of them are very tech-savvy, some are less, and some are not tech-savvy at all.

T

Tech-readiness Tech-readiness could be referred to as an individual's ability to safely use contemporary technologies in order to be benefited by using the technology.

Trade-in allowances Money subtracted from the sales price of a new product in exchange for the customer's current product, which is to be replaced.

Transparency in business Transparency in business involves honest communication about a company's practices, policies, and impacts, allowing stakeholders and consumers to make informed decisions.

Trendjacking Trendjacking is the art of inserting a brand message into popular conversations by creating content around those topics or events to connect with your audience and draw attention to your products and services.

Triple bottom line The triple bottom line framework posits that businesses should also focus on social and environmental concerns rather than purely profit. Instead of one bottom line, there should be three—people, planet, and profit.

U

User-Generated Contents In web dynamics, the contents, e.g., the original post that is shared on social media by a brand, may have been created by the brand's marketing team, and the contents (e.g., the like, share, or comments) that are generated by the brand's followers on social media are known as user-generated contents as word-of-mouth for the brand.

Unethical concerns Unethical concerns in branding arise when companies engage in deceptive practices, exploitation of labor or resources, or disregard for social and environmental responsibilities, leading to a loss of trust and reputation among consumers.

Uniform delivered pricing Charging the same price to customers regardless of their distance from the seller and the shipping costs.

Unpaid consumers Unpaid consumers in digital environments refer to individuals who provide valuable data, content, or engagement to online platforms without receiving monetary compensation, raising ethical questions about data privacy, fair exchange, and the commercialization of user-generated content.

Up-Selling Exploring opportunities to convince customers to purchase the same product in more volume in order to increase the product's overall sales.

Upstream channel partner A channel partner that is further away from the firm's customers than the firm is—it usually supplies materials, machinery, finance, or other resources.

V

Value Value is "an anticipated outcome of any sort of planned and organized activity. The activity could be derived from monetary, psychic, or physical resources. The more the outcome meets initial anticipation, the more the possibility of win-win outcomes or value optimization for all involved stakeholders."

Value anticipation Value anticipation is customers' own interpretation about a particular product's utility before they purchase and consume it, and how they think of what they may need to sacrifice and what utility they may obtain in return for their sacrifice after they consume it.

Value proposition Value proposition is a comprehensive plan of designing, delivering, and evaluating value to satisfy the customers' and non-customer stakeholders' needs, wants, and expectations.

Variable costs Costs that increase as more products are manufactured and sold.

Veblen effect Veblen effect describes the phenomenon where the demand for a product increases as its price rises, often due to the product's perceived exclusivity or status symbol value rather than its inherent utility or quality.

Vertical integration The process of a firm acquiring its downstream channel intermediaries by buying them.

Vision Vision is establishing a long-term target as a lighthouse that can help us accurately navigate toward our destination, which we want to reach within a targeted period.

Viral marketing Viral marketing through social media is a popular digital marketing strategy that helps marketers to reach out to an enormous number of people based on an appealing message in less possible time and promotional cost.

Volume discounts Discounts offered in return for bulk orders or buying high volumes over time.

Volume production Volume production is the manufacturing and production of large quantities of products to meet high demand.

VRIN competency VRIN competency of a brand enables its market offering to proffer value to satisfy customers' needs in a way that is rare among the competitive brands and make that value inimitable and non-substitutable by the competitive brands in order to sustain the brand's competitive advantage.

W

Waste generation Waste generation is the creation of redundant or discarded materials during the production, consumption, or disposal of products.

Wholesalers Channel partners located downstream of the originator (manufacturer) and upstream of the retailers. They deal with manufacturer deliveries, manage stock, break bulk, offer variety, make retailer deliveries, undertake administrative tasks, and share risk with suppliers.

Z

Zone pricing Charging customers different prices according to the geographical zone in which they are located

Index

D

E

F

G

H

I

K

L

M

N

O

P

R

S

T

U

V

W

Z

The manufacturer's authorised representative in the EU is Springer
Nature Customer Service Centre GmbH, Europaplatz 3, 69115 Heidelberg,
Germany. If you have any concerns regarding our products, please
contact ProductSafety@springernature.com

Printed and bound by CPI Group (UK) Ltd, Croydon, CR0 4YY
01/10/2026
02225956-0020